ITALIAN PHYSICAL SOCIETY

PROCEEDINGS

OF THE

INTERNATIONAL SCHOOL OF PHYSICS « ENRICO FERMI »

COURSE XLIV

edited by CH. SCHLIER
Director of the Course

VARENNA ON LAKE COMO
VILLA MONASTERO
JULY 29th - AUGUST 10th - 1968

Molecular Beams and Reaction Kinetics

1970

ACADEMIC PRESS • NEW YORK AND LONDON

SOCIETA' ITALIANA DI FISICA

RENDICONTI
DELLA
SCUOLA INTERNAZIONALE DI FISICA « ENRICO FERMI »

XLIV Corso

a cura di CH. SCHLIER
Direttore del Corso

VARENNA SUL LAGO DI COMO
VILLA MONASTERO
29 LUGLIO - 10 AGOSTO 1968

Fasci molecolari e cinetica delle reazioni

1970

ACADEMIC PRESS • NEW YORK AND LONDON

ACADEMIC PRESS INC.
111 Fifth Avenue
New York 3, N. Y.

United Kingdom Edition
Published by
ACADEMIC PRESS INC. (London) Ltd.
Berkeley Square House, London W. 1

Library of Congress Catalog Card Number: 71-79893

PRINTED IN ITALY

INDICE

D. R. HERSCHBACH – Reactive collisions of thermal neutral systems.

J. ROSS and E. F. GREENE – Elastic scattering of reactive systems.

F. S. ROWLAND – Hot-atom chemistry. - I

F. S. ROWLAND – Hot atom chemistry. - II

A. HENGLEIN – Kinematics of ion-molecule reactions.

G. G. Volpi – Gas-phase proton-transfer reactions.

R. S. Berry – Chemiionization.

Preface.

CH. SCHLIER
Universität Freiburg - Freiburg i. Br.

Looking back to the days when the lectures presented in this book were given to a young international audience, I cannot help to remember the small guide to Villa Monastero, which you get at the entrance. The German version most adequately tells tourists what they will find there: « ein Fleckchen der Wonnen » - « a hamlet of bliss ». Perhaps tourists heading for Milano, Firenze, and Roma may forget this place soon. I am sure that this will not be the case with the participants of the 44th Course of the Enrico Fermi School, who could enjoy the « ambiance » of Varenna for two weeks, discussing physics and non-physics, swimming, and working hard to prepare lectures or preliminary drafts of what is printed here.

I have tried to depict my personal view of the topics treated in this course and their importance in the pages called « Introduction ». Here follows a historic remark and a handful of acknowledgements: A first school on « Reaction Kinetics » was held in Irvine, California in January, 1967. Since that time, the repetition of a similar school at some place in Europe and perhaps having a series of such schools alternately in Europe and the USA was discussed between D. L. BUNKER and F. S. ROWLAND of the University of California at Irvine, and D. BECK and myself at Freiburg. I am much indebted to these three colleagues for advice, which helped me in the preparation of the course and its schedule.

I think I am speaking for all who attended as students or lecturers if I express my sincerest thanks to the Italian Physical Society and the sponsoring organizations of the Enrico Fermi School for making this Course possible.

It is a pleasure for me to thank Professor G. GERMANÀ of the Italian Physical Society for his and his staff's experienced organization of the environment in which the course was held.

I am obliged to Professor SCOLES of Genova not only for his first idea of having a molecular beam school at Varenna, but also for his services as « Secretary » of the course, including the clearing-up of many linguistic difficulties of a non-Italian-speaking German like myself.

Last not least, I probably express a general feeling of the audience if I thank all lecturers and seminar speakers for their excellent job.

Introduction.

CH. SCHLIER

Universität Freiburg - Freiburg i. Br.

When Norman F. RAMSEY published his book « Molecular Beams » [1], in 1956, a « typical molecular beam » apparatus [2] was necessarily a magnetic resonance machine. The situation has changed since then. At about the time Ramsay's book was in press, TAYLOR and DATZ [3] published their pioneering experiment in which the chemical reaction $K+HBr \rightarrow KBr+H$ was studied by molecular beam techniques. Now, less than 15 years later, most molecular beam experiments are probably not of the R. F. spectroscopy type but are performed to explore scattering processes. Molecular beam techniques have already set new standards for the accuracy with which intermolecular potentials can be determined. The application of the molecular beam technique to reactive scattering, *i.e.* to the study of simple chemical reactions, has already produced information about the dynamics of some of these reactions, which is far more detailed than what people have imagined in the otherwise so fertile pre-war years.

Yet it is not straightforward, to solve the old problems of reaction kinetics with this new tool. This statement remains true, even if we restrict ourselves to homogeneous reactions in the gas phase, and to reactions involving small molecules. (Evidently, reactions of big aggregates will probably always need more « chemical » concepts to describe their complexity, than the « physical » pictures for which we look in small systems.) Still, old classics of reaction kinetics have not been solved by the beam technique, as is the case for $HI+HI \rightarrow I_2+H_2$ [4]. But these beam techniques are quickly developing. « Chemical accelerators » [5], *i.e.* methods to do scattering experiments at energies between 0.5 and 10 eV, are being put into service. Better detectors are about to end the « alkali age ». So we can expect a wealth of information to be produced within the next few years.

Under these circumstances, the opportunity to teach the new techniques imbedded in a general course in reaction kinetics to a larger group of people from different laboratories was most welcome, and I hope that this book will continue in the dissemination of what has been taught at Varenna. I should be happy if it could contribute to bridge the gap which still exists between those dedicated to (so to speak) « classical » reaction kinetics, not yet convinced of the power of beam techniques, and those who are so much beam-minded, that they forget the existence of all other tools. At the same time at least in Germany, and probably in some other parts of Europe, this gap only too often coincides with that between « chemists » and « physicists », and I feel strongly towards filling it with true « chemical physicists ».

There is another point to make, and I now take the liberty to speak of my own field, atomic and molecular physics. This field has yielded a wealth of data about atomic levels and structures in the last decades, together with many nuclear and more fundamental data obtained with atomic methods. What are we about to do next? In my view, somewhat different of that of a recent survey [6], it is the atomic *processes* and their dynamics (*not* only their kinetics) that we are to explore in the next decade. (It may not be pure chance, that the notion of society as a « process » has gained so much importance in the current left-oriented sociological discussion. But to discuss this, is rather a homework problem).

Concentrating again on molecular processes, I tend to see the tasks of the chemist and the physicist merging their effort towards common problems like the following: the chemist can question the physicist's faith in formal solutions, asking for the models underlying them and for those numbers which can be compared with experimental data. We must continuously require that a quantitative theory has to provide numbers, not only an easily-written T-matrix element, in order to be regarded as finished; if that is impossible a qualitative model may be more useful in understanding what happens.

On the other hand the physicists must be allowed to constantly inquire about the phenomenologically defined concepts of the chemists. It is, of course, legitimate, *e.g.*, to define « complex » by the phenomenon of a reaction having a forward-backward symmetric cross section. On the other hand, construction of a model theory in which a « complex » is a system of atoms, which is so strongly coupled that its energy is equilibrated among the various degrees of freedom, is quite an other thing. Therefore, $complex_1$ is all but identical to $complex_2$.

Finally, I will make some remarks on what has been included in this volume and what some of the readers may miss. Elastic and inelastic scattering is, of course, the natural starting point of a discussion of atomic and molecular processes. The subject of reactions has been extended to include ionic reactions and reactions of the electron (excitation transfer): the first topic because of its invaluable contribution to the exploration of reaction dynamics at energies above the thermal range, where neutral beams are very difficult to produce;

the second topic because several theoretical models are similar for atomic and electronic transfer. The next group of lectures treats nonbeam methods and the connections between beam and nonbeam results. Case studies of H^+- and H-transfer are included. A final group of lectures tries to point out where theory stands today. A heavy weight has been given to trajectory studies (and to what comes before: computation of energy surfaces) because these methods, though numerical, have yielded so much insight in the dynamics of small systems.

Some people may miss lectures on the technical aspects of molecular beam experiments. After some consideration, the conviction that lectures on constructional details are liable to be boring has succeeded in keeping them off the schedule. The interested reader is, *e.g.*, referred to ref. [7].

REFERENCES

[1] N. F. Ramsey: *Molecular Beams* (Oxford, 1956).
[2] *loc. cit.*, chapter I.2.
[3] E. H. Taylor and S. Datz: *Journ. Chem. Phys.*, **23**, 1711 (1955).
[4] S. B. Jaffé and J. B. Anderson: *Journ. Chem. Phys.*, **49**, 2859 (1968).
[5] R. Wolfgang: *Scientific American*, **219**, No. 4, 44 (1968).
[6] V. Hughes: *Physics Today*, **22**, No. 2, 33 (1969).
[7] H. Pauly and J. P. Toennies: in *Methods of Experimental Physics*, edited by L. Marton, vol. **7** A (New York and London, 1968), p. 227.

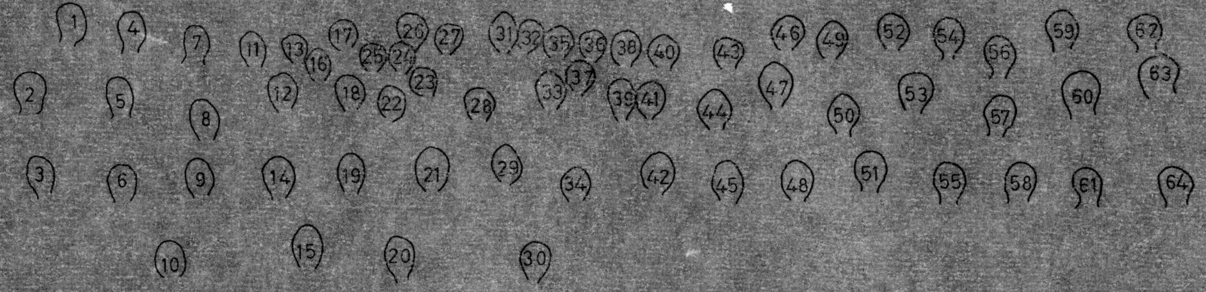

1. C. Schmidt
2. G. Lempert
3. D. Chang
4. R. Düren
5. A. Niehaus
6. G. Scoles
7. F. Torello
8. M. G. Dondi
9. F. Baede
10. V. Kempter
11. U. Buck
12. P. Eckelt
13. T. Rose
14. A. Van der Meulen
15. R. M. Yealland
16. A. Schultz
17. E. Teloy
18. A. Galli
19. G. G. Volpi
20. H. Haberland
21. R. Ross
22. M. Lenzi
23. U. Landman
24. N. Gershon
25. C. J. Chapman
26. H. U. Mittmann
27. C. Kaplinsky
28. R. Gengenbach
29. D. Herschbach
30. G. Aniansson
31. H. Hotop
32. J. A. Vliegenthart
33. H. Schlumbohm
34. Ch. Schlier
35. A. Ding
36. J. Alper
37. G. Bosse
38. K. Laemann
39. B. Schimpke
40. I. Johansson
41. J. W. Bredewout
42. F. Rowland
43. D. Beck
44. K. Schügerl
45. Berry
46. M. Berrondo
47. M. Von Seggern
48. Conroy
49. U. Koller
50. J. P. Treguier
51. D. Bunker
52. W. D. Held
53. A. Saplakoglu
54. I. Leksell
55. R. J. Cross
56. G. Müller
57. J. Schoettler
58. F. A. Gianturco
59. A. Henglein
60. L. Holmlid
61. T. Nenner
62. R. Haerten
63. H. G. Wagner
64. J. Grosser

SOCIETÀ ITALIANA DI FISICA

SCUOLA INTERNAZIONALE DI FISICA « E. FERMI »

XLIV CORSO - VARENNA SUL LAGO DI COMO - VILLA MONASTERO - 29 Luglio - 10 Agosto 1968

Collision Mechanics.

D. Beck

Physikalisches Institut der Universität Freiburg - Freiburg i. B.

In this course we shall be interested in molecular collision processes. They are the elementary step in a number of rather different phenomena, such as transport properties, relexation times or the rate of a chemical reaction. Therefore, there is one common concept to which all of these phenomena can be reduced: the appropriate collision cross-section $d\sigma$. Let us begin with a few remarks on this concept.

1. – Definition of the collision cross-section in the laboratory system (LS).

To define $d\sigma$ we consider an idealized experiment (cf. Fig. 1): A number Z_b of target particles B is somehow confined to a small volume $d\tau = df\,dl$. They are assumed to be stationary and dilute. A stream of particles A with uniform velocity $\boldsymbol{v}_a$ and current density $\boldsymbol{j}_a$ impinges. The number of collisional « events » per second may be written as

$$dN = j_a Z_b\, d\sigma \tag{1a}$$

$$= N_a n_b\, dl\, d\sigma \tag{1b}$$

$$= n_a n_b v_a\, d\tau\, d\sigma\,. \tag{1c}$$

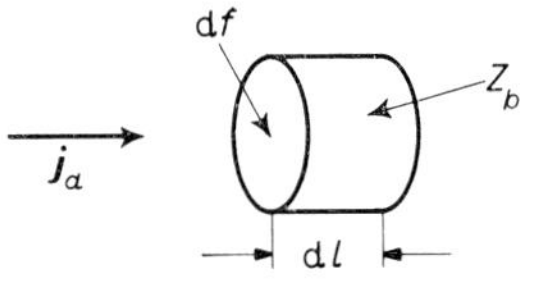

Fig. 1.

Here N_a is the number of projectiles entering $d\tau$ per second, n_a, n_b are the respective number densities. The target is dilute if $n_b\, dl\, d\sigma \ll 1$ (no multiple scattering, two body collisions). If species B is not stationary, but has uniform velocity $\boldsymbol{v}_b$ there is an additional factor v/v_a on the right-hand side of (1a, b), $v = |\boldsymbol{v}_a - \boldsymbol{v}_b|$ being the relative speed, while v_a in (1c) is replaced by v.

The quantity $d\sigma$ so defined will not be meaningful unless we specify the event which we observe. Talking about an *ideal* experiment on a collision of the type

$$A + B \to C + D\,,$$

the *complete* specification will be the list of the maximum number of observables in the initial as well as the final states of the collision partners. Let i stand for one set of eigenvalues of all observables for the internal motion of A, $\boldsymbol{v}_a$ for its (nonrelativistic) velocity and let j, k, l, $\boldsymbol{v}_b$, $\boldsymbol{v}_c$, $\boldsymbol{v}_d$ be correspondingly defined for B, C, D respectively. The specification of the observed event may be indicated as

$$\text{(2)} \qquad \mathrm{d}\sigma(k, l, \boldsymbol{v}_c, \boldsymbol{v}_d | i, j, \boldsymbol{v}_a, \boldsymbol{v}_b) \, .$$

As linear momentum

$$\boldsymbol{P} = \text{const} \, ,$$

and total energy

$$E_{\text{tot}} = \text{const} \, ,$$

are conserved, the number of parameters in (2), *i.e.* the number of experimentally observed quantities, may be reduced not impairing the completeness of the specification. Depending on the choice of the eliminated parameters (4 at the most) there exists a number of different equivalent specifications. One of them is most often used:

$$\text{(3)} \qquad \mathrm{d}\sigma(k, l, \Theta, \Phi | i, j, \boldsymbol{v}_a, \boldsymbol{v}_b) \, .$$

Here the angles Θ, Φ specify the direction of one of the reaction products with respect to a direction given in the initial preparation, say of C with respect to A as indicated in Fig. 2.

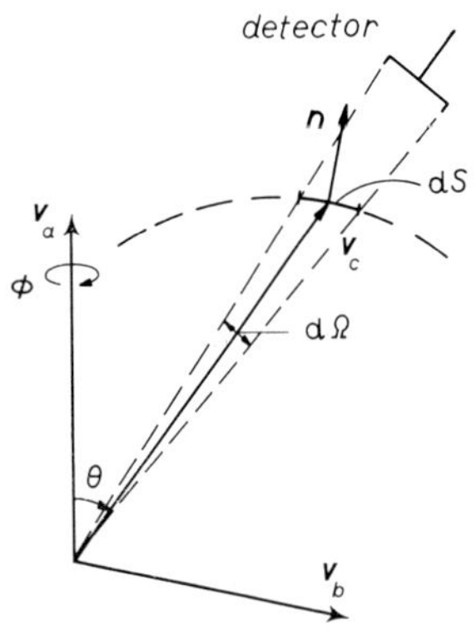

Fig. 2.

Note that conservation of angular momentum does not in general reduce a specification because the orbital angular momentum of the relative motion cannot be controlled. However, considering only nonrelativistic velocities, conservation of mass,

$$M_{\text{tot}} = \text{const} \, ,$$

is valid and may be useful if rearrangement processes can occur and a complete specification like (3) is experimentally not feasible.

Quantum mechanics implies a statistical notion of the cross-section. Classically, this may be considered as a consequence of our ignorance of the relative phases of the translational and internal motions of the partners and their translational angular momentum. Even if the initial state is prepared in a δ-function manner a continuum of final translational states will be populated. We can factor

$$\text{(4)} \qquad \mathrm{d}\sigma = f\cdot \mathrm{d}\boldsymbol{v}_c\,\mathrm{d}\boldsymbol{v}_d\,, \quad \mathrm{d}\boldsymbol{v}_c \equiv v_c^2\,\mathrm{d}v_c\,\mathrm{d}\Omega_c\,, \quad \text{etc.}$$

Again the conservation laws permit several simpler but equally general notations. The specification of a 2-dimensional volume element in the final-velocity space of the system is sufficient. Pictorially, $\boldsymbol{v}_d$ being tied to $\boldsymbol{v}_c$ by linear momentum conservation, the end point of the vector $\boldsymbol{v}_c$—by energy conservation—is restricted to move on some surface S in the velocity space of particles C (Fig. 2). f may be considered to be different from zero only on this surface. The notation (3) refers to detection of all particles C within the solid angle element $\mathrm{d}\Omega = \mathrm{d}\Omega_c$ spanned by the detector as seen from the interaction region $\mathrm{d}\tau$. If, instead of (4), we factor according to this situation we obtain the definition of the differential partial cross-section I_L in the laboratory reference frame (LS)

$$\text{(5)} \qquad \mathrm{d}\sigma = I_L(k, l, \Theta, \Phi | i, j, \boldsymbol{v}_a, \boldsymbol{v}_b)\,\mathrm{d}\Omega\,.$$

From Fig. 2 $\mathrm{d}\Omega$ is seen to be

$$\text{(6)} \qquad \mathrm{d}\Omega = \frac{\cos(\boldsymbol{n}\cdot\boldsymbol{v}_c)}{v_c^2}\,\mathrm{d}S\,, \qquad \boldsymbol{n} \text{ perpendicular to } S.$$

Often a different, less ideal situation may arise: k, l, one or both are not determined at all or so closely spaced that they cannot be resolved in an experiment. As a result $\mathrm{d}\boldsymbol{v}_c$ will be effectively 3-dimensional. Employing velocity analysis of C we can at best measure a signal according to a definition different from (5), namely

$$\text{(7)} \qquad \mathrm{d}\sigma = J_L(\boldsymbol{v}_c | i, j, \boldsymbol{v}_a, \boldsymbol{v}_b)\,\mathrm{d}\boldsymbol{v}_c\,\mathrm{d}\Omega\,.$$

As Prof. HERSCHBACH pointed out to me it is generally necessary to use this definition in the analysis of reactive scattering (see also Sect. **3**·3).

2. – Cross-section in the center-of-mass system (CMS).

The notation (5) of $\mathrm{d}\sigma$ specifies the collision completely in an instrumental way. It does contain redundant information. We know that, whatever the interaction is, it depends on relative coordinates only. Conservation of linear momentum and mass guarantees that the center of mass velocity $\boldsymbol{C}$ is conserved.

The process is independent from it. To account for this we transform

$$\boldsymbol{v}_a - \boldsymbol{v}_b = \boldsymbol{v} ,$$

$$\boldsymbol{v}_c - \boldsymbol{v}_d = \boldsymbol{v}' ,$$

$$(1/M)(\boldsymbol{P}_a + \boldsymbol{P}_b) = (1/M)(\boldsymbol{P}_c + \boldsymbol{P}_d) = \boldsymbol{C} ,$$

where $\mathrm{d}\boldsymbol{v}_c\,\mathrm{d}\boldsymbol{v}_d = 1\cdot\mathrm{d}\boldsymbol{v}\,\mathrm{d}\boldsymbol{C}$, and for $\boldsymbol{C}$ fixed by the initial preparation

$$\mathrm{d}\boldsymbol{v}_c = \left(\frac{m_d}{M}\right)^3 \mathrm{d}\boldsymbol{v} = 1\cdot\mathrm{d}\boldsymbol{u}_c .$$

An observer stationed in the (moving) center of mass describes the situation indicated by (5) in the following way: From (1) he determines the same number $\mathrm{d}\sigma$ as is found in the LS

$$\mathrm{d}\sigma(\mathrm{LS}) = \mathrm{d}\sigma(\mathrm{CMS}) . \tag{8}$$

Measuring particle velocities $\boldsymbol{u}_a$ etc. with

$$\boldsymbol{u}_a - \boldsymbol{u}_b = \boldsymbol{v}; \quad \boldsymbol{u}_c - \boldsymbol{u}_d = \boldsymbol{v}' ,$$

(Fig. 3) he writes the conservation theorems

$$m_a\boldsymbol{u}_a + m_b\boldsymbol{u}_b = m_c\boldsymbol{u}_c + m_d\boldsymbol{u}_d = 0 ,$$

$$\frac{\mu}{2}v^2 + E_i + E_j = \frac{\mu'}{2}v'^2 + E_k + E_l .$$

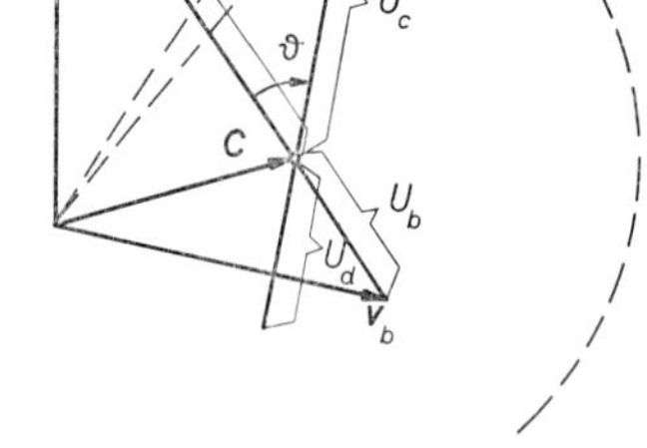

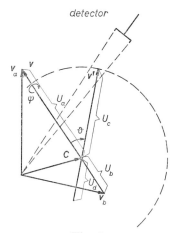

Fig. 3.

μ, μ' are the reduced masses before and after the collision, E_i etc. the total internal energies of particles A etc. in quantum states i etc. In the ideal, fully specified situation, evidently

$$v' = \mathrm{const} \rightarrow u_c = \mathrm{const} .$$

Graphically, the end point of the vector $\boldsymbol{u}_c$ is free to move on a sphere centered around the tip of $\boldsymbol{C}$. This sphere is the surface on which f in (4) or I_L in (5) are different from zero.

Let θ, φ indicate the direction of $\boldsymbol{u}_c$ as in Fig. 3. With the same reasoning as for (3) and (5) the observed quantity can now be written

$$\mathrm{d}\sigma = I(k, l, \theta, \varphi | i, j, \boldsymbol{v})\,\mathrm{d}\omega , \quad \mathrm{d}\omega = \mathrm{d}\omega_c = \frac{\mathrm{d}S}{u_c^2} . \tag{9}$$

I, the differential partial cross-section in the CMS, also gives a complete description of the process. This description is of greater generality than (5) because it is independent of the apparatus *i.e.* of $\boldsymbol{v}_a$, $\boldsymbol{v}_b$.

3. – CMS-LS conversion.

To state an experimental result in a general way we have to transform it to the CMS. This can be done for almost any actual situation starting from (5), (6), (8) and (9).

3·1. *Differential partial cross-section.* – In the fully specified case we have from (8)

$$(10)\qquad \begin{cases} I_L \mathrm{d}\Omega = I \mathrm{d}\omega\,, \\ I = I_L \dfrac{\mathrm{d}\Omega}{\mathrm{d}\omega} = I_L \dfrac{u_c^2}{v_c^2} \cos(\boldsymbol{u}_c, \boldsymbol{v}_c)\,. \end{cases}$$

Along with this « intensity » transformation the LS scattering angles Θ, Φ have to be converted into θ, φ. The transformation depends on the initial preparation of the system and the change of masses and translational energy in the process. Depending on the same parameters as well as the scattering angles the Jacobian factor in (10) may make I markedly different from the measured I_L. For the general type of process assumed here the transformation formulae are rather lengthy but their calculation is straightforward [1].

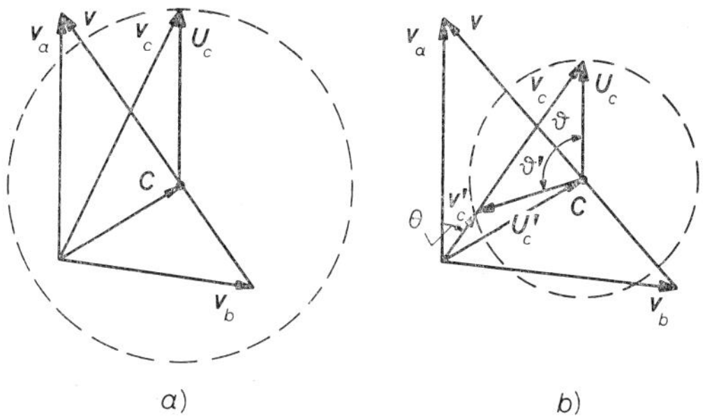

Fig. 4.

For the situation in which k, l are not determined, with a definition of the CMS cross-section similar to (7), that is

$$(11)\qquad \mathrm{d}\sigma = J(\boldsymbol{u}_c|i, j, \boldsymbol{v})\, \mathrm{d}u_c\, \mathrm{d}\omega\,,$$

we have the intensity transformation [1b]

$$(12)\qquad J = J_L \frac{u_c^2}{v_c^2}\,.$$

The Jacobian factor may be strongly different from the one given in (10).

The transformation CMS $\rightarrow$ LS will always be unambiguous. However, LS $\rightarrow$ CMS, the one we need to convert experimental results according to (10) or (12) will be unambiguous only if

$$u_c > C \,,$$

as can best be seen graphically from Fig. 4*a*, *b*. In Fig. 4*b* at a LS angle Θ an intensity corresponding to two CMS angles, θ, θ' is detected. In such cases there exist « forbidden » regions in the LS within which no scattering is observed. If no velocity selection is employed ($v_c \neq v_c'$!) only an approximate intensity conversion will be possible by the fact that θ and θ' may be widely separated and there may be reason to assume $I(\theta)$ and $I(\theta')$ to be very different so that the smaller contribution can be neglected.

3'2. *Integral partial cross-section.* – It is defined by

$$\sigma(k, l|i, j, \boldsymbol{v}) = \int\limits_{\substack{\text{all final}\\ \text{translational}\\ \text{states}}} \mathrm{d}\sigma = \int\limits_{4\pi} I\,\mathrm{d}\omega = \int\limits_{4\pi} I_L \mathrm{d}\Omega \,.$$

The integration is over the 2-dimensional volume in velocity space to which the endpoints of the vectors $\boldsymbol{v}_c$ and $\boldsymbol{u}_c$ are restricted in the ideal situation. σ is independent of the co-ordinate system.

3'3. *Differential, partially summed cross-section.* – The internal-state analysis assumed in the fully specified situation might be difficult in practice. More likely there will be no state selection k, l but velocity selection for the detected product C. The detector at Θ, Φ measures a signal proportional to

$$\sum_{k,l}{}' I_L(k, l, v_c, \Theta, \Phi|i, j, \boldsymbol{v}_a, \boldsymbol{v}_b)\,\mathrm{d}\Omega_c = \Delta n \cdot \bar{I}_L(v_{c0}, \Theta, \Phi|i, j, \boldsymbol{v}_a, \boldsymbol{v}_b)\,\mathrm{d}\Omega_c \,.$$

In the CMS this will be equal to

$$\sum_{k,l}{}' I\,\mathrm{d}\omega_c = \Delta n \cdot \overline{I\,\mathrm{d}\omega_c} \,.$$

The $\sum'$ runs over Δn pairs k, l such that v_c will be in the range Δv_c around v_{c0} transmitted by the velocity selector. $\bar{I}_L$ and $\overline{I\,\mathrm{d}\omega_c}$ are averages in Δv_c. In some cases (*e.g.* inelastic collisions) Δn may be known. If $\mathrm{d}\omega_c$, θ and φ vary little in Δv_c they can be taken as constant and evaluated at v_{c0}. We can evaluate an average cross-section in the CMS just as in (10)

$$\bar{I}(u_{c0}, \theta, \varphi|i, j, \boldsymbol{v}) \approx \bar{I}_L(v_{c0}, \Theta, \Phi|i, j, \boldsymbol{v}_a, \boldsymbol{v}_b)\,\frac{\mathrm{d}\Omega_c}{\mathrm{d}\omega_c} \,.$$

There are, however, other cases (reactive collisions with highly excited products, dissociative or ionization processes for which k, l includes continuous contributions) where the enumeration of states may be not interesting or impossible. Then

$$\sum_{k,l}' \ldots = \bar{I}_L \frac{\Delta n}{\Delta v_c} \Delta v_c \, \mathrm{d}\Omega_c = \overline{I \, \mathrm{d}\omega_c} \frac{\Delta n}{\Delta u_c} \Delta u_c \, ,$$

and in the limit of infinitesimally small Δv_c, setting $I_L(\mathrm{d}n/\mathrm{d}v_c) = J_L$ and $I(\mathrm{d}n/\mathrm{d}u_c) = J$ the transformation law (12) results for the J_L, J.

4. – Examples.

A few simple examples may serve to illustrate the above considerations. Several more complicated examples will be discussed by Prof. WAGNER and HERSCHBACH in their lectures on « Rate Constants and Reaction Cross-Sections » and « Kinematic Analysis of Reactive Scattering ».

4·1. *Differential elastic-scattering cross-section from a crossed molecular beam experiment.* – For elastic scattering the specification reduces to

$$\mathrm{d}\sigma = I_L(\Theta, \Phi | \boldsymbol{v}_a, \boldsymbol{v}_b) \, \mathrm{d}\Omega = I(\theta | v) \, \mathrm{d}\omega \, .$$

A spherically symmetric interaction has been assumed. Physically the process does not depend on the direction of $\boldsymbol{v}$. We need it only to define θ. Also, I is independent of φ, *i.e.* in the CMS the scattering has azimuthal symmetry around the $\boldsymbol{v}$ direction.

Note: this symmetry holds for *any* interaction if the relative orientation of the collision partners is random.

The detector signal is

$$S = \mathrm{d}N_a(\Theta, \Phi | \boldsymbol{v}_{a0}, \boldsymbol{v}_{b0}) = n_a n_b \, \mathrm{d}\tau \, \mathrm{d}\Omega \int \mathrm{d}\boldsymbol{v}_a \mathrm{d}\boldsymbol{v}_b f_a f_b v I \frac{\mathrm{d}\omega}{\mathrm{d}\Omega} \, .$$

$f_a = f(\boldsymbol{v}_a)$, $f_b = f(\boldsymbol{v}_b)$ are known distribution functions, $\boldsymbol{v}_{a0}$, $\boldsymbol{v}_{b0}$ are nominal velocities of the species A and B for which an instrumental definition is possible. For simplicity we neglect the finite size of the interaction region $\mathrm{d}\tau$ and the detector $\mathrm{d}\Omega$.

In practice one often finds a very weak dependence of the Jacobian $\mathrm{d}\omega/\mathrm{d}\Omega$ on $\boldsymbol{v}_a$ and $\boldsymbol{v}_b$ in their respective ranges given by f_a, f_b. With a negligible error it may be taken out of the integral and evaluated for representative values $\boldsymbol{v}_{a0}$, $\boldsymbol{v}_{b0}$. If in the experiment one beam is velocity-selected while the other is thermal (see Fig. 5) a more careful analysis shows that this procedure will be accurate to first order if $\boldsymbol{v}_{a0}$ is taken as the most probable velocity of the selected beam while $\boldsymbol{v}_{b0}$ is the *average* velocity of the thermal beam.

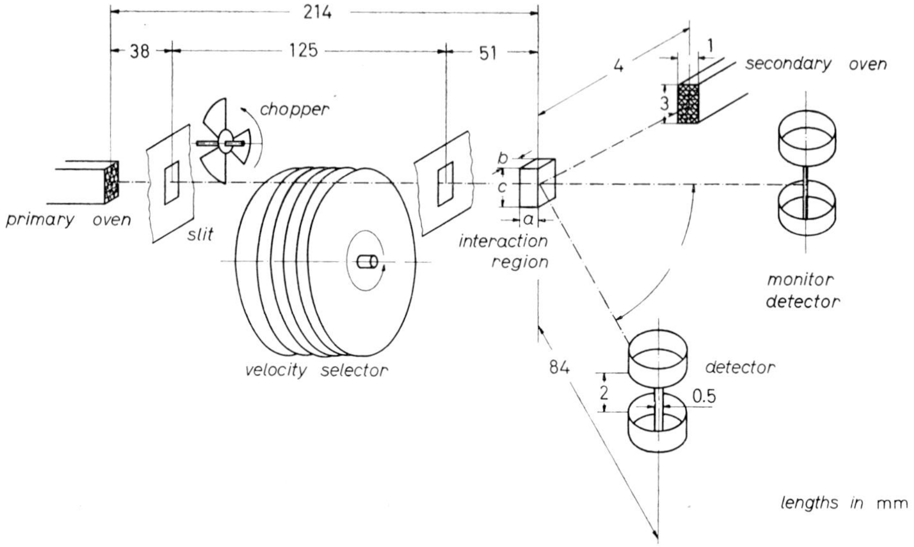

Fig. 5.

The result of the experiment is usually quoted as

$$I_{\text{eff}}(\theta_0|v_0) = \left[\int \mathrm{d}\boldsymbol{v}_a \,\mathrm{d}\boldsymbol{v}_b f_a f_b v\right]^{-1} \cdot \text{const} \cdot \left.\frac{\mathrm{d}\Omega}{\mathrm{d}\omega}\right|_{\boldsymbol{v}_{a0}\boldsymbol{v}_{b0}} \cdot S(\Theta, \Phi|\boldsymbol{v}_{a0}, \boldsymbol{v}_{b0}) \,,$$

where the nominal velocities are used to convert the angles. To the same order of accuracy

$$v_0 \approx v_{a0}\left[1 + \frac{1}{2}\frac{\overline{v_{b0}^2}}{v_{a0}^2}\right],$$

if the beams are crossed at a right angle.

I_{eff} will be reasonably close to I depending on the quality of the initial preparation and the choice of the system (unselected partner slow and heavy). For the distribution of the relative speed v half-widths as narrow as 2% have been reported [2].

4·2. *Integral elastic-collision cross-section from a beam experiment.* – The experiment is usually done by measuring the attenuation of one beam by the other at high angular resolving power $\delta\Omega$. The signal is

$$S(\boldsymbol{v}_{a0}, \boldsymbol{v}_{b0}) = -\Delta N_a = \int\limits_{4\pi-\delta\Omega} \mathrm{d}N_a =$$

$$= n_a n_b \,\mathrm{d}\tau \left[\int \mathrm{d}\boldsymbol{v}_a \,\mathrm{d}\boldsymbol{v}_b f_a f_b v\sigma(v) - \int\limits_{\delta\Omega} \mathrm{d}\Omega \int \mathrm{d}\boldsymbol{v}_a \,\mathrm{d}\boldsymbol{v}_b f_a f_b v I \frac{\mathrm{d}\omega}{\mathrm{d}\Omega}\right].$$

Here the second term accounts for particles with deflections so small that they still reach the detector. It is evaluated as outlined in Sect. **4**.1 using a small-angle approximation to I and $\mathrm{d}\omega/\mathrm{d}\Omega$. High resolving power means the second term is negligible compared to the first. If this is true the result may be quoted as

$$\sigma_{\text{eff}}(v_0) = \left[n_a n_b \,\mathrm{d}\tau \int \mathrm{d}\boldsymbol{v}_a \,\mathrm{d}\boldsymbol{v}_b f_a f_b v\right]^{-1} \cdot S .$$

For details see [3], [4], [5], but note that this definition of σ_{eff} differs from the one adopted in [3] and [4].

4.3. *Transport properties.* – A comparison may be interesting. The shear viscosity coefficient to first order is given by

$$\eta = \frac{5\pi}{32\sqrt{2}} \frac{m\bar{v}_a}{\sigma_\eta},$$

$$\sigma_\eta = \frac{1}{32}\left(\frac{m}{2kT}\right)^4 \int\limits_v \left[\int\limits_{4\pi} I(\theta|v) \sin^2\theta \,\mathrm{d}\omega\right] v^7 \exp\left[-\frac{mv^2}{4kT}\right] \mathrm{d}v .$$

I is averaged over a wide v distribution. In addition the weighting factor $\sin^2\theta$ damps out the contributions from small and large angles. This restricts the obtainable information about the interaction to a limited range in distance r of the partners during the collision. This loss can only be regained by large variations of temperature. The coefficient of thermal conductivity also includes $\sin^2\theta$, while the diffusion coefficient has $(1 - \cos\theta)$ as a weighting factor.

5. – Classical calculation of the elastic-scattering cross-section.

If the interaction can be represented by a spherically symmetric, static potential $V(r)$ it is a simple and well known matter to predict the differential scattering cross-section $I(\theta|v)$ using classical mechanics. From Fig. 6 with b as the impact parameter of the incoming projectile we have by « conservation of particles »: $I(\theta|v)\,d\omega = b\,db\,d\varphi$. From this

$$I \sin\theta = b \left|\frac{db}{d\theta}\right| , \tag{13}$$

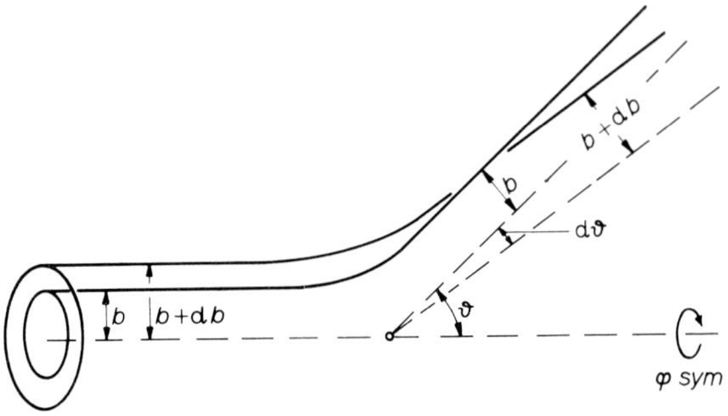

Fig. 6.

where the φ symmetry has been used and the $|...|$ sign is taken in order to have some freedom in defining the deflection angle. Its dependence from impact parameter and relative speed is given by

$$\theta = \theta(b, v) = \pi - 2\int_{r_c}^{\infty} \frac{b\,dr}{r^2[1 - V(r)/E - b^2/r^2]^{\frac{1}{2}}} , \tag{14}$$

$\mu = m_a m_b/(m_a + m_b)$ reduced mass,

$E = (\mu/2)v^2$ initial relative kinetic energy,

$r_c =$ distance of closest approach, determined by $1 - \dfrac{V(r_c)}{E} - \dfrac{b^2}{r_c^2} = 0$.

Except for the simplest potentials like the Coulomb-($\sim r^{-1}$), the pure polarization-($\sim r^{-4}$) or the van der Waals-potential ($\sim r^{-6}$), θ and I have to be calculated numerically.

Note. If r_m is some scaling distance of the potential θ depends only on the reduced impact parameter $\beta = b/r_m$, and $I \sin\theta = r_m^2 \beta |d\beta/d\theta|$. This means that r_m can only be inferred from an absolute measurement of the cross-section but *not* from its angle dependence.

Let us review the results of the classical analysis for two particularly important potential models.

5˙1. *Hard spheres* (Fig. 7). – The radii of the colliding species be R_a, R_b, $R = R_a + R_b$.

Deflection function:

$$b = R \sin\alpha = R \cos\theta/2 \ .$$

Cross-sections:

$$I(\theta|v) = \frac{1}{\sin\theta}\, b \left|\frac{\mathrm{d}b}{\mathrm{d}\theta}\right| = \frac{R^2}{4}, \text{ independent of } \theta \text{ and } v.$$

$$\sigma(v) = \pi R^2, \text{ as expected .}$$

The quantum-mechanical result, even in its high velocity limit is different:

$$\sigma(v \to \infty) = 2\pi R^2 \ .$$

Explicit calculation shows that the additional contribution to the cross-section is confined to small angles in the forward direction, cf. Fig. 8. This has a close

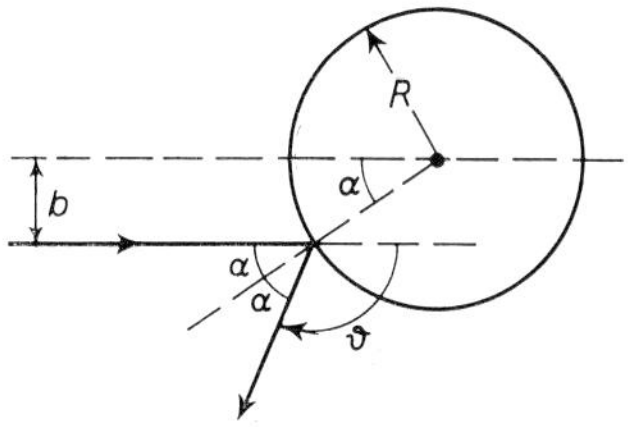

Fig. 7.

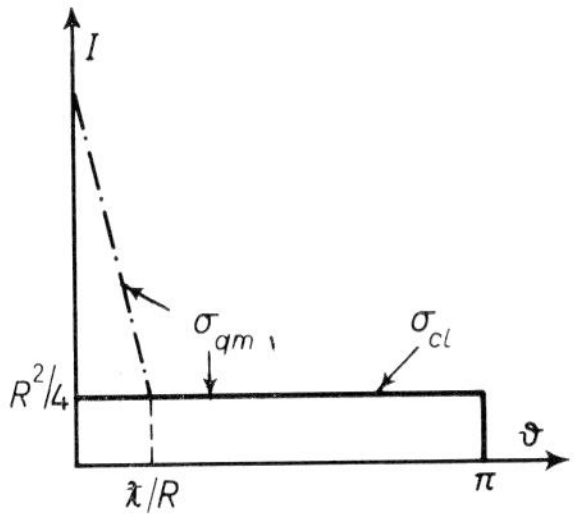

Fig. 8.

parallel in optics: in the shadow behind a reflecting object on the axis it is bright due to diffraction. In a distance large compared to R^2/λ behind the object just a slight decrease of the light wave amplitude will be reminiscent of the shadow. It accounts for the light scattered into all directions by the obstacle.

5˙2. *Interatomic potential.* – Figure 9 shows a potential which is qualitatively realistic for atomic collisions. The analysis of the scattering from such an interaction is interesting because it shows the features of superposition, rainbow and glory scattering, and orbiting. These have become valuable diagnostic tools to judge qualitatively the course of more complicated inelastic and reactive interactions.

The qualitative behavior of the deflection function is given in Fig. 9*b*). Its main features are:

a) a positive and a negative branch as the impact parameter varies, due to net repulsion and net attraction, respectively, in the collision,

b) a minimum at $\theta_r(b_r)$ due to the beginning influence of the repulsive potential core on the particle trajectory,

c) a zero for finite impact parameter corresponding to cancellation of large attractive and repulsive forces in the collision.

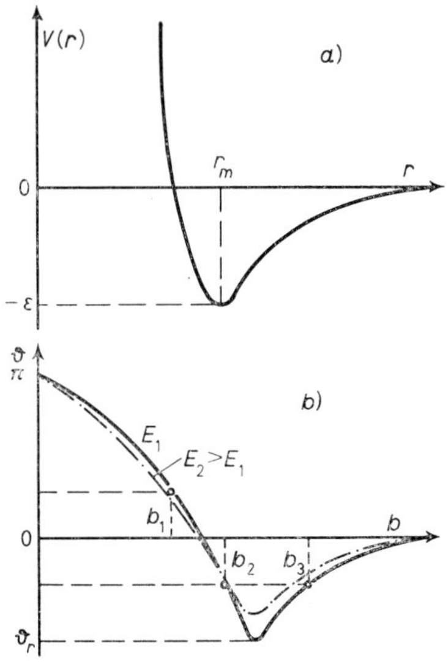

Fig. 9.

The consequences of this type of deflection function are

$$(13a)\quad a)\qquad I\sin\theta=\sum_{i=1}^{3} b_i\left|\frac{\mathrm{d}b}{\mathrm{d}\theta}\right|_i.$$

Three impact parameters contribute to a given observation angle $\theta<\theta_r$ (see Fig. 9*b*)). $b_3|\mathrm{d}b/\mathrm{d}\theta|_3$ is obviously the largest contribution. As it drops with increasing observation angle the cross-section also drops.

b) At θ_r

$$I\sin\theta\sim[\theta-\theta_r]^{-\frac{1}{2}},$$

due to the negative branch of $\theta(b)$. Because of the close analogy of this singularity to the beautiful phenomenon in the scattering of light from water droplets it is called rainbow scattering. In Fig. 10 the third branch of the deflection function—here given as $\theta=\theta(\alpha)$—shows the extremum which is responsible for the meteorological main rainbow.

As the minimum in $\theta(b)$ is strongly energy-dependent (see Fig. 9*b*) the classical singularity in I shifts in angle when the energy is changed (rainbow colours).

Fig. 11 qualitatively depicts the angular dependence of $I\sin\theta$ as inferred from (13*a*). As can be seen from (14)

$$\theta_r=\theta_r\left(\frac{E}{\varepsilon}\right).$$

For often used model potentials such as the Lennard-Jones 12-6 the rainbow trajectory has an r_c little larger than the minimum distance r_m and just slightly changing with collision energy

$$1.06 r_m < r_c < 1.16 r_m .$$

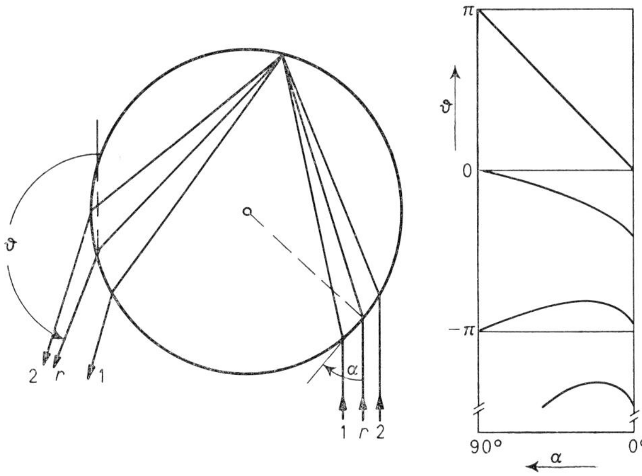

Fig. 10.

c) Classical mechanics will always give a divergent cross-section at $\theta = 0°$ for a potential of unlimited range due to the contribution from large impact parameters. Hidden under this is another interesting feature of the cross-section caused by the zero of $\theta(b)$. Here b as well as $|db/d\theta|$ are finite. As $\sin\theta \to 0$, $I \to \infty$, the classical glory scattering singularity. It is a geometrical effect which should occur whenever $\theta(b)$ passes through $0, \pi, 2\pi, \ldots$. Again, for the scattering of light it may be seen as strong backscattering around the shadow of one's plane on the clouds underneath.

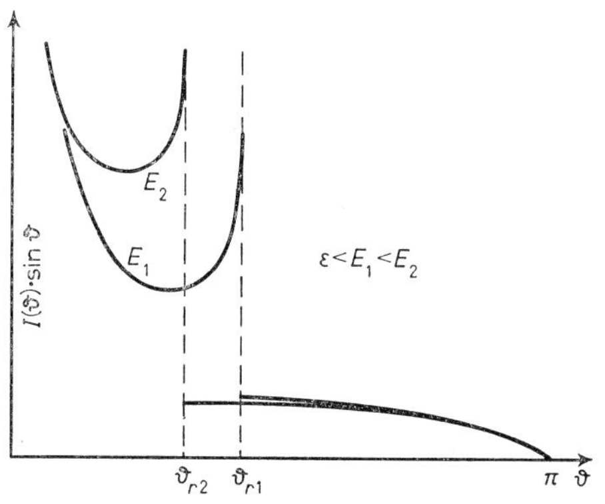

Fig. 11.

All classical singularities discussed above are replaced by finite maxima of the cross-section in a quantum-mechanical treatment if $V(r) \sim r^{-(3+\varepsilon)}$ asymptotically.

d) For $\theta > \theta_r$ only one impact parameter contributes. I is smaller than to the left of θ_r (dark side of meteorological rainbow) and approximates const$\cdot\sin\theta$ the better the more the repulsive potential core resembles a hard sphere.

e) As the collision energy is lowered $\theta_r(b_r)$ drops moving the rainbow to larger observation angles until this phenomenon disappears: $\theta(b_r) \to \infty$, the partners orbit around each other. The orbiting condition occurs for the trajectory having an r_c coinciding with the maximum in the effective potential

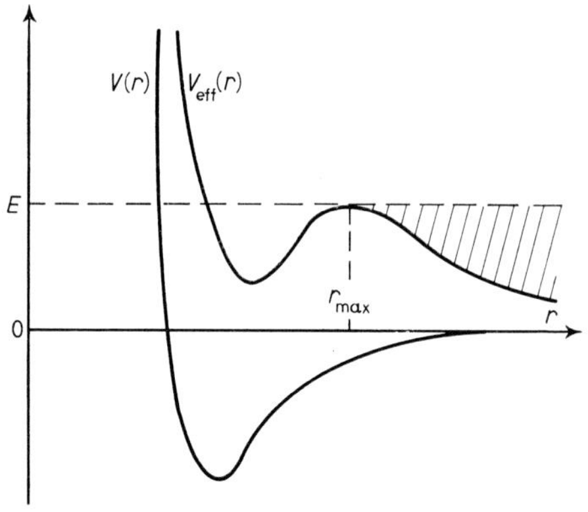

Fig. 12.

$$V_{eff}(r) = V(r) + E\frac{b^2}{r^2}$$

caused by the centrifugal barrier, see Fig. 12.

$$\frac{\partial V_{eff}}{\partial r} = 0 \to b(r_{max}) ,$$

$$1 - \frac{V(r_c)}{E} - \frac{b^2}{r_c^2} = 0 \to b(r_c) .$$

Equating the b's and dropping the indices of r yields

$$1 - \frac{V(r)}{E} - \frac{r}{2E}\frac{\partial V}{\partial r} = 0 , \tag{15}$$

the largest zero of which is $r_{c\ \mathrm{orbit}}$. Inspection shows that there exists no real root of (15) for $E > E^* = \max\left(V + (r/2)(\partial V/\partial r)\right)$. E^* is uniquely defined by the potential. The limiting orbit is given by $r^*_{c\ \mathrm{orbit}}$ to be determined from $\partial^2 V/\partial r^2 = -(3/r)(\partial V/\partial r)$. Orbiting will occur whenever the attraction is stronger than r^{-2}.

* * *

Per l'aiuto datomi nella preparazione del lavoro ringrazio caldamente i dottori M. G. Dondi e F. Torello.

REFERENCES

[1] *a*) R. K. B. Helbing: *Journ. Chem. Phys.*, **48**, 472 (1968); **50**, 4123 (1969); *b*) See also Prof. Herschbach's lecture on *Kinematic analysis of reactive scattering*.

[2] U. Buck and H. Pauly: *Zeits. Naturf.*, **23** a, 475 (1968).

[3] K. Berkling, R. Helbing, K. Kramer, H. Pauly, Ch. Schlier and P. Toschek: *Zeits. Phys.*, **166**, 406 (1962).

[4] F. v. Busch, H. J. Strunck and Ch. Schlier: *Zeits. Phys.*, **199**, 518 (1967).

[5] D. Beck and H. J. Loesch: *Zeits. Phys.*, **195**, 444 (1966).

Elastic Scattering of Nonreactive Atomic Systems.

D. BECK

Physikalisches Institut der Universität Freiburg - Freiburg i. Br.

1. – Elastic scattering and the atomic interaction.

The very simplest process to occur in the encounter of two atomic or molecular particles is the elastic scattering. It is a change in the state of the translational motion of the system only. Hence, it will provide information on the forces which act between the particles and which we shall assume to be represented by an intermolecular potential. It also serves to define familiar concepts like the mean free path in a gas, the coefficient of thermal conductivity, etc. Moreover, in the framework of the « optical model » it has been used as a basis to state a number of properties of very complicated interactions such as chemically reactive collisions [1].

In this lecture we shall begin with a reminder of the formal treatment of elastic scattering, and a review of the validity range of approximate methods in Sect. **2**. The JWKB and semi-classical methods which have been useful for quantitative work will be presented in some detail in Sect. **3** and **4**. We shall restrict the discussion to the characteristic features of the scattering from a spherically symmetric interaction potential as it is encountered for atomic partners in S-ground states. This entails several limitations. The application of our results to molecular partners will loose its quantitative significance with increasing deviation of the interaction from spherical symmetry. Collision energies should generally be in the thermal range or slightly above in order to conveniently observe the characteristic properties of atomic scattering. Also, results from beam experiments will virtually be the only reference to compare and confirm theoretical predictions. Finally, in Sect. **5** we shall make a few remarks on how potential data may be actually extracted from the scattering.

We shall not touch on the rather complex relations between the scattering and macroscopic properties of a gas. The point of major interest—the acquisition of information about the intermolecular potential—will thus be covered in a particularly significant aspect in our program, but very incompletely in general.

Let me, therefore, mention as an introduction what the major sources of information are. Much of our present knowledge about the potential comes from observations on macroscopic phenomena such as virial coefficients, transport properties of gases, the Joule-Thomson effect, compressibility and crystal data, optical line broadening and shifts. These methods and data obtained from them have been discussed in the literature [2]. Since about 1960 there is also an increasing number of contributions from beam experiments. They supplied, indeed, a very detailed insight into the scattering process, and—in spite of a rather limited number of systems investigated hitherto—turned out some useful experimental potential information, probably the most reliable to date. The largest body of data from this source is on alkali-rare gas and alkali-Hg interactions. Rare gas and H_2-rare gas as well as alkali-alkali systems have been investigated to some extent in addition to a large number of diverse observations primarily on alkali-molecule systems [3-5].

Besides experimental observations there are theoretical approaches, *a priori* calculations as well as semi-empirical methods. The former are considerably difficult to do. The variation energy has to be carried to very high accuracy to obtain a significant number for the desired interaction energy which is a small fraction of the total energy. Electron correlation must be carefully considered. Results, quantitatively useful in our context, are available for H-H [6] at all distances of interest, very accurate, and for He-He between 0.03 to 28 eV, probably accurate [7, 8]. Semi-empirical methods mainly predicting the strength of the Van der Waals force yield reliable values, often good to 10 % or better for a number of systems primarily rare gas-rare gas, rare gas-alkali and alkali-alkali partners [9, 10].

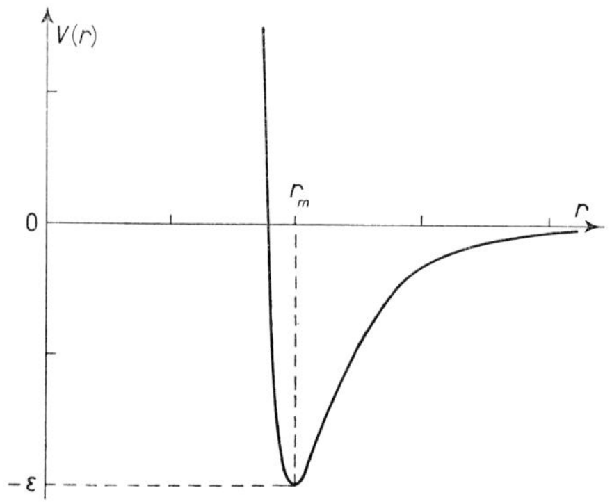

Fig. 1. – Interaction potential.

To give a rough summary it is found (see Fig. 1) that the interaction will be nonmonotonic, attractive at large, repulsive at small distances, and spherically symmetric for atoms in S-ground states [11]. For chemically inert partners the depth of the well ε may vary by 2 to 3 orders of magnitude, a representative number being 10^{-2} eV. The minimum distance r_m varies by hardly more than a factor of two, and is typically about 4 Å. For neutral atoms and nonpolar molecules in their ground states and $r \gtrsim 2r_m$

$$V(r) \approx -\frac{C^{(6)}}{r^6},$$

the induced dipole-induced dipole interaction, is a good approximation [12]. The repulsive interaction is not satisfactorily known except for H-H and He-He. It is often approximated by either an inverse power of r (Lennard-Jones (LJ) n-6 type potential models) or by an exponential term $e^{-\alpha r}$ (Buckingham exp [α-6] type models). There is no convincing argument for one to be generally superior to the other. Experiments are often insensitive to the repulsive part of the potential, or too limited in energy range and can, then, be explained by either assumption. Evidence in favour of an exp-repulsion in particular cases is:

a) The above-quoted He-He results may be closely approximated by an exp-repulsion between 4 and 28 eV.

b) Experiments on high-energy scattering, high temperature thermal conductivity, compressibility and viscosity of He-He can reasonably accurately be explained by an exp-repulsion. This is a significant finding because it holds in a remarkably wide energy range $0{,}01 \leqslant V(r) \leqslant 25$ eV [5*a*, *b*, 13].

c) Experiments on transport properties in gases often indicate a « soft » interaction left of the minimum. To account for it the exp-repulsion is reported to be superior to r^{-n} [14].

All potential data derived from experimental observations should be used with some caution. The general inversion problem is not solved. Instead, a fitting procedure must be used. A model interaction is chosen, *e.g.* the LJ 12-6 model

$$V(r) = \varepsilon\left[\left(\frac{r_m}{r}\right)^{12} - 2\left(\frac{r_m}{r}\right)^{6}\right],$$

which is represented by some functional form of the independent variables of the problem, as for this lecture simply r, the partner distance. It also contains a small number of parameters for flexibility, ε and r_m in the example. The model is varied until some observable quantity calculated from it agrees with the respective experimental result to within the latter's error. The model variation is brought about by parameter variation. Potential information so obtained are numerical values for the parameters but *not* the respective quantities of the real interaction. Consequently, a parameter value may come out different, but still give a fit within experimental error if the functional form or other parameters of the model are changed: parameters are model-dependent. For critical use it is not sufficient to know the uncertainty they carry due to experimental errors. Some indication of their model-dependence would also be very desirable. Obviously, no statement of any generality will be possible. There is no convincing way out of this situation except increasing the experimental accuracy and amount of different data on the same system. Nevertheless, there are features of the scattering which are more sensitive to a certain property of the interaction than to others. To the degree that the respective potential property is predominantly determined by a single parameter this will come out weakly model-dependent if it is evaluated from the sensitive feature of the scattering [15]. Regarding their significance for the real interaction it is, therefore, possible to distinguish « safe » and « unsafe » parameters. This quality will depend on the feature of the scattering from which the parameter is evaluated as well as on the parametrization of the model. Unfortunately, only recently has there been paid closer attention to this problem. We shall point out a few more specific details in Sect. **5**.

2. – Formal solutions and validity of approximate methods [16].

Let the interaction be represented by a single, static and spherically symmetric potential $V(r)$. The elastic-scattering problem is simply to find the solution of

$$\left\{-\frac{\hbar^2}{2m}\left[\Delta_r+\frac{1}{r^2}\Delta_{\vartheta\varphi}\right]-E+V(r)\right\}\psi(\boldsymbol{r})=0$$

for

$$E=\frac{\hbar^2k^2}{2m}>0\,,$$

which asymptotically behaves as

$$\psi_{\boldsymbol{k}}(\boldsymbol{r})\sim\exp[i\boldsymbol{k}\cdot\boldsymbol{r}]+\frac{f(\boldsymbol{k}',\boldsymbol{k})}{r}\cdot\exp[ikr]\,,\qquad |\boldsymbol{k}'|=|\boldsymbol{k}|,$$

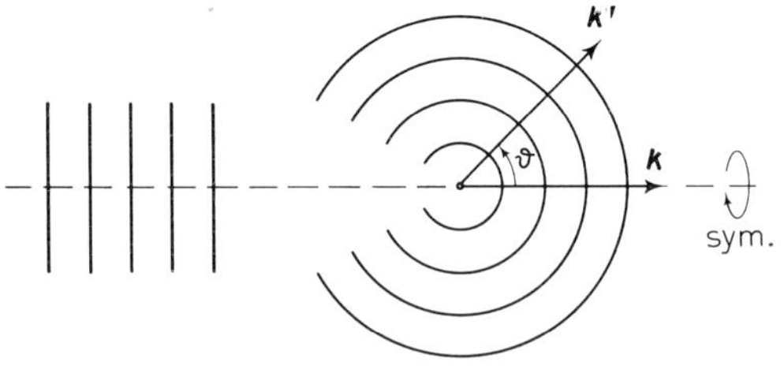

Fig. 2. – Asymptotic structure of the wave field for a scattering process. The plane wave passes over the scatterer and proceeds to right.

i.e. which describes the field sketched in Fig. 2. Two useful formal solutions are

$$(1)\qquad f(\boldsymbol{k}',\boldsymbol{k})=f(\vartheta,k)=\sum_{l=0}^{\infty}(2l+1)\,P_l(\cos\vartheta)\cdot\frac{\exp[2i\eta_l]-1}{2ik}\,,$$

where the azimuthal symmetry of $f(\boldsymbol{k}', \boldsymbol{k})$ for a spherically symmetric interaction is explicitly used, and

$$f(\boldsymbol{k}', k) = -\frac{1}{4\pi}\int \exp[-i\boldsymbol{k}'\cdot\boldsymbol{r}]\, U(r)\cdot\psi_{\boldsymbol{k}}(\boldsymbol{r})\,\mathrm{d}\boldsymbol{r}\,, \tag{2}$$

where $U(r) = (2m/\hbar^2)V(r)$.

Both (1) and (2) essentially restate the content of the Schrödinger equation including the boundary condition and symmetry, but do not actually solve the problem. In (1) the asymptotic phase differences η_l between outgoing spherical waves with and without $V(r)$ have to be found for each contributing angular momentum l. They are obtained by integration of the radial Schrödinger equation. (2) shows that a knowledge of the solution of Schrödinger's equation in that region of space will be sufficient where $U(r)$ differs from zero. In (2) $U(r)$ need not have spherical symmetry which makes it a suitable starting point for more complicated problems [17, 18].

For a realistic interaction at collision energies of practical interest (from 10^{-2} eV on up) numerical integration procedures must in general be used. As this will often be a considerable effort let us briefly review the nature of several approximation methods and indicate their range of validity. To do so we shall describe the interaction by two parameters V_0 and R_0 of which the former is a representative absolute measure of the strength of the interaction, the latter stands for its representative linear dimension. In a spherical Gaussian potential such parameters would have a particularly obvious meaning.

2·1. *Born's approximation.* – In (2) let

$$\psi_{\boldsymbol{k}}(\boldsymbol{r}) \to \exp[i\boldsymbol{k}\cdot\boldsymbol{r}]\,. \tag{3}$$

This leads to

$$f(\boldsymbol{k}', \boldsymbol{k}) = -\frac{1}{4\pi}\int \exp[i(\boldsymbol{k}-\boldsymbol{k}')\cdot\boldsymbol{r}]\cdot U(r)\,\mathrm{d}\boldsymbol{r}\,,$$

the well known Born scattering amplitude. By (3) we neglect the deviation of $\psi_{\boldsymbol{k}}(\boldsymbol{r})$ from the plane wave throughout the range of $U(r)$. We might interpret this as assuming the potential to exert a small action on the incoming wave, or

$$V_0\frac{2R_0}{v} \ll \hbar\,, \tag{4}$$

v being the particle velocity. (4) is a sufficient condition for Born's approximation to be valid. In (1) it implies that *all* $\eta_l \ll 1$ which will never be true in cases we are interested in. However, for some l it might be true. The advantage of the sum in (1) is that better approximations may be used where Born's assumption tends to fail.

2·2. JWKB *approximation.* – This method applies under conditions which are very often true in atomic collisions. There are two variations of it which are not always clearly distinguished in the literature although their range of validity is considerably different for 3-dimensional problems. One variation [17] has the validity range of the semi-classical treatment of scattering which we shall describe later at some detail. Here, we mean the other which —broader in range and identical to the first under semi-classical conditions— proceeds as follows: It is assumed that the *local* wave length $\lambda\!\!\!^{-}_r$ of the motion changes slowly within the range of the potential

$$\lambda\!\!\!^{-}_r \left|\frac{\mathrm{d}\lambda\!\!\!^{-}_r}{\mathrm{d}r}\right| \ll \lambda\!\!\!^{-}_r \quad \text{or} \quad \lambda\!\!\!^{-}_r \left|\frac{\mathrm{d}V}{\mathrm{d}r}\right| \ll T_r\,. \tag{5}$$

With this assumption the radial Schrödinger equation is approximately integrated to find the phase shifts η_l. (1) is used to obtain the scattering amplitude. Note that the approximation is introduced in *one* dimension and exact angular functions are used in this procedure.

(5) is a necessary condition. It is not sufficient because the *local* kinetic energy T_r may become zero. For the moment we shall exclude this problem of the classical turning point and the possibility of tunnelling from the discussion and come back to it later. For the representative interaction

$$\left|\frac{\mathrm{d}V}{\mathrm{d}r}\right| \sim \frac{V_0}{R_0}$$

and the JWKB condition (5) reads

$$\frac{\lambda\!\!\!^{-}}{R_0} \ll \frac{E}{V_0} \qquad \text{for } E \gg V_0\,, \tag{5a}$$

$$\frac{\lambda\!\!\!^{-}_r}{R_0} \ll 1 \qquad \text{for } E \lessgtr V_0\,. \tag{5b}$$

E is the energy at infinite partner separation and $\lambda\!\!\!^{-}$ the corresponding de Broglie wave length. The turning-point problem being excluded (5*b*) will be physically interesting only for attractive interactions. Both conditions mean that the particle wave length be « sufficiently » small compared to the linear dimension of the potential everywhere except for the vicinity of the classical turning point.

2.3. *Classical approximation.* – A classical trajectory cannot be reasonably defined unless the uncertainty of the scattering angle $\Delta\vartheta$ is small compared to 1. From the uncertainty relation at best we have

$$\Delta\vartheta \sim \frac{\Delta p_\perp}{p_\parallel} \sim \frac{\hbar}{R_0}\frac{1}{p_\parallel} = \frac{\lambda\!\!\!^-}{R_0}\,.$$

So

$$\frac{\lambda\!\!\!^-}{R_0} \ll 1 \tag{6}$$

is necessary. It is sufficient if for the representative scattering angle ϑ we also have

$$\frac{\Delta\vartheta}{\vartheta} \ll 1\,.$$

Obviously, this will become stringent if forward scattering is dominant, *i.e.* at high energies. Now,

$$\vartheta \sim \frac{\int K_\perp \mathrm{d}t}{p_\parallel} \sim \frac{V_0}{R_0}\cdot\frac{2R_0}{v}\cdot\frac{1}{p_\parallel} \sim \frac{V_0}{E} \qquad \text{for } E \gg V_0\,.$$

This requires

$$\frac{\vartheta}{\Delta\vartheta} \sim \frac{V_0}{E}\frac{R_0}{\lambda\!\!\!^-} \sim 2\frac{V_0 R_0}{\hbar v} \gg 1\,. \tag{6a}$$

2.4. *High-energy approximation.* – Although we shall not use it in this lecture the high-energy approximation should be mentioned. Its formal appearance makes it particularly useful for application to nonspherically symmetric potentials [17-19].

$$\frac{V_0}{E} \ll 1 \qquad \text{and} \qquad \frac{\lambda\!\!\!^-}{R_0} \ll 1 \tag{7}$$

are its simultaneous sufficient conditions which may often be true in heavy-particle collisions at not too large angles.

To summarize: The action parameter brings out clearly that the Born and the classical approximations are opposite extremes which never overlap. The JWKB and high-energy approximations place an entirely different restriction on the action. From (5*a*) and (7)

$$\left(\frac{R_0}{\lambda\!\!\!^-}\right)^2 \gg \frac{2V_0 R_0}{\hbar v} \qquad \text{for } E \gg V_0\,. \tag{8}$$

In atomic collisions virtually always $R_0/\lambda\!\!\!^{-} \gg 1$. Thus, these methods bridge the gap between the former approximations and may well overlap both of them. With regard to the JWKB version sketched above it seems worth mentioning that it is *not* restricted to semi-classical conditions. Nevertheless, situations in which the action obeys (8), but considerably exceeds 1 are quite common in atomic collisions.

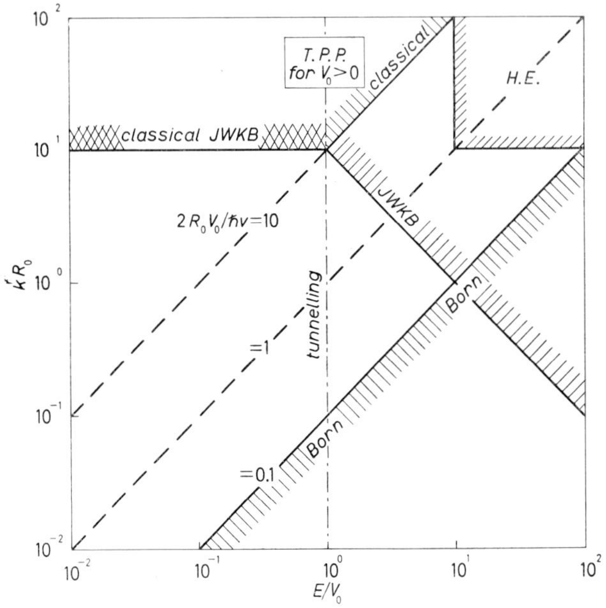

Fig. 3. – Survey of the validity of approximate methods. The interaction is given by its representative strength V_0 and its representative linear dimension R_0. Effects due to the centrifugal potential are not included. For $V_0 > 0$ tunnelling will occur for $E \leqslant V_0$ but will be particularly important at $2V_0R_0/\hbar v \sim 1$. For $V_0 = E$ the turning point problem (T.P.P.) will arise in the JWKB method.

Figure 3 surveys the above results (*). The values 10 or 0.1 for the important scattering parameters are arbitrarily chosen as large or small compared to 1, respectively. The degree of convergence of the various approximations towards the respective limiting case requires detailed analysis. It may not be uniform in terms of the scattering parameters used here [20].

To be a little more specific about the applicability to atomic collisions let the well depth ε of a potential like that in Fig. 1 stand for V_0 and the equilib-

(*) Many papers use abbreviations for the parameter combinations which are involved here (identifying $R_0 = r_m$ and $V_0 = \varepsilon$): $r_m/\lambda\!\!\!^{-} = kr_m = A$, $E/\varepsilon = K$, $2r_m\varepsilon/\hbar v = D$ (Editor).

rium distance r_m for R_0. Table I lists these quantities for the systems He-He and K-Xe. A typical collision energy is 0.1 eV. For this the relevant scattering parameters are also given. For He-He the JWKB method is appropriate. The action parameter indicates that the scattering will be predominantly at the Born limit. The classical approximation is expected to fail. For K-Xe, on the other hand, the scattering is again far in the JWKB regime, but approaching classical conditions.

TABLE I. – *Scattering parameters for two atomic systems at a collision energy of* 0.1 eV.

	$\varepsilon \cdot 10^{14}$ (erg)	$r_m \cdot 10^8$ (cm)	$r_m/\lambda\!\!\!^-$	E/ε	$2\varepsilon r_m/\hbar v$
He-He	0.14	3.0	29	112	0.26
K-Xe	2.1	5.2	200	7.7	26

At angles below $\Delta\vartheta \sim \lambda\!\!\!^-/r_m$ we know that the scattering will be ruled by the uncertainty relation and will never be classical, regardless of what happens at larger angles. One consequence of this is the removal of a dramatic failure to which a classical treatment would lead: For potentials asymptotically approaching zero it would, obviously, give unbounded differential cross-sections at zero angle, divergent integral cross-sections being a consequence. Actually, as we shall indicate later, cross-sections remain finite under the following conditions:

$$I(\vartheta) \text{ finite for } \vartheta \to 0 \quad \text{if} \quad V(r) \sim r^{-(3+\delta)} \text{ asymptotically,}$$

$$\sigma \quad \text{finite} \quad \text{if} \quad V(r) \sim r^{-(2+\delta)} \text{ asymptotically,}$$

where $\delta > 0$. Note that the condition for the integral cross-section is just the one for an (attractive) potential to hold a finite number of bound states [21].

In a partial-wave treatment (1), for both our systems, a substantial number of angular momenta will significantly contribute to the scattering. By semi-classical equivalence this number is roughly given by

$$|\boldsymbol{l}| \sim \hbar l \sim mv \cdot r_m \to l \sim \frac{r_m}{\lambda\!\!\!^-}. \tag{9}$$

In (1) the action (in units of $\hbar$) of the potential on the wave is expressed by the phase shifts η_l. Intuitively, one may argue, a representative phase shift will be of the order of the action parameter. It is noteworthy that for K-Xe a majority of phase shifts will be large compared to one. For He-He a majority will be small. By inspection of (1) we see that these partial waves will constructively interfere in the forward direction. Hence, we may expect the scattering to be heavily concentrated at small angles for He-He.

3. – The JWKB approximation.

3·1. *The* JWKB *phase shift.* – Knowing the nature of approximate methods, let us proceed to quantitative work. Very rarely have exact numerical solutions of the Schrödinger equation been obtained. In such cases the Numerov method was found to be more than three times faster than the Runge-Cutta integration [22]. Instead, the JWKB method as outlined in Sect. **2·1** and the semi-classical approximation to be discussed in Sect. **4** have been of greatest practical importance. At the range and accuracy of present experiments the former is fully satisfactory. It does, however, involve the calculation of a generally large number of phase shifts and the summation according to (1). The semi-classical method is substantially simpler. Except for very light systems its accuracy will be better than 10 % [23]. For the majority of experiments reported to date it was felt to be sufficient. The outstanding merit of this method lies, however, in the insight it affords into how the scattering comes about if a large number of partial waves contribute.

As to the JWKB method its accuracy will, of course, be primarily a matter of the phase shifts [22*b*]. Let us briefly review how they are obtained. The radial Schrödinger equation

$$\left[\frac{1}{r^2}\frac{\partial}{\partial r}\left(r^2\frac{\partial}{\partial r}\right)+k^2-U(r)-\frac{l(l+1)}{r^2}\right]u_l(r)=0$$

in view of (5) is solved by

$$u_l(r)=\frac{1}{kr}\,a_l(r)\exp\left[\frac{i}{\hbar}S_l(r)\right] \tag{10}$$

iteratively to first order in a_l and S_l. This yields Jeffrey's approximation

$$^{\text{JWKB}}u_l(r)=\frac{\text{const}}{kr}\cdot\frac{1}{[k(r)]^{\frac12}}\sin\left[\frac{\pi}{4}+\int_{r_c}^{r}k(r')\,\mathrm{d}r'\right]. \tag{11}$$

From this we read the phase shift

$$^{\text{JWKB}}\eta_l=\lim_{r\to\infty}\left[\int_{r_c}^{r}k(r')\,\mathrm{d}r'-\int_{{}^0r_c}^{r}{}^0k(r')\,\mathrm{d}r'\right], \tag{12}$$

where

$$k(r)=\left[k^2-U(r)-\frac{(l+\frac12)^2}{r^2}\right]^{\frac12},\qquad k(r_c)=0,$$

$${}^0k(r)=\left[k^2-\frac{(l+\frac12)^2}{r^2}\right]^{\frac12},\qquad {}^0k({}^0r_c)=0,$$

are the local wave numbers with and without the interaction $U(r)$. The second term on the right of (12) may be integrated to give an alternative form of the phase shift

$$(12a)\qquad {}^{\text{JWKB}}\eta_l = \frac{\pi}{4} + \frac{l\pi}{2} - kr_c + \int_{r_c}^{\infty}[k(r) - k]\,\mathrm{d}r\,.$$

Note that according to LANGER [24] the $l(l+1)$ of the Schrödinger equation has been replaced by $(l+\frac{1}{2})^2$. This often leads to increased accuracy of ${}^{\text{JWKB}}\eta_l$ [20]. A general reason is that the local wave number integral in the force-free ${}^{\text{JWKB}}u_l(r)$ may be analytically integrated. It takes on the correct asymptotic behaviour—which is known to be

$${}^{0}u_l(r) = j_l(r) \sim \frac{1}{kr}\sin\left[kr - \frac{l\pi}{2}\right]—$$

only with Langer's modification.

Figure 4 *a*) shows the deviation of ${}^{\text{JWKB}}\eta_l$ from the exact numerically integrated phase shift $\eta_{l\text{ex}}$ for a LJ 12-6 model potential. The scattering parameters

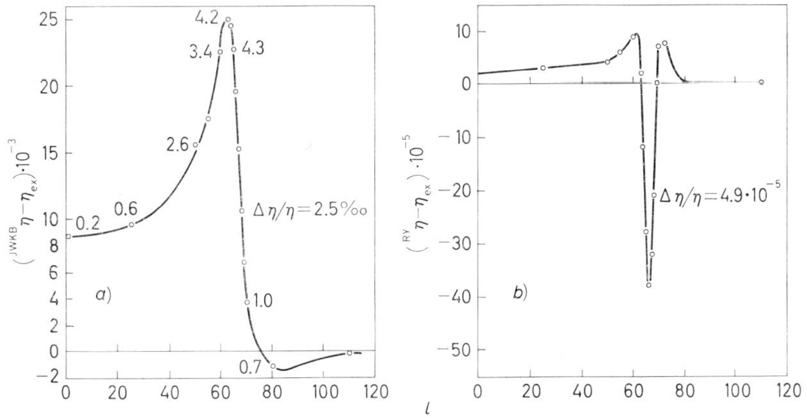

Fig. 4. – *a*) Deviation of the JWKB approximation to the phase shifts from the exact values [22*a*]. The numbers along the curve give the fractional deviation in ‰. A LJ 12-6 model interaction has been assumed. The scattering parameters are: $K = E/\varepsilon = 3$; $A = r_m/\lambda\!\!\!^{-} = 54.7$; $B = A^2/K = 10^3$. *b*) Deviation of the Rosen-Yennie phase shifts from the exact values. Parameters are as in Fig. 4 *a*. Note the units on the ordinate scale.

are in the JWKB regime; but for the majority of systems (except the lightest ones) they will be even more favourable. The agreement is good, indeed. The largest error does *not* occur for $l \to 0$ as is often argued. The JWKB phase shift does not in general converge uniformly towards the exact one. For the majority of atomic scattering problems it may be interesting to note that the convergence effectively does not hinge on condition (5) but on the turning point or « connection » problem of the JWKB method [25, 26]. It raises an independent and different requirement for the validity of the ${}^{\text{JWKB}}\eta_l$. We mentioned that (5) never holds at a classical turning point r_c as $T_r \to 0$. Jeffrey's solution (11) diverges at this point because of the factor $[k(r)]^{-\frac{1}{2}}$. This needs some interpretation: Jeffrey's solution does *not* solve the radial Schrödinger equation in the vicinity of $r = r_c$. Instead, it is constructed as the analytic continuation of an exact solution at the turning point and its immediate vicinity. Now, the exact solution continued by (11) is obtained for a *linear* approximation to the (effective) *potential* at r_c. This is an inherent feature of the JWKB method whose consequences are: *a*) The characteristic « start phase » $\pi/4$ along with the lower limit r_c of the wave number integral in (11). The ${}^{\text{JWKB}}\eta_l$ are completely insensitive to the potential in the classically forbidden region. *b*) A further necessary validity requirement for the ${}^{\text{JWKB}}\eta_l$. In the classically allowed region near the turning point there must be some r-interval in which both (5) *and* the linear approximation to the (effective) potential are true. This may be simply shown to require analytically

$$\left.\frac{|U''_{\text{eff}}|^{\frac{1}{2}}}{(U'_{\text{eff}})^2}\right|_{x=0} \ll \lambda\!\!\!^{-} , \tag{13}$$

with $U_{\text{eff}} = U + (l + \frac{1}{2})^2/r^2$, and the derivatives to be taken with respect to $x = k(r - r_c)$. It is really (13) that accounts for the order of magnitude and the overall behaviour of $\Delta\eta_l$ in Fig. 4*a* and almost everywhere in the range of scattering parameters pertinent in atom-atom collisions. Figure 5 shows the maximum deviation throughout the region of interest. In terms of the order-of-magnitude considerations of Sect. 2 condition (13) reads

$$\left[\frac{\hbar^2}{2mV_0 R_0^2}\right]^{\frac{1}{2}} = \frac{1}{B^{\frac{1}{2}}} \ll 1 ,$$

m being the reduced mass of the partners, and B the «capacity» of the potential, a quantity related to the number of bound states which an attractive well can hold. Note that this simplified version of (13) is closely borne out on a relative scale

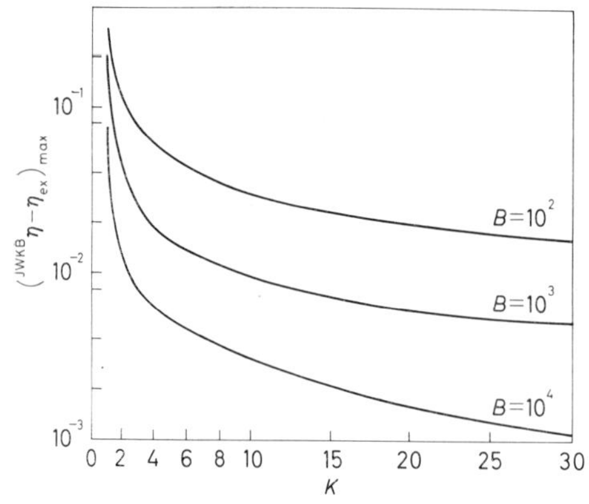

Fig. 5. – Maximum error, of ${}^{\text{JWKB}}\eta_l$ corresponding to the peak of the curve in Fig. 4*a*. The majority of experiments to date had parameters $3 \leqslant K \leqslant 200$ $100 < B \leqslant 10\,000$.

by the data of Fig. 5. *c*) The JWKB approximation breaks down if the effective potential U_{eff} has zero slope at r_c. A general reason for this to happen corresponds to classical orbiting [27]. It is brought about by every potential which is attractive and falls off faster than r^{-2} asymptotically. It occurs for a particular l-value if the energy is below a critical limit which is characteristic for the potential, $E \leqslant 0.80 \cdot \varepsilon$ for the LJ 12-6 model. Again, Fig. 5 clearly indicates this feature. As the point of zero tangent is a maximum in U_{eff} tunnelling will occur for collision energies below the top of the U_{eff} barrier. The ${}^{\text{JWKB}}\eta_l$ do *not* account for this as is implied by (11) or the statement under *a*) [28]. Potential scattering resonances due to virtual bound levels will, thus, be absent from JWKB calculations [28, 20]. This is, however, not a serious drawback in practice. As long-range forces produce the barrier, tunnelling turns out to be very inefficient except for the first few l-values just below the top of the barrier. Unless hydrogen or helium is a partner, and sufficiently low-energy collisions can be produced there is little hope to observe tunnelling effects in atomic scattering [29].

3'2. *The Rosen-Yennie phase shift.* – There have been many attempts to improve on the JWKB approximation. Very convincing work has been done by ROSEN and YENNIE [30]. They account from the beginning for the three-dimensional problem using the spherical Bessel function instead of (10) in their ansatz:

$$U_l(r) = a_l(r) \cdot j_l(S(r)) + b_l(r) \cdot j_l'(S(r)) \, .$$

j_l' is the derivative with respect to r. They obtain

$${}^{\text{RY}}\eta_l = {}^{\text{JWKB}}\eta_l + \operatorname{arc\,tg} b_l(\infty)$$

with

$$b_l(\infty) = -\frac{1}{24}\int\limits_{r_c}^{\infty}\frac{\mathrm{d}r}{r}\left[k^2 - U(r) - \left(\frac{l+\frac{1}{2}}{r}\right)^2\right]^{-\frac{1}{2}} \cdot \frac{\mathrm{d}}{\mathrm{d}r}\left\{\frac{4(k^2 - U) - 5rU' - r^2 U''}{2(k^2 - U) - rU'}\right\},$$

defined only for $l > l_{\text{orbit}}$ in case $k^2 < k^2_{\text{orbit}}$ ($U' = \partial U/\partial r$ etc.). The typical improvement in accuracy illustrated by Fig. 4*b* is remarkable. The computing time for η_l is but two times that for ${}^{\text{JWKB}}\eta_l$. Apart from orbiting conditions there is virtually no need any more for numerical phase integrations.

3'3. *The Jeffreys-Born phase shift.* – This is the Born limit of the phase shifts

$$(14) \qquad {}^{\text{JB}}\eta_l = -\frac{1}{\hbar v}\int\limits_{{}^0 r_c}^{\infty}\frac{V(r)\,\mathrm{d}r}{[1 - ((l+\frac{1}{2})/kr)^2]^{\frac{1}{2}}};$$

valid if

$$|U(r)| \ll \left(\frac{l+\frac{1}{2}}{r}\right)^2 \qquad \text{for all } r \geqslant r_c \approx {}^0r_c\,.$$

Relating the phase shift linearly to the potential it is even simpler than the ${}^{\text{JWKB}}\eta_l$. It may be obtained from these as the first term of an expansion in $f(V(r))/E$ [31]. Therefore, it will not be surprising that the ${}^{\text{JB}}\eta_l$ are the same as the high-energy phase shifts [17]

$$\text{(14a)} \qquad {}^{\text{HE}}\eta_l = -\frac{1}{2\hbar v}\int_{-\infty}^{+\infty} V\left(\sqrt{(l+\tfrac{1}{2})^2+z^2}\right)\mathrm{d}z = {}^{\text{JB}}\eta_l\,.$$

This is easily shown by noting that ${}^0r_c$ is really the classical impact parameter $b = \lambda\!\!\!^{-}(l+\frac{1}{2})$, and $z^2 = r^2 - b^2$. Although we shall not discuss the high-energy approximation the interpretational aspect of (14a) with respect to (14) is noteworthy: ${}^{\text{JB}}\eta_l$ is essentially the action of the potential on the wave integrated over a straight-line trajectory.

Both (14) and (14a) suggest that if

$$\text{(15)} \qquad V(r) = \frac{C}{r^s}\,, \qquad s > 2\,,$$

there will be an (energy-dependent) l-value above which the ${}^{\text{JB}}\eta_l$ become a valid approximation. For (15) and a few other simple forms they can be integrated analytically:

$$\text{(16)} \qquad {}^{\text{JB}}\eta_l = -f(s)\cdot\frac{C}{\hbar v}\cdot\frac{1}{[\lambda\!\!\!^{-}(l+\frac{1}{2})]^{s-1}}\,, \qquad f(s) = \frac{\sqrt{\pi}}{2}\cdot\frac{\Gamma((s-1)/2)}{\Gamma(s/2)}\,.$$

The asymptotic behaviour of the potential determines the forward scattering amplitude and the integral cross-section. The statements at the end of Sect. **2** about their relations can be easily proven by expansion for small η_l in (1) and (26), replacement of sums over l by integrals, and the use of (16). Becaues of their simplicity the ${}^{\text{JB}}\eta_l$ will often be useful for purposes of rough qualitative interpretation.

Having seen the power and restrictions of the JKWB method we know how to get reliable cross-sections for a large number of atom-atom scattering problems. The superposition (1) of all the phase-shifted partial waves to bring about the scattering is, however, still a little obscure. It will be worth-while to make it more transparent.

4. – The semiclassical approximation.

This method, previously existing in fragments only, was developed and applied in two fascinating articles by FORD and WHEELER in 1959 [20, 32]. Its range of applicability is different from that of the JWKB method. The action parameter is required to be larger than 1, thus tending towards classical conditions.

4'1. *Basic idea.* – On the basis of the partial-wave treatment the method may be defined by adopting the following approximations:

1) $\eta_l \to {}^{\text{JWKB}}\eta_l$; we drop the index JWKB henceforth. In the semicalssical method phase shifts are always used in this approximation.

2) $P_l(\cos\vartheta) \simeq \dfrac{\sin[(l+\frac{1}{2})\vartheta + \pi/4]}{[(\pi/2)(l+\frac{1}{2})\cdot\sin\vartheta]^{\frac{1}{2}}}, \qquad \text{for } \sin\vartheta \gtrsim \dfrac{1}{l}.$

3) $\sum\limits_l \to \int \mathrm{d}l$

4) Evaluation of integrals by the method of stationary phase or other appropriate techniques.

From (8) we see that these assumptions are consistent: If the ${}^{\text{JWKB}}\eta_l$ are correct and large, *i.e.* under semi-classical conditions, many partial waves do, indeed, contribute to the scattering. By the approximation to the Legendre polynomials angles ϑ close to 0 or π are excluded from the following. We may, thus, use that $\sum\limits_l (2l+1)P_l = 0$ for $\vartheta \neq 0, \pi$ and obtain from (1) with the adopted approximations the semiclassical scattering amplitude

$$f_{\text{sc}}(\vartheta, E) = -\frac{\lambda\!\!\!^{-}}{[2\pi\sin\vartheta]^{\frac{1}{2}}}\int\limits_0^\infty (l+\tfrac{1}{2})^{\frac{1}{2}}\{\exp[i\varphi_+] - \exp[i\varphi_-]\}\,\mathrm{d}l \tag{17}$$

with

$$\varphi_\pm = 2\eta(l) \pm (l+\tfrac{1}{2})\vartheta \pm \frac{\pi}{4}\,. \tag{17a}$$

$\eta(l)$ is considered to be a continuous function of l. (17) is a very illuminating form. According to our assumptions both the $2\eta(l)$ as well as the $(l+\frac{1}{2})\vartheta$ terms may be large compared to one. Both exponents $\varphi_\pm$ can be expected to show a large variation with l. The integrand of (17) will, thus, be a rapidly oscillating function of l. A negligible contribution to the scattering amplitude results. A quite different situation arises if for some l either exponent becomes stationary

$$\frac{\partial \varphi_\pm}{\partial l} = 0 \qquad \text{for } l = l_\vartheta,\ i.e.\ 2\frac{\partial \eta}{\partial l} = \mp \vartheta\,. \tag{18}$$

Let us keep the observation angle ϑ fixed for a moment and use it to label the stationary l (which apparently depends on ϑ). From (18) we take that for a given ϑ either φ_+ or φ_-, but not both, may become stationary. Figure 6 shows the situation for a purely attractive potential, the $\eta(l)$ being positive and monotonically decreasing. Here φ_+ will be stationary.

Now, there will be a nonnegligible contribution to f_{sc} arising from the l_ϑ partial wave and some neighbourhood of it. We expand $\eta(l)$ about l_ϑ to second order

$$\varphi_\pm(l) = 2\eta(l_\vartheta) - 2\eta'(l_\vartheta)\cdot [l_\vartheta + \tfrac{1}{2}] + 2\eta'(l_\vartheta)[l + \tfrac{1}{2}] + \eta''(l_\vartheta)[l - l_\vartheta]^2 \pm (l + \tfrac{1}{2})\vartheta \pm \frac{\pi}{4}\,. \tag{19}$$

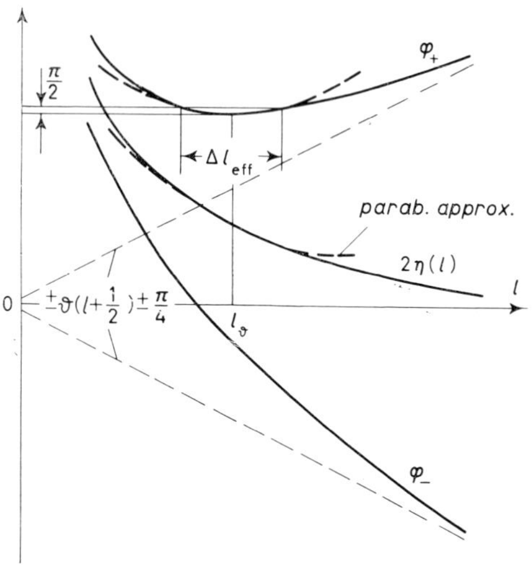

Fig. 6. – How a point of stationary phase comes about for an attractive potential.

Two terms cancel because of (18). We neglect the nonstationary branch, φ_- in the example of Fig. 6, insert the stationary one into f_{sc}, take all constant terms including the slowly varying $(l + \frac{1}{2})^{\frac{1}{2}}$ at l_ϑ out of the integral and obtain

$$f_{sc} \sim \int_{-\infty}^{+\infty} \exp\,[i\eta''\cdot(l - l_\vartheta)^2]\, dl\,.$$

Consistent with the semi-classical assumptions it causes negligible error to extend the lower limit of the integral from 0 to $-\infty$. The resulting Gauss-Fresnel integral is easily evaluated to give the final result

$$f_{sc}(\vartheta, E) = \lambda\left[\frac{l + \frac{1}{2}}{2|\eta''|\cdot\sin\vartheta}\right]_{l_\vartheta}^{\frac{1}{2}} \cdot \exp\,[i\beta]\,, \tag{20}$$

$$\beta = \left[2\eta(l) - 2\eta'(l)\cdot[l + \tfrac{1}{2}] - \left(2 - \frac{\eta''}{|\eta''|} - \frac{\eta'}{|\eta'|}\right)\frac{\pi}{4}\right]_{l_\vartheta}\,. \tag{20a}$$

This line of arguments clearly reveals the important feature of the scattering at semi-classical conditions. According to (1) *all* angular momenta contribute to the scattering at a given observation angle. Almost all of them, however, will be destructively interfering. There will be *one* l—related to the angle by (18)—and its immediate neighbours for which the interference is constructive, thus generating intensity at this angle. We have a situation reminiscent of the eikonal approximation of optics. Also we can see how the classical limit —in mechanics or optics—with its *sharp* one-to-one correspondence of intensity at a given angle to impact parameter comes about. A few further comments are:

1) The main contribution to the Gauss-Fresnel integral comes from an l range

$$\Delta l_{\text{eff}} = \left[\frac{\pi}{|\eta''|}\right]^{\frac{1}{2}} \tag{21}$$

centered around l_ϑ. For f_{sc} in (20) to be quantitatively useful the parabolic approximation (19) must represent the actual phase shifts correctly within this interval, and there must be many l within it. For systems like K-Xe this will in general be true. Naturally, (19) will become incorrect somewhere. If this is outside Δl_{eff} it matters little because destructive interference has set in.

2) The stationary-φ condition (18) establishes the relation between observation angle and contributing l-value through the derivative of the phase shift with respect to l. Exactly this relation

$$2\frac{\partial}{\partial l}\,{}^{\text{JWKB}}\eta(l) = \vartheta(l) \tag{22}$$

is well known to exist between the JWKB phase shifts and the classical deflection function $\vartheta(l)$. (Other than the observation angle ϑ the latter may have either sign, thus distinguishing between repulsive and attractive deflections $\vartheta(l) > 0$ and < 0 respectively.) This indicates again the consistency of this phase choice with the semi-classical argumentation.

3) Apart from the phase β there is no difference in the semi-classical scattering amplitude (20) whether the φ_+, *i.e.* $\eta' < 0$, or the φ_- branch, *i.e.* $\eta' > 0$, of (17) become stationary.

4) Other than in the JWKB method f_{sc} is very simply obtained once the phase shifts are known.

4·2. *Implications.*

1) Classical cross-section. From the semi-classical equivalence relationship for the angular momentum we have for the classical impact parameter

$$b = \lambda\!\!\!^{-}(l + \tfrac{1}{2})\,.$$

The stationarity condition for φ may be written in classical terms

$$2\lambda \frac{\partial \eta(b)}{\partial b} = \vartheta(b) \,. \tag{22a}$$

From (20) the semi-classical differential cross-section is

$$I_{sc}(\vartheta, E) = |f_{sc}|^2 = \frac{\lambda(l+\frac{1}{2})}{2|\eta''|\cdot \sin\vartheta} = \frac{b}{|\partial\vartheta(b)/\partial b|\cdot\sin\vartheta} = I_{cl}\,,$$

which turns out to be exactly the classical result. This may not be surprising but we can see the implications: The parabolic approximation to the phase shifts must be valid, and there must be just *one* stationary l for each ϑ.

2) Interference. There may be more than one point of stationary phase. Figure 7 shows the two simplest cases where two l values, l_1 and l_2, are stationary

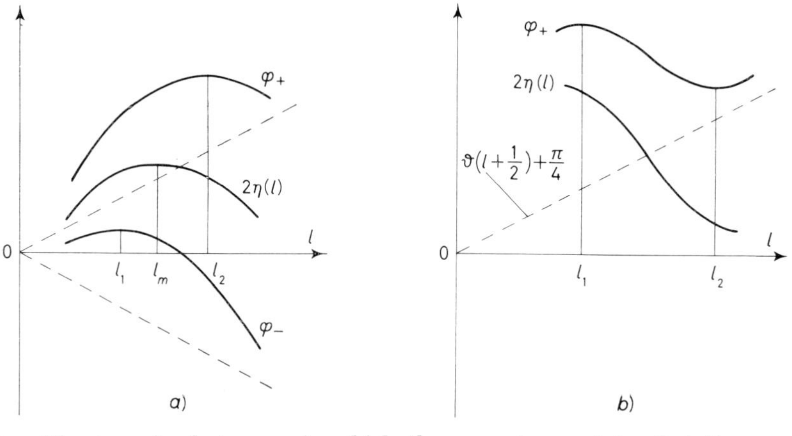

Fig. 7. – The two simplest cases in which there are two points of stationary phase. *a*) An extremum is present in $\eta(l)$. The stationary l belong to different branches. *b*) An inflection point is present. The stationary l occur in the same branch.

for a given ϑ. In case *a*) $\eta(l)$ exhibits an extremum. For a maximum, $l_1 < l_2$ will be in the φ_- branch, l_2 in the φ_+ branch. Inspection of (22) indicates a repulsive deflection function for l_1, an attractive deflection function for l_2. Classically, we are talking about two different trajectories which must have passed the scatterer on opposite sides because one is attracted, the other repelled, and yet both are deflected into the same observation angle. In case *b*) there is a point of inflection in $\eta(l)$. Both l_1 and l_2 are in the *same* branch. In Fig. 7 it is assumed to be the φ_+ branch, *i.e.* attractive deflection. Classically, an inflection point corresponds to an extremal deflection (rainbow scattering) which must lead to equal deflection for pairs of impact parameters, one being smaller (l_1) the other larger (l_2) than that for extremal deflection.

Let us assume that the l-values of stationary φ are widely separated so that their effectively contributing l ranges $\Delta l_{\text{eff}\,1}$ and $\Delta l_{\text{eff}\,2}$ do not overlap. A little reflection shows that we can carry out the procedure described above separately for each stationary point:

$$f_{\text{sc}}(\vartheta) = f_{\text{sc}\,1}(\vartheta) + f_{\text{sc}\,2}(\vartheta)\,.$$

This leads to interference. The cross-section is no longer classical, *i.e.* the sum of two intensity contributions $I_{\text{cl}\,1} + I_{\text{cl}\,2}$ but

(23) $$I_{\text{sc}}(\vartheta) = I_{\text{cl}1}(\vartheta) + I_{\text{cl}\,2}(\vartheta) + 2(I_{\text{cl}\,1}I_{\text{cl}\,2})^{\frac{1}{2}}\cdot\cos(\beta_2 - \beta_1)\,,$$

where

$$\beta_2 - \beta_1 = \begin{cases} 2\eta(l_2) - 2\eta(l_1) + (l_2 + l_1 + 1)\vartheta - \dfrac{\pi}{2} \\[2ex] 2\eta(l_2) - 2\eta(l_1) + (l_2 - l_1)\vartheta + \dfrac{\pi}{2} \end{cases}$$

for the cases of Fig. 7*a*) and 7*b*) respectively.

Semiclassically, the phases β are expected to be large. As the angle is changed they may vary by many multiples of π. Thus, the interference term in (23) will produce an oscillation of the differential cross-section about the classical value. The angular wavelength of the oscillation tends to be distinctly different in the two prototype cases because

(24) $$\frac{\mathrm{d}(\beta_2 - \beta_1)}{\mathrm{d}\vartheta} = \begin{cases} l_2 + l_1 + 1 & \text{for } a\,, \\ l_2 - l_1 & \text{for } b\,. \end{cases}$$

If l_1 and l_2 are comparable in magnitude the oscillation in case a will be much more rapid than for b. As it is intimately related to the existence of an inflection point in $\eta(l)$ this latter type of interference is called the supernumerary rainbow oscillation.

In both cases (23) is incorrect if the two stationary l-values come too close to each other so that the nonoverlapping l-band assumption fails. From (18) we see that this will occur in the forward direction for case a, and in the vicinity of a classical rainbow angle for b.

4·3. ***Rainbow scattering.*** – If we get close to a classical rainbow angle (23) fails on two counts: The effective l-bands will overlap, and the parabolic phase shift approxition (19) will be insufficient: at the inflection point $\eta''=0$.

Using a cubic expansion of the phase shifts about the inflection point at

$l = l_r$ the semi-classical scattering amplitude (17) may again be reduced to a known integral:

$$f_{sc}(\vartheta, E) = \lambda \left[\frac{2\pi(l_r + \frac{1}{2})}{\sin\vartheta}\right]^{\frac{1}{2}} \cdot (\eta''')^{-\frac{1}{3}} \cdot \mathrm{Ai}\left(\frac{\pm(\vartheta - \vartheta_r)}{(\eta''')^{\frac{1}{3}}}\right) \cdot \exp[i\beta_r] \qquad \text{for } \varphi_\pm,$$

$$\beta_r = 2\eta(l_r) \pm (l_r + \tfrac{1}{2})\vartheta \begin{cases} +\dfrac{5\pi}{4} \\ -\dfrac{\pi}{4} \end{cases} \qquad \text{for } \varphi_\pm.$$

Here

$$\mathrm{Ai}(x) = \frac{1}{2\pi}\int_{-\infty}^{+\infty} \exp[ixu + \tfrac{1}{3}iu^3]\, du$$

is the Airy function, and $\eta'''(l_r)$ may have either sign which has to be included in β_r to give the total phase of the rainbow scattering amplitude.

(25) $$I_{sc}(\vartheta) \cdot \sin\vartheta \sim \mathrm{Ai}^2$$

is plotted as a function of $x = (\vartheta - \vartheta_r/(\eta''')^{\frac{1}{3}})$ in Fig. 8. The sharp spike of classical mechanics [27] at the rainbow angle $\vartheta_r\,(x = 0)$ is lost. The parabolic

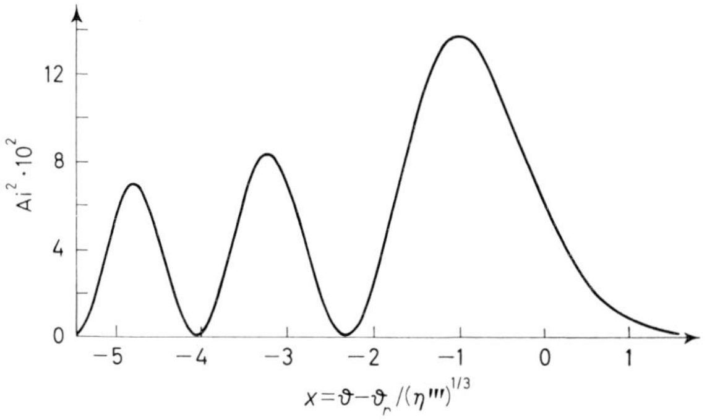

Fig. 8. – Rainbow scattering in the semi-classical approximation.

phase approximation failing, classical and semi-classical results bear little resemblance. The oscillation on the « bright » left side of the main rainbow peak at ϑ_r is, of course, the supernumerary rainbow pattern discussed in the preceding paragraph case *b*. Here its two stationary points are included from the outset in the cubic phase shift approximation. Al long as it is correct the Airy function oscillations show the interference of two effective l-band separating from an original coincidence as we go from the rainbow angle away towards its left.

4'4. *Integral cross-section.* – The colaescing l-band scattering in case *a*) brings up the question of forward scattering for which the f_{sc} of (17) is not the appropriate starting point. Closely related but of greater relevance to experimental studies are integral cross-sections. Let the conditions again be semi-classical: Many l contribute and the action, *i.e.* a representative phase shift, is large compared to 1. The integral cross-section

$$\sigma(E) = 4\pi\lambda\!\!\!^{-2} \sum (2l+1)\sin^2\eta_l \tag{26}$$

becomes

$$\sigma_{sc}(E) = 2\pi\lambda\!\!\!^{-2}\int_0^\infty (2l+1)[1-\cos 2\eta(l)]\,\mathrm{d}l\,. \tag{27}$$

Again we may expect the cos term to oscillate except in two cases: 1) The $\eta(l)$ may become stationary. The simplest such case will be an extremum as in Fig. 7*a*). 2) For large l $\eta(l)\to 0$, and the cos term approaches 1 monotonically. Figure 9 shows the phase shifts which are qualitatively characteristic

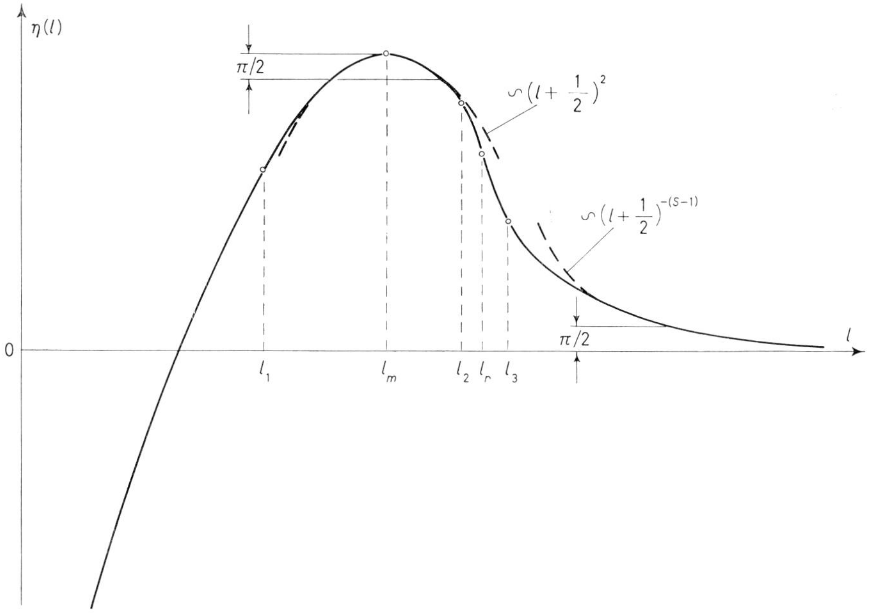

Fig. 9. – Phase shifts qualitatively correct for an interaction potential as in Fig. 1. The integral cross-section is sensitive to the indicated regions at the maximum and in the Jeffreys-Born regime at large l. Here valid approximations are needed (dashed lines). Outside of these regions the l contribute in a « random » fashion. Also shown are the three l values of stationary phase contributing to the intensity at an observation angle a little smaller than the classical rainbow angle.

for an interaction as in Fig. 1. The repulsive interaction at small r produces the negative phase shifts at small l, the attractive interaction leads to positive $\eta(l)$ for intermediate l. Like the interaction at large distances the $\eta(l)$ tend to zero at large l. In the JWKB approximation the curve may be quantitatively obtained by (22) from the classical deflection function discussed in « *Collision Mechanics* » 5·2. It exhibits both conditions under which the cos term fails to oscillate.

In evaluating the integral cross-section it is necessary to use valid approximations to the $\eta(l)$ in the nonoscillating l ranges. The oscillation regions will be very insensitive to the $\eta(l)$. Substitutes may be used provided they also vary rapidly with l.

For large l we take [33]

$$\eta(l) \to {}^{\text{JB}}\eta(l)$$

and

$$V(r) \to -\frac{C^{(s)}}{r^s}\,.$$

Of course, for ${}^{\text{JB}}\eta(l) \geqslant 1$ the JB phase shifts become incorrect but this is just where the cross-section tends to be insensitive to their exact value. Use of the approximate potential is only permitted if phase shifts for the real and the approximate potential do not deviate from each other until they reach values > 1. Obviously, $(\varepsilon r_m/\hbar v) \gg 1$ is the condition for this to be true. For $l \sim l_m$ we expand

$$\eta(l) \simeq \eta(l_m) - \tfrac{1}{2}|\eta''|_{lm}\cdot(l - l_m)^2\,.$$

This must approximate the actual phase shifts accurately within $\Delta l_{\text{eff}} = [\pi/|\eta''|_{lm}]^{\frac{1}{2}}$ centered at l_m because it again leads to a Fresnel integral. Inserting into (27) yields

$$\sigma_{\text{sc}}(E) = 2\pi\lambda\!\!\!^{-2}\left\{\int_0^\infty (2l+1)\cdot[1-\cos 2\,{}^{\text{JB}}\eta(l)]\,\mathrm{d}l - \right.$$

$$\left. -\int_{-\infty}^{+\infty}(2l+1)\cos\left[2\eta(l_m) \pm |\eta''|_{l_m}\cdot(l-l_m)^2\right]\mathrm{d}l\right\} \qquad \text{for } \eta''(l_m) \gtrless 0\,.$$

Both approximations vary rapidly with l outside their validity range (see Fig. 9). The lower limit of the second integral may safely be extended to $-\infty$. The integrals can be analytically evaluated [33] to give the integral cross-section

$$(28)\qquad \sigma_{\text{sc}}(E) = g(s)\cdot\left(\frac{C^{(s)}}{\hbar v}\right)^{2/(s-1)} - \lambda\!\!\!^{-2}\,\frac{2\pi^{\frac{3}{2}}(l_m+\frac{1}{2})}{|\eta''|^{\frac{1}{2}}_{l_m}}\cdot\cos\left[2\eta(l_m) \pm \frac{\pi}{4}\right], \qquad \eta''(l_m) \gtrless 0\,,$$

where

$$g(s) = \frac{2\pi^2}{\Gamma((s+1)/(s-1))\cdot(s-1)\cdot\sin\pi/(s-1)}\cdot\sqrt{\pi}\cdot\frac{\Gamma((s-1)/2)}{\Gamma(s/2)}\,.$$

v is the relative partner velocity, and for the dispersion force $s=6$, $g(s) = 8.083$ [34].

For atomic partners the first term in (28) due to the long-range interaction is the dominant one. It decreases monotonically with the velocity. Note that it is the only expression we have given which relates a feature of the interaction—its dependence at large r—directly to the cross-section. In all other cases phase shift properties are used to express cross-sections. This increases the uncertainty and labour in extracting potential information immensely. The second term in (28)—called the glory undulations [20 ,36]—is due to concurrent attractive-repulsive interactions. It oscillates because of the velocity-dependence of $\eta(l_m)$ and for $\eta''(l_m) < 0$ has extrema at

$$(29) \qquad \eta(l_m) = (N - \tfrac{3}{8})\pi\,, \qquad N = \begin{cases} 1, \quad 2, \quad 3 \;\dots \text{for maxima,} \\ 1.5,\ 2.5,\ 3.5 \dots \text{for minima,} \end{cases}$$

counting from high to low velocities. For increasing v $\eta(l_m) \to 0$, the oscillation dies out. This means, of course, that $\varepsilon r_m/\hbar v$ has dropped below 1 and the long-range term also becomes incorrect due to the approximate potential for which it was derived. Even in the forward direction the scattering is now predominantly from the repulsive core. In Fig. 10 $\sigma(v)$ is plotted as a result of a full JWKB

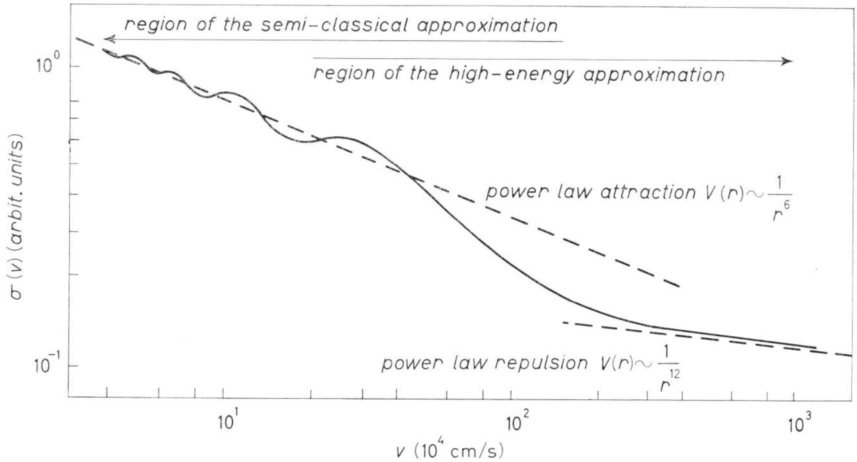

Fig. 10. – Integral cross-section for a LJ 12-6 model and a wide range of the action parameter $2\varepsilon r_m/\hbar v$. For large action the potential well dominates the cross-section. Two contributing l-bands produce the glory undulation. For small action the repulsive potential core dominates.

calculation [4]. It avoids the above approximations thus being capable to account for σ in a wide velocity range containing the oscillation region where (28) is a good approximation, the « great fall » region where $\varepsilon r_m/\hbar v$ passes through 1 and the high-velocity region where a result similar to the first term of (28) takes over, calculated for the repulsive interaction instead of the attractive one.

As $v\rightarrow 0$ isotropic s-wave scattering occurs. From Levinson's theorem [35] we known that

$$\lim_{v\to 0} \eta\,(l=0) = n_0 \pi\,,$$

where n_0 is the number of bound states with zero angular momentum which the potential well is capable to hold. The number of maxima in the total cross section must be finite. BERNSTEIN shows that it is nearly *equal* to n_0 [36]. Another interesting relation exists between the glory undulations and the rainbow structure [37]: Each time the integral cross-section passes through a glory maximum the differential forward scattering cross-section $I(0)$ also takes on a maximum. Then, at decreasing velocity, $\sigma(v)$ moves on to the next glory minimum while the $I(0)$ maximum swings out to a finite angle to become observable as a new rainbow maximum. Thus, at a velocity where the N-th glory maximum is observed in $\sigma(v)$ we shall find $N-1$ rainbow maxima at finite angles in the angular distribution.

5. – Experimental results and extraction of potential information.

In accurate work the experimentally determined cross-section or certain features thereof are calculated in JWKB or RY approximation on the basis of an assumed model interaction containing free parameters for flexibility. By parameter variation a fit is obtained to the experiment and available theoretical data (mainly Van der Waals' constants $C^{(6)}$). Once the parameters are fixed the potential is used to predict other features of experimental cross-sections. In this way a check is obtained as to how physically realistic the model may be. This is a laborious process. The insight into the relation of specific potential properties and features of the cross-section is not always clearly apparent. Numerical methods must often be used. To characterise briefly the present experimental accuracy and interpretational experience we may say: *a*) As model parameters the two important quantities, the well depth ε and the equilibrium distance r_m, may be obtained with errors close to 1%. This is true for 2 parameter models, such as LJ 12-6. Other parameters—if present—should be kept fixed. *b*) Compared to the respective quantities of the real interaction, model parameter values are expected to be significant to within $\pm 10\%$. This guess is based on the model-dependence of ε and r_m for a number of models with very different analytic form and parametrization.

It is necessary that the models to be compared fit glory *and* rainbow data as well as rapid oscillations (see below) and Van der Waals constants $C^{(6)}$ for one system. *c*) For a number of systems it has been definitely shown that the standard models LJ n-6 or $\exp[\alpha\text{-}6]$ are insufficient. More flexible models have been proposed. They are expected to afford a more realistic representation of the interaction because they fit more different data on the same system. In a forthcoming article by SCHLIER these things are discussed in more detail [38].

Often the amount and quality of available data or the purpose of the investigation may make a semi-classical data evaluation fully adequate. It is useful to obtain first-order data for more elaborate calculations. It is invaluable in making the relation between cross-sections and the potential more transparent. Therefore, we shall proceed on this basis in pointing out a few details about the actual data extraction. The errors quoted along with a potential property are considered typical for a careful experiment, but are based on a JWKB data evaluation.

5·1. *Total cross-sections.*

1) Average velocity dependence: From the first term of (28) averaged over the undulations

$$\sigma_{\text{av}} \sim v^{-2/(s-1)} \to s \pm 5\% \,.$$

The slope of a log-log plot of σ *vs.* v yields the « asymptotic » r-dependence of the potential. « Asymptotic » should not be taken literally. σ is sensitive to an r-range which—velocity-dependent— extends down to $r \approx 1.4 r_m$ as long as σ is in the undulation region. The main contribution arises from $r \approx 2r_m$. Figures 11 and 12 show the first data of this kind on K-N_2 [39] as well as a more recent measurement on K-Hg [40] which yields $s = 6.3 \pm 5\%$.

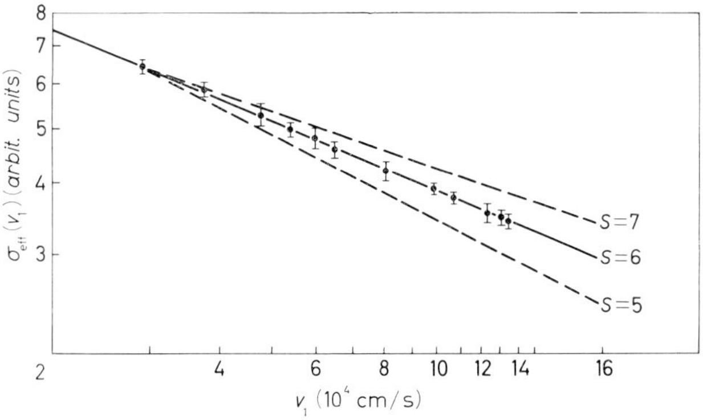

Fig. 11. – $\sigma(v)$ for K-N_2. First direct experimental verification of the r^{-6} dispersion-force law.

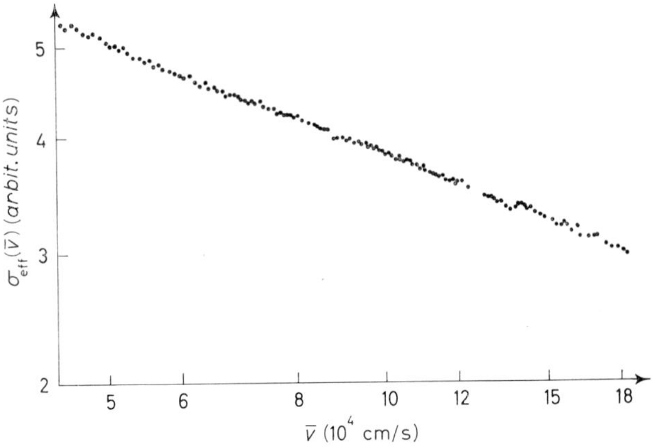

Fig. 12. – $\sigma(v)$ for K-Hg. Due to very large action the glory undulations are very rapid. They are washed out because of the experimental velocity inhomogeneity.

2) Absolute total cross-sections: After elimination of the glory effect by averaging or numerical correction absolute values of total cross-sections at a known velocity yield the Van der Waals constants $C^{(6)}$.

$$\sigma_{av} = \text{const}\left(\frac{C^{(6)}}{v}\right)^{\frac{2}{5}} \to C^{(6)} \pm 20\,\% \,.$$

The agreement with theoretical predictions is generally good for alkali-rare gas systems [9, 41], for alkali-alkali systems it is still poor [10].

3) Extrema velocities: For exploitation it is necessary to know the velocity-dependence of the maximum phase shift. Inspection of (12) and (12*a*) shows that an expansion of the JWKB phase shift must be of the form

$$\eta(l_m) = \frac{2\varepsilon r_m}{\hbar v}\sum_{v=0}^{\infty}\frac{a_v}{K^v}\,, \qquad K = \frac{E}{\varepsilon}\,.$$

The first term is the JB maximum phase shift. Its coefficient a_0 may be calculated analytically for LJ 12-6 or exp [α-6] models, higher coefficients are numerically obtained. From (29) we have for an extremum

$$\left(N - \frac{3}{8}\right)\pi = \frac{2\varepsilon r_m}{\hbar v_N}\left(a_0 + \frac{a_1}{K} + \dots\right) \begin{matrix} \nearrow \varepsilon r_m \pm 3\,\%\,, \\ \searrow \varepsilon \pm 25\,\%\,, \end{matrix}$$

N, v_N being taken from the experiment a plot of N *vs.* v_N^{-1} yields a curve which at high v_N, *i.e.* large K, becomes a straight line whose slope gives the product εr_m. The deviation from the limiting behaviour independently yields ε. Figure 13

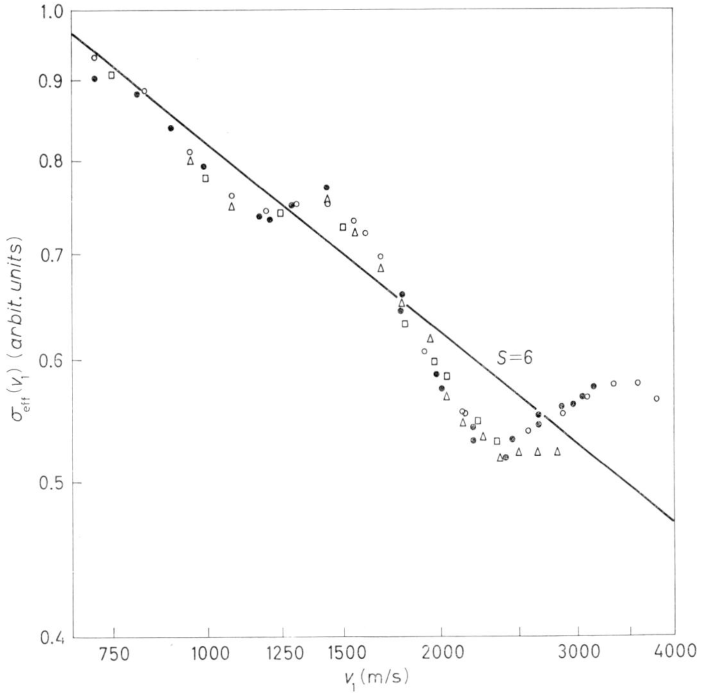

Fig. 13. – Glory undulations for Li-Xe.

shows the experiment on Li-Xe by ROTHE and coworkers in which the glory undulations were discovered [42]. Figure 14 again is a more recent measurement on K-Xe for which the quoted accuracy is 2 % in εr_m [40]. Agreement

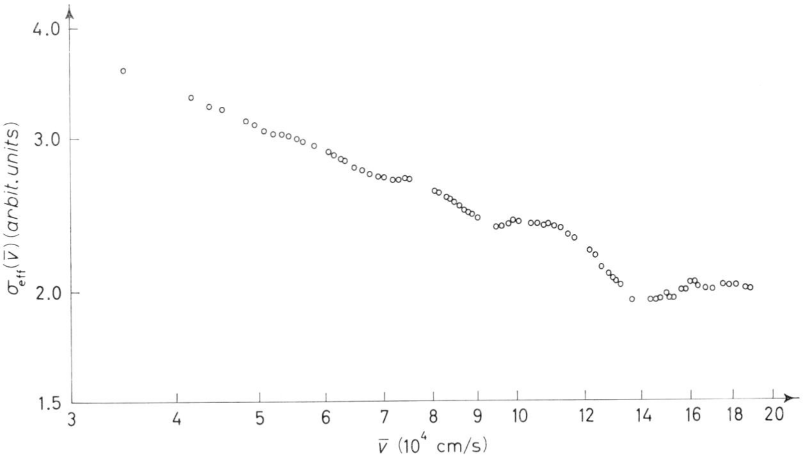

Fig. 14. – Glory undulations for K-Xe. For low velocities they are increasingly washed out due to experimental velocity spread. The last high-velocity undulation is virtually undiminished.

to 1 % was found in an independent experiment in Prof. Schlier's laboratory. Note that other than under 1) and 2) parameters evaluated here and below refer to a given model. Errors quoted indicate the accuracy of experiment and data evaluation *not* including any kind of model-dependence of the parameters.

The model-dependence of the εr_m, caused by the coefficient a_0, is weak. To get some idea, using LJ 8-6 instead of LJ 12-6 changes εr_m by about -10%. The εr_m products are among the most accurately obtainable data from scattering experiments.

4) Glory amplitudes: In Fig. 15 the data of Fig. 14 are plotted with the monotonic v-dependence normalized out. The dashed line is a JWKB calculation which fits the extrema velocities but disagrees with respect to the ampli-

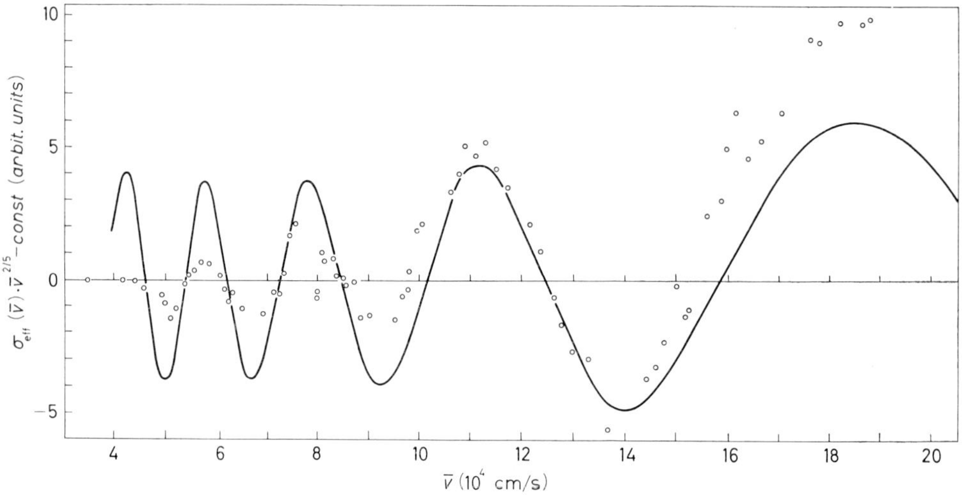

Fig. 15. – K-Xe undulations compared with a model calculation which fits the extrema velocities. Low-velocity discrepancies are not significant. The experimental high-velocity undulation is too *large*. This has been seen for many systems. It indicates an insufficency of the model.

tudes (only the deviation at high v is significant). This type of disagreement was found to persist for all potentials of the LJ or exp α families, *i.e.* for all values of the shape parameters n or α. From (28) we know that the amplitudes depend on $\eta''(l_m)$. There is no simple, accurate relation of η'' to the potential. Crudely,

$$\eta''(l_m) \to V''(r_m) .$$

V'' is to be considered an effective curvature which characterizes the well over the r-range of the lowest few bound states. New models were chosen capable of a variation of the width of the well at constant long-range interaction. Increasing the well width over that of LJ 12-6 DÜREN, RAABE and SCHLIER obtained very good agreement of which Fig. 16 gives an example for K-Ar [43].

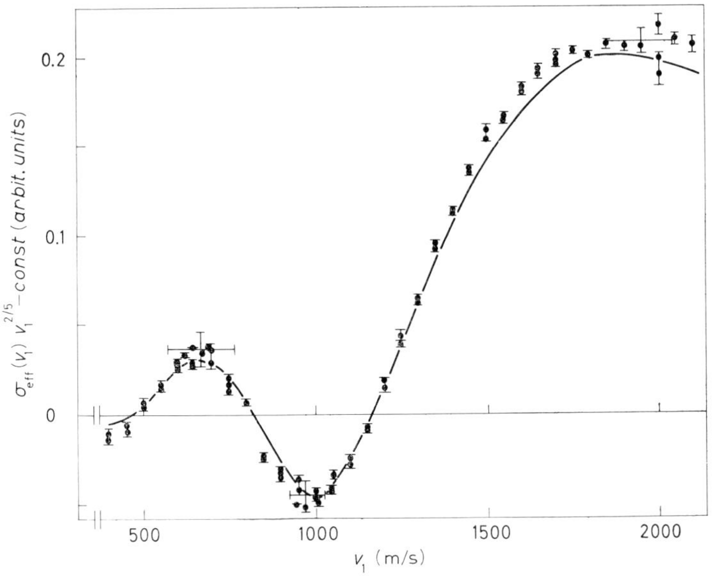

Fig. 16. – K-Ar undulations. The full line is calculated from an improved model. The computation also takes care of the experimental velocity averaging.

5.2 *Rainbow scattering.* – Mainly the angular positions of rainbow extrema have been used to obtain potential data.

1) Main Rainbow maximum: In the semi-classical approximation due to (22a) the classical deflection function $\vartheta(b)$ retains its key role in explaining the scattering. Specifically, the inflection point of the $\eta(l)$ of Fig. 9 which causes the rainbow in atomic scattering is in the φ_+ branch. It corresponds to a minimum in $\vartheta(b)$, the rainbow angle ϑ_r which we already mentioned in « *Collision Mechanics* » 5.2. By an expansion of $\vartheta(b)$ one obtains

$$\vartheta_r = \sum_{v=1}^{\infty} \frac{b_v}{K^v} \qquad \text{for } K = \frac{E}{\varepsilon} > K_{\text{orbit}},$$

$$b_1 = -2.04\,, \qquad b_2 = -0.211 \qquad \text{for LJ 12-6.}$$

For not too low reduced energies K, ϑ_r varies as E^{-1}. According to (25) or Fig. 8 it is to be found as the point where the intensity on the dark side of the rainbow has fallen to 43.9 % of the main maximum intensity. In atomic scattering the rainbow sits on the repulsive-core scattering due to the positive branch of $\vartheta(b)$ (see « *Collision Mechanics* » **5·2**, Fig. 11 at large angles). In a semi-classical analysis it must be separated from the latter. This is usually done by linear extrapolation of the repulsive-core scattering to smaller angles. Measuring ϑ_r at known E yields the well depth

$$\vartheta_r \to \varepsilon \pm 3\%$$

again model-dependent. Typically, b_1 changes by -12% in going from LJ 8-6 to LJ 12-6. The frequent quotation « ϑ_r measures the well depth » must be referred to a given model. It may, however, be shown that there exists a virtually model-independent relation of b_1 and the quantity $(r(\mathrm{d}V/\mathrm{d}r))_{\max}$. This is not surprising because the extremal deflection ϑ_r should be closely related to the maximum force, *i.e.* the slope of the potential at its inflection point. For application to chemically reactive systems it is worthwhile to keep in mind: Rainbow scattering reveals the existence of a *repulsive* deviation from long-range forces.

2) Rainbow structure including supernumeraries: As the supernumerary rainbow structure is an interference pattern angular distances $\Delta\vartheta$ between extrema must contain information about the ratio $r_m/\lambda\!\!\!^{-}$. More adequately, we suspect from (24*b*) and Fig. 7 that the linear potential dimension to be used instead of r_m may be some small r-interval near the inflection point of $V(r)$. The explicit first-order dependence on $\eta'''(l_r)$, *i.e.* the curvature of the deflection function at its minimum, also suggest this. Unfortunately, this length cannot easily be defined uniquely. Nevertheless it will be some model-dependent fraction of r_m and we obtain

$$\Delta\vartheta \to r_m \pm 8\%$$

as expected strongly model-dependent. A change from LJ 8-6 to LJ 12-6 changes r_m so obtained by 25% [44].

Figures 17, 18, 19 present three examples of measured rain-

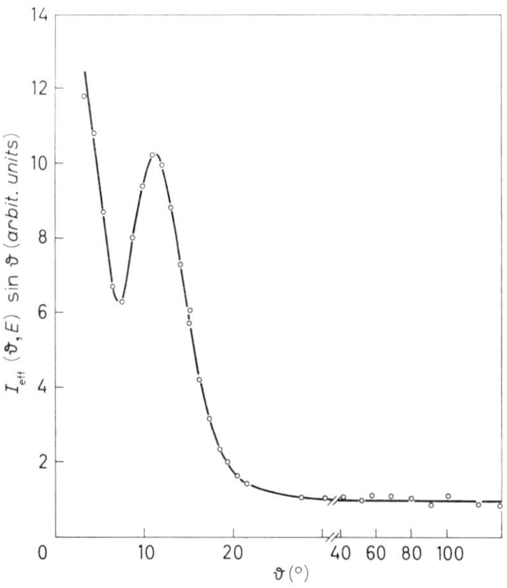

Fig. 17. – Early example of rainbow scattering for K-Kr. The flat, almost constant repulsive core scattering at large angles is nicely seen. $E = 0.056$ eV.

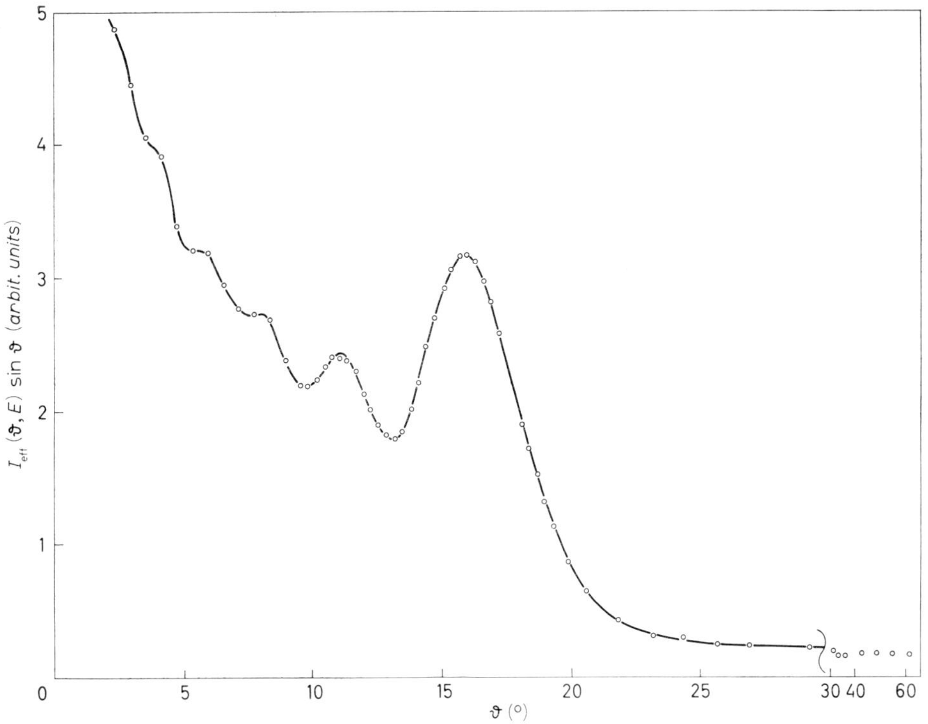

Fig. 18. – Raibow scattering for K-Hg showing supernumeraries. The larger rainbow angle at a higher collision energy indicates a potential well considerably deeper than for K-Kr. $E = 0.31$ eV.

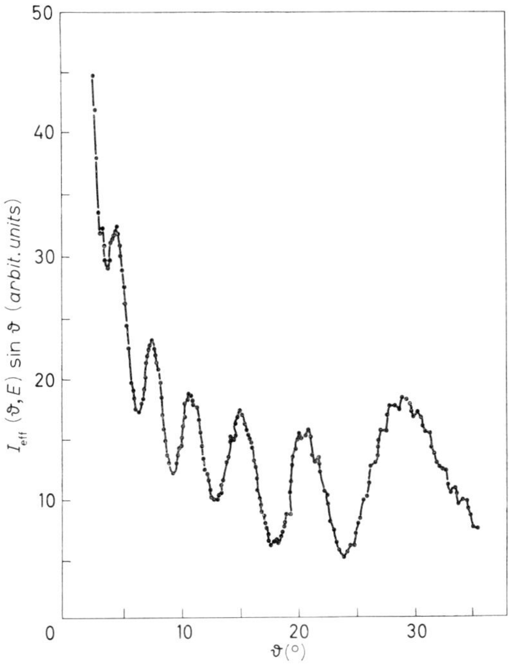

Fig. 19. – A very well resolved rainbow structure for Na-Hg. $E =$ $= 0.195$ eV.

bow scattering. In the K-Kr experiment it was first seen [45], the later K-Hg measurement shows the supernumerary oscillation clearly [46], and in the recent Na-Hg experiment it is fully resolved [47] including definite indications of the rapid oscillation pattern to be discussed below.

5'3. *Rapid oscillation pattern.* – Classically there were three impact parameters b contributing to the intensity at some observation angle $\vartheta < \vartheta_r$ (see « *Collision Mechanics* » **5'2**), but just one for $\vartheta > \vartheta_r$. From (22*a*) we recognize the former as the three l values of stationary $\varphi_\pm$ which the $\eta(l)$ of Fig. **9** will produce for all $\vartheta < 2|\eta'(l_r)|$. Two of them, l_2 and l_3, are of type b and give rise to the supernumerary rainbow oscillation. l_1 and either of the former are of type a. Simplifying the triple interference but correct to first order we may consider the interference of l_1 with the arithmetic mean $\bar{l} = \frac{1}{2}(l_2 + l_3)$. From (24*a*) we see that it will generate an oscillation considerably more rapid than the rainbow structure. Again, the angular wavelength of this rapid oscillation

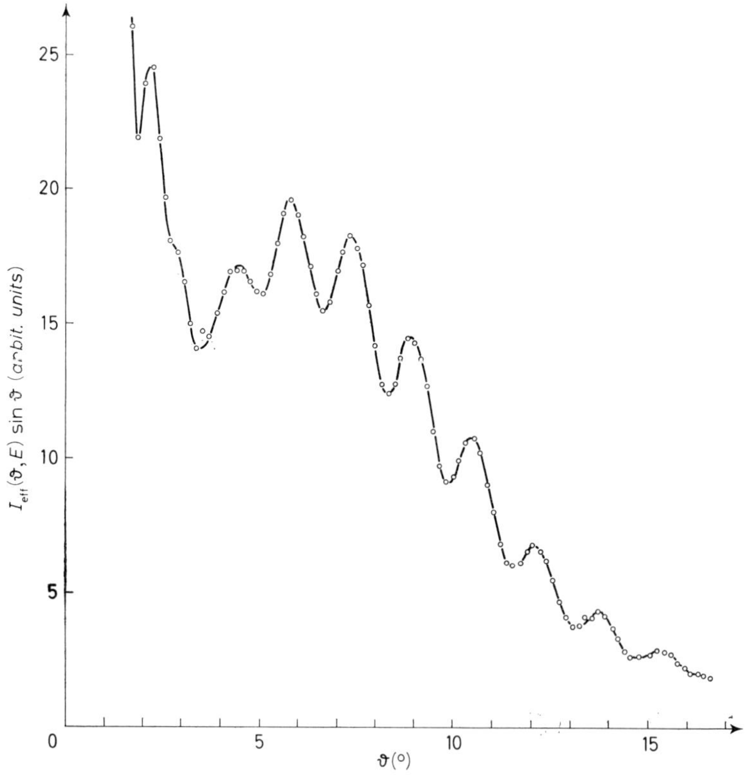

Fig. 20. – The rapid oscillation pattern superimposed on the main rainbow maximum for K-Ar. $E = 0.050_5$ eV.

must carry information about r_m. Using (24a) we have

$$\Delta\vartheta \simeq \frac{2\pi}{l_1+l} = \frac{\lambda}{r_m}\cdot\frac{2\pi}{\beta_1+\beta} \to r_m \pm 2\,\%\,.$$

The reduced impact parameters $\beta_i = b_i/r_m$ contribute to the intensity at the angle ϑ, where the oscillation is observed. They are simply read from the deflection function which depends on β rather than on b and r_m separately as we pointed out in « *Collision Mechanics* » **5**.

It has been shown that the r_m values from the rapid oscillations are very little model-dependent for a large class of potentials [48, 37]. Such values can, thus, be expected to be much closer to the r_m of the real interaction than all other data derived from scattering except s and $C^{(6)}$. This interesting feature may be related to the attractive-repulsive type oscillation for potentials with a small ratio of width of well to distance of well from origin [49]. Moreover, the r_m from this source may be obtained very accurately using the JWKB method. In Fig. 20 a reasonably well resolved rapid oscillation pattern is seen for K-Ar [49]. It is superimposed on the main rainbow peak which is the only rainbow maximum to be expected for this system and energy. Beyond the maximum the oscillation clearly tends to die out as we expect. Here, only one stationary l-value is responsible for the intensity and the cross-section becomes classical. For this experiment the error of r_m is entirely determined by the accuracy with which the collision energy can be measured.

* * *

The assistance of Drs. M. G. Dondi and F. Torello in preparing a first draft of this lecture is gratefully acknowledged.

REFERENCES

[1] See this volume p. 86.

[2a] J. O. Hirschfelder, C. F. Curtis and R. B. Bird: *Molecular Theory of Gases and Liquids* (New York, 1954).

[2b] R. G. Breene: *The Shift and Shape of Spectral Lines* (Oxford, 1961).

[2c] J. O. Hirschfelder, Ed.: *Intermolecular Forces*, in *Adv. Chem. Phys.*, vol. **12** (New York, 1967).

[3] R. B. Bernstein and J. T. Muckermann: in *Adv. Chem. Phys.*, edited by J. O. Hirschfelder, vol. **12** (New York, 1967), p. 389.

[4] H. Pauly and J. P. Toennies: in *Adv. At. Mol. Phys.*, edited by D. R. Bates and J. Estermann, vol. **1** (New York, 1965), p. 195.

[5a] J. AMDUR and J. E. JORDAN: in *Adv. Chem. Phys.*, vol. **10**, *Molecular Beams*, edited by J. ROSS (New York, 1966).
[5b] J. E. JORDAN and J. AMDUR: *Journ. Chem. Phys.*, **46**, 165 (1967).
[6a] W. KOLOS and L. WOLNIEWICZ: *Journ. Chem. Phys.*, **41**, 3663 (1964).
[6b] W. KOLOS and L. WOLNIEWICZ: *Journ. Chem. Phys.*, **49**, 404 (1968).
[7] P. E. PHILLIPSON: *Phys. Rev.*, **125**, 1981 (1962).
[8] See this volume p. 349: Prof. CONROY advised me after this lecture that he was able to reproduce Phillipson's results closely using a different method. Thus, there can be little doubt that the theoretical data on He-He are correct.
[9] A. DALGARNO and W. D. DAVISON: in *Adv. At. Mol. Phys.*, edited by D. R. BATES and J. ESTERMANN, vol. **2** (New York, 1966), p. 1.
[10] A. DALGARNO and W. D. DAVISON: *Mol. Phys.*, **13**, 479 (1967).
[11] See SCHLIER *et al.*: *Zeits. Phys.*, **168**, 81 (1962), for the first experiment on deviations from spherical symmetry for atomic partners not both in S-states.
[12] Beyond distances comparable to the optical wave length which corresponds to excited states of the partners the dispersion interaction changes to an r^{-7} law. This is due to retardation in the coupling electromagnetic field; see: A. B. G. CASIMIR and D. POLDER: *Phys. Rev.*, **73**, 360 (1948); P. R. FONTANA and R. B. BERNSTEIN: *Journ. Chem. Phys.*, **41**, 1431 (1964). These distances are too large to be of interest in atomic scattering. Below $r \sim 2r_m$ higher-order dispersion or electrostatic terms, as well as first-order forces may begin to contribute appreciably to the interaction.
[13] To see how close different estimates of the He-He interaction come to each other see: D. E. BECK: *Mol. Phys.*, **14**, 311 (1968); J. N. MURRELL and G. SHAW: *Mol. Phys.*, **15**, 325 (1968).
[14] E. A. MASON, *et al.*: *Journ. Chem. Phys.*, **22**, 169, 522 (1954).
[15] To give a simple example let us consider the evaluation of the well depth ε from rainbow scattering using the LJ 12-6 model. As was pointed out in *Collision Mechanics* the angular position ϑ_r at which rainbow scattering appears is solely determined by ε in the classical limit. It is strictly independent of the other model parameter r_m. Now, for an actual system classical mechanics is but an approximation. Therefore, we may expect the angular location of rainbow scattering to yield ε almost, but not fully independent of a choice of r_m.
[16] See standard text books on quantum mechanics.
[17] R. J. GLAUBER: *Lectures Theor. Phys.*, **1**, 315 (1959).
[18] L. I. SCHIFF: *Phys. Rev.*, **103**, 443 (1956).
[19] G. MOLIÈRE: *Zeits. Naturforschg.*, **2** a, 133 (1947).
[20] K. W. FORD and J. A. WHEELER: *Ann. of Phys.*, **7**, 259 (1959).
[21] L. D. LANDAU and E. M. LIFSHITZ: *Quantum Mechanics* (New York, 1959), p. 50.
[22a] K. H. KATERBAU: Diplom Thesis, Freiburg, unpublished (1965).
[22b] R. P. MARCHI and C. R. MUELLER: *Journ. Chem. Phys.*, **38**, 740 (1963).
[23] E. HUNDHAUSEN and H. PAULY: *Zeits. Phys.*, **187**, 305 (1965).
[24] R. E. LANGER: *Phys. Rev.*, **51**, 669 (1937).
[25] B. S. JEFFREYS: in *Quantum Theory*, I, edited by D. R. BATES, Chap. 7 (New York, 1961).
[26] H. A. KRAMERS: *Zeits. Phys.*, **39**, 828 (1926).
[27] See lecture on *Collision Mechanics*.
[28] K. W. FORD, D. L. HILL, M. WAKANO and J. A. WHEELER: *Ann. of Phys.*, **7**, 239 (1959), derive a mixing formula which joins JWKB solutions to the left

and to the right of a parabolic barrier to each other. In this way they obtain a phase shift which varies smoothly through the top of the barrier.

[29] W. C. STWALLEY, A. NIEHAUS and D. R. HERSCHBACH: *Proc. V Intern. Conf. Phys. Electr. At. Coll.*, (Leningrad, 1967), p. 639, very probably observed resonance scattering for the H-Hg system. This appears to be the only evidence from scattering experiments so far. See, however, R. B. BERNSTEIN: *Phys. Rev. Lett.*, **16**, 385 (1966).

[30] M. ROSEN and D. R. YENNIE: *Journ. Math. Phys.*, **5**, 1505 (1964).

[31] $f=(r^2V-r_c^2V(r_c))/(r^2-r_c^2)$ is obtained by rewriting the integrand of (12*a*) in terms of r_c instead of $(l+\frac{1}{2})$. For another derivation of (14) see N. R. MOTT and H. S. MASSEY: *Theory of Atomic Collisions* (Oxford, 1965), p. 89, 635.

[32] K. W. FORD and J. A. WHEELER: *Ann. of Phys.*, **7**, 287 (1959).

[33] L. D. LANDAU and E. M. LIFSHITZ: *Quantum Mechanics* (New York, 1959), p. 416.

[34] R. B. BERNSTEIN and K. H. KRAMER: *Journ. Chem. Phys.*, **38**, 2507 (1963), estimate an accuracy to 0.5% for the first term in (28) if $V\sim r^{-12}$. For the second term errors smaller than 2% were found if $\eta''(l_m)$ is replaced by an effective curvature within the contributing l-band and for $r_m/\lambda\leqslant 160$, $\varepsilon r_m/\hbar v\geqslant 3$ (D.B. unpublished). Thus, the approximation (28) is really astonishingly good.

[35] N. LEVINSON: *Kl. Danske Videnskab. Selskab. Mat.-Fys. Medd.*, **25**, No. 9 (1949).

[36] R. B. BERNSTEIN: *Journ. Chem. Phys.*, **38**, 2599 (1963).

[37] U. BUCK and H. PAULY: *Zeits. Phys.*, **208**, 390 (1968).

[38] CH. SCHLIER: *Ann. Rev. Phys. Chem.*, **20**, 191 (1969).

[39] H. PAULY: *Zeits. Naturforsch.*, **15** A, 277 (1960).

[40] D. BECK and H. L. LOESCH: *Zeits. Phys.*, **195**, 444 (1966).

[41] Experimental values prior to 1964 were mostly in error due to wrong pressure calibrations, see *e.g.* H. G. BENNEWITZ and H. D. DOHMANN: *Vak. Tech.*, **1**, 8 (1965).

[42] E. W. ROTHE, P. K. ROOL, S. M. TRUJILLO and R. H. NEYNABER: *Phys. Rev.*, **128**, 659 (1962).

[43] R. DÜREN, G. P. RAABE and CH. SCHLIER: *Zeits. Phys.*, **214**, 410 (1968).

[44] The quadratic approximation to the deflection function at its minimum on which the semi-classical method is based becomes invalid for angles too far away from ϑ_r. Semi-classical r_m values from the supernumerary rainbow structure may, therefore, be substantially in error. The full JWKB method is to be preferred here. For the LJ 8-6 and LJ 12-6 models tabulated extrema angles are available [23].

[45] D. BECK: *Journ. Chem. Phys.*, **37**, 2884 (1962).

[46] D. BECK, H. J. DUMMEL and U. HENKEL: *Zeits. Phys.*, **185**, 19 (1965).

[47] U. BUCK and H. PAULY: *Zeits. Naturforsch.*, **23** a, 475 (1968).

[48] P. BARWIG, U. BUCK, E. HUNDHAUSEN and H. PAULY: *Zeits. Phys.*, **196**, 343 (1966).

[49] R. KRÄMER, H. FÖRSTER and D. BECK: Freiburg, unpublished.

Vibrationally and Rotationally Inelastic Scattering.

R. J. Cross, jr.
Yale University - New Haven, Conn.

1. – Introduction.

In order to treat molecular collisions it is necessary to consider inelastic processes involving transitions between vibrational and rotational quantum states. Unlike the problem of atom-atom elastic scattering there is no exact or simple numerical solution to the problem, and one must resort to approximate procedures. The wide variety of experiments in the field has led to an equally wide variety of approximate theories, none of them, unfortunately, useful for all cases of interest. In this article we will first treat the formal quantum-mechanical theory of inelastic scattering in order to obtain the basic equations to be solved. We will then sketch several approximate methods of solution and various experimental techniques.

2. – Formal theory of inelastic scattering.

Schrödinger's equation for the scattering of two molecules may be written as

$$\left[H_{\text{int}} - \frac{\hbar^2}{2\mu}\nabla^2 + V_0(r) - E\right]\Psi(r, \theta, \varphi, \boldsymbol{r}_{\text{int}}) = -\Delta V \Psi, \tag{1}$$

where H_{int} is the Hamiltonian for the free rotation and vibration of *both* molecules. The potential is divided for convenience into two parts, a spherically symmetric part, $V_0(r)$, and the remaining part ΔV which contains the dependence on the internal molecular co-ordinates; E is the total energy of the system, (r, θ, φ) is the vector joining the centers of mass of the molecules, and $\boldsymbol{r}_{\text{int}}$ are the collective internal co-ordinates of both molecules.

Equation (1) is specific for each particular problem. To obtain a more general equation and to show the transition more specifically we use the substitution,

$$\Psi(r,\theta,\varphi,\boldsymbol{r}_{\text{int}})=\sum_{nlm} r^{-1}u_{nlm}(r)\,Y_{lm}(\theta,\varphi)\,\psi_n(\boldsymbol{r}_{\text{int}})\,, \tag{2}$$

where ψ_n is the solution of $H_{\text{int}}\;\psi_n=E_n\psi_n$, and n stands for all of the various internal quantum numbers of both molecules. Introducing (2) into (1) and making some convenient substitutions yields,

$$\left[\frac{\mathrm{d}^2}{\mathrm{d}r^2}-\frac{l(l+1)}{r^2}+k_n^2-U_0(r)\right]\sum_{nlm}u_{nlm}Y_{lm}\psi_n=\Delta U\sum_{nlm}u_{nlm}Y_{lm}\psi_n\,, \tag{3}$$

where $U_0=2\mu V_0/\hbar^2$, $\Delta U=2\mu\Delta V/\hbar^2$, and $k_n^2=2\mu(E-E_n)/\hbar^2$. Since ψ_n and Y_{lm} form complete sets of functions in $\boldsymbol{r}_{\text{int}}$ and (θ,φ), we can multiply both sides of (3) by $Y_{l'm'}\psi_{n'}$ and integrate over $\boldsymbol{r}_{\text{int}}$, θ, and φ to generate the series of coupled equations,

$$\left[\frac{\mathrm{d}^2}{\mathrm{d}r^2}-\frac{l'(l'+1)}{r^2}+k_{n'}^2-U_0(r)\right]u_{n'l'm'}=\sum_{nlm}U_{n'nl'lm'm}u_{nlm}\,, \tag{4}$$

where

$$U_{n'nl'lm'm}=\int Y_{l'm'}^*\psi_{n'}^*\,\Delta U\,Y_{lm}\psi_n\,\mathrm{d}r_{\text{int}}\sin\theta\,\mathrm{d}\theta\,\mathrm{d}\varphi\,. \tag{5}$$

In matrix form we can write (4) as,

$$\Omega\boldsymbol{u}=\boldsymbol{U}\boldsymbol{u}\,. \tag{6}$$

To obtain expressions for the scattering cross-sections from (4) we must examine the asymptotic solutions of (4) and construct a wave function with the proper boundary conditions. At large r, (4) becomes,

$$\mathrm{d}^2u_{n'l'm'}/\mathrm{d}r^2=-k_{n'}^2u_{n'l'm'}\,. \tag{7}$$

It is convenient to write the solution as

$$u_{n'l'm'}\sim k_{n'}^{-\frac{1}{2}}\left[A_{n'l'm'}\exp[-ik_{n'}r+il'\pi/2]-B_{n'l'm'}\exp[ik_{n'}r-il'\pi/2]\right]. \tag{8}$$

The first term is an incoming spherical wave (traveling toward $r=0$), and the second term is an outgoing wave. The scattering or S-matrix is defined by the relationship between the two amplitudes [1],

$$B_{n'l'm'}=\sum_{nlm}S_{n'nl'lm'm}A_{nlm}\,, \tag{9}$$

or $\boldsymbol{B}=\boldsymbol{S}\boldsymbol{A}$ in matrix notation.

When (9) is substituted into (8) to eliminate $\boldsymbol{B}$, and the result is introduced into (2) we obtain an asymptotic expression for Ψ,

$$(10)\qquad \Psi \sim \sum_{\substack{nlm\\ n'l'm'}} r^{-1} k_{n'}^{-\frac{1}{2}} \{A_{n'l'm'}\,\delta_{n'n}\,\delta_{l'l}\,\delta_{m'm} \exp[-ik_{n'}r + il'\pi/2] - $$
$$- S_{n'nl'lm'm} A_{nlm} \exp[ik_{n'}r - il'\pi/2]\}\, Y_{l'm'}\psi_{n'}\,.$$

Physically, we will have a beam of state-selected molecules which we will assume travels along the z-axis toward the scattering center. Emerging from the scattering center will be scattered molecules in all accessible states. Thus, Ψ is given asymptotically by

$$(11)\qquad \Psi \sim \psi_0 \exp[ik_0 z] + \sum_{n'} r^{-1} f_{n'0}(\theta, \varphi) \exp[ik_{n'}r]\psi_{n'}\,,$$

where the first term is a plane wave representing the beam, and the second term is the scattered wave weighted by an amplitude factor f and by r^{-1} since the scattered intensity, proportional to $\Psi^*\Psi$, must decrease as r^{-2}. At this point let us note that there are states for which $E_n > E$ and for which k_n is therefore imaginary. These states (often referred to as closed channels) may contribute to Ψ in the interaction region, but, as we can see from (11), the scattered amplitude must decrease as $\exp[-\varkappa r]$ $(k = i\varkappa)$ and so do not contribute to the final scattered wave.

Since (10) and (11) must be equal we can obtain $\boldsymbol{A}$ by equating the terms corresponding to incoming waves. Equating the terms corresponding to outgoing waves gives $f_{n'0}$ in terms of $\boldsymbol{S}$. To show this we expand $\exp[ikz]$ and f_{n0} in terms of spherical harmonics [1],

$$(12)\qquad \Psi \sim \sum_{l} i^l [4\pi(2l+1)]^{\frac{1}{2}} (2ik_0 r)^{-1} Y_{l0}(\theta, \varphi)\psi_0 [\exp[ikr - il\pi/2] -$$
$$- \exp[ik_0 r + il\pi/2]] + \sum_{n'l'm'} r^{-1} C_{n'l'm'} \exp[-il'\pi/2]\, Y_{l'm'}(\theta, \varphi) \exp[ik_{n'}r]\psi_{n'}\,.$$

Note that in the expansion of $\exp[ikz]$, $m = 0$; there can be no component of angular momentum along the initial velocity. The term in (12) proportional to $\exp[-ikr]$ gives A_{nlm}, and by using (9) we can relate C to the S-matrix giving,

$$(13)\qquad f_{n'0}(\theta, \varphi) = (4\pi)^{\frac{1}{2}} [2i(k_0 k_n)^{\frac{1}{2}}]^{-1} \sum_{l} i^l (2l+1)^{\frac{1}{2}} \cdot$$
$$\cdot \left\{ \sum_{l'm'} i^{-l'} [S_{n'0l'lm'0} - \delta_{n'0l'lm'0}]\, Y_{l'm'}(\theta, \varphi) \right\}.$$

The matrix $[\boldsymbol{S} - \boldsymbol{I}]$ appearing in (13) is known as the transmission or T-matrix. Note that there are several phase conventions used in defining $\boldsymbol{S}$ and $\boldsymbol{T}$. The differential cross-section is given by [2]

$$I_{n'0}(\theta, \varphi) = (k_{n'}/k_0)|f_{n'0}(\theta, \varphi)|^2 . \tag{14}$$

Note that $|f|^2$ relates the number densities or probabilities of the scattered and incident beams. The cross-section relates the two fluxes, and the ratio of $k_{n'}/k_0$, merely converts the number density to flux. By integrating I_{n0} over angles the partial cross-section for the transition from 0 to n' can be obtained as,

$$Q_{n'0} = \pi k_0^{-2} \sum_{l_1 l_2 l' m'} [(2l_1 + 1)(2l_2 + 1)]^{\frac{1}{2}} i^{l_1 - l_2} T_{n'0l'l_1m'0} T^*_{n'0l'l_2m'0} . \tag{15}$$

Several authors use a basis set of wave functions that couples the orbital and rotational angular momenta [5]. For the simple case of an atom and a diatomic molecule the radial wave functions for the two representations are related by the unitary transformation $u'_{JMjl} = \sum_{m_1 m} C(jlJ; m_1 m M) u_{jm_1 lm}$ where C is the Clebsch-Gordan or vector-coupling coefficient and J and M are the quantum numbers associated with the total angular momentum.

There are two important properties of the S-matrix [2]. It is unitary. From (9) we see that the probability of a transition between nlm and $n'l'm'$ is proportional to $|S_{n'nl'lm'm}|^2$. The unitarity condition $\sum_i |S_{ij}|^2 = 1$ means that the sum of all transition probabilities is unity, *i.e.* particles are conserved. From the Hermitian property of H one can show that S is also a symmetric matrix [2]. Thus the cross-section for the process $nlm \to n'l'm'$ is proportional to the cross-section for $n'l'm' \to nlm$. This is the law of microscopic reversibility.

3. – Approximate methods.

We now look at several approximate methods for solving the coupled equations (4).

3·1. *Numerical methods.* – With high-speed digital computers it is possible to solve (4) by truncating the coupled set and using any one of a number of numerical methods. One very fast method has been developed by LIGHT *et al.* [6] and is effectively a generalization of the WKB approximation. Up to about fifty coupled equations can be handled by this method depending on the particular problem.

3·2. *Integral equations.* – Several approximate methods can be derived by expressing (4) in terms of a set of integral equations.

We can write (4) as

$$\Omega(r) u(r) = f(r) . \tag{16}$$

We define the Green's function, $G(r, r')$, for the operator Ω by the equation

$$\Omega(r)\, G(r, r') = \delta(r, r') \;. \tag{17}$$

Then $u(r)$ can be obtained by convolution as

$$u(r) = \int G(r, r') f(r')\, dr' , \tag{18}$$

as can easily be verified by substituting (18) into (16). The Green's function for Ω can be expressed in terms of the two linearly independent solutions, u_a^0 and u_b^0, of the related problem of elastic scattering, $\Omega u^0 = 0$ [7],

$$G(r, r') = \tfrac{1}{2} i\, u_a^0(r_<)\, u_b^0(r_>) \;, \tag{19}$$

where $r_< = \min(r, r')$ and $r_> = \max(r, r')$. The solutions u^0 are chosen so that u_a^0 is well behaved at the origin. The integral equation corresponding to (4) is

$$\boldsymbol{u}(r) = \boldsymbol{u}_a^0(r) + \int G(r, r')\, \boldsymbol{U}(r')\, \boldsymbol{u}(r')\, dr' \;. \tag{20}$$

The values of k and l used to evaluate G and u_a^0 are those corresponding to the final state, *i.e.* the one specified by the left-hand side of (20).

3·3. *Distorted-wave approximation.* – We can construct a perturbation approximation from (20). The unperturbed wave function, the undistorted wave, is u_a^0, the elastically scattered wave function. The perturbation is ΔV or $\boldsymbol{U}$. If we substitute $u_a^0(r')$ for $\boldsymbol{u}$ in the integrand of (20) we obtain the first-order distorted-wave approximation. The S-matrix is given by [8]

$$S_{n'n\, l'l\, m'm} = \exp\left[i\eta_0(k, l) + i\eta_0(k', l')\right]\Big[\delta_{n'n}\delta_{l'l}\delta_{m'm} + \\ + \frac{1}{2} i \int_0^\infty u_a(r, k', l')\, U_{n'n\, l'l\, m'0}(r)\, u_a^0(r, k, l)\, dr\Big] , \tag{21}$$

where η_0 is the phase shift corresponding to V_0 and (k', l') and (k, l) are the final and initial values of k and l, respectively.

3·4. *Born approximation.* – A somewhat more drastic but simpler approximation is the first Born approximation which is a perturbation treatment based on the initial plane wave as the unperturbed wave function. This may be derived from the distorted-wave approximation by setting V_0 equal to zero and including the whole potential in ΔV. In this case u_a^0 is given by the spherical Bessel function,

$$u_a^0(r) = 2ik^{\frac{1}{2}} r j_l(kr) \tag{22}$$

and η_0 is, of course, zero.

Neither the Bornn or the distorted-wave approximation is particularly good for molecular inelastic scattering. The Born approximation breaks down for elastic scattering except at very small angles. In most cases of interest ΔV is so large that second-order and higher-order effects are necessary, and the transition probabilities predicted by the Born and distorted-wave approximations are incorrect—they are often larger than one, which is physically unrealistic.

3·5. *Infinite-order distorted-wave approximation* [8]. To include higher-order effects we can continue the process used in obtaining (21). If we substitute the wave function for the first-order distorted-wave approximation for $\boldsymbol{u}$ in (20) we can generate a second-order term and so on. We obtain

$$\boldsymbol{u} = \boldsymbol{u}_a^0 + \int G\boldsymbol{U}\boldsymbol{u}_a^0 \, dr' + \int_0^\infty dr' \int_0^\infty dr'' \, G(r, r')\boldsymbol{U}(r')G(r', r'')\boldsymbol{u}_a^0(r'') + \dots \, . \tag{23}$$

The values of k and l used in calculating each G are specified by the subscripts on the $\boldsymbol{U}$ or $\boldsymbol{u}$ which bracket G. The convergence properties of (23) are not simple, but it does not converge for all possible $\boldsymbol{U}$. As it stands (23) is very difficult to evaluate and not very useful, but with some further approximations it can be reduced to a simple form.

3·6. *The semi-classical approximation* [8]. – We now consider a method to simplify (23) which is valid for the case where the internal levels of the colliding molecules are closely spaced. We use the WKB approximation to evaluate u_a^0 and u_b^0,

$$u_a^0 = 2i(\hbar/p_r)^{\frac{1}{2}} \sin\left[\hbar^{-1}\int_{r_c}^{r} p_r \, dr + \pi/4\right], \tag{24}$$

where

$$p_r = [2\mu(E - V_0) - (l + \tfrac{1}{2})^2\hbar^2/r^2]^{\frac{1}{2}} \, . \tag{25}$$

u_b^0 is similar to u_a^0 except that the sine is replaced by exp $i[\;]$, and r_c is the classical turning point. We keep the changes in E and l if they occur in the exponential or sine, but we neglect the changes in evaluating $p_r^{\frac{1}{2}}$ and r_c (which limits the theory to closely spaced levels). Two types of terms occur in the evaluation of (23). There are rapidly oscillating terms with a frequency given by

$(p_r + p'_r)/\hbar$ and slowly oscillating terms with a frequency given by $(p_r - p'_r)/\hbar$. If we keep only the slowly oscillating difference terms, we obtain [8]

$$\boldsymbol{S} = \exp[2i\boldsymbol{\eta}] = \sum_{n=0}^{\infty} (2i\boldsymbol{\eta})^n/n!\,, \tag{26}$$

where

$$\boldsymbol{\eta} = \eta_0(k', l')\,\boldsymbol{I} - (2\hbar)^{-1} i^{\Delta l} \int_{-\infty}^{\infty} \Delta\boldsymbol{V}(r(t)) \exp[i\omega t] \exp[-i\Delta l\theta]\,\mathrm{d}t + i(2\hbar)^{-2} i^{\Delta l} \cdot \int_{-\infty}^{\infty} \mathrm{d}t_1 \int_{-\infty}^{t_1} \mathrm{d}t_2 [\Delta V(t_1) \exp[i\omega t_1] \exp[-i\Delta l\theta_1], \Delta V(t_2) \exp[i\omega t_2] \exp[-i\Delta l\theta_2]] + \ldots, \tag{27}$$

where $\omega = \Delta E_{\text{internal}}/\hbar$, $\Delta l = l' - l$, and the integrals are over the classical trajectory specified by V_0 and with the initial conditions that $\theta\,(t = -\infty) = \pi$, $z = -vt$, $(x^2 + y^2)^{\frac{1}{2}} = b$, the impact parameter.

In two important limiting cases the third and successive terms in (27) involving the commutators of $V(t_1)$ and $V(t_2)$ vanish. These limits are the sudden limit where $\omega = 0$, *i.e.* the energy level separation is negligible compared to the kinetic energy, and the classical limit of large rotational or vibrational quantum numbers where ω and $\boldsymbol{U}$ are independent of j, m, and n and dependent only on the changes in the various quantum numbers. In the classical limit ω is the frequency of the classical motion; thus the sudden limit corresponds to the case of fixed molecular orientation. In both limits there exists a representation where the internal wave function can be described as a narrow wave packet during the collision.

By considering the rapidly oscillating sum terms in (23) additional terms in $\boldsymbol{\eta}$ can be obtained which correct for the deviation of the true trajectory from the unperturbed trajectory specified by V_0 [8]. Thus the semi-classical approximation is valid if the trajectory is close to the unperturbed trajectory, the energy levels are closely spaced, and the problem is close to the classical or sudden limit. It was designed to treat rotationally inelastic scattering where, in most cases, the anisotropy of the potential is large enough so that the Born and distorted-wave approximations are not valid but small enough that the trajectory is close to the unperturbed trajectory. In most cases the rotational quantum numbers are large enough that the classical limit applies. For the case of small quantum numbers the levels are often spaced closely enough for the sudden limit to apply. There are cases, however, where none of the above procedures will work.

3·7. *Diagonalization of the S-matrix.* – The existence of a wave-packet representation for the sudden and classical limits which resembles a classical description suggests that we might simplify the problem by transforming the S-matrix into this representation. This turns out to be the case, and in this representation the theory can be simplified in a manner similar to the semi-classical theory of elastic scattering.

We must first eliminate the orbital angular momentum from the scattering expressions. It is evident from (13) and (15) that we must always sum over l and m. By using asymptotic expressions for the spherical harmonics in (5) and by using closure relations for the sums over l and m one can show that [8]

$$\sum_{\substack{\Delta l \\ \Delta m}} i^{-\Delta l-\Delta m} S_{n'n\,l'l\,m'm} \exp[i\Delta m\varphi] = S'_{n'n}(b, \varphi) = \exp[2i\eta'] , \tag{28}$$

where

$$\eta'_{n'n}(b, \varphi) = \eta_0(b) - (2\hbar)^{-1}\int_{-\infty}^{\infty} \Delta V_{n'n}(r, \theta, \varphi) \exp[i\omega t]\, dt + \dots . \tag{29}$$

As before, the integration is over the trajectory specified by V_0. The trajectory depends on the impact parameter $b = l'/k' = (x^2 + y^2)^{\frac{1}{2}}$ *and* on the initial azimuthal angle $\varphi = \operatorname{arctg}(y/x)$.

In the case of the sudden limit a narrow stationary wave packet centered at r' can be constructed by using the closure relation [9]

$$\delta(r - r') = \sum_n \psi_n(\boldsymbol{r})\, \psi_n^*(\boldsymbol{r}') . \tag{30}$$

In the sudden limit we can diagonalize S' by the transformation

$$S(\boldsymbol{r}')\, \delta(\boldsymbol{r} - \boldsymbol{r}') = \sum_{nn'} \psi_{n'}(\boldsymbol{r})\, S'_{n'n}\, \psi_n^*(\boldsymbol{r}') . \tag{31}$$

The new S-matrix is diagonal in terms of the continuous quantum numbers $\boldsymbol{r}'$. This sudden limit has been used for a long time [10]. In the classical limit the transformation is more difficult. In this case [11]

$$S(\boldsymbol{\varphi}) = \prod_j \sum_{\Delta n_j} S'_{\Delta n_j} \exp[i\Delta n_j \varphi_j] . \tag{32}$$

The product is over all the quantum numbers. Each quantum number is related to a generalized momentum; φ_i is the value of the conjugate co-ordinate at $t = 0$. In both limits the diagonal phase shift matrix is given by [11],

$$\eta = \eta_0 - (2\hbar)^{-1}\int_{-\infty}^{\infty} \Delta V(t)\, dt , \tag{33}$$

where now ΔV varies in time as r, θ, and $\boldsymbol{r}_{\text{int}}$ change in time.

In the diagonal representation we can simplify (5) by using an asymptotic expression for the spherical harmonic. The use of (28) converts (13) into a simple two-dimensional integral in b and φ which can be evaluated by the method of stationary phase to give [12],

$$(34)\qquad f(\theta',\varphi') = b^{\frac{1}{2}}|\sin\theta'\,(\partial(\theta',\varphi')/\partial(b,\varphi))|^{-\frac{1}{2}}\exp[2i\eta - ik'b\theta'\cos(\varphi-\varphi')]\,i^{p},$$

where η, b, and φ are evaluated at the point of stationary phase,

$$(35)\qquad \theta' = +2k^{-1}[(\partial\eta/\partial b)^2 + b^{-2}(\partial\eta/\partial\varphi)^2]^{\frac{1}{2}},$$

$$(36)\qquad \varphi' = \varphi + \operatorname{arctg}[(\partial\eta/\partial\varphi)/b(\partial\eta/\partial b)]\,.$$

The phase factor $p=0$ if the Jacobian in (34), $J<0$; $p=1$ if $J>0$, $\partial^2\eta/\partial b^2>0$; $p=-1$ if $J>0$, $\partial^2\eta/\partial b^2<0$. For a spherically symmetric potential, $\varphi'=\varphi, \varphi+\pi$, and (34) reduces to the common elastic expression for $f(\theta',\varphi')$; (35) reduces to the usual relation between the phase shift and scattering angle. By reversing the transformation (31) or (32), one can transform back to the original representation to give $f_{n'n}(\theta',\varphi')$.

3·8. *Effect of anisotropy on total scattering measurements.* – In the case of atom-atom scattering much information can be obtained from scattering measurements about the intermolecular potential. It is natural, therefore, to ask what happens to these measurements in the case of atom-molecule scattering where the potential is not spherically symmetric. If the anistropy is large, obviously one can say little about the scattering without a complete analysis of the resulting inelastic scattering. If the anisotropic term is small, $V = V_0(r) + a\Delta V$, we can perform a perturbation treatment; η, θ', and φ' all depend on a to first order. Yet if we average over all orientations, we find that the *classical* differential cross-section and the quantal total cross-section (not including the glory undulations) is identical to the cross-section calculated from V_0 except for a small term which is second order in a. The effect of the anisotropy is quite severe on quantal interference effects such as the glory oscillations in the total cross-section [13]. In this case the phase shift is $\eta_0+\Delta\eta$, and even though $\Delta\eta$ may be small compared to η_0, it may be comparable to π. Thus at a given velocity one orientation may correspond to a maximum and another to a minimum in the cross-section. When an average is taken over orientations, the two will cancel giving a severely reduced amplitude.

3·9. *Vibrationally inelastic scattering.* – The problem of vibrationally inelastic scattering at thermal energies is a difficult one. The energy level spacing is comparable to or larger than the incident kinetic energy, and only a few transitions are possible. However, every system with vibrational energy levels must also have rotational energy levels which are closely spaced compared to the vibrational levels. Thus there will be rotational transitions concurrent with every vibrationally inelastic event, and the two should be considered together. Unfortunately, the approximations which work for a few widely spaced levels fail for the rotational part of the problem, and the approximations valid for a large number of closely spaced levels will fail for the vibrational part. There is currently no way of treating both halves of the problem accurately. The most common approach is to approximate the problem by the case of a collinear collision of an atom and a diatomic molecule thereby eliminating the rotational part of the problem and the anisotropy of the vibrational part of the potential. While this is a crude approximation at best, it is better than nothing and will give a rough idea of the magnitude of the cross-section and its dependence on the various initial conditions. A numerical solution of the coupled equations (5) has been done by SECREST and JOHNSON [14]. They compare their results to a number of approximate treatments and give a detailed formulation of the problem.

4. – Experimental.

There are a variety of methods for obtaining information about inelastic processes. This topic has been reviewed extensively by GORDON, KLEMPERER, and STEINFELD. [4]

4·1. *Spectroscopic methods.* – In a low-pressure gas a spectrum will consist of many sharp lines whose width is determined by the lifetime of the states involved in the transition. At higher pressures the molecules undergo inelastic collisions, which shorten the lifetime of the states, and, by the uncertainty principle, the uncertainty in energy and hence the width of the line must increase. The pressure-dependence of the linewidth is a measure of the inelastic scattering [15]. A more recent technique is microwave-microwave double resonance [16]. If a gas is subjected to a strong microwave signal at the frequency of a particular transition the populations of the two states involved in the transition will change. Since these states are coupled to all other states by inelastic scattering, the change in population will cause corresponding changes in the populations of all the levels, which will depend on the strength of the coupling. These changes can be measured by the absorption of a second weaker microwave signal.

When a molecule such as I_2 is irradiated by a narrow atomic spectral line a transition occurs to a single rotational-vibrational level of an electronically excited state. The excited molecules then undergo vibrationally and rotationally inelastic collisions then undergo a spectroscopic transition to the ground electronic state. From the emitted fluorescence spectrum one can tell what inelastic transitions were made by the excited molecules [17].

4·2. *Relaxation methods* [18]. – If the translational temperature of a system is changed by some means such as a shock wave, the rotational and vibrational temperatures will also change because they are coupled to the translational motion by inelastic collisions. The rate of change of rotational or vibrational temperature rise is usually expressed as relaxation time and is determined by inelastic scattering. Vibrational relaxation times can be measured from a Schlieren photograph of a shock wave in a gas. Rotational relaxation can be measured by studying the expansion of gas through a supersonic nozzle [19]. The absorption of high-frequency sound waves is also determined by inelastic scattering [4]. Here, the perturbation is periodic rather than sudden. Nuclear spins are weakly coupled to molecular rotation. Nuclear relaxation times obtained from gas-phase NMR experiments give information on the change in rotational angular momentum [20].

4·3. *Beam methods.* – Rotationally inelastic scattering has been measured by passing a beam of polar molecules such as CsF or TlF through a specially constructed inhomogeneous electric field which selects a single rotational state of the beam. This state-selected beam is then scattered by a gas, passed through a second state selector tuned to a different state, and finally detected [21]. Vibrationally inelastic scattering has been measured by scattering a beam of high-energy alkali atoms or ions from H_2. Since the mass of the alkali atom is larger than that of H_2, the atoms or ions scattered at 180° in the center-of-mass system are scattered at 0° in the laboratory system. By measuring the flight time of the scattered ions, one can measure the energy transferred from translation to vibration [22]. Rotational and vibrational energy transfer can also be measured by scattering a velocity-selected beam from a crossed beam and then measuring the velocity distribution of the scattered beam [23].

4·4. *Some general results.* – All experiments and theories show that rotationally inelastic scattering is very probable. Several rotational quanta are usually transferred in each collision. The inelastic cross-sections are even larger for polar molecules due to the large anisotropy of the dipole-dipole force [21]. Vibrationally inelastic scattering is much less likely. However, fluorescence experiments on I_2 indicate that more than one vibrational quantum can be transferred on a single collision [17]. Vibrational energy transfer seems to be accompanied by a larger amount of rotational transfer than if no vibrational transfer occurred [17].

* * *

The author would like to thank Drs. W. D. Held, L. Homlid, and I. Leksell for their help in the preparation of this manuscript.

REFERENCES

General references:

[1] N. F. Mott and H. S. Massey: *The Theory of Atomic Collisions* (London, 1965), see especially p. 369-394.
[2] E. H. S. Burhop: *Quantum Theory I*, edited by D. R. Bates(New York, 1961).
[3] K. Takayanagi: *Advances in Atomic and Molecular Physics*, edited by D. R. Bates and I. Estermann (New York, 1965).
[4] R. G. Gordon, W. Klemperer and J. I. Steinfeld: *Ann. Rev. Phys. Chem.*, **19**, 215 (1968).

Specific references:

[5] A. M. Arthurs and A. Dalgarno: *Proc. Roy. Soc.*, A **256**, 540 (1960); R. B. Bernstein, A. Dalgarno, H. Massey and I. C. Percival: *Proc. Roy. Soc.*, A **274**, 427 (1963); W. D. Davison: *Disc. Faraday Soc.*, **33**, 71 (1962).
[6] S. K. Chan and J. C. Light: *Journ. Chem. Phys.*, **49**, 86 (1968); R. G. Gordon: *Journ. Chem. Phys.*, **51**, 14 (1969).
[7] P. M. Morse and H. Feshbach: *Methods of Theoretical Physics I* (New York, 1953), p. 826.
[8] R. J. Cross: *Journ. Chem. Phys.*, **48**, 4838 (1968).
[9] R. J. Cross: *Journ. Chem. Phys.*, **47**, 3724 (1967).
[10] L. I. Schiff: *Phys. Rev.*, **103**, 443 (1956); R. J. Glauber: *Phys. Rev.*, **100**, 242 (1955); K. Alder and A. Winther: *Kgl. Danske videnskab. Selskab. Mat., Fys. Medd.*, **32**, n. 8 (1960); K. H. Kramer and R. B. Bernstein: *Journ. Chem. Phys.*, **40**, 200 (1964); **44**, 4473 (1966); R. W. Fenstermaker and R. B. Bernstein: *Journ. Chem. Phys.*, **47**, 4417 (1968); J. P. Toennies: *Zeits. Phys.*, **193**, 76 (1966).
[11] R. J. Cross: *Journ. Chem. Phys.*, **49**, 1753 (1968).
[12] R. J. Cross: *Journ. Chem. Phys.*, **50**, 1036 (1969).
[13] R. E. Olson and R. B. Bernstein: *Journ. Chem. Phys.*, **49**, 162 (1968); R. J. Cross: *Journ. Chem. Phys.*, **49**, 1976 (1968); R. E. Olson and R. B. Bernstein: *Journ. Chem. Phys.*, **50**, 246 (1969).
[14] D. Secrest and B. R. Johnson: *Journ. Chem. Phys.*, **45**, 4556 (1966).
[15] P. W. Anderson: *Phys. Rev.*, **76**, 647 (1949).
[16] A. P. Cox, G. W. Flynn and E. B. Wilson: *Journ. Chem. Phys.*, **42**, 3094 (1965); R. G. Gordon: *Journ. Chem. Phys.*, **46**, 4399 (1967).
[17] J. I. Steinfeld and W. Klemperer: *Journ. Chem. Phys.*, **42**, 3475 (1965); H. P. Broida and T. Carrington: *Journ. Chem. Phys.*, **38**, 136 (1963).
[18] H. G. Wagner: This volume, p. 62.
[19] D. R. Miller and R. P. Andres: *Journ. Chem. Phys..*, **46**, 3418 (1967).
[20] R. G. Gordon: *Journ. Chem. Phys.*, **44**, 228 (1966).
[21] J. P. Toennies: *Zeits. Phys.*, **182**, 257 (1965).
[22] J. Schöttler and J. P. Toennies: *Zeits. Phys.*, **214**, 472 (1968); P. F. Dittner and S. Datz: *Journ. Chem. Phys.*, **49**, 1969 (1968).
[23] D. Beck: unpublished results.

Relaxation Methods.

H. Gg. Wagner

University of Bochum - Bochum

1. – Introduction.

In thermal equilibrium the population of the energy levels of one degree of freedom, *e.g.* of the vibrational levels in a diatomic gas, can be described by an equilibrium distribution function. The various degrees of freedom are also in equilibrium with each other. In chemical reactions, in nozzle flow, and other processes this equilibrium within one and between various degrees of freedom can be disturbed. It is re-established by energy transfer processes. Except for the case of flow of vibrational energy in large molecules the energy exchange occurs via inelastic collisions. The energy exchange by radiation is negligible under normal conditions.

Calculations of thermodynamic data in gases show that specific heats obtained under the assumption that the various degrees of freedom are practically independent from each other are in very good agreement with experimental values (as long as the density is sufficiently low). Therefore one may expect that the energy transfer processes between the various degrees of freedom can be characterized by various time constants. Measurements have confirmed that various types of energy transfer processes in a system can be described by characteristic times which often are orders of magnitude apart from each other.

Important energy transfer processes in gases are the exchange of translational energy from one molecule to the other, and that between translation and rotation, translation and vibration, and vibration and vibration. Most of the information about these processes has been obtained from relaxation measurements.

In such an experiment the equilibrium state of the system is slightly disturbed, either by a step function or by a periodic disturbance. One observes then the reaction of the system to that disturbance, *i.e.* the re-establishment of the equilibrium.

Such a process can be described on the basis of thermodynamics of irreversible processes [1, 2], as long as the deviation from equilibrium remains sufficiently small. This means that the large number of different inelastic collisions which are necessary, *e.g.*, for vibrational-translational energy transfer in a gas are characterized by one (or a few) relaxation times τ. One does not look at a single process but at the whole system. For a closed homogeneous system, one has

$$T\,\mathrm{d}S = \mathrm{d}U + P\,\mathrm{d}V + \sum_r A_r\,\mathrm{d}\xi_r\,, \tag{1}$$

S = entropy of the system,
T = temperature,
U = internal energy,
P = pressure,
V = volume,
μ_k = chemical potential,
ν_{kr} = stoichiometric coefficient,
$A_r = -\sum_k \nu_{kr}\mu_k$ = affinity,

ξ_r is some variable of concentration, *e.g.* the partial density or a reaction progress variable. The rate of reaction is

$$w_r = \frac{\mathrm{d}\xi_r}{\mathrm{d}t} = \sum_s a_{rs}A_s \tag{2}$$

(with $\sum A_r w_r \geqslant 0$, the determinant $|a_{ik}| > 0$, and $a_{rs} = a_{sr}$). For simplicity the energy transfer process shall be described for one variable only. Then

$$\frac{\mathrm{d}\xi}{\mathrm{d}t} = aA \qquad \text{with } a > 0\,.$$

In equilibrium there is $A_e = 0$ and $\xi = \bar{\xi}$. A can be expanded as $A = (\xi - \bar{\xi})\cdot$ $\cdot(\partial A/\partial\xi)_{\alpha,\beta}$, where α, β are the thermodynamic variables used to describe the system. Stability conditions require $(\partial A/\partial\xi)_{\alpha,\beta} < 0$. This gives a first order differential equation, the relaxation equation,

$$\frac{\mathrm{d}\xi}{\mathrm{d}t} = -\frac{1}{\tau_{\alpha\beta}}(\xi - \bar{\xi})\,, \tag{3}$$

where $1/\tau_{\alpha\beta} = -a(\partial A/\partial\xi)_{\alpha,\beta}$. $\tau_{\alpha\beta}$ is the relaxation time for α = const and β = const. An important point here is that one has to specify α, β, *e.g.* as S, V or P, T, and that the relaxation time $\tau_{\alpha\beta}$ depends on this specification. The case of several processes with various characteristic times τ_i can be treated in similar fashion.

If the equilibrium quantity $\bar{\xi}$ is changed periodically so that the variation is $\Delta\bar{\xi}_{eq} = \alpha e^{iwt}$ then one easily obtains from the relaxation equation that the quantity ξ varies about the static equilibrium value as

$$(4) \qquad \xi = \frac{\Delta\bar{\xi}_{eq}}{1 + iw\tau} .$$

Let ξ be the population in the excited state with energy ε. Then the energy content of that state is $E = \varepsilon\xi$. Under equilibrium conditions there is $\Delta E = \varepsilon\Delta\bar{\xi}_{eq} = c'\Delta T_{eq}$ where c' is the equilibrium specific heat contribution of that state and ΔT_{eq} the variation of temperature.

A periodic variation with frequency w of the state of a substance therefore causes an effective contribution of the state under consideration to the specific heat of the system, which is

$$(5) \qquad c_{eff} = \frac{c'}{1 + iw\tau} .$$

The case of a step function disturbance need not be treated here.

Methods for the determination of relaxation times by periodical or stepwise variation of the thermodynamic properties of a system are [3-5]:

1) Measurement of sound velocity dispersion and sound absorption.

2) Shock waves.

3) Impact tube.

The range accessible to each of these methods is determined by the magnitude of the corresponding specific heat contribution. They are normally useful for the investigation of relaxation of low-lying energy levels.

There is a second group of methods which do not utilize a variation of thermodynamic properties like pressure, density, and temperature. They are based on the generation of an overpopulation in certain energy levels, the disappearance of which is observed. This overpopulation can be achieved by flash photolysis, all kinds of radiolysis, chemical activation and others. These methods allow the excitation also of high-energy levels.

2. – Experimental methods (group I).

We will give a brief review of the applicability and limitations of various experimental methods and cite some recent results.

2·1. *Sound dispersion and absorption.* – The measurement of sound velocity and sound absorption for the determination of relaxation times is a very important method. For an ideal gas with $pV^{\gamma}=P_0V_0^{\gamma}$ and $V=1/\varrho$ one obtains for the velocity of sound, C_0,

$$C_0^2=\gamma\frac{p}{\varrho}=\frac{RT}{M}\left(1+\frac{R}{C_V}\right). \tag{6}$$

(M=molecular weight, C_V=specific heat for V=const, $R=C_p-C_V$, $\gamma=C_p/C_V$.)

Two factors are neglected here:

1) Normally a gas does not behave as an ideal gas. Except for special cases this does not cause a large effect [4]. In the case of ethylene for example the deviation of the real from the ideal sound velocity at 1 atm and room temperature is less than 0.5 %.

2) Due to heat conduction, friction and diffusion (transport processes!) the sound waves are attenuated roughly proportional to the square of the sound frequency. The absolute value of this absorption coefficient however is small [4] and can be measured independently.

As already mentioned the specific heat C_V of a gas consists of the contribution of the various degrees of freedom: translation, rotation and vibration. Therefore

$$C_{V_0}=C_{V\,\text{trans}}+C_{V\,\text{rot}}+C_{V\,\text{vib}}.$$

For low-frequency sound waves ($w\tau\ll1$), all the contributing internal degrees of freedom are able to follow the change in density and temperature caused by the sound wave (see above). Therefore all terms of C_{V_0} contribute and the sound velocity is

$$C_0^2=\frac{RT}{M}\left(1+\frac{R}{C_{V_0}}\right). \tag{7}$$

But when the vibrational degrees of freedom are no longer able to follow the variations of density and temperature, due to the fact that the transfer of energy from translation into vibration and *vice versa* becomes too slow ($w\tau\gg1$), then the vibrations are no longer excited and do no longer contribute to the specific heat. The expression for the sound velocity at high frequencies, C_∞, is

$$C_\infty^2=\frac{RT}{M}\left(1+\frac{R}{C_{V\,\text{trans}}+C_{V\,\text{rot}}}\right)=\frac{RT}{M}\left(1+\frac{R}{C_{V\infty}}\right). \tag{8}$$

The sound velocity at high frequencies $w\tau\gg1$ has a higher value than at low frequencies ($w\tau\ll1$).

The transition between these two limits is then determined by

$$C^2 = \frac{RT}{M}\left(1 + \frac{R}{Cv_\infty + C'_v}\right), \tag{9}$$

where

$$C'_v = \frac{Cv_{\text{vib}}}{1 + i w \tau}.$$

The sound velocity contains an imaginary term which means that absorption takes place, in the transition region between C_0 and C_∞.

For the experimental determination of sound velocity and absorption in connection with relaxation processes it is convenient to change conditions by varying the pressure because $\tau \sim P^{-1}$. Therefore the independent variable is w/P. As an experimental tool, the ultrasonic interferometer can be used for relaxation times between 10^{-3} and 10^{-8} s. Temperatures around room temperature are easily accessible. The method becomes difficult to apply at temperatures above 1000 °K. An example for the application of this method is shown in Fig. 1. Part *a*) of the figure shows the sound velocity plotted as a function of sound frequency for pure CS_2. The experimental procedure was as mentioned above. At low and at high frequencies the sound velocity is practically constant, corresponding to the two values C_0 and C_∞. From the transition region (inflection point w_{infl}) one easily obtains the relaxation time for the energy

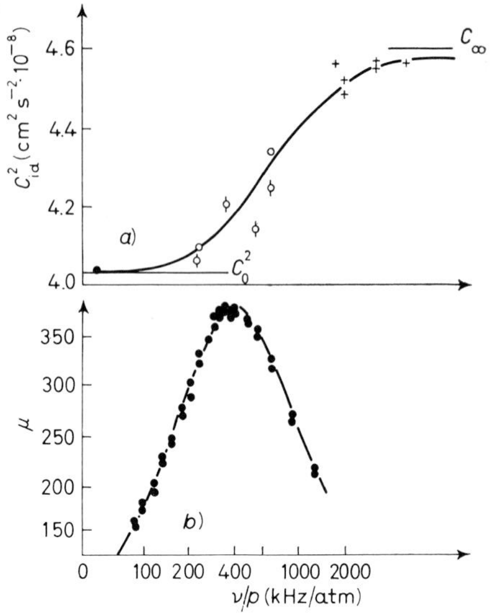

Fig. 1. – Sound velocity, C_{id}, and sound intensity absorption coefficient per wavelength, μ, for CS_2 [6] as a function of frequency ν over pressure P. Data of 1*a*) from W. T. RICHARDS and I. A. REID: *Journ. Chem. Phys.*, **2**, 193 (1934), 1*b*) from F. A. ANGONA: *Journ. Acoust. Soc. Am.*, **25**, 1116 (1953). *a*) + 451 kHz; ϕ 94 kHz; ○ 92 kHz; ● 9 kHz. *b*) ● experimental data. ——— theoretical curve.

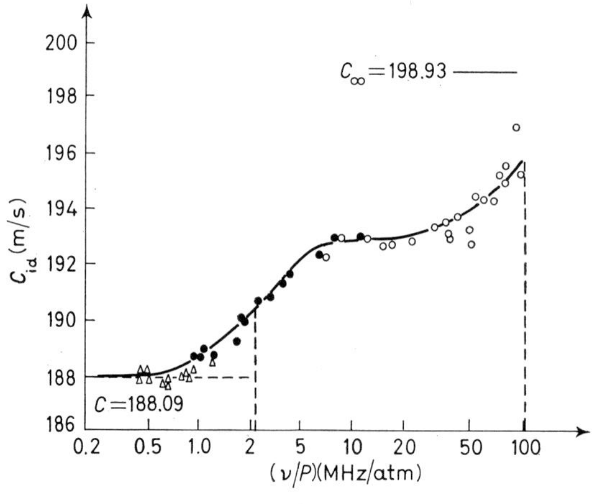

Fig. 2. – Sound velocity dispersion in methylene chloride vapour at 30 °C as a function of ν/P [7]. ○ at 1.9760 MHz, ● at 0.399 36 MHz, and △ at 0.203 38 MHz.

transfer process as $\tau = C_{V_0}/C_{V\infty} \cdot w_{V_{\text{infl}}}$. The process studied is the vibrational-translational energy exchange.

In the region where the sound velocity changes from the low to the high value one expects increased sound absorption. The shape of the absorption curve can easily be obtained from formula (9) above. The absorption coefficient for CS_2 in the dispersion region is shown in Fig. 1 *b*) as a function of w/P. Absorption measurements are rather difficult. Most authors therefore prefer the determination of the velocity dispersion.

In many cases the collisional deactivation of a molecular vibration with a frequency above that of the lowest normal mode takes place through the vibration with the lowest frequency. Sometimes however the relaxation times for two excitation processes are rather close together. This shows the sound velocity dispersion curve for CH_2Cl_2 in Fig. 2. The first increase of sound velocity corresponds to a translational-vibrational energy transfer, the second is attributed to a complex collisional energy transfer process, in which an oscillator with frequency ν_2 is deactivated so that part of the energy goes into ν_1 the rest into, *e.g.*, translation. This type of process can take place if $\nu_2 > 2\nu_1$.

2·2. *Shock waves.* – When a shock front passes through a gas the state of that gas is changed very rapidly, transition regions being only a few mean free paths wide. For the investigation of energy transfer processes shock waves are applied in various ways [8]. From measurements of the thickness of the shock front (or better of the profile of the shock front) one obtains information about the very fast translation-rotation energy transfer processes. If degrees of freedom, *e.g.* vibrations, cannot follow the rapid change of state in the shock front one can observe how they relax behind the front. This relaxation can be observed by measuring density profiles behind the shock front using, *e.g.*, a Mach-Zehnder interferometer. In this case one needs comparatively large density variations during the relaxation process in order to achieve reasonable accuracy for the relaxation times.

If the shock wave is stationary the state behind the shock front (index 1) is related to the state ahead (index 0) by the Hugoniot curve

$$\int_0^1 C_V \mathrm{d}T = -\frac{1}{2}(p_1 - p_0)\left(\frac{1}{\varrho_1} + \frac{1}{\varrho_0}\right)$$

and by the equation for the Rayleigh line (v_0 = shock velocity)

$$p_1 - p_0 = -\frac{\varrho_0}{\varrho_1}(\varrho_0 v_0^2) + \varrho_0 v_0^2 \,.$$

These two equations form a relation between ϱ_1 and C_V. The effective specific heat immediately behind the shock front is C_{v0} and $C_{v\infty}$ the one in thermal equilibrium. Therefore the relaxation process consists in a transition from C_{v0} to $C_{v\infty}$, which is related to a corresponding density increase. At the same time, however, pressure and temperature change as well, and they can also be used for the measurement of the relaxation process. The fact that all state variables change as a function of time complicates the evaluation of reliable relaxation time values.

If the species whose relaxation is to be investigated is infrared-active and present only in small quantities, such that during the relaxation process the state variables behind the shock front remain practically constant, one can obtain very well defined relaxation times by measuring the infrared emission of that species as a function of time.

Shock waves are especially well suited for measurements at elevated temperatures, up to many thousand degrees and for relaxation times between 1 and 100 μs.

Figure 3 shows a plot of relaxation times (given as pressure times relaxation time) as function of temperature (as $T^{-\frac{1}{3}}$) for various pure and mixed systems. They represent the energy transfer process between translation and vibration. The parameter O_2-Ar means that the curve represents the vibrational relaxation of O_2 with Ar as collision partner.

Shock waves can also be applied to the investigation of other energy transfer processes, *e.g.* that between vibrations of different molecules. It was often observed that small amounts of polyatomic impurities in a diatomic gas can drastically reduce its relaxation times. This is attributed to energy transfer between the vibration of the diatomic and the vibrations of the polyatomic species, since the polyatomic species show a rather rapid vibrational-translational energy exchange. Another energy transfer process of the type vib-vib is the resonance or near resonance transfer between oscillators with similar frequencies.

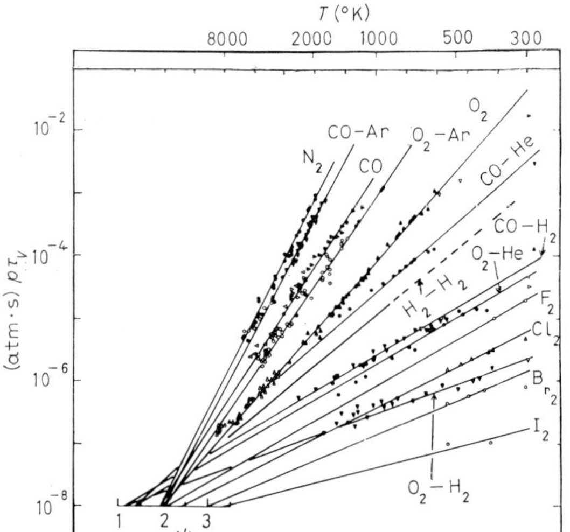

Fig. 3. – Relaxation times for various gases and gas mixtures as a function of temperature, taken from [9] except H_2—H_2 from [10] and [18].

For example [11] NO—N_2 ($\Delta\omega$=456 cm^{-1}), NO—CO ($\Delta\omega$=

$= 267\ \text{cm}^{-1}$), CO_2—N_2 ($\Delta\omega = 18\ \text{cm}^{-1}$ for the asymmetric stretch mode ν_3 of CO_2). TAYLOR, CAMAC and FEINBERG [12] investigated these processes in shock waves. In the case of CO—NO they observed signals which correspond to the rather fast vib-trans relaxation of NO. After a short time the relaxation process continued at a lower rate which is due to the vib-vib transfer from NO to CO. These measurements showed, that an addition of 3 % NO reduced the relaxation time of CO at 1000 °K by a factor of 10.

Such a process can be described by the following mechanisms (a star means vibrational excitation):

$$1)\qquad A+A^* \rightleftarrows A+A\,,$$

$$2)\qquad B+B^* \rightleftarrows B+B\,,$$

$$3)\qquad A^*+B \rightleftarrows A+B\,,$$

$$4)\qquad B^*+A \rightleftarrows B+A\,,$$

$$5)\qquad A^*+B \rightleftarrows B^*+A\,.$$

If reaction 5 is negligible the two species A and B will exhibit various relaxation times and different vibrational temperatures. If reaction 5 is very fast compared to the other reactions it couples the two species A and B close together and the system has one relaxation time and one vibrational temperature. If the vibration-vibration coupling is intermediate A and B will have slightly different vibrational temperatures, but after an induction period only one relaxation time. This intermediate case is realized in the systems NO—CO, CO_2—N_2 and CO—N_2. It shows, that a vibration-vibration energy transfer process can control the vibrational relaxation, particularly if there is one component present in low concentration with rapid translation-vibration relaxation.

Other possibilities of using shock waves to investigate relaxation processes exist, they were, *e.g.*, also applied to the investigation of the relaxation of excited electronic states.

2·3. *Other methods.* – Two other methods should be briefly mentioned: the impact tube [13] and the experimentally rather complicated spectrophone [14]. Both methods are not as widely used as ultrasonic interferometers or shock waves.

3. – Experimental methods (group II).

A good deal of information about energy transfer processes, especially about those including higher vibrational levels, comes from flash photolysis experiments [4]. The decay of the overpopulation is monitored, *e.g.*, by kinetic spectroscopy. A few examples may illustrate this method.

If BrCN in the reaction vessel is flashed it dissociates [15] into Br and CN. The CN radical will be electronically excited by the same flash. Its radiative or collisional decay to the electronic ground-state causes an overpopulation in the higher vibrational states ($v \geqslant 6$) of CN. The deactivation by collision can be easily observed by taking absorption spectra at various time intervals.

The overpopulation in certain states can also be generated by other methods like electron impact or X-rays. In some cases chemical activation is quite useful, *e.g.* for the production of O_2 which is vibrationally excited [16] up to quantum states near $v = 20$.

$$O_3 + O \rightarrow O_2 + O_2 \qquad (v \leqslant 20):$$

In the relaxation experiments utilizing flash photolysis and kinetic spectroscopy one generally obtains more information than in sound dispersion or shock wave experiments because the population of various states is measured directly as a function of time. On the other hand, there are certain restrictions: in general only the spectra of small molecules can be used to measure the population of certain vibrational or rotational states. The time resolution of the method is limited towards short times by the duration of the photolysis flash. The application of other methods, *e.g.* the *Z*-pinch or the laser as photolysis flash, could bring marked improvements.

Lasers have been successfully used for the investigation of energy exchange processes of low-lying vibrational levels. The overpopulation in the state of interest can either be achieved by a single laser pulse or by periodic illumination. The population and the change of population of that state can be monitored, *e.g.*, by a characteristic infra-red emission.

YARDLEY and MOORE [17] in their experiments on energy transfer in methane used a He-Ne Laser (10 mW at 3.39 μm) which was mechanically chopped (10^2 to 10^4 Hz). Its light excited the ν_3 mode of CH_4 (2 947.9 cm^{-1}). Part of the excitation is rapidly transferred to ν_4. A Cu-doped Ge-detector measured the infrared fluorescence of ν_3 and ν_4. Since the radiative lifetimes (0.04 and 0.4 s) are much longer than the nonradiative ones, the fluorescence intensity

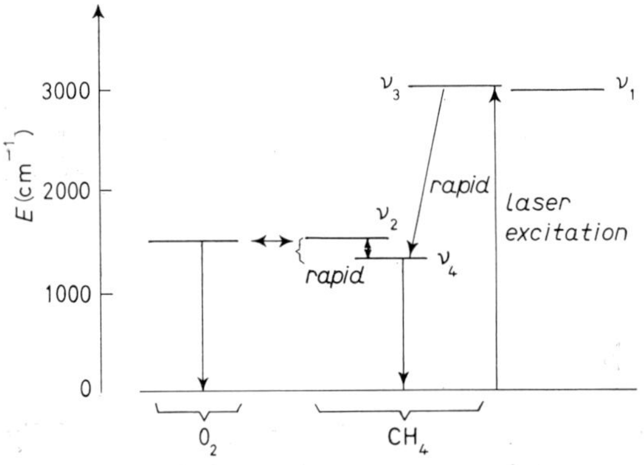

Fig. 4. – Energy level diagram for CH_4 and O_2 [17].

is proportional to the modulated populations of the excited vibrational states.

Besides the vibrational-vibrational energy transfer in CH_4, the authors studied the vibrational energy exchange between O_2 and CH_4. This is a good example for the influence of small amounts of polyatomic impurities on relaxation times. Figure 4 shows an energy level diagram for methane and oxygen. The authors found that the vibrational energy exchange CH_4—CH_4 is fast. The transition from ν_3 to ν_4 needs about 60 collisions. About 440 collisions are required to transfer one quantum of vibrational excitation from O_2 to CH_4. In order to transfer one vibrational quantum in a vibrational-translational process in pure oxygen more than 10^7 collisions are required (all data for room temperature). The same process in pure methane takes place much faster, it needs about 10^4 collisions. These figures explain the strong influence of CH_4 on the relaxation of O_2. There are other substances which are still more effective as impurities than CH_4 in O_2, as can be seen in the following Table. It gives the vibrational deactivation of CO_2 by various collision partners M. α is the efficiency of partner M relative to that of $M = CO_2$ (room temperature data again).

M	CO_2	He	H_2	N_2	CO	H_2O	CH_3OH	$C_6H_6CH_3$
α	1	20	200	40	223	10^3	$2\cdot10^3$	$3.4\cdot10^3$

Stimulated Raman scattering was applied by DE MARTINI and DUCUING to relaxation measurements in H_2 at room temperature [18]. A Q-switched ruby laser beam was focussed through a lens into the cell containing H_2 under pressure, where it produced stimulated Raman scattering (about 1% of the molecules excited). An ordinary laser was used as probing source. The authors obtained $P\cdot\tau_{H_2\text{-}H_2} \approx 10^{-3}$ atm s at 300 °K. This value fits very well into the extrapolation of shock tube measurements (see Fig. 3).

A very powerful method for the investigation of energy exchange processes should be mentioned though it is not a relaxation method. It uses the competition between two reactions. Species A is excited (reaction 1) and can be deactivated in two processes 2) and 3)

$$\begin{aligned} &1) \quad && A + h\nu \longrightarrow A^*, \\ &2) \quad && A^* \xrightarrow{\tau_e} A + h\nu, \\ &3) \quad && A^* + M \xrightarrow{k_3} A + M. \end{aligned}$$

In the stationary state $d[A^*]dt \approx 0$ and the emitted light intensity I is given by

$$I_0/I = 1 + k_3\tau_e\cdot[M],$$

where τ_e is the known radiative lifetime of A^* and I_0 the emitted intensity for $[M] \to 0$. From a plot of I^{-1} against the concentration of the collision partner **M** (Stern-Vollmer plot) one obtains the rate constant k_3 for the energy exchange process and the relaxation time $\tau_3 = k_3^{-1}$.

The excitation can be performed in many different ways. MILLIKAN [19] used the competition between collisional and radiative deactivation of excited CO to determine the efficiency of the collision partners ortho-H_2, H_2, D_2, HD, He and others for the deactivation of CO ($v=1$). The CO was excited by the radiation of CH_4—O_2 flames. POLANYI [20] used the process $Hg(6\,^3P) + CO(^1\Sigma) \to Hg(6\,^1S_0) + CO(^1\Sigma, v \leqslant 11)$ for the excitation of CO. Other authors excited the molecules by electric discharges [21]. It need not be discussed that electronic fluorescence spectra can be used for investigations of energy transfer processes with electronically excited species (see [4]).

4. – Some results.

The results of relaxation measurements are relaxation times and their dependence on collision partner, pressure, and temperature.

The measurements show, that translational relaxation needs only a few collisions. Energy transfer between translation and rotation is somewhat slower. For H_2, rotational relaxation at room temperature requires about 300 collisions, for D_2 about 200. For most other species the relaxation for trans-rot transfer requires about 10 collisions or less.

Data for vibrational relaxation have been given in Fig. 3. An excellent survey of data can be found in [4]. Generalizing somewhat one can say the following: (Z_{10} is the average number of collisions necessary to deactivate the first vibrational level).

1) Z_{10} increases with increasing oscillator energy;

2) Z_{10} decreases with increasing temperature;

3) lighter partners are generally more efficient than heavier ones;

4) normally triatomic species relax faster than diatomic ones, and non-linear triatomic molecules faster than linear ones, cf. the following numbers:

Species	CO_2	COS	SO_2	H_2O
Lowest vibrational quantum (cm^{-1})	672	527	519	1595
Z_{10} (at room temperature)	$5 \cdot 10^4$	10^4	$5 \cdot 10^2$	10^2

5) larger molecules generally relax faster. In families such as C_nH_{2n} Z_{10} decreases with increasing n;

6) impurities may have large influence in reducing the relaxation times, *e.g.* H_2O, alcohol a.o.

The examples given above were mainly concerned with the transfer of vibrational into translational or vibrational energy. These energy exchange processes are of great practical importance, but it is clear, that they are not the only ones which can be investigated with the relaxation methods described.

It must, however, not be forgotten that relaxation times respectively rate constants for the energy transfer processes have a meaning similar to the meaning of rate constants for chemical reactions (see the other lectures of the author). They do not give direct information about the inelastic collision processes and the cross-sections of such processes.

5. – Theoretical considerations.

We will at first estimate under which conditions nonadiabatic energy transfer is likely to happen in a collision. This will be the case if the perturbation of the quantum states of the colliding system, as given by the uncertainty principle, is of the same order of magnitude as the energy difference ε between two adjacent states of the system. Efficient collisional (de)excitation will therefore happen, if $\varepsilon\tau_{\text{coll}} \approx \hbar$, where τ_{coll} is the collision time. This is usually put equal to a/v, where a is a characteristic interaction distance and v the mean velocity of the encounter.

So we have

$$\varepsilon a/v\hbar \approx 1$$

as the condition for efficient energy transfer, while adiabaticity makes the transfer unlikely if $\varepsilon a/v\hbar \gg 1$. Under normal conditions (room temperature, $a \approx 1$ Å) this means that ε has to be smaller than $(10 \div 20)\ \text{cm}^{-1}$ for efficient transfer of energy. In general, therefore, electronic-translational energy exchange ($\varepsilon \approx 10\,000\ \text{cm}^{-1}$) and vibrational-translational energy transfer ($\varepsilon \approx (200 \div 3\,000)\ \text{cm}^{-1}$) will be inefficient, whereas rotational-translational transfer (H_2: $\varepsilon \approx 60\ \text{cm}^{-1}$, I_2: $\varepsilon \approx 0.04\ \text{cm}^{-1}$) or resonant vibration-vibration transfer should be quite efficient. This is in general born out by the data (*).

More detailed theories of energy transfer have been constructed on various

(*) It should, however, be kept in mind, that an otherwise adiabatic collision may be locally nonadiabatic. In that case asymptotically $\varepsilon a/v\hbar \gg 1$, but locally $\varepsilon' a'/v\hbar \approx 1$ is possible, giving high transition probabilities. Cf. also R. S. Berry's lectures (Editor).

levels: classical, semiclassical, and quantum-mechanical ones. They are often based on the following assumptions [3, 4]:

1) the internal states of *one* molecule only are considered,

2) the number of internal states treated is limited, *e.g.* rotation is neglected while treating vibrational transfer, and the system is idealized, *e.g.* to be a harmonic oscillator,

3) one treats only head-on collisions, so that there is only one relative co-ordinate,

4) the dynamics of the encounter is considered to be unaffected by energy transfer.

The procedure is as follows:

a) the collision is described in terms of an interaction potential and the relative kinetic energy E of approach at infinite distance to obtain an expression for the time-dependent perturbing force $F(t)$,

b) then an expression is derived for the transition probability P induced by $F(t)$ as a function of the collision parameter b. Finally one integrates over all values of E and b.

Most treatments are either adiabatic or nearly adiabatic. If the potential is represented by an exponential $e^{-\alpha r}$, the parameter a in the adiabaticity condition $\varepsilon a/v\hbar \gg 1$ has to be taken as $a \approx 1/\alpha$. In that case one would expect that the transition probability is a certain function of $\varepsilon/\alpha v$ only, *i.e.* $P(v) = f(\varepsilon/\alpha v)$. In the well known treatment of LANDAU and TELLER [22] one gets (putting $\varepsilon = h\nu$) $f = \exp[-4\pi^2 \nu/\alpha v]$. A detailed treatment of various methods is given *e.g.* in the books of COTTRELL [3] or STEVENS [4]. Here we will only briefly discuss a classical approach for the vibrational-translational energy transfer given by RAPP [23]. He considered the case shown in Fig. 5. The potential function is taken to be

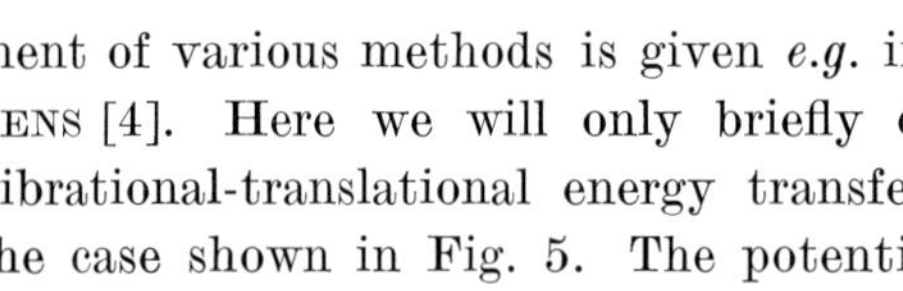

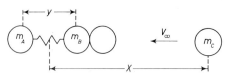

Fig. 5. – Model for head-on collisions of AB and C.

$$V(x, y) = E_\infty \exp\left[-\alpha\left(x - \frac{m_A}{m_A + m_B}\, y\right)\right].$$

The approximate disturbing force is then

$$F(t) = \frac{\partial}{\partial y} V(y, t) = \alpha E_\infty \frac{m_A}{m_A + m_B} \operatorname{sech}^2 \frac{\alpha v_\infty t}{2}.$$

The resulting transition probability per collision from the first excited vibrational state to the ground state becomes (with velocity symmetrization)

$$P_{10} = 2\left(\frac{2\pi}{3}\right)^{\frac{1}{2}} \frac{m_C(m_A^2 + m_B^2)}{2m_A m_B(m_{AB} + m_C)} \left(\frac{\varepsilon}{kT}\right)^{\frac{1}{6}} \left(\frac{\varepsilon}{h\nu}\right) \exp\left[-\frac{3}{2}\left(\frac{\varepsilon}{kT}\right)^{\frac{1}{3}} + \frac{h\nu}{2kT} + \frac{\varepsilon_a}{kT}\right],$$

where $\varepsilon = 16\pi^4\mu\nu^2/\alpha^2$, $\mu = \mu_{AB,C}$, and ε_a is the well depth of a possible additional attractive potential.

The temperature-dependence of the transition probability is governed by the term $T^{-\frac{1}{3}}$ in the exponent. Figure 3 shows that the temperature-dependence of measured relaxation times for small molecules can be quite well described by that expression, but absolute values of τ are not fitted as well.

In the semi-classical treatment one takes $F(t)$ from the classical formula, and uses then time-dependent perturbation theory. The result is the same as in the classical case. Moreover, the classical result agrees also with the correct quantum-mechanical calculation if quantization is introduced properly.

* * *

The author acknowledges the collaboration of U. Buck, R. Gengenbach and M. v. Seggern in preparing a first draft of this lecture.

REFERENCES

[1] J. Meixner: *Acustica*, **2**, 101 (1952); J. Meixner and H. G. Reik: *Handbuch der Physik*, vol. **3**/2, edited by S. Flügge (Berlin, 1959), p. 479.
[2] R. Haase: *Thermodynamik der irreversiblen Prozesse* (Darmstadt, 1963).
[3] T. L. Cottrell and J. C. McCoubrey: *Molecular Energy Transfer in Gases* (London, 1961).
[4] B. Stevens: *Collisional Activation in Gases* (London, 1967).
[5] K. F. Herzfeld and T. A. Litovitz: *Absorption and Dispersion of Ultrasonic Waves* (New York, 1959).
[6] K. F. Herzfeld and V. Grifting: *Journ. Phys. Chem.*, **61**, 844 (1967).
[7] D. Sette, A. Busala and J. C. Hubbard: *Journ. Chem. Phys.*, **23**, 787 (1955).
[8] J. P. Toennies and E. F. Greene: *Chemical Reactions in Shock Waves* (London, 1964); Ye. V. Stupochenko, S. A. Losev and A. I. Osipov: *Relaxation in Shock Waves* (Berlin, 1967).
[9] R. C. Millikan and D. R. White: *Journ. Chem. Phys.*, **39**, 3209 (1963).
[10] J. H. Kiefer and R. W. Lutz: *Journ. Chem. Phys.*, **44**, 668 (1966).
[11] G. Herzberg: *Spectra of Diatomic Molecules* (Princeton, 1950).
[12] R. L. Taylor, M. Camac and R. M. Feinberg: *Eleventh Symposium on Combustion* (Pittsburgh, 1967).

[13] A. Kantrowitz: *Journ. Chem. Phys.*, **14**, 150 (1946).
[14] See M. E. Jacox and S. H. Bauer: *Journ. Phys. Chem.*, **61**, 833 (1957).
[15] N. Basco, J. E. Nicholas, R. G. W. Norrish and W. H. J. Vickers: *Proc. Roy. Soc.*, A **272**, 147 (1963).
[16] N. Basco and R. G. W. Norrish: *Canad. Journ. Chem.*, **38**, 1769 (1960).
[17] J. T. Yardley and C. B. Moore: *Journ. Chem. Phys.*, **49**, 1111 (1968); **48**, 14 (1968).
[18] F. De Martini and J. Ducuing: *Phys. Rev. Lett.*, **3**, 117 (1966).
[19] R. C. Millikan: *Journ. Chem. Phys.*, **38**, 2855 (1963).
[20] G. Karl and J. C. Polanyi: *Journ. Chem. Phys.*, **38**, 271 (1963).
[21] A. N. Terenin and H. G. Nenimin: *Acta Physicochim. URSS*, **16**, 257 (1942).
[22] L. Landau and E. Teller: *Phys. Zeit. Sowjetunion*, **10**, 34 (1936).
[23] D. Rapp: *Journ. Chem. Phys.*, **32**, 735 (1960).

Reactive Collisions of Thermal Neutral Systems (*).

D. R. HERSCHBACH

Harvard University, Department of Chemistry - Cambridge, Mass.

PART A - Kinematics.

This group of lectures deals with a discussion of reactive scattering at thermal energies. Unlike the case of elastic scattering, there is no « royal road » from the Schrödinger equation to measurable quantities. In this part we consider the kinematic analysis in studying reactive scattering. The following part will deal with some prototype modes of reaction dynamics.

It is convenient to adopt some brief labels which classify reactive scattering experiments according to the degree of control of analysis achieved. Somewhat facetiously we can describe as « primitive experiments » those in which there is no velocity selection in either the primary, the secondary, or the product beams. All that is measured is the intensity of the product as a function of the scattering angle. In « not so primitive experiments » the velocity distribution of the products is measured. Finally in what may be called « millennial experiments » we envision an ideal apparatus which is capable of high angular resolution, velocity selection of all beams, sharp definition of all states, and complete specification of all internal quantum numbers. Even for alkali reactions these millennial experiments remain somewhat far off (if not far-fetched). Thus we will emphasize especially how much can be extracted even from primitive experiments, since for some time to come we must rely on such experiments as the field moves « beyond the alkali age ».

(*) Due to circumstances out of control of the editor, these lectures cannot be presented here. Instead, Part A gives a short account of kinematics, based on lecture notes prepared by J. ALPER, R. M. DÜREN, H. HABERLAND and C. SCHMIDT, while Part B gives an annotated bibliography of reactive beam scattering experiments, which will allow the reader to fill up the gap in this book. The Editor.

The kinematic analysis is based solely on conservation laws, which yield results that are independent of the model used to describe the reaction dynamics. For primitive experiments these conservation laws are expecially important since the information available is so rudimentary. The basic reaction to be studied will be of the form $A + BC \rightarrow AB + C$ where A, B, C may be atoms or molecules.

1. – The conservation of energy.

Let E, E' denote the kinetic energies of the relative motions of the reactants and products respectively; W, W' the internal energies of the reactants and products. In primitive experiments E, E', W, W' are taken as the mean or most probable thermal energies. If ΔD_0 denotes the difference in the dissociation energy of AB and BC (measured from zero point levels), the energy balance requires

$$E + W + \Delta D_0 = E' + W' .$$

2. – The conservation of linear momentum.

The relations involved in the conservation of linear momentum may be pictured by means of Newton Diagrams illustrated, *e.g.* in Fig. 3 and 4 of ref. [A1] in this book.

Notation: $\boldsymbol{v}_a$, $\boldsymbol{v}_b$ are the velocities of the reactant beams, the center of mass moves with velocity

$$\boldsymbol{c} = \frac{1}{M}(m_a \boldsymbol{v}_a + m_b \boldsymbol{v}_b) ,$$

where M is the total mass $m_a + m_b$, $\boldsymbol{v}_r$ is the relative veloity $\boldsymbol{v}_a - \boldsymbol{v}_b$, and $\boldsymbol{u}_a$, $\boldsymbol{u}_b$ are the velocities of beams a and b in the center of mass system.

$$u_a = \frac{m_b}{M} v_r , \qquad u'_c = \frac{m_d}{M} v'_r ,$$

$$u_b = -\frac{m_a}{M} v_r , \qquad u'_d = -\frac{m_c}{M} v'_r .$$

The Newton diagrams represent graphically the transformation from the lab to the center of mass (c.m.) system. The usefulness of the Newton diagram in analysing scattering experiments is illustrated in Fig. 4 *a*), *b*) of ref. [A1]. If the velocity of a product is of magnitude $u'_c < c$ in the c.m. system, its position is confined to a circle of radius u'_c centered about the endpoint of $\boldsymbol{c}$. In the lab system this means that the product is confined to a limited range of angles about $\boldsymbol{c}$. We may note here that the intensity at one deflection angle in the lab system arises from scattering into two different angles in the c.m. system. Figure 4 *a*) illustrates the case of $u'_c > c$. There is a one-to-one correspondence between scattered angle in the lab system and the c.m. system. Thus if we measure the intensity in the lab system at an angle Θ we know that the corresponding angle in the c.m. system is ϑ.

The mathematical details of the transformation from the lab to the c.m. system can be found in ref. [A2, A2*a*, A3*a*].

3. – The conservation of intensity.

The intensity of the product beam is measured in the lab system. However for comparison with theoretical models we need to evaluate the intensity in the c.m. system. The transformation law for the intensities is determined by two relations:

$$I_{\text{lab}}(\Theta, \Phi, v)\,\mathrm{d}v\,\mathrm{d}\Omega = I_{\text{c.m.}}(\vartheta, \varphi, u)\,\mathrm{d}u\,\mathrm{d}\omega$$

and

$$v^2\,\mathrm{d}v\,\mathrm{d}\Omega = u^2\,\mathrm{d}u\,\mathrm{d}\omega\,.$$

The first relation merely states that all the particles which emerge in a specified element $\mathrm{d}u\,\mathrm{d}\omega$ in the c.m. system appear in the corresponding element $\mathrm{d}v\,\mathrm{d}\Omega$ in the lab system. The second relation follows from inspection of the lab-c.m. transformation and is discussed in ref. [A1, A3, A4]. Thus we have

$$I_{\text{lab}}(\Theta, \Phi, v) = \left(\frac{v}{u}\right)^2 I_{\text{c.m.}}(\vartheta, \varphi, u)\,.$$

The Jacobian factor $(v/u)^2$ weights strongly the contribution from reactions in which the c.m. product velocity u'_c is small and hence may severely distort the lab « image » of the c.m. distribution.

Note that the Jacobian $(u/v)^2$ differs from that derived in Prof. Beck's lecture in that it lacks a factor of $\cos(u, v)$. It is readily shown [3] that for the case in which the product c.m. velocity is fixed the proper Jacobian

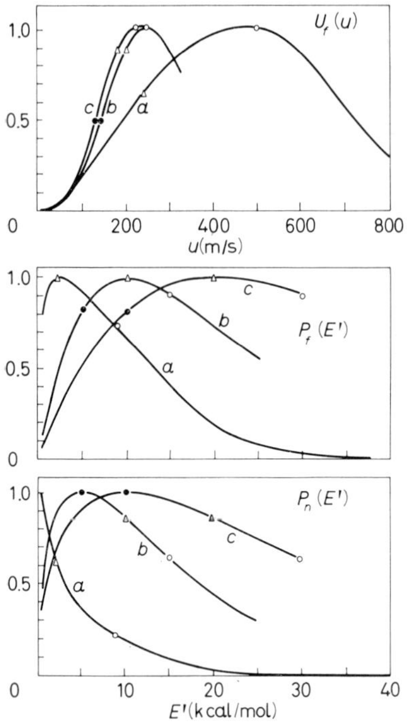

Fig. 1.

again becomes

$$\frac{v^2}{u^2 \cos(u, v)}$$

regardless of the magnitude of u.

Up to this point intensity has always referred to flux, u_f, as a function of velocity u'. For all calculations involving cross-sections this is the appropriate form. We can transform the flux distribution as a function of velocity to a flux distribution as a function of energy $P_f(E')$ by dividing by u'. Finally we transform to number density as a function of energy by another division by u':

$$U_f(u') = uP_f(E') = u^2 P_n(E')$$

We have ignored the normalization factors that enter into this equation. The relationship among these quantities is illustrated in Fig. 1 of ref. [A5].

References to Part A:

[A1] D. Beck: First lecture in this book.

[A2] R. K. B. Helbing: *Journ. Chem. Phys.*, **48**, 472-477 (1968); **50**, 4123 (1969).

[A2*a*] H. Pauly and J. P. Toennies: in *Methods of Experimental Physics* (ed. by L. Marton), vol. **7** A (New York and London, 1968), p. 306.

[A3] T. T. Warnock and R. B. Bernstein: *Journ. Chem. Phys.*, **49**, 1878-1886 (1968).

[A3*a*] F. A. Morse and R. B. Bernstein: *Journ. Chem. Phys.*, **37**, 2019-2027 (1962).

[A4] W. B. Miller, S. A. Safron and D. R. Herschbach: *Disc. Faraday Soc.*, **44**, 108-122 (1968).

[A5] E. A. Entemann and D. R. Herschbach: *Disc. Faraday Soc.*, **44**, 289-292 (1967).

Part B - **Bibliography of reactive beam scattering.**

The following references include most of the papers on reactive neutral thermal beam scattering which appeared since 1966. Elastic scattering of reactive systems, and ion-molecule reactions are adequately treated in the lectures by Ross and Greene, respectively Henglein in this volume.

As an introduction to the topic the following three reviews are recommended:

[B1] D. R. HERSCHBACH: *Reactive scattering in molecular beams*, *Adv. Chem. Phys.*, **10**, 319-393 (1966).
[B2] J. P. TOENNIES: *Molecular beam studies of chemical reactions*, in *Chemische Elementarprozesse*, ed. by H. HARTMANN (Berlin, 1968), p. 157-218.
[B3] J. P. TOENNIES: *Molecular beam investigations of bimolecular reactions*, *Ber. Bunsenges. Phys. Chem.*, **72**, 927-949 (1968).

At the moment, for the non-specialist, ref. [B2] will probably give the best overall view. A much more detailed view as of 1967 gives

[B4] *Discussions of the Faraday Society*, vol. **44** (1968), entitled: *Molecular Dynamics of the Chemical Reactions.*

Some twenty-odd original papers have appeared in the field since 1966. Of these, more than $\frac{2}{3}$ belong to the « Alkali age », describing reactions of alkali atomic beams with compounds containing either halogen or oxygen. The rest of the papers is concerned with reactions of halogen atomic beams, and in one case of a HI beam. The dawn of the new post-alkali eon is, however, there, and such non-alkali beams have been used in at least 5 laboratories.

Another classification was mentioned in Part A of this exposé: About 60 % of all experiments published since 1966 are « primitive » experiments, *i.e.* they employ velocity selection neither in the reactant beams nor in the product beam. 20 % are « not so primitive », employing product beam velocity selection, 20 % use oriented primary beams. Of course, it is not essential how « primitive » the experiments are, one should rather look how detailed and precise the results are.

One wishes to know, in the first place, the angular distribution of reaction products in the center-of-mass system, $\sigma_{\text{c.m.}}(\vartheta)$, in the second place the distribution of the available energy into translation and internal energies of the products. In the third place only, there are more detailed questions, *e.g.* the energy distribution into different internal degrees of freedom, the correlation of the internal energy of a product molecule with its direction, and the dependence of the reaction cross-section on the relative orientation or the energy of the reactants.

The first two problems, determination of $\sigma_{\text{c.m.}}(\vartheta)$ and $P(E')$, are handicapped by the lack of the information necessary to transform from the lab to the c.m. system. The reactant beams have broad Maxwellian velocity distributions, and the product energy E' is not unique even at fixed incoming energy E. Intensity considerations forbid velocity selection in both primary beams, and only in a few cases [B5, B12, B13, B16] one such beam has been velocity-selected. It is general usage to do the lab → c.m. transformation with the most probable incoming beam velocity only.

In all experiments up to now, additional assumptions have been used. The most common is to presuppose the factorization of angular and energy distribution of the products

$$\sigma_{\text{c.m.}}(E; E', \vartheta) \sim \sigma(E; \vartheta)\cdot P(E; E'). \tag{1}$$

This is not sufficient for «primitive» experiments, where an often used approximation is that of a single product energy only (*e.g.* [B6]):

$$P(E; E') = \delta(E - E'_0(E)) \tag{2}$$

More stringent postulates have been used, *e.g.* the analytic form of $\sigma(\vartheta)$ or $P(E')$ with adjustable parameters has been assumed, cf. [B17].

Another, most simple approach is to compare the lab distributions with the predictions of certain models transformed to the lab system. This is done, *e.g.* in ref. [B5] and in almost all preliminary accounts of new experiments. If the c.m. → lab transformation of the predictions of the model is done properly, this procedure has the big advantage of not introducing approximations into the comparison of experiment with model, its disadvantage being its model-dependence from the beginning. In those experiments where velocity selection of the products is employed and enough angles are scanned (« not so primitive experiments ») relatively complete information about $\sigma(E', \vartheta)$ can be obtained. It has been possible—still assuming the validity of eq. (1), however—to construct complete « flux contour maps », *i.e.* an intensity map in a (E', ϑ)-polar-coordinate system. Examples are given in ref. [B7, B12, B13], these maps could also have been plotted in the c.m. system and would then look similar to Fig. 17 of Prof. Henglein's paper in this book. (Note that in ion-molecule reactions velocity selection is essentially complete, and assumption (1) has not to be used. Note further that this Figure, which is from p. 141 of [B4] gives an intensity rather than flux contour map.)

Further discussion of the experiments, which has as its aim to shed some light on the internal dynamics of the collisional encounter, is generally done by comparing the c.m. distribution with model calculations. On the one hand, these are trajectory studies which are treated in other parts of this book. On the other hand, simple phenomenological models have been used: « rebound » reactions have been opposed to « stripping » mechanisms, and « direct » processes have been discriminated against « complex » formation.

A discussion of the rebound *vs.* stripping mechanisms can be found in [B1], later experiments showed that the extreme cases are much more seldom realized as was thought a few years ago. Stripping or more precisely « spectator stripping » is discussed extensively in [B5], and—applied to ion reactions—in Prof. Henglein's lecture in this volume. This model has been modified to include pre- and post-collisional polarization forces by WOLFGANG and coworkers see p. 130 f. of [B4]. For a somewhat related discussion one should also consult ref. [B27], especially from chapter VII on.

While reactions classified as of the « rebound » or « stripping » type are generally thought to be « direct », *i.e.* the molecules make a single approach and then the newly formed ones recede, other reactions have been found [B16, B17] which apparently form a more or less long-lived « complex » before decaying into products or back into the elastic channel. Long-lived complexes must have forward-backward symmetry in their $\sigma(\vartheta)$, they may or may not adequately be described by a statistical model (see Prof. Bunker's lecture on « *Compound State Approaches* » in this volume), since a lifetime sufficiently long to produce angular symmetry of decay must not necessarily be long enough for energetic equilibrium.

In what follows I list the more important papers, where much more detailed information can be found. The order is that of chemical species participating in the reaction.

Reactions of alkali atoms with halogen molecules:

[B5] R. E. MINTURN, S. DATZ and R. L. BECKER: *Alkali atom-halogen molecule reactions in molecular beams: the spectator stripping model, Journ. Chem. Phys.*, **44**, 1149-1159 (1966).

[B6] J. H. BIRELY and D. R. HERSCHBACH: *Reactive scattering in molecular beams: velocity analysis of* KBr *formed in the* $K+Br_2$ *reaction, Journ. Chem. Phys.*, **44**, 1690-1701 (1966).

[B7] T. T. WARNOCK, R. B. BERNSTEIN and A. E. GROSSER: *Internal energy of reaction products by velocity analysis. - III: Center-of-mass angular distribution and product excitation function for* $K+Br_2$, *Journ. Chem. Phys.*, **46**, 1685-1693 (1967).

[B8] J. H. BIRELY, R. R. HERM, K. R. WILSON and D. R. HERSCHBACH: *Molecular beam kinetics: reactions of* K, Rb *and* Cs *with* Br_2 *and* I_2, *Journ. Chem. Phys.*, **47**, 993-1004 (1967).

[B9] R. GRICE and P. B. EMPEDOCLES: *Molecular beam kinetics: reaction of* K, Rb, Cs *with* Cl_2, *Journ. Chem. Phys.*, **48**, 5352-5357 (1968).

[B10] D. D. PARRISH and R. R. HERM: *Possible mass effect in alkali-atom reactions: crossed beam studies of* $Li+Cl_2$, ICl, Br_2, $SnCl_4$, PCl_3, *Journ. Chem. Phys.*, **49**, 5544-5545 (1969).

In this connection, the following reference gives a non-beam determination of an integral reactive cross-section:

[B10*a*] D. C. BRODHEAD, P. DAVIDOVITS and S. A. EDELSTEIN: *Cross-sections for the alkali metal-halogen molecule reactions:* Cs *with* I_2, *Journ. Chem. Phys.*, **51**, 3601-3603 (1969).

Reactions of alkali atoms with halogen hydrides:

[B11] L. R. Martin and J. L. Kinsey: *Crossed molecular beam reactions of tritium bromide*, *Journ. Chem. Phys.*, **46**, 4834-4838 (1967).

[B12] C. Riley, K. T. Gillen and R. B. Bernstein: *Polar (velocity-angle) flux contout maps for* KBr *from the crossed-beam reactions* K+HBr, DBr: *evidence for both forward and backward (c.m.) scattering*, *Journ. Chem. Phys.*, **47**, 3672-3674 (1967).

[B13] K. T. Gillen, C. Riley and R. B. Bernstein: *Reactive scattering of* K *by* HBr, DBr *in crossed molecular beams: angular and velocity distributions of* KBr *in laboratory and c.m. systems*, *Journ. Chem. Phys.*, **50**, 4019-4033 (1969).

Reactions of alkali atoms with other halogen compounds (see also [B10]):

[B14] K. R. Wilson and D. R. Herschbach: *Molecular beam kinetics: transition between rebound and stripping mechanism in reactions of alkali atoms with polyhalide molecules*, *Journ. Chem. Phys.*, **49**, 2676-2683 (1968).

[B15] R. J. Gordon, R. R. Herm and D. R. Herschbach: *Molecular beam kinetics: magnetic deflection analysis of scattering of alkali atoms from polyhalide molecules*, *Journ. Chem. Phys.*, **49**, 2684-2691 (1968).

Reactions of alkali atoms with oxides or alkali chlorides (these are the two cases where complex behaviour has been found for the first time):

[B16] D. O. Ham and J. L. Kinsey: *Long lived complexes in molecular collisions*, *J. Am. Chem. Soc.*, **48**, 939-940 (1968).

[B17] W. B. Miller, S. A. Safron and D. R. Herschbach: *Exchange reactions of alkali atoms with alkali halides: a collision complex mechanism*, *Disc. Faraday Soc.*, **44**, 108-122 (1968).

A special group of experiments are neither «primitive» nor «not so primitive» in that *primary beam state selection* is employed. By electrical polarization of CH_3I or CF_3I the dependence of the reactivity on the side of attack of an alkali atom against these molecules is determined:

[B18] R. J. Beuhler, R. J. Bernstein and K. H. Kramer: *Observation of the reactive asymmetry of* CH_3I. *Crossed beam study of the reaction of* Rb *with oriented* CH_3I, *Journ. Amer. Chem. Soc.*, **88**, 5331 (1966).

[B19] P. R. Brooks and E. M. Jones: *Reactive scattering of* K *atoms from oriented* CH_3I *molecules*, *Journ. Chem. Phys.*, **45**, 3449-3450 (1966).

[B20] R. J. Beuhler and R. B. Bernstein: *Reactive asymmetry of* CH_3I: *crossed beam study of the reaction of* Rb *with oriented* CH_3I *molecules*, *Chem. Phys. Lett.*, **2**, 166 (1968).

[B21] R. R. Brooks: *Molecular beam reaction of* K *with oriented* CF_3I. *Evidence for harpooning?*, *Journ. Chem. Phys.*, **50**, 5031-5032 (1969).

A growing but smaller group of experiments uses halogen atomic beams to react with halogen molecules. The atoms are produced by thermal dissociation, the products are detected with an electron bombardment ionizer and mass spectrometer.

[B22] Y. T. LEE, J. D. MCDONALD, P. R. LEBRETON and D. R. HERSCHBACH: *Molecular beam kinetics: evidence for short range attraction in halogen atom-molecule exchange reactions*, *Journ. Chem. Phys.*, **49**, 2447-2448 (1968).

[B23] D. BECK, F. ENGELKE and H. J. LOESCH: *Reactive Streuung in Molekularstrahlen*: $Cl+Br_2$, *Ber. Bunsenges. Phys. Chem.*, **72**, 1105-1107 (1968).

[B24] J. B. CROSS and N. C. BLAIS: *Crossed molecular beam kinetics: energy disposal in the reaction* $Cl+Br_2 \rightarrow ClBr+Br$, *Journ. Chem. Phys.*, **50**, 4108-4109 (1969).

[B25] Y. T. LEE, P. R. LEBRETON, J. D. MCDONALD and D. R. HERSCHBACH: *Molecular beam kinetics: evidence for preferred geometry in interhalogen exchange reactions*, *Journ. Chem. Phys.*, **51**, 455-456 (1969).

Apart from hot-atom reactions and ion-molecule reactions adequately treated elsewhere in this book, there is only a single experiment so far published using the « nozzle beam » technique developed in the last decade to produce beams at higher-than-thermal energies. Unfortunately only a negative result on the reaction HI+HI has been obtained:

[B26] S. B. JAFFE and J. B. ANDERSON: *Molecular beam study of the* HI *reaction*, *Journ. Chem. Phys.*, **54**, 1057-1064 (1969).

[B27] P. J. KUNTZ, M. H. MOK and J. C. POLANYI: *Distribution of reaction products. - V: Reactions forming an ionic bond*, M+XC(3D), *Journ. Chem. Phys.*, **50**, 4623-4652 (1969).

Elastic Scattering of Reactive Systems.

J. Ross (*)
MIT - Cambridge, Mass.

E. F. Greene
Brown University - Providence, R. I.

Elastic scattering in reactive systems may yield information not only on interactions as in nonreactive systems [1] but also about the chemical reaction. In this lecture we shall discuss briefly a theory for elastic scattering in reactive systems, the so-called optical model of reactions [2], some measurements, and their interpretation which leads to information on reaction probabilities, reaction threshold conditions, total reaction cross-sections and their energy, dependence, and on potential surfaces of reaction in the entrance channel. The results obtained from nonreactive scattering in reactive systems and those obtained from measurements on the products of the reaction tend to supplement each other. For the few systems for which comparisons have been made the results agree within experimental uncertainties.

1. – Optical model.

Consider an effective two-body potential

$$V = V_0 - iV_1$$

with a real, V_0, and imaginary, iV_1, component. The flux density j in quantum theory (ref. [3], p. 10-12) is

$$j = \frac{\hbar}{2i\mu}[\psi^* \nabla\psi - \psi\nabla\psi^*] , \tag{2}$$

(*) The lecture was given by J. Ross.

where μ is the reduced mass of the colliding system in the nonreactive channels ($\mu = m_A m_B/(m_A + m_B)$ for the reactive system $A+B \Rightarrow C+D$), and ψ is the wave function. The continuity equation holds for the density $\varrho \equiv \psi^* \psi$

$$\frac{\partial \varrho}{\partial t} + \nabla \cdot j = 0 \,, \tag{3}$$

so that for elastic scattering from a real potential

$$\int_{\mathscr{V}} \nabla \cdot j \, d\mathscr{V} = 0 \,, \qquad \int_{\mathscr{V}} \frac{\partial \varrho}{\partial t} d\mathscr{V} = 0 \,. \tag{4}$$

These equations simply state conservation of particles within a volume $\mathscr{V}$. All particles, say those in a given quantum state A, which enter the scattering, region $\mathscr{V}$ leave that region still in state A. If the potential is complex, however then we have

$$\int_{\mathscr{V}} \nabla \cdot j \, d\mathscr{V} = -\int_{\mathscr{V}} \frac{2}{\hbar} V_1 \psi^* \psi \, d\mathscr{V} = \int_{\mathscr{V}} \frac{\partial \varrho}{\partial t} d\mathscr{V} \,. \tag{5}$$

We still have conservation of mass, of course, but the flux density of A out of the scattering region is less than the flux density entering that region. Some of the particles are absorbed, that fraction which has reacted or has been scattered inelastically.

If the interaction potential is complex, then the phase shifts are complex. It is instructive to look at the asymptotic form of the radial wave function for the l-th partial wave after scattering, which may be written as

$$u_l(r) \sim C_l \sin\left(kr - \frac{l\pi}{2} + \eta_l\right) , \tag{6}$$

where C_l is a constant, k is the magnitude of the wave vector, and η_l is the phase shift. In terms of incoming and outgoing radial waves eq. (6) is

$$u_l(r) \sim \frac{i}{2} C_l \exp[-i\eta_l]\left[\exp\left[-i\left(kr - \frac{l\pi}{2}\right)\right] - \exp[2i\eta_l] \exp\left[i\left(kr - \frac{l\pi}{2}\right)\right]\right] . \tag{7}$$

If η_l is real, then $|\exp[2i\eta_l]|^2$ is unity and the incoming and outgoing flux densities will be the same. If η_l is complex, $\eta_l = \eta_l^0 + i\xi_l$, then the amplitude of the outgoing wave function is decreased by a factor $\exp[-2\xi_l]$, the density corresponding to the outgoing wave function is decreased by a factor $\exp[-4\xi_l]$, and hence the probability of absorption of the l-th partial wave from the elastic channel due to inelastic and reactive collision is

$$P_l = 1 - \exp[-4\xi_l]\,. \tag{8}$$

The total cross-section for reaction (and inelastic collisions) is

$$\sigma_R(E) = \frac{\pi}{k^2}\sum_l (2l+1)(1-\exp[-4\xi_l])\,, \tag{9}$$

and for elastic collisions is

$$\sigma(E) = \frac{\pi}{k^2}\sum_l (2l+1)|1-\exp[2i\eta_l]|^2\,. \tag{10}$$

The scattering amplitude as a function of the relative scattering angle is

$$f(\chi) = \frac{1}{2ik}\sum_{l=0}^{\infty}(2l+1)(\exp[2i\eta_l]-1)\,P_l(\cos\chi) \tag{11}$$

for η real or complex.

A semi-classical analysis [4-7] of the scattering amplitude for complex phase shifts leads to a useful result. We replace the summation in eq. (11) by integration, use the asymptotic formula for $P_l(\cos\theta)$ [7] in the limit $l\chi \gg 1$ and use the semi-classical relation $\beta = (l+\frac{1}{2})k^{-1}$ to obtain

$$f(\chi) = -(2\pi\sin\chi/k)^{-\frac{1}{2}}[I^+\exp[\tfrac{1}{4}i\pi] - I^-\exp[-\tfrac{1}{4}i\pi]][1+O(h)]\,, \tag{12}$$

with

$$I^\pm = \int_0^\infty \{[1-P(\beta)]\beta\}^{\frac{1}{2}}\exp[2i\eta^0(\beta)\pm i\beta k\chi]\,d\beta\,. \tag{13}$$

We repeat this procedure for the real part of the phase shift η_l^0, for which the scattering amplitude is $f_0(\chi)$ and the elastic differential scattering cross-section $\sigma_0(\chi) = |f_0(\chi)|^2$. The analogues of eqs. (12), (13) are

$$f_0(\chi) = -\left(\frac{2\pi\sin\chi}{k}\right)^{-\frac{1}{2}}\left[I_0^+\exp\left[\frac{i\pi}{4}\right] - I_0^-\exp\left[-\frac{i\pi}{4}\right]\right][1+O(h)]\,, \tag{14}$$

$$I_0^\pm = \int_0^b \beta^{\frac{1}{2}}\exp[2i\eta^0(\beta)\pm i\beta k\chi]\,d\beta\,. \tag{15}$$

The evaluation of the integrals $I^\pm$ by the method of stationary phase [8] for analytic absorption functions $P(\beta)$ leads to

$$(16)\qquad I^{\pm} = (2\pi \sin\chi/k)^{\frac{1}{2}}[1-P(b)]^{\frac{1}{2}}[\sigma_0(\chi)]^{\frac{1}{2}} \exp[2i\eta^0(b) \pm ibk\chi \pm i/4\pi]\cdot \cdot\left[1 \pm \frac{i\sigma_0(\chi)\sin\chi}{2kb\{[1-P(b)]b\}^{\frac{1}{2}}}\left(\frac{d^2\{[1-P(\beta)]\beta\}^{\frac{1}{2}}}{d\beta^2}\right)_{\beta=b}\right][1+O(h)]\,,$$

where b is the classical impact parameter and $|\eta_0''(b)|$ is $\frac{1}{2}kb[\sigma_0(\chi)\sin\chi]^{-1}$. Similar evaluation of the integrals $I_0^{\pm}$ and the use of the results in eq. (16) yields the desired relation

$$(17)\qquad P(Eb) = 1 - \frac{\sigma(E\chi)}{\sigma_0(E\chi)}$$

provided the condition

$$(18)\qquad [1+O(h)]\frac{\sigma_0(\chi)\sin\chi}{4kb[1-P(b)]}\{P''(b) + \tfrac{1}{2}[P'(b)]^2[1-P(b)]^{-1} + [P'(b)/b]\} \ll 1$$

holds for the second term in eq. (16). The applicability of eq. (17) limited by this condition can be tested with the following simple example. Suppose the form of the reaction probability is

$$(19)\qquad P(b) = \tfrac{1}{2}c[1-\operatorname{erf}[(b-b_0)/\alpha]]\,,$$

with $\operatorname{erf}(x)$ being the error function defined for positive and negative values of x. If we choose $c=1$, $b_0=3$ Å, $\alpha=1$ Å (arbitrarily but probably reasonably), we find the condition (18) to give $k \gg 1$ Å^{-1} for $b=2$ Å, $\chi=\pi/2$, and $\sigma_0=\frac{1}{4}b_0^2$. For the experiments to be discussed below k is in the range $(20\div 50)$ Å^{-1} so that the condition is satisfied.

It is usually difficult to obtain information about the real part of the phase shift, η_l^0, necessary for the evaluation of $\sigma_0(\chi)$ in eq. (17), since η_l^0 depends on both V_0 and V_1 in the complex potential. However, a perturbation development [5] shows that in first order η_l^0 depends only on V_0 provided the stronger condition

$$(20)\qquad |P'(b)| \ll 2k(\pi-\chi)[1-P(b)]$$

holds.

A number of calculations [9-11] have been made of the differential elastic scattering cross-section with complex phase shifts, $\eta_l = \eta_l^0 + i\xi_l$. Usually W.K.B. phase shifts are calculated for η_l^0 and an assumed Lennard-Jones or exp-six potential; simple models are used for ξ_l or P_l. Figure 1 shows the results of a calculation made by Hundhausen and Pauly [9] for a step function for P_l

$$P_l = \begin{cases} 0\,, & L \leqslant l < \infty\,, \quad \xi_l = 0\,, \\ 1\,, & 0 < l < L\,, \quad \xi_l = \infty\,, \end{cases} \tag{21}$$

where L is the threshold angular-momentum quantum number. If L is zero, only elastic scattering occurs and there is no absorption. For small L, the small-angle elastic scattering is not affected by reaction but the large-angle scattering is. As L increases the effect of absorption becomes noticeable at smaller angles until for very large L the prominent rainbow maximum no longer appears. Minor modifications [10, 11] in the form of the absorption function do not alter the general features of the cross-sections in Fig. 1.

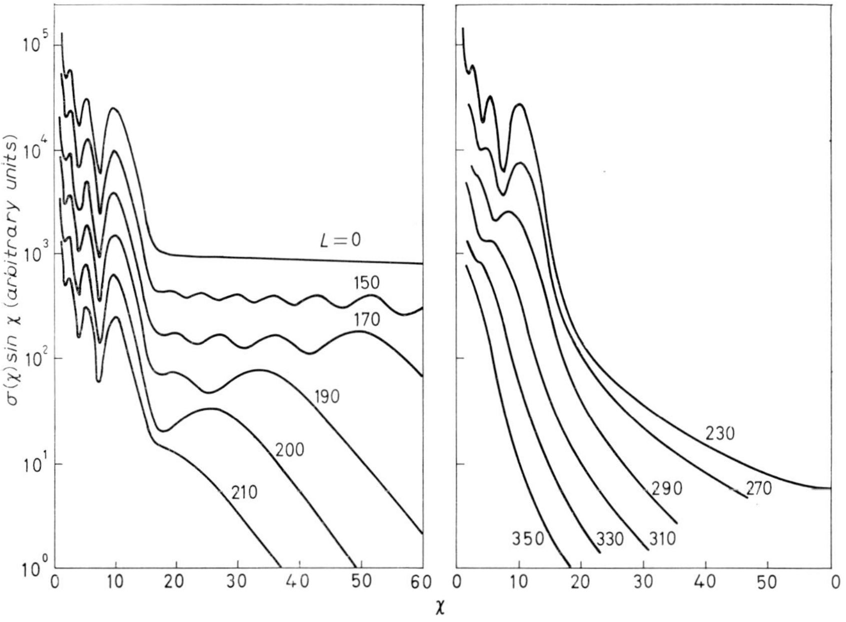

Fig. 1. – Differential elastic-scattering cross-section calculated with complex phase shifts, eq. (21), and various threshold angular momentum quantum numbers L. From HUNDHAUSEN and PAULY, ref. [9].

The semi-classical limit of the total reaction cross-section, eq. (9), is

$$\sigma_R(E) = 2\pi \int P(b,\, E)\, b\, \mathrm{d}b\,. \tag{22}$$

In principle both the real and imaginary component of the effective two-body optical potential may be calculated from the actual potentials of interaction in the reacting system [12, 13]. The presence of reactive channels, inelastic channels, and anisotropies of the potentials in the elastic channels are all represented by the optical potential.

2. – Measurements.

All the measurements cited here were made in a molecular-beam apparatus [14] in which a beam of alkali-metal atoms is velocity-selected and then made to collide with a thermal beam of a nonreactive or reactive species, Fig. 2.

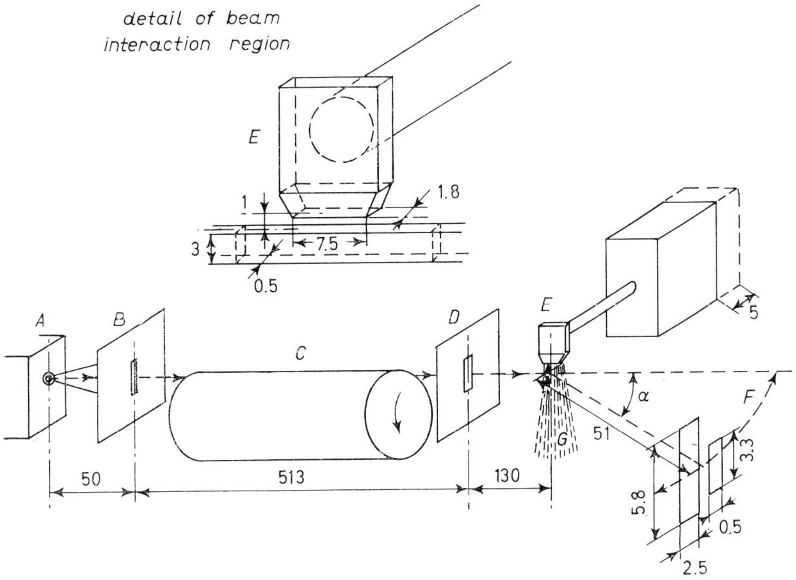

Fig. 2. – Schematic drawing of molecular beam apparatus, from ref. [14*a*]. Dimensions in mm.

The scattered alkali beam is not velocity-analysed. The measurements are plotted either with the ordinate proportional to a laboratory scattering cross-section $\sigma(\bar{E}, \alpha)\sin\alpha$ *vs.* laboratory scattering angle α, (eq. (1), ref. [14*a*]) or with the ordinate proportional to $(\bar{E}\alpha)^{\frac{1}{3}}\alpha\sigma(\bar{E}\alpha)\sin\alpha$ *vs.* $\bar{E}\alpha$, eq. (1), ref. [11] as suggested in ref. [15]. $\bar{E}$ is the effective initial relative kinetic energy. The latter method of representation has the advantage of nearly superimposing angular distributions at different energies. Transformation to center-of-mass co-ordinates leaves the general shape of the angular distribution practically unaltered.

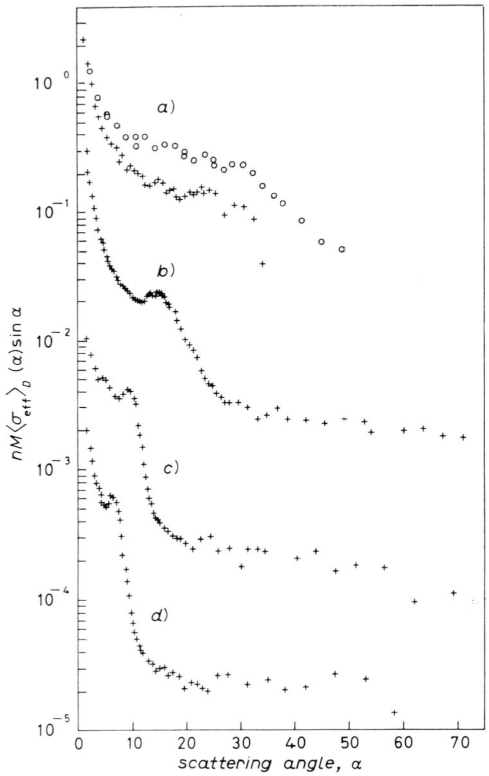

Fig. 3. – $K+C(CH_3)_4$, differential laboratory cross-section for nonreactive scattering, multiplied by const · sin α, for four relative energies, *vs.* laboratory angle in degrees. From ref. [16]. The subscript D denotes an average over the dimensions of the detector. *a*) $\bar{E} = 1.30$ kcal/mol: + run 1, $n = 1000$; ○ run 2, $n = 2 \cdot 10^3$; • run 3, $n = 4 \cdot 10^3$; *b*) $\bar{E} = 2.18$ kcal/mol, run 4, $n = 100$; *c*) $\bar{E} = 3.71$ kcal/mol, run 5, n = 10, *d*) $\bar{E} = 5.36$ kcal/mol, run 6, $n = 1$.

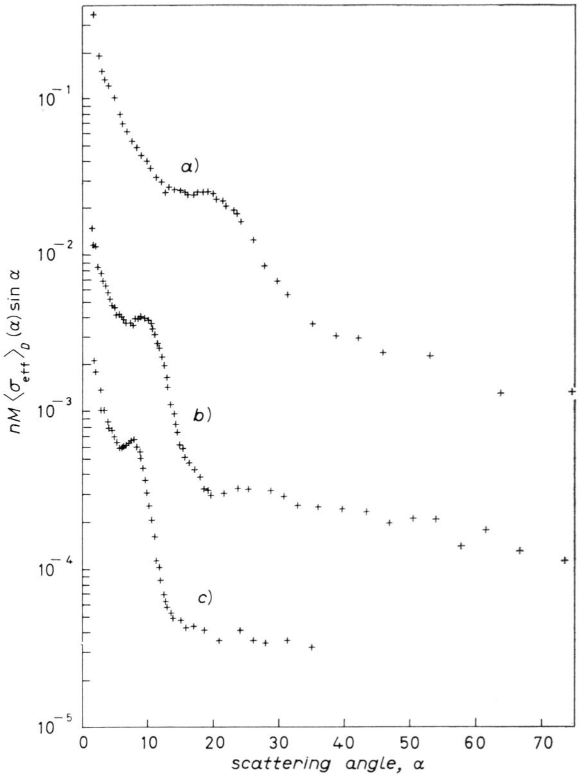

Fig. 4. – $K+C_6H_{12}$ (cyclohexane), differential laboratory cross-section for nonreactive scattering, multiplied by const · sin α, for three relative energies, *vs.* laboratory angle in degrees. From ref. [16]. *a*) $\bar{E} = 2.25$ kcal/mol, run 10, $n = 100$; *b*) $\bar{E} = 4.31$ kcal/mol, run 11, $n = 10$; *c*) $\bar{E} = 5.60$ kcal/mol, run 12, $n = 1$.

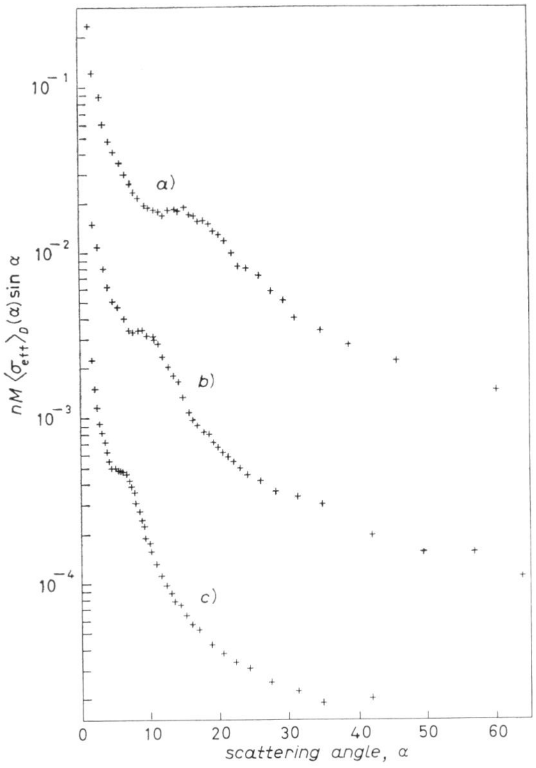

Fig. 5. – $K + C_6H_6$, differential laboratory cross-section for nonreactive scattering, multipled by const·sin α, for three relative energies, *vs.* laboratory angle in degrees. From ref. [16]. *a)* $\bar{E} = 2.22$ kcal/mol, run 7, $n = 100$; *b)* $\bar{E} = 3.65$ kcal/mol, run 8, $n = 10$; *c)* $\bar{E} = 5.48$ kcal/mol, run 9, $n = 1$.

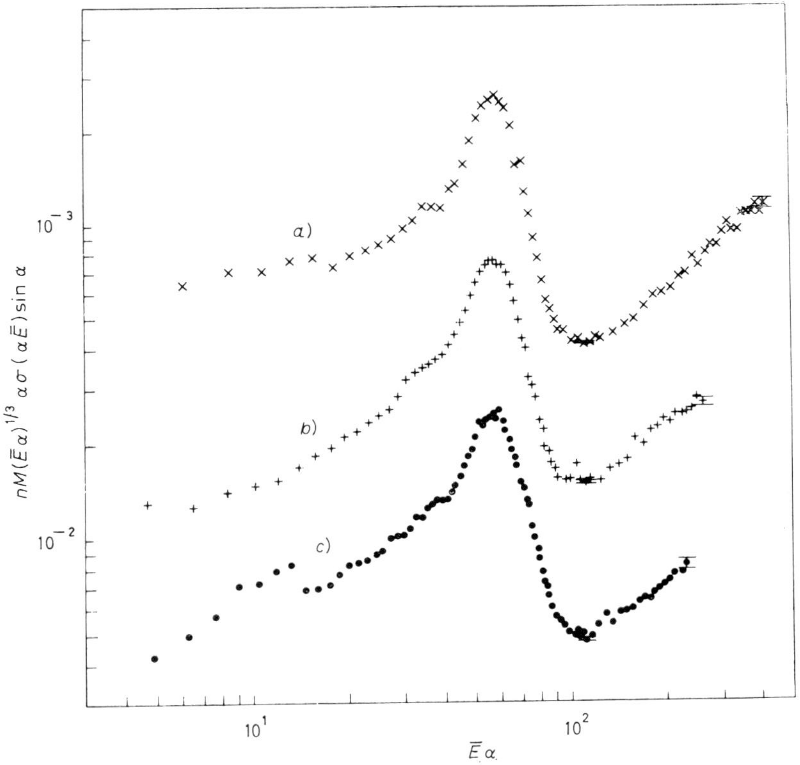

Fig. 6. – $K + C_{10}H_{16}$ (adamantane), differential laboratory cross-section for nonreactive scattering multiplied by $nM(\bar{E}\alpha)^{\frac{1}{3}}\alpha\sin\alpha$, *vs.* relative energy, $\bar{E}$ (kcal mole^{-1}) × laboratory scattering angle (degrees). From ref. [11]. *a)* $\bar{E} = 6.70$ kcal/mol, run 3, $n = 10$; *b)* $\bar{E} = 5.18$ kcal/mol, run 2, $n = 3$; *c)* $\bar{E} = 3.86$ kcal/mole, run 1, $n = 1$.

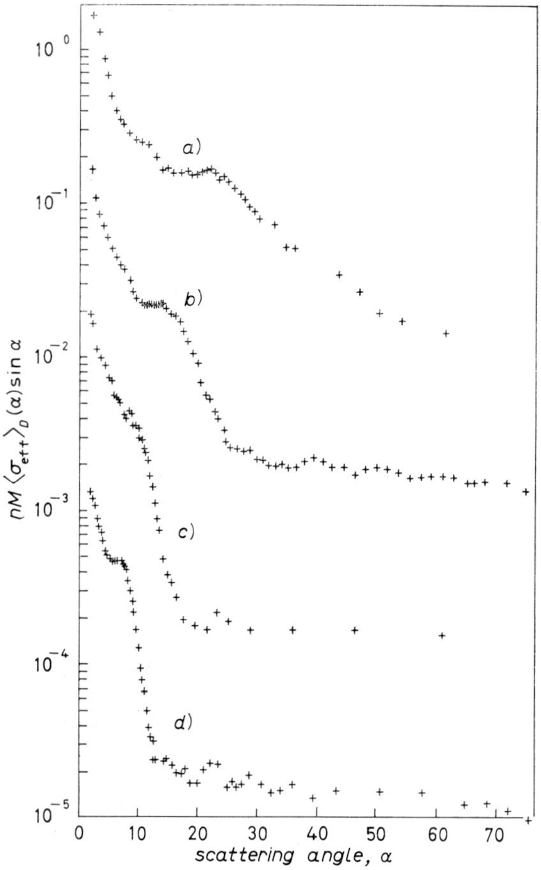

Fig. 7. – $K+SiCl_4$, differential laboratory cross-section for nonreactive scattering, multiplied by const·sin α, for four relative energies, *vs.* laboratory angle in degrees. From ref. [16]. *a*) $\bar{E} = 2.06$ kcal/mol, run 27, $n = 1000$; *b*) $\bar{E} = 3.36$ kcal/mol, run 28, $n = 100$; *c*) $\bar{E} = 5.35$ kcal/mol, run 29, $n = 10$; *d*) $\bar{E} = 6.94$ kcal/mol, run 30, $n = 1$.

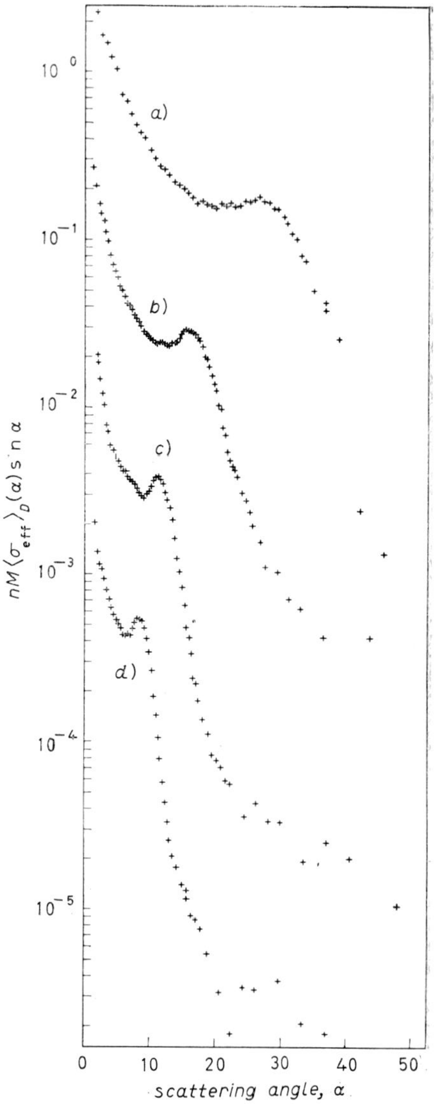

Fig. 8. – $K+CCl_4$, differential laboratory cross-section for nonreactive scattering, multipled by const·sin α, for four energies, *vs.* laboratory angle in degrees. From ref. [16]. *a*) $\bar{E} = 2.04$ kcal/mol, run 23, $n = 1000$; *b*) $\bar{E} = 3.31$ kcal/mol, run 24, $n = 100$; *c*) $\bar{E} = 4.90$ kcal/mol, run 25, $n = 10$; *d*) $\bar{E} = 6.83$ kcal/mol, run 26, $n = 1$.

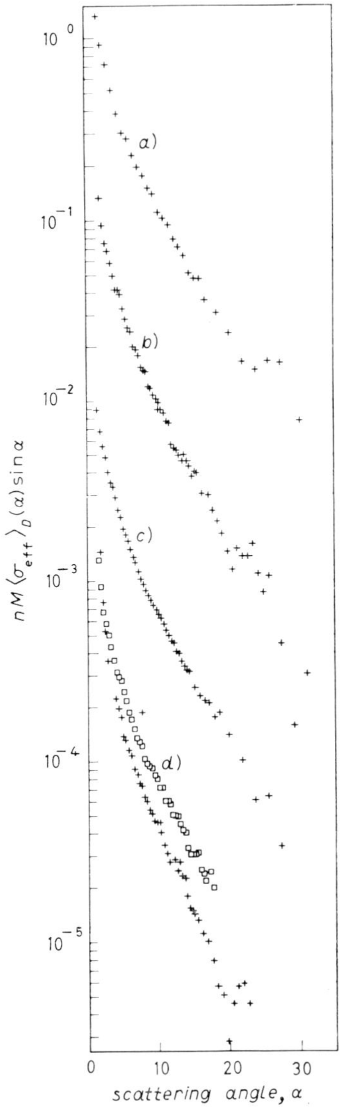

Fig. 9. – K+$SnCl_4$, differential scattering cross-section for nonreactive scattering, multiplied by const·sin α, for four relative energies, *vs.* laboratory angle in degrees. From ref. [16]. *a*) $\bar{E}$ = 2.15 kcal/mol, run 31, *n* = 1000; *b*) $\bar{E}$ = 3.86 kcal/mol, run 32, *n* = 100; *c*) $\bar{E}$ = 5.67 kcal/mol, run 33, *n* = 10; *d*) $\bar{E}$ = 7.83 kcal/mol, *n* = 1: + run 34, □ run 35.

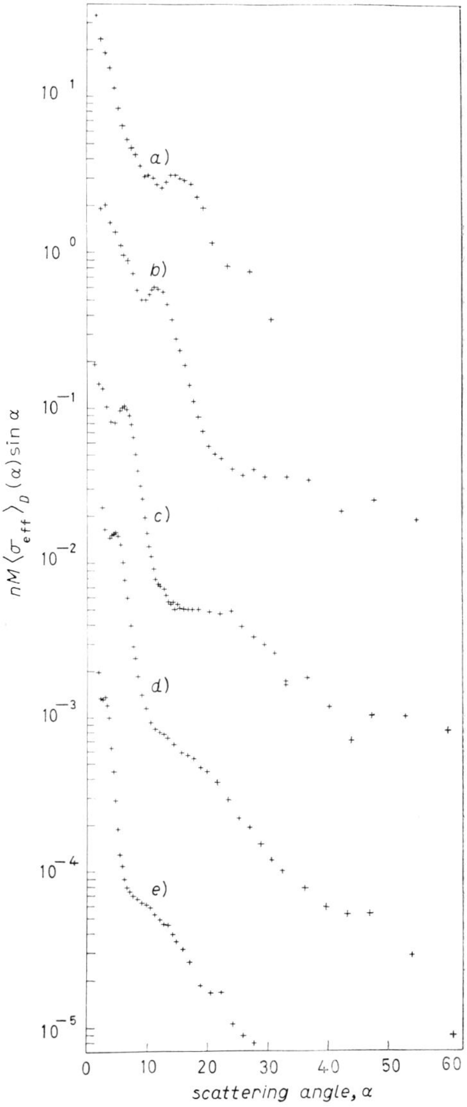

Fig. 10. – K+SF_6, differential laboratory cross-section for nonreactive scattering, multiplied by const·sin α, for five relative energies, *vs.* laboratory angle in degrees. From ref. [16]. *a*) $\bar{E}$ = 1.42 kcal/mol, run 36, $n = 10^4$; *b*) $\bar{E}$ = 2.02 kcal/mol, run 37, $n = 10^3$; *c*) $\bar{E}$ = 3.57 kcal/mol, run 38, *n* = 100; *d*) $\bar{E}$ = 5.21 kcal/mol, run 39, *n* = 10; *e*) $\bar{E}$ = 6.76 kcal/mol, run 40, *n* = 1.

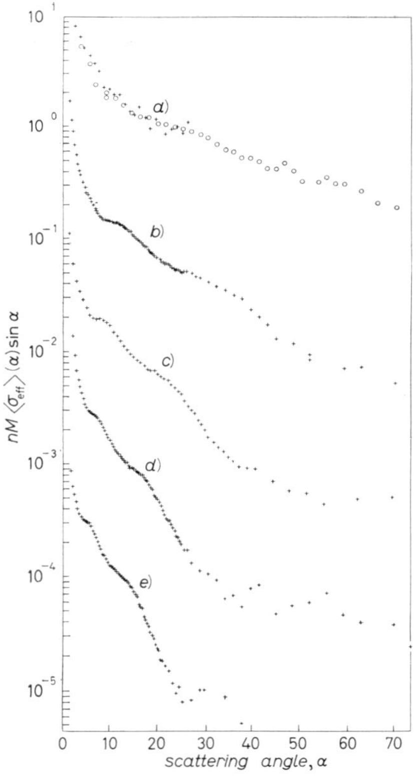

Fig. 11. – $K + CH_3I$, differential laboratory cross-section for nonreactive scattering, multiplied by const·sin α, for five relative energies, *vs.* laboratory angle in degrees. From ref. [16]. *a*) $\bar{E} = 1.42$ kcal/mol, $n = 10^4$: + run 17, ○ run 18; *b*) $\bar{E} = 2.99$ kcal/mol, run 19, $n = 10^3$; *c*) $\bar{E} = 4.17$ kcal/mol, run 20, $n = 100$; *d*) $\bar{E} = 5.55$ kcal/mol, run 21, $n = 10$;

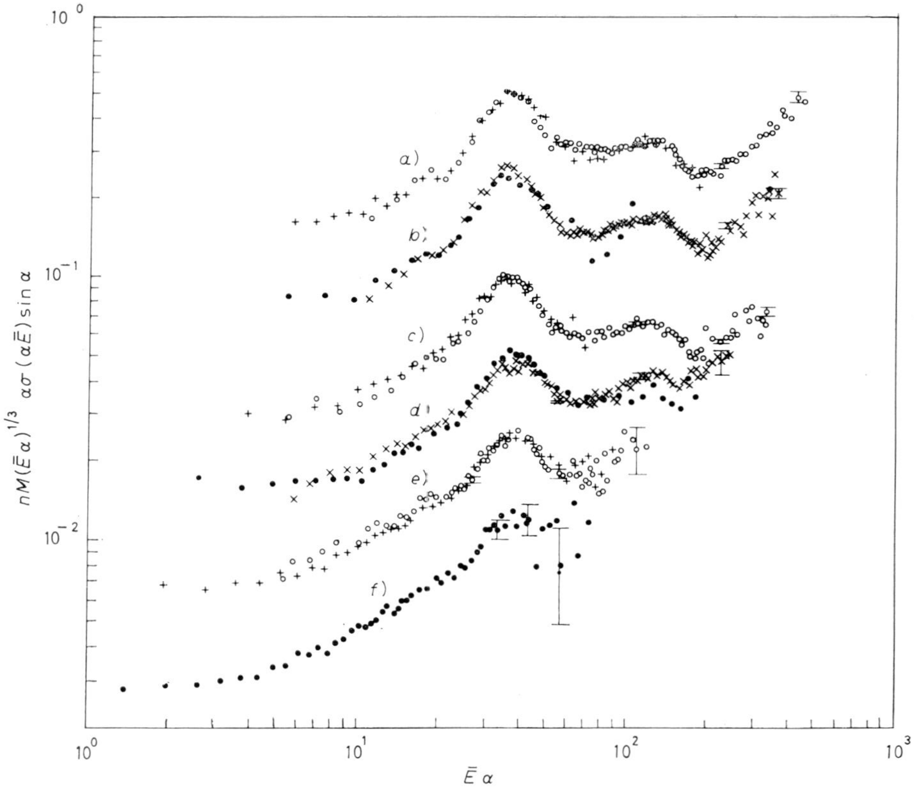

Fig. 12. – $K + (CH_3)_3CBr$, differential laboratory cross-section for nonreactive scattering multiplied by $nM(\bar{E}\alpha)^{\frac{1}{3}}\alpha\sin\alpha$, *vs.* relative energy, $\bar{E}$(kcal mol^{-1})×laboratory scattering angle (degrees). From ref. [11]. *a*) $\bar{E} = 6.27$, $n = 40$: ○ run 14, + run 13; *b*) × $\bar{E} = 5.15$, $n = 20$, run 12; • $\bar{E} = 5.51$, $n = 20$, run 11; *c*) $\bar{E} = 4.14$, $n = 8$: ○ run 10, + run 11; *d*) – $\bar{E} = 2.97$, $n = 4$, run 8; • $\bar{E} = 2.96$, $n = 4$, run 7; *e*) – $\bar{E} = 2.29$, $n = 2$, run 6; $\bar{E}$ + $= 2.22$, $n = 2$, run 5; *f*) $\bar{E} = 1.59$, $n = 1$, run 4.

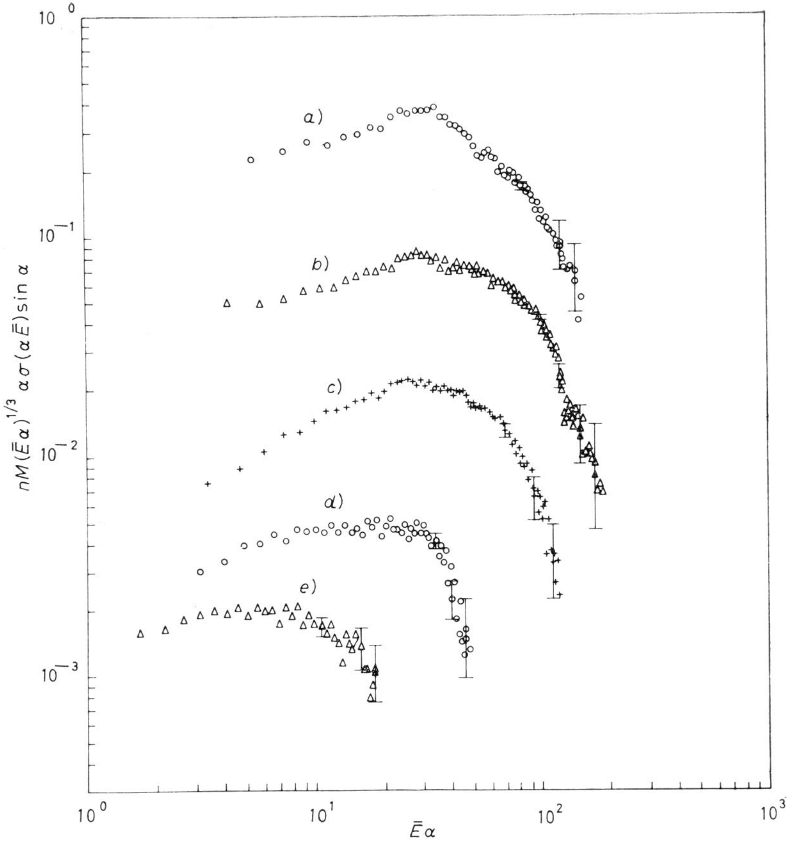

Fig. 13. – $K+Cl_2$, differential laboratory cross-section for nonreactive scattering mulplied by $nM(\bar{E}\alpha)^{\frac{1}{3}}\alpha\sin\alpha$, *vs.* relative energy, $\bar{E}$(kcal mol^{-1}) × × laboratory scattering angle (degrees). From ref. [11]. *a*) $\bar{E}=5.66$, $n=100$, run 22; *b*) $\bar{E}=4.39$, $n=20$, run 21; *c*) $\bar{E}=3.55$, $n=6$, run 20; *d*) $\bar{E}=2.37$, $n=3$, run 19; *e*) $\bar{E}=1.29$, $n=1$, run 18.

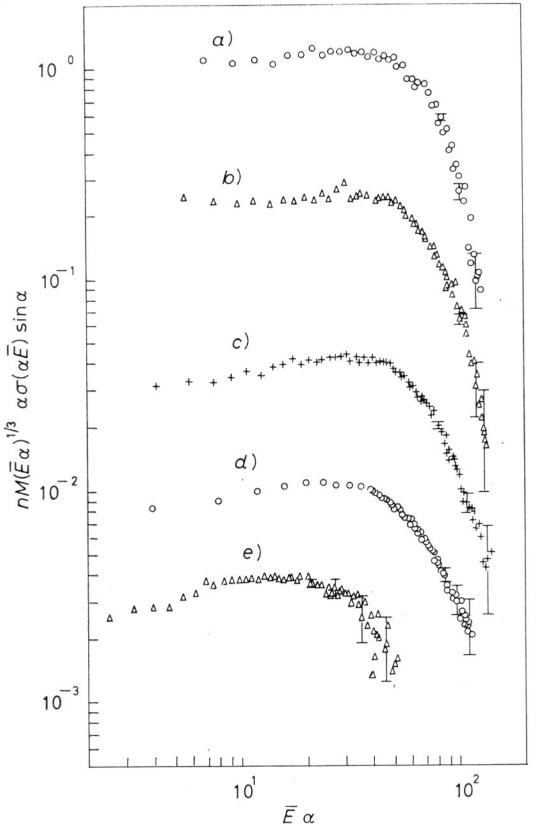

Fig. 14. – $K+Br_2$, differential laboratory cross-section for nonreactive scattering multiplied by $nM(\bar{E}\alpha)^{\frac{1}{3}}\alpha\sin\alpha$, *vs.* relative energy, $\bar{E}$(kcal mol^{-1}) × laboratory scattering angle (degrees). From ref. [11]. *a*) $\bar{E}=6.87$, $n=500$, run 27; *b*) $\bar{E}=5.67$, $n=100$, run 26; *c*) $\bar{E}=4.58$, $n=10$, run 25; *d*) $\bar{E}=3.62$, $n=3$, run 24; *e*) $\bar{E}=2.04$, $n=1$, run 23.

No corrections have been made for the remaining distributions, such as the residual velocity distribution in the primary beam due to the range of speeds transmitted by the mechanical velocity selector, and the speed and angular distribution in the secondary beam.

Figures 3, 4, 5, 6 show measurements of the differential scattering cross-section for some nonreactive systems [11, 16]. The scattering of K from the nearly spherically symmetric $C(CH_3)_4$ or $C_{10}H_{16}$ (adamantane) has essentially the same characteristics as that of K from rare gases [17]. The resolution of the experiment is sufficient to show the rainbow scattering and at times a supernumerary rainbow, but it is not sufficient to reveal the fine structure due to interferences from all three branches of the deflection function [1]. A comparison of the systems $K+C(CH_3)_4$, $K+C_6H_{12}$, and $K+C_6H_6$, however, shows

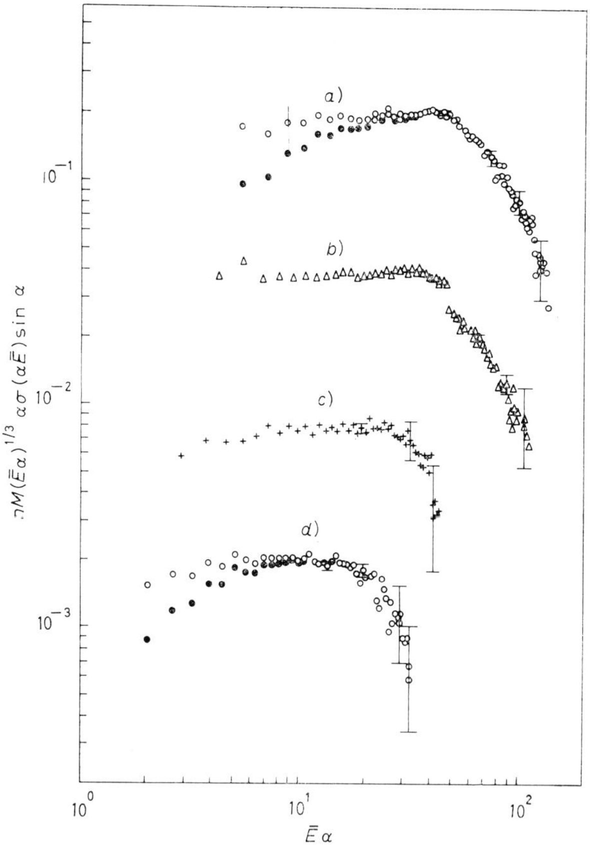

Fig. 15. – $K+I_2$, differential laboratory cross-section for nonreactive scattering multi plied by $nM(\bar{E}\alpha)^{\frac{1}{3}}\alpha\sin\alpha$, *vs.* relative energy, $\bar{E}$(kcal mol^{-1}) × laboratory scattering angle (degrees). From ref. [11]. The filled circles on runs 28 and 31 show results not corrected for the nonzero height of the detector. *a*) $\bar{E}=4.54$, $n=100$, run 31; *b*) $\bar{E}=3.54$, $n=20$, run 30; *c*) $\bar{E}=2.41$, $n=4$, run 29; *d*) $\bar{E}=1.71$, $n=1$, run 28.

that the rainbow scattering for $K+C_6H_6$ is less clearly visible. Anisotropic interactions and possibly inelastic transitions are likely causes for such deviations.

In Fig. 7-15 are shown representative angular distributions of the nonreactive scattering of K from reactive compounds [11, 16].

3. – Interpretation.

3·1. *Potential parameters.* – In a number of systems, for instance $K+SiCl_4$, $K+CCl_4$, $K+SF_6$, $K+C(CH_3)_3Br$, the small-angle scattering seems unaffected by the possibility of chemical reactions. The angular distributions appear similar to those in Fig. 1 for $L < 230$. We shall interpret the nonreactive scattering in such cases as elastic scattering. Inelastic cross-sections which lead to substantial changes in internal energy are usually smaller than elastic ones, while inelastic scattering which produces small energy changes generally also causes small angular-momentum changes. In either case, the effect on the angular distribution may be expected to be small. Further experiments are necessary to tell how serious the simplification of neglecting inelastic collisions is, but here, in any case we make it. Moreover, we shall interpret the elastic scattering with a simple spherically symmetric potential. That may be good enough for $K+SnCl_4$, or $K+SF_6$, but is not likely to be sufficient for $K+CH_3I$ where the rainbow pattern seems much distorted.

Potential parameters in an assumed form of potential (we shall use the exp-six potential, $\alpha = 12$, eq. (1), ref. [14*b*]) may be obtained from a semi-classical analysis of the rainbow scattering (see Method A of Sect. **4**, ref. [14*a*], as an example) or preferably form the angular positions of the rainbow and supernumerary rainbows (see ref. [9]; Sect. **4** of ref. [11]). Values of potential parameters for a number of systems, are listed in Table I (*)

In evaluating and using the information in Table I it is useful to remember the assumptions and approximations made in the analysis. The potential parameters are not independent of the assumed form of potential, not even for atom-atom scattering. The unreasonably low values of the size parameter r_m for $K+C_6H_{12}$, C_6H_6, $C_{10}H_{16}$, $SiCl_4$ obtained from the rainbow structure is probably due to a distortion and broadening of that structure brought about by anisotropic components in the potential and possibly inelastic scattering.

3·2. *Probability of reaction and threshold conditions.* – The nonreactive scattering in reactive systems differs markedly from that in nonreactive systems.

(*) The exp-six potential is $V(x) = \varepsilon[6/(\alpha-6)\exp(\alpha(1-x)) - \alpha/(\alpha-6)x^{-6}]$; $x = r/r_m$. The extraction of potential parameters is similar to that described in D. BECK's lecture on *Elastic scattering of nonreactive atomic systems*, Sect. **5** b (this Volume, p. 15) (Ed.).

TABLE I. – *Potential parameters (in exp-six potential, $\alpha = 12$) for some reactive and nonreactive systems, and threshold conditions for reactive systems.* The results are deduced from measurements made in the range of energies $(1 \div 7)$ kcal/mole ([a]).

System	Reference	ΔE_0^0 ([b]) (kcal/mol)	ε (kcal/mol)	r_m (Å)	Threshold energy (kcal/mol)	Reduced threshold distance in units of r_m
$K+C(CH_3)_4$	[16]		0.53	5.1		
$K+C_6H_{12}$	[16]		0.66	3.3		
$K+C_6H_6$	[16]		0.59	2.0		
$K+C_{10}H_{16}$	[11]		{ 0.76	4.4 }		
			0.69 ([c])	6.2 ([c])		
$K+SiCl_4$	[16]	-14	0.59	2.4		
$K+CCl_4$	[16]	-31	0.69	5.5	~ 0	0.88
$K+SF_6$	[16]	-44	0.31	6.3	0.15	0.86
$K+CH_3I$	[16]	-22	0.51		0.16	0.87
$K+(CH_3)_3CBr$	[11]	-33?	{ 0.48	2.4 }	0.47	0.84
			0.44 ([c])	5.6 ([c])		
$K+HCl$	[14]	$+1$	0.45	3.2	0.55	0.85
$K+HBr$	[14a]	-4	0.57	4.2	~ 0	0.89
$K+HI$	[14]	-6	0.67	4.2	0.2	0.87

([a]) Errors in the parameters are difficult to estimate. We guess that a range of $\pm 5\,\%$ for ε and $\pm 10\,\%$ for r_m may cover reproducibility and variations with energy, or the commonly used forms of the potential.

([b]) Standard change of energy of reaction at 0 °K.

([c]) Determined from the angular position of rainbows and supernumerary rainbows (see text Sect. 3·1).

For some systems, like $K+CCl_4$, $K+SF_6$, $K+(CH_3)_3CBr$, the differences occur at angles larger than the rainbow angle and consist of a much lower intensity of scattered species for the reactive systems. The decreased intensity is due to absorption, and the angular distribution is similar to that in Fig. 1 for $L \sim 230 \div 190$. For other systems, like $K+Cl_2$, $K+Br_2$, $K+I_2$, $K+SnCl_4$, the differences extend further toward small scattering angles: there is not a sign of rainbow scattering. The angular distribution is now somewhat like that in Fig. 1 for $L \sim 350$.

Where rainbow scattering occurs and the onset of reaction is evident, as in $K+(CH_3)_3CBr$ for instance, an interpretation of the measurements by application of eq. (17) can be made. Potential parameters for an assumed form of the potential are obtained from the rainbow region and, where possible, from supernumerary rainbows. The scattering cross-section $\sigma_0(\chi)$, which should be calculated with the real part of a phase shift dependent on both the real (V_0) and imaginary (V_1) part of an optical potential, is approximated by $\sigma_0(\chi)$,

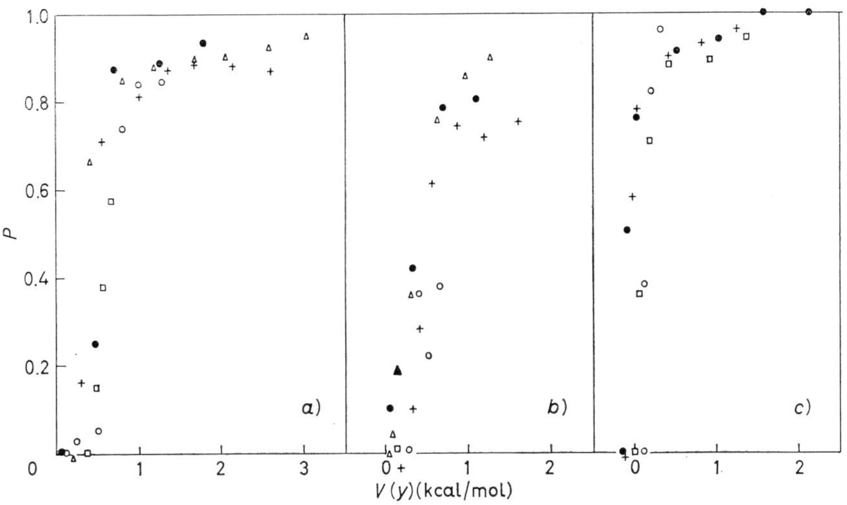

Fig. 16. – Probability of reactions *vs.* potential energy at the distance of closest approach of reactants at several relative energies $\bar{E}$(kcal/mol) for K+CH_3I, K+SF_6, K+CCl_4: From ref. [16]. *a*) K+CH_3I: □ $\bar{E}=1.42$, ○ $\bar{E}=2.99$, + $\bar{E}=4.17$, △ $\bar{E}=5.55$, ● $\bar{E}=6.73$; *b*) K+SF_6:) ○ $\bar{E}=2.02$, + $\bar{E}=3.57$, △ $\bar{E}=5.21$, ● $\bar{E}=6.76$; *c*) K+CCl_4 ○ $\bar{E}=2.04$, + $\bar{E}=3.31$, □ $\bar{E}=4.90$, ● $\bar{E}=6.83$.

defined by the calculation with V_0 only [18]. As a first guess, V_0 is taken to be the potential obtained from the rainbow region. In Fig. 16, 17, 18 are shown probabilities of reaction as a function of reduced impact parameter or potential energy at the distance of closest approach of reactants. The elastic-scattering angle, the impact parameter, and the potential energy at the distance of closest approach are related to each other in classical dynamics and can be calculated for two-body scattering with a given potential.

The probabilities of reaction for K+SF_6 and K+CCl_4 are very close to step functions as given by the simple model eq. (21) and as required by the collision theory of hard spheres. The limiting values of the probabilities of

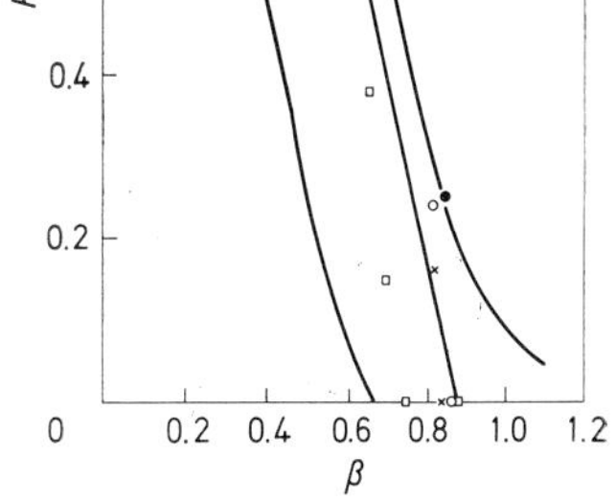

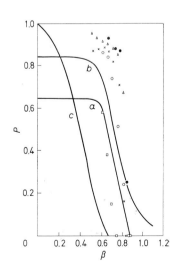

Fig. 17. – K+CH_3I, probability of reaction *vs.* reduced impact parameter for several relative energies, in kcal/mole. The solid lines are classical calculations for three potential surfaces (L. M. RAFF and M. KARPLUS: *Journ. Chem. Phys.*, **44**, 1212 (1966)). From ref. [16]. □ $\bar{E}=1.42$, ○ $\bar{E}=2.99$; × $\bar{E}=4.17$, △ O $=5.55$, ● $\bar{E}=6.73$.

reaction for the sequence $K+HBr$, $K+CH_3Br$, $K+(CH_3)_3CBr$ are in qualitative agreement with the steric factors to be expected for this sequence [19]. The origin of the energy-dependence of the plateau values of the probability of reaction in $K+(CH_3)_3CBr$ is not known.

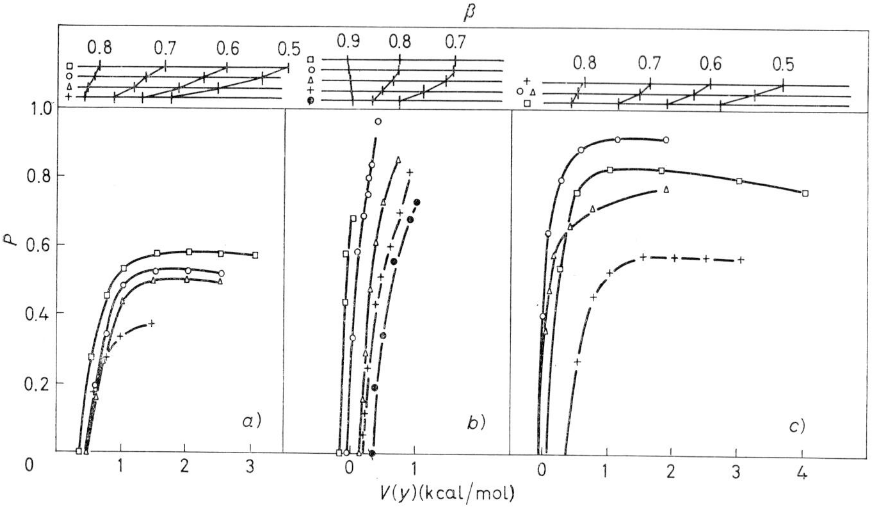

Fig. 18. – The probability of reaction P *vs.* the potential energy at the distance of closest approach $V(y)$; $\bar{E}$ in kcal mol^{-1}. The separate abscissa scales above the figure show the values of the reduced impact parameter β for each curve. *a*) $K+(CH_3)_3CBr$. □ $\bar{E}=6.28$, ○ $\bar{E}=5.15$, △ $\bar{E}=4.14$, + $\bar{E}=2.97$. *b*) Cs+HBr: • $\bar{E}=1.74$, + $\bar{E}=2.53$ △ $\bar{E}=3.50$, ○ $\bar{E}=4.67$, □ $\bar{E}=6.03$. *c*) Comparison for several different bromine compounds at $\bar{E}=4$ kcal mol^{-1}. The numbers in parentheses are reduced relative kinetic energies, K. From ref. [11]. ○—○—○ HBr (8.7), □—□—□ CH_3Br (12.9), △—△—△ DBr (8.7), +—+—+ $(CH_3)_3CBr$ (13.2).

Threshold energies and distances of each reactive system are available from the variation of the probability of reaction with the potential energy at the distance of closest approach, and this information is listed in Table I for a few systems.

The probability of reaction for the endothermic system [14*b*] K+HCl is shown in Fig. 19. The threshold of reaction is apparently just the endothermicity of the reaction, and at threshold the product KCl is formed in as low an internal state as the conservation laws permit. The rotational state of KCl is near that one for which the rotational angular momentum is equal to the total angular momentum of the system, but the vibrational state is the ground state. As the potential energy at the distance of closest approach is increased a second increase in the probability of reaction is observed where the threshold is expected for the production of KCl in the first vibrational state.

In cases where rainbow scattering is observed but the onset of reaction occurs in that same region, as in $K+CCl_4$ or $Cs+HBr$ the interpretation may be ambiguous [11]. Figura 20, shows the results of calculatons made with real phase shifts (Fig. 20 *a*)) and with the absorption function

$$(23)\qquad P(l)=\begin{cases} 1-\exp[-4\xi_l]\,, & \\ 0\,, & L+\Delta\leqslant l<\infty\,,\\ \left[\frac{3}{4}\left(\frac{l-L-\Delta}{\Delta}\right)^2+\frac{1}{4}\left(\frac{l-L-\Delta}{\Delta}\right)^3\right], & L-\Delta<l<L+\Delta\,,\\ 1\,, & 0<l<L-\Delta\,,\end{cases}$$

where L and Δ are parameters which govern the location and range in which that function rises from zero to unity. In the calculations for Fig. 20 *b*), L and Δ are chosen as indicated by small vertical bars on the deflection functions (inserts on the figures) at positive values of the relative scattering angle χ for each of the three reduced relative kinetic energies $K(=E/\varepsilon)$, and in the calculations for Fig. 20 *c*) $L\pm\Delta$ are chosen as indicated by the vertical bars at negative values of the deflection function. The point is that the angular distribution in the rainbow region is essentially the same at each K for the very different threshold conditions *b*) and *c*). Differences are noticeable only in the supernumerary rainbow structure.

In the interpretations described above the choice for the real potential V_0 has been arbitrary, and the absorption functions obtained depend on that choice. The fact that the threshold conditions deduced from the absorption functions are,

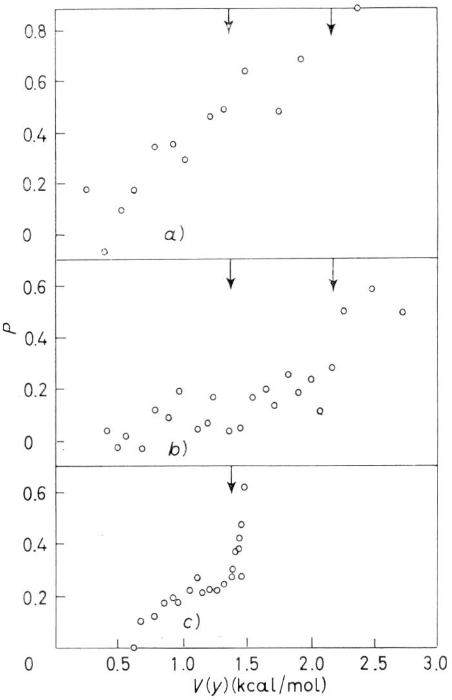

Fig. 19. – The probability of reaction *vs.* the potential energy at the distance of closest approach of the reactants $K+$ $+HCl$, for the three runs with the best signal-to-noise ratio. The arrows indicate the value of the abscissa corresponding to the measured threshold of the reaction plus one and two vibrational quanta of KCl. From ref. [14*b*]. *a*) $\bar{E}=4.53$ kcal/mol, run 9; *b*) $\bar{E}=2.88$ kcal/mol, run 6; *c*) $\bar{E}=1.41$ kcal/mol, run 2.

to within the probable likely experimental error, independent of initial relative kinetic energy leads to some belief in the utility of the results. This is a long way from having a unique answer; the problem is similar to, but still more complicated than, determing potentials from elastic scattering in two-particle systems. As an alternative to the procedure just outlined it is possible to proceed by choosing an absorption function, arbitrarily but hopefully

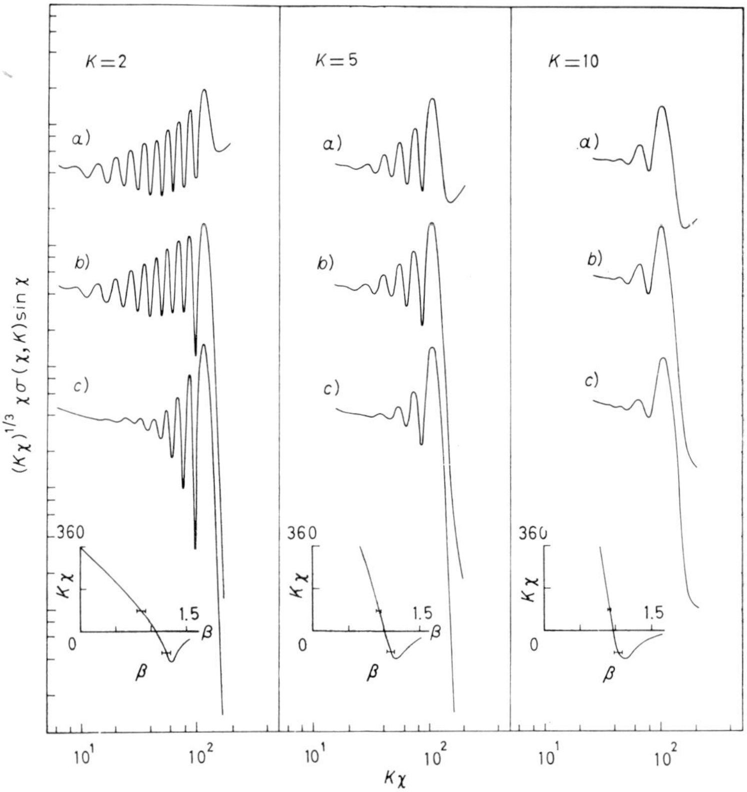

Fig. 20. – Differential scattering cross-sections multiplied by $(K\chi)^{\frac{1}{3}}\chi\sin\chi$ *vs.* $K\chi$ for the 3 indicated relative kinetic energies K, χ in degrees. *a*) No reaction, *b*) reaction thresholds at positive values of deflection functions, see inserts and text, *c*) reaction thresholds at negative values of deflection function. From ref. [11].

reasonably, and then deducing some information about the real part of the potential, V_0. This inverse procedure seems to be the more reasonable to use for systems like $K+Cl_2$, Br_2, I_2 for which no rainbow scattering is observed. If we follow the previous procedure and choose an exp-six potential for V_0 $K+I_2$, for instance, then the threshold distances of the absorption function obtained from the measurements vary considerably with relative kinetic energy.

It seems preferable to take the second course and assume a simple-step function for the absorption function at a fixed distance invariant with relative kinetic energy, and then to vary V_0 to fit the measurements. This method [11, 20] is illustrated in Fig. 21. The absorption function is taken to be a step function which is unity at all distances smaller than the smallest distance at which the effective potential $V_{eff} = V + Eb^2/r^2$ has a maximum (rotational barrier). Thus if the kinetic energy for a given impact parameter b is less than the maximum in the effective potential elastic scattering occurs; if the kinetic energy

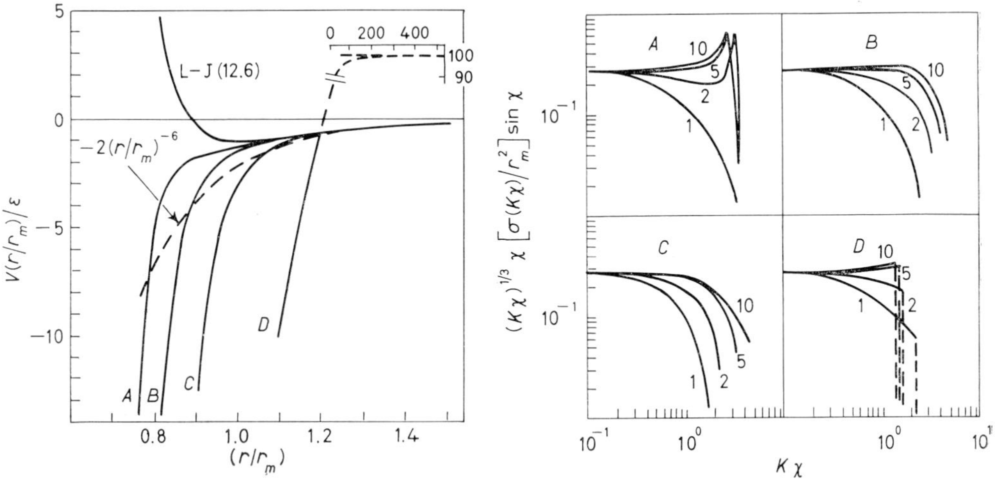

Fig. 21. – Two-body potentials used as model for the interaction of K with halogens and TCNE, eqs. (3) and (4), and the resulting reduced differential cross-sections for several reduced energies, K. Reaction is assumed to occur abruptly when a system passes over the angular-momentum barrier. Angles in radians. From ref. [11].

exceeds the maximum then reaction occurs. Potential D is a Van der Waals potential at large distances which changes abruptly to a Coulomb potential corresponding to an electron jump at a given separation of reactants [21]. Potentials A, B, and C are also Van der Waals potentials at large distances but change to more attractive interactions gradually.

A comparison of the elastic differential scattering cross-section calculated with the potentials A, B, C, D, Fig. 21, and the measurements, Fig. 13, 14, 15, shows qualitative agreement for potentials B, C, in particular with regard to to the variation of the angular distribution with energy. The opposite of the observed variation is predicted by potential D. Hence it seems that a model of a potential corresponding to a gradual electron transfer, rather than an electron jump at one distance, is in better agreement with the measurements.

3·3. *Total reaction cross-section.* – The total reaction cross-section can be calculated from the probability of reaction by means of eq. (22). The variation of the total reaction cross-section with relative energy, in reduced units, is shown in Fig. 22 for some reactive systems. The curve for $K+Br_2$ is calculated from a model outlined in the previous Section for a potential like V_B.

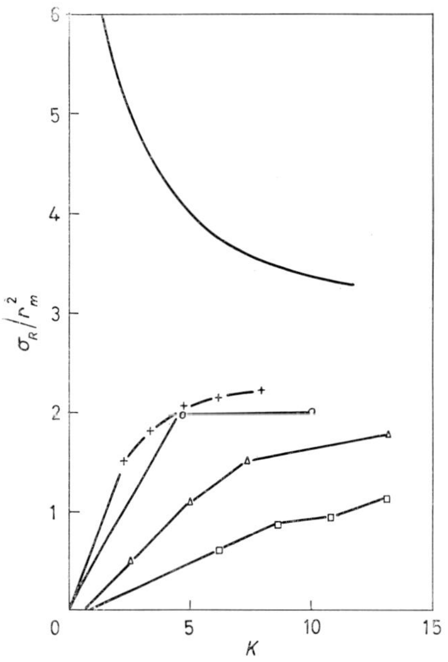

Fig. 22. – The total reaction cross-section obtained from an analysis of the nonreactive scattering of K by several bromine compounds *vs.* reduced relative kinetic energy $K=E/\varepsilon$. The curve for $K+Br_2$ is calculated from a model discussed on p. 105 above and in Subsect. **4**.B.4 of ref. [11] for a potential V_B (see Fig. 21). From ref. [11]. —— $K+Br_2$, + $Cs+HBr$, ○ $K+HBr$, △ $K+CH_3Br$, □ $K+(CH_3)_3CBr$.

4. – Conclusions.

The nonreactive scattering in reactive systems studied so far is interpretable with an optical model and leads to information about the chemical reaction. A number of simplifications have been made, such as neglecting the influence of inelastic collisions and anisotropic forces.

For some of the systems reviewed chemical reaction occurs only for small reduced impact parameters so that the rainbow scattering is unaffected by reaction. For these cases, where comparison is possible, the angular distribution of the products is peaked backwards in the center of mass system. For other systems reaction occurs at large reduced impact parameters and here the distribution of the products is peaked forwards in the center of mass system. This correlation holds for reactions proceeding by a direct interaction mechanism [23]. For the first group reduced total reaction cross-sections are, as expected, smaller than for the second group.

* * *

We thank Drs. HAERTEN, MÜLLER and SCHÖTTLER for preparing a preliminary draft of the manuscript from the lecture.

REFERENCES

[1] D. Beck: this volume, p. 15.

[2] R. M. Eisberg and C. E. Porter: *Rev. Mod. Phys.*, **33**, 190 (1961); P. E. Hodgson: *The Optical Potential of Elastic Scattering* (Oxford, 1963); L. D. Landau and E. M. Lifshitz: *Quantum Mechanics* (London, 1958).

[3] N. F. Mott and H. S. W. Massey: *The Theory of Atomic Collisions*, III ed. (London, 1965).

[4] C. Nyeland and J. Ross: *Journ. Chem. Phys.*, **49**, 843 (1968).

[5] J. L. J. Rosenfeld and J. Ross: *Journ. Chem. Phys.*, **44**, 188 (1966).

[6] *a*) H. Y. Sun and J. Ross: *Journ. Chem. Phys.*, **46**, 3306 (1967); *b*) F. J. Smith, E. A. Mason and J. T. Vanderslice: *Journ. Chem. Phys.*, **42**, 3257 (1965).

[7] K. W. Ford and J. A. Wheeler: *Ann. of Phys.*, **7**, 259 (1959).

[8] A. Erdelyi: *Asymptotic Expansions* (New York, 1956).

[9] E. Hundhausen and H. Pauly: *Zeits. Phys.*, **187**, 305 (1965).

[10] R. B. Bernstein and R. D. Levine: *Journ. Chem. Phys.*, **49**, 3872 (1968).

[11] E. F. Greene, L. F. Hoffman, M. W. Lee, J. Ross and C. E. Young: *Journ. Chem. Phys.*, **50**, 3450 (1969).

[12] K. M. Watson: *Adv. Theor. Physics*, edited by K. A. Brueckner, vol. **1** (New York, 1965), p. 115, and references therein.

[13] B. C. Eu and J. Ross: *Disc. Faraday Society*, No. 44, 39 (1967).

[14] The apparatus and the method of measurement are described in *a*) J. R. Airey, E. F. Green, K. Kodera, G. P. Reck and J. Ross: *Journ. Chem. Phys.*, **46**, 3287 (1967); *b*) M. Ackerman, E. F. Greene, A. L. Moursund and J. Ross: *Journ. Chem. Phys.*, **41**, 1183 (1964); *c*) D. Beck, E. F. Greene and J. Ross: *Journ. Chem. Phys.*, **37**, 2895 (1962).

[15] E. Everhart: *Phys. Rev.*, **132**, 2083 (1963); F. T. Smith, R. P. Marchi and K. G. Dedrick: *Phys. Rev.*, **150**, 79 (1966); E. A. Mason: *Journ. Chem. Phys.*, **26**, 667 (1957).

[16] J. R. Airey, E. F. Greene, G. P. Reck and J. Ross: *Journ. Chem. Phys.*, **46**, 3295 (1967).

[17] *a*) D. Beck and H. Dummel: private communication; H. Dummel: Ph. D. Thesis, Freiburg (1968); *b*) D. Beck, H. Dummel and U. Henkel: *Zeits. Phys.*, **185**, 19 (1965); *c*) P. Barwig, U. Buck, E. Hundhausen and H. Pauly: *Zeits. Phys.*, **196**, 343 (1966).

[18] For further discussion of the theory on this point see ref. [5].

[19] For a possible explanation of the apparent anomaly in this sequence, K+DBr, see ref. [11].

[20] R. Anderson and D. R. Herschbach: private communication.

[21] D. R. Herschbach: *Adv. Chem. Phys.*, **10**, 319 (1966).

[22] E. F. Greene, A. L. Moursund and J. Ross: *Adv. Chem. Phys.*, **10**, 135 (1966).

[23] D. R. Herschbach: this volume, p. 77.

Hot-Atom Chemistry - I.

F. S. Rowland

University of California - Irvine, Cal.

1. – The nature of hot-atom chemistry.

The central problem in hot-atom chemistry is the study of chemical reactions over the whole range of energies for which such reactions are possible. We have already seen the importance of reactions near threshold for thermally activated systems; however, thermal systems do not provide information about processes occurring at energies much above the threshold energy simply because only a negligible proportion of such reactions occurs at higher energies. The distribution in collision energies for thermal species reacting with a 1.0 eV threshold is shown by the shaded area in Fig. 1 for tritium atoms in thermal equilibrium at 273 °K. An increase in the temperature of the tritium atoms will result in a substantial increase in the magnitude of the shaded area, but the distribution of energies of the tritium atoms at reaction will continue to be concentrated in the range just above the threshold energy. Hot-atom chemistry is thus concerned with experimental and theoretical approaches to the probability of reaction, the mechanisms, and the energetics of reactions over the entire chemical reaction range, and especially for the uninvestigated range well above threshold energies [1-5].

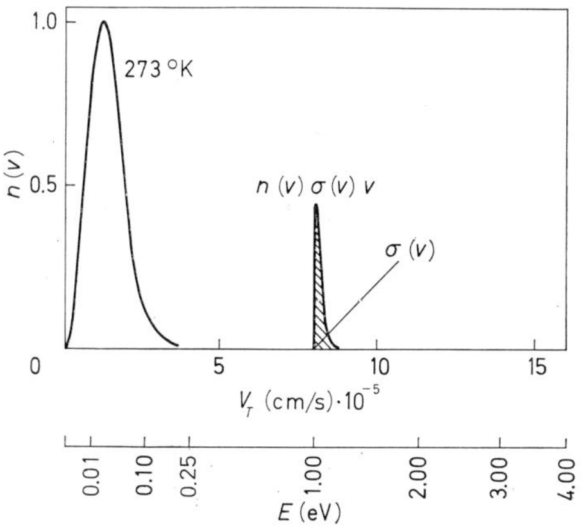

Fig. 1. – Cross-product of reaction cross-section and Maxwellian distribution. Tritium atoms at 273 °K with 1 eV reaction threshold.

2. – Hot-atom techniques. Nuclear recoil, radio gas chromatography.

Experimental solutions for the investigation of hot-atom chemistry problems are found through the development of energetic sources of atoms *not* in equilibrium with the surroundings. Several types of such sources are available, and can generally be classified into three categories, placed in descending order of utility to-date: a) nuclear reactions, b) photochemical sources, and c) beam methods. The nuclear-reaction sources generally produce atoms with energies which are essentially « infinite » on the scale of chemical-reaction energies. An excellent example of the energies involved, and a nuclear reaction used extensively in the studies to be discussed later, is the formation of recoil tritium from thermal neutron reactions with ^{3}He. The nuclear energetics for this reaction can be directly calculated from the masses of the various particles involved in the reaction:

Mass ^{3}He	= 3.016030	^{3}H	= 3.016049
Mass ^{1}n	= 1.008665	^{1}H	= 1.007825
	4.024695		4.023874
	4.023874		

$$0.000821 \text{ a.m.u.} \times 931.48 \text{ MeV/a.m.u.} = 0.765 \text{ MeV exoergic}$$

$$\frac{1.007825}{4.023874} \times 0.765 = 0.192 \text{ MeV for } ^3\text{H}$$

The excess energy of the nuclear reaction appears as kinetic energy of the products, and is distributed in reciprocal ratio of the masses—the tritium atom is thus formed with 192 000 eV of energy, and is, of course, far above the (4 ÷ 5) eV of the typical C—H bond. Atoms produced in this manner must eventually lose their excess energy, enter the chemical reaction range from higher energy, and effectively sample the characteristic chemical reactions over the *entire* range of possible energies. Such a reaction source thus gives the very real advantage that *all* possible hot reactions should be occurring in the system, and are potentially observable under appropriate experimental conditions [1-5].

The tritium atoms formed in this reaction are so energetic that they actually are in the ionized +1 state for most of their decelerating path in the medium; neutralization occurs in the later stages, the −1 state also appears and disappears, but the tritium atoms react in the neutral state. Neutralization in high ionization potential media, *e.g.* helium or neon, is not always assured, and experimental verification of the neutral state assumption must be sought in those systems. Similarly, the atoms exist in excited electronic states during much of the deceleration process, but concentrate more and more in lower states as the energy decreases, and react in the ground state without electronic excitation.

Nuclear reaction sources have one other important characteristic—the nuclear reactions are chosen to produce a radioactive atom with a convenient half-life, and the radioactivity of the chemical product permits subsequent identification of the atom in its final chemical form. In this way, one is certain through the radioactivity that each such atom was one of the original high-energy atoms available in the system. The high sensitivity of tracer analysis also permits the detection of the high-energy atoms even though the total fractions of atoms so involved may only be 10^{-10}. This high sensitivity for detection also permits successful experiments after short nuclear irradiations before the build up of any significant amount of radiation damage.

Appropriate nuclear reactions are available not only for hydrogen, but also for fluorine (^{18}F), chlorine (^{38}Cl, ^{39}Cl), carbon (^{11}C), and most of the other elements whose atomic reactions are of current research interest [1-5]. One conspicuously difficult element is oxygen, for which the longest lived radioactive isotope has a half-life of only two minutes, and which therefore has severe problems of rapid analytical technique.

The analytical technique now used in the great majority of hot-atom chemistry experiments by nuclear techniques is *radio gas chromatography* [6, 7].

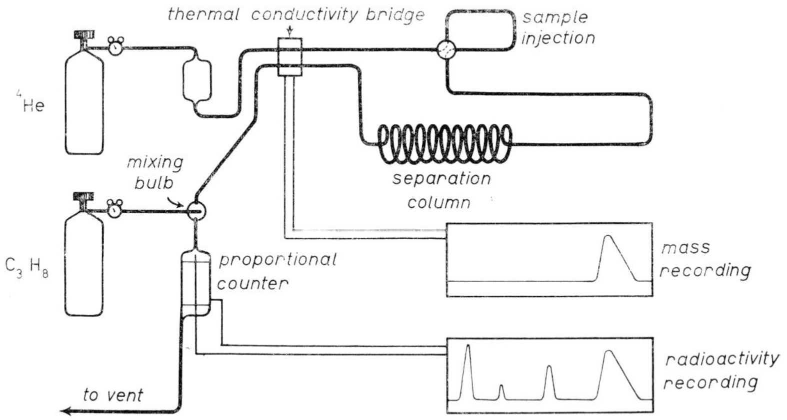

Fig. 2. – Block diagram of radio gas chromatographic apparatus.

The standard separation technique of gas chromatography has been coupled with a radioactivity detector to produce chromatograms of the distribution of radioactivity in irradiated samples, as illustrated in Fig. 2. Figure 3 illustrates a typical radio gas chromatogram for the reactions of energetic ^{38}Cl with CH_3Cl, in the presence of *cis*-CHCl=CHCl and O_2 as scavenger molecules [8]. The scavenger molecules are intended to remove thermalized species from the system in such a way that one can readily distinguish their reaction products from those which were initiated by *energetic* atoms. As with most radio gas chromatograms, two separate records are made, the primary one concerned with the distribution of radioactivity, and the secondary measuring the macroscopic distribution of material in the system. In Fig. 3, the lower line illustrates the macroscopic record from this sample, and indicates that the sample

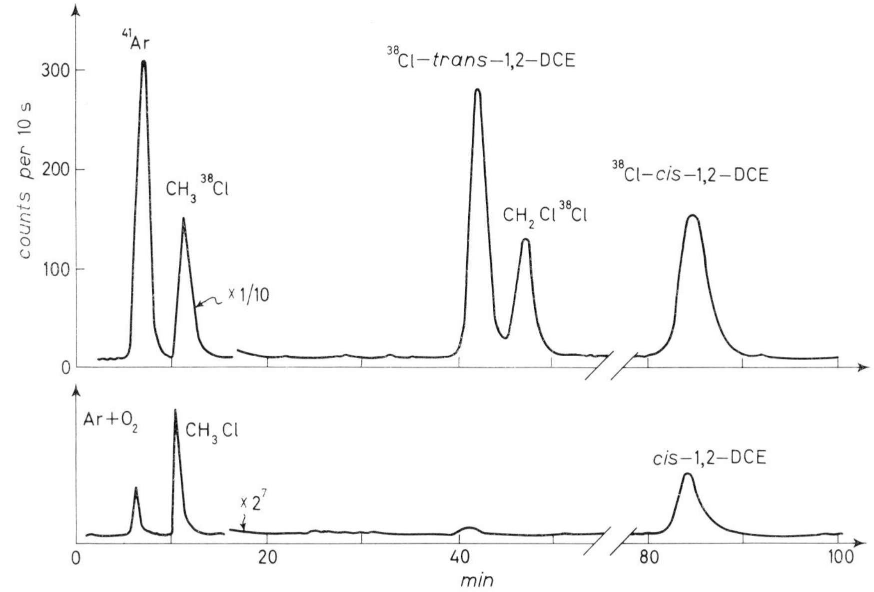

Fig. 3. – Radio gas chromatogram of radioactive products from reaction of ^{38}Cl with CH_3Cl. Lower line: macroscopic composition; Upper line: radioactivity distribution.

contained largely CH_3Cl, with small amounts of argon and *cis*-CHCl=CHCl. The energetic, radioactive ^{38}Cl ($t_{\frac{1}{2}} = 37.3$ min) is formed by the $^{37}Cl(n, \gamma)^{38}Cl$ nuclear reaction on the chlorine atoms in the mixture.

The upper line shows the radioactivities observed in the same sample: $CH_3{}^{38}Cl$ from the substitution of ^{38}Cl for Cl; $CH_2Cl^{38}Cl$ from the substitution of ^{38}Cl for H in CH_3Cl; both *cis*- and *trans*-CHCl=$CH^{38}Cl$ from the reactions of *thermal* ^{38}Cl with *cis*-CHCl=CHCl; and ^{41}Ar from the nuclear reaction $^{40}Ar(n, \gamma)\ ^{41}Ar$. From absolute yield measurements, monitored through the ^{41}Ar production, the ^{38}Cl radioactivity observed accounts for only 7.5% of the total produced in the sample at one atmosphere pressure. The remainder has gone into $H^{38}Cl$, into higher-boiling products after thermal scavenging by CHCl=CHCl, or into oxygenated species. While the products clearly formed by hot reactions are $CH_3{}^{38}Cl$ and $CH_2Cl^{38}Cl$, $H^{38}Cl$ may be formed by either thermal or hot reactions, and some of the primary products may be lost for measurement through decomposition of excited polyatomic molecules, such as $CH_3{}^{38}Cl^*$ or $CH_2Cl^{38}Cl^*$.

Another characteristic radio gas chromatogram is shown in Fig. 4, in this case for the reactions of energetic tritium atoms in a mixure of CH_4 and CD_4 [9]. The parent molecules are separated on the gas chromatographic column, as

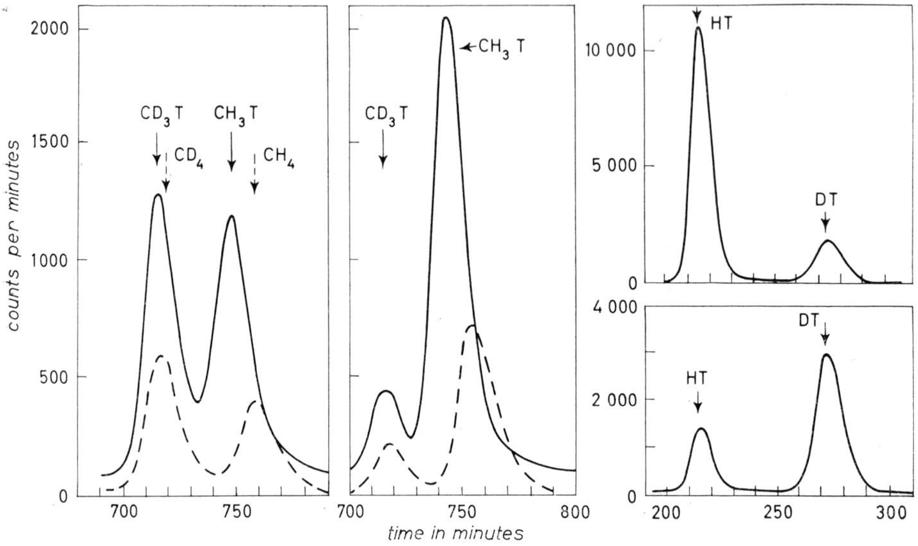

Fig. 4. – Radio gas chromatographic separation of isotopic methanes and hydrogens. Two different samples are illustrated for each.

shown by the lower trace, while the radioactive substitution products, CH_3T and CD_3T, are also separated as shown in the upper trace. From the relative peak heights, one can see that substitution of tritium into CH_4 seems to be easier then into CD_4—a reactive isotope effect of about 1.31 ± 0.03. The other trace in Fig. 4 shows the separation of HT and DT, formed by the abstraction reaction in the same mixtures. The abstraction from a given molecule or from particular locations in a molecule can be identified through the selective use of deuterated or partially-deuterated molecules.

3. – Types of hot-atom reaction.

The reactions observed for hot tritium atoms fall into three general classes, as observed from the study of reactions with more than one hundred parent molecules. These three classes are [1-5]:

a) abstraction of hydrogen,

$$T^* + RH \rightarrow HT + R; \tag{1}$$

b) substitution for an atom or radical,

$$T^* + RX \rightarrow RT^* + X; \tag{2}$$

c) addition to unsaturated π-bonded systems,

$$T^* + {>}C{=}C{<} \rightarrow {>}CT{-}\dot{C}^*{<} \tag{3}$$

The substitution reaction has been written in the very general form of replacement of X, since tritium replacement is observed not only for H and D, but also for other atoms, such as F, Cl, Br, I, and for polyatomic radicals, such as OH, NH_2, CH_3, C_2H_5, etc. The reaction products from reactions (2) and (3) are labeled with asterisks to indicate that these molecular species are often observed to have substantial amounts of internal excitation energy subsequent to the primary hot reaction—sufficient to permit secondary decomposition or isomerization reactions unless the excitation energy is very rapidly removed or reduced by collision with another molecule of the medium. Such secondary reactions will be considered in detail in the second lecture on hot atom chemistry.

The abstraction reaction might also be generalized in principle to include the removal of F, Cl, etc., to form TF, TCl, etc., and such reactions may very well occur. However, the rapid intermolecular exchange of hydrogen atoms (including tritium) prohibits the isolation without atomic exchange of TF, TCl, HTO, or NH_2T, and almost invariably will mix the radioactivity of these products with thermal tritium atoms which have reacted with scavenger molecules such as O_2 or Br_2. Studies of hot atom reactions have therefore usually been limited to those chemical systems for which the isotopic integrity of the products is maintained from formation through analysis. Fortunately, atoms bonded to carbon do not usually undergo ready isotopic exchange, permitting the convenient use of a very large number of organic reactant molecules.

The qualitative reactions found for hot halogen atoms parallel those found for tritium—abstraction, substitution, addition to the double bond—although the quantitative yields reflect the large differences among the atoms in mass and chemical behavior [4]. The substitution of ^{38}Cl for H and for Cl have already been illustrated in Fig. 3; addition to the double bond of CHCl=CHCl is primarily by thermal atoms in Fig. 3, but occurs by a similar path for more energetic ^{38}Cl atoms; abstraction of H to form $H^{38}Cl$ gives problems both in analysis and in distinguishing hot reactions from thermal, and is usually not measured.

Many studies have also been carried out with polyvalent species, with ^{11}C (or ^{14}C) as the most prominent example [10, 11]. The isolation of individual steps of reaction is difficult for atoms which must form several new bonds (the free ^{11}C atom must form four C—H bonds to become CH_4), and information about hot atoms is much harder to obtain. However, free carbon atoms, unlike hydrogen or halogen atoms, are very difficult to produce by other chemical methods, and most of our present knowledge about the chemical reactions of carbon atoms, hot or thermal, has been obtained through study of the hot carbon atoms formed by nuclear recoil.

4. – Other techniques for hot-atom chemistry: photochemistry, beams.

Two other experimental techniques that are now applied to hot atom studies are *a*) photochemical, and *b*) accelerated beams. The photochemical method involves the photolysis of a suitable target molecule with ultraviolet light of short enough wavelength to furnish an excess of energy over that required for simple decomposition of the molecule. The method has been applied especially with the hydrogen halides (HBr, HI, DBr, DI, TBr) as the source of hydrogen atoms [12-15].

The photolysis of HBr with 1849 Å light from a mercury arc lamp ejects H atoms with about 3 eV kinetic energy [12]; the use of DBr or TBr leads to D or T atoms of similar energy. This photochemical system has been used in two distinct modes of operation as illustrated in the following set of equations:

$$DX \xrightarrow{h\nu} D^* + X \qquad \text{(photolysis)},$$

$$D^* + RH \rightarrow HD + R \qquad \text{(hot abstraction)},$$

$$D^* + RH \rightarrow RD + H \qquad \text{(hot substitution)},$$

$$D^* + RH \rightarrow RH + D \qquad \text{(nonreactive scattering)}.$$

In one method of experimentation, the scavenger molecule is also the target molecule, DX. In the other, the scavenger is the molecular halogen, X_2. The subsequent scavenger reactions for each are shown in the boxes:

Radical	Scavenger	
	DX	X_2
R	RD	RX
H	HD	HX
D	D_2	DX

When the same molecule serves as both target and scavenger, hot abstraction cannot be distinguished from hot substitution since both sequences form one molecule each of HD and RD. However, thermalized D atoms give a distinguishable product, D_2, and the total hot reaction, abstraction plus substitution, can be obtained by comparison of the yield of HD (or RD) with D_2. Experimentally, the ratio of HD/D_2 is extrapolated to zero concentration of DX to eliminate the contribution of D_2 from the hot abstraction reaction of D^* with DX [12-14].

When X_2 is used as scavenger, the initial hot reactions lead to easily distinguishable products, but the thermalized atoms react to reform the originally photolyzed molecule, DX. In this method of operation, the ratio of HD/RD can be determined, but the total hot yield is not measured [15].

The first experiments used the same molecule as target and scavenger (HX or DX) and measured total hot yields in a number of systems. In our own experiments, we have utilized TBr as the target, have observed both abstraction and substitution, and have crudely estimated total yields by comparison with the earlier experiments (*).

The initial energy of the hydrogen atoms in such experiments can be controlled by the choice of the wavelength of light in photolysis, and by variation of the target molecule between DBr and DI [12-14]. This variation in initial energy has been applied by KUPPERMANN and WHITE [14] in the measurement of the threshold of the reaction $D + H_2 \rightarrow \rightarrow HD + H$. A measurement of the threshold for the substitution of T for D in CD_4 is shown in Fig. 5 [16].

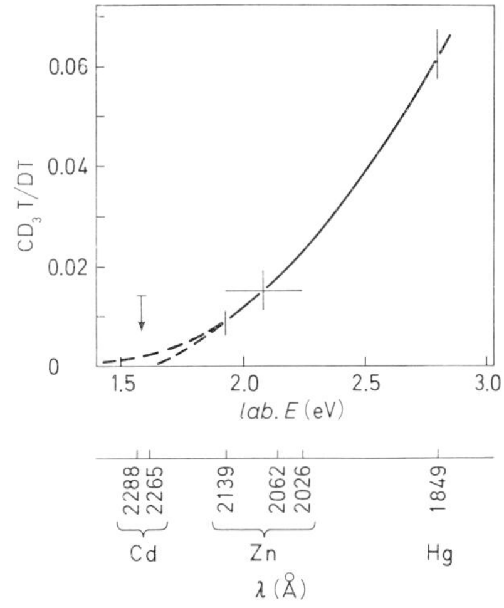

Fig. 5. – Threshold measurement for the substitution of hot tritium atoms for D in CD_4.

The third general method of acceleration that can be used is the beam method. This method has just been developed to a useful level by MENZINGER and WOLFGANG, who have accelerated both T_2^+ and T^+ into a target of solid cyclohexane, and have observed the tritiated products by radio gas chromatography [17]. The beam can be varied in energy between 200 eV and a lower limit in the vicinity of 1 eV. The yield of reaction products in their

(*) Independent measurement of total light absorbed, which could lead to estimates of number of molecules photolyzed and to total hot yield, is not yet possible since the ultra-violet absorption spectrum of TBr has not been measured.

experiments rises rapidly in the energy range from $(1 \div 10)$ eV, and more slowly above 10 eV, as shown in Fig. 6 for T^+. The reaction yields are almost flat above 20 eV, suggesting that most of the hot reactions occur below that energy. MENZINGER and WOLFGANG have also observed the threshold for the hot reaction for the production of hexenes in this system as about 5 eV.

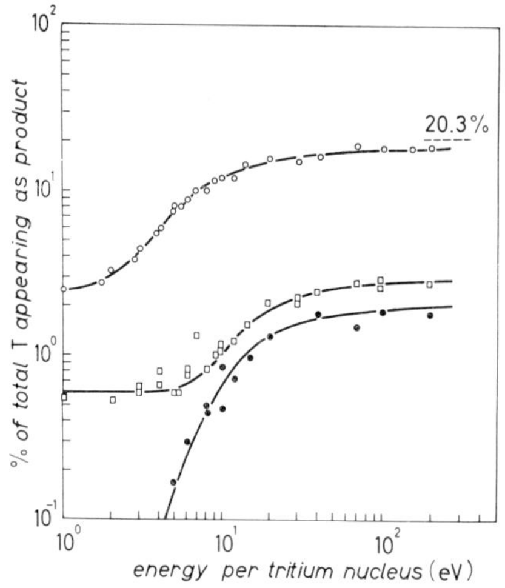

Fig. 6. – Reaction yields *vs.* initial tritium atom energy, T^+ beam into solid cyclohexane Ref. [17]. ○ Cyclohexane-T, □ n-hexane-T, • hexenes-T.

The beam method with a solid target offers considerable promise as a further tool for hot atom reactions. Since beams with « thick » targets are multiple collision systems without any possibility for angular analysis, their advantage lies in the control of the initial atom energy over the whole chemical range. The beams are controlled as ions, and the formation of neutrals is assumed to occur through charge exchange in the target material. Control of the beam energy is not as accurate in the energy range below 3 eV as in the photochemical process, but the beams operate satisfactorily over the whole energy range. The nonzero reaction yields at the lowest energy suggest that some complications still exist in this system, perhaps incomplete neutralization of the T^+ beam.

Crossed-beam, single collision experiments will be limited for some time to nonreactive scattering and to measurement of diatomic product yields, since secondary decomposition of primary substitution products will probably be very extensive in most systems at the low pressures required for crossed apparatuses [18]. Further reference to such decomposition will be made in the second lecture on hot atom chemistry.

5. – Chemical reaction range.

A hypothetical probability of reaction curve is illustrated in Fig. 7 (*). The number of collisions made by an energetic atom in the chemical reaction

(*) The basic description of hot reactions is ultimately done in terms of reaction cross-sections. So far, however, multiple collision experiments do not have any measurement of time involved in them—the atoms are formed, undergo one or many collisions, and finally react with measurement only of the chemical fate, and not of the transit

range is determined by the energy losses in nonreactive scattering collisions, both elastic and inelastic—if the energy losses are large, then fewer collisions are possible in the reactive range, and the total hot yield will be smaller. Since the energy losses in scattering processes in the $(5 \div 10)$ eV range are generally not known, interpretations of the significance of percentage hot yields are made in terms of (probability of reaction) × (average number of collisions), and not in terms of either measured separately. Numerous experiments in nuclear recoil tritium systems indicate that the total hot yields for reaction with organic molecules fall into the $(30 \div 80)\%$ range, and represent an important fraction of the total atoms formed [1-5].

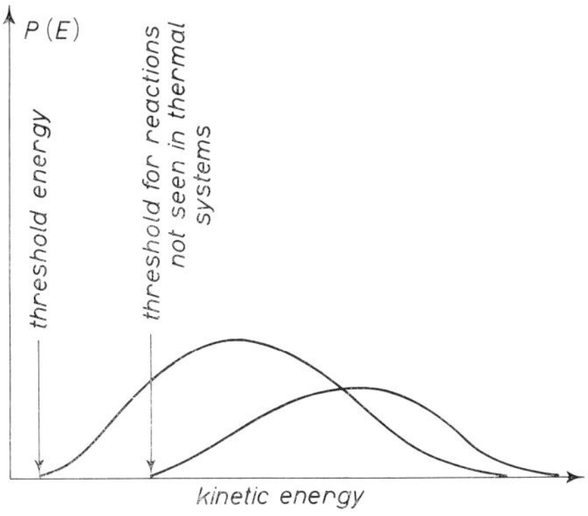

Fig. 7. – Hypothetical graph of reaction probability *vs.* energy for a system with two observable reaction products.

Many questions can be asked about such hot reactions, including such topics as their cross-sections, energy thresholds, time scales, and detailed dynamics. Several of these topics will be briefly illustrated in this first lecture, although most of the considerations of energetics will be reserved for the second lecture on hot atom chemistry.

I should also insert at this point that my selection of topics is not chosen randomly from all hot atom research, but is very much oriented toward those facets which seem to me of special interest—it will come as no surprise, I'm sure, that the facets of special interest are often ones on which my research group has worked, and the illustrations will often be made from our own work. If I wished to defend these choices on impersonal grounds, I would indicate that *a*) thermal hydrogen atom reactions are better understood by far than other atomic reactions, thus giving a wide, solid background for hot tritium studies; *b*) the thermal reactions of hydrocarbons are better understood than those of other classes of molecules, thereby favoring experiments with hydrogen or carbon atoms; *c*) the reactions of univalent species, capable of full satisfaction of bonding requirements in a single interaction, are inherently simpler to understand then the reactions of polyvalent species, which may require formation

---

time between formation and final reaction. Under these conditions, the probability of reaction, defined as the ratio of reactive scattering to total scattering, is a convenient variable for considering hot reactions. Since no angular measurements have been made either, the scattering cross-sections are all total cross-sections, integrated over angle and averaged over the thermal distribution of internal energy states.

of several bonds; and *d*) experiments in the gaseous and liquid phases are freer from additional constraints than those in crystalline materials.

Eventually, hot atom chemists will certainly want to attempt a complete understanding of the reactions of polyvalent species, and of reactions in crystals. For the present, however, the experiments with energetic tritium atoms in fluid phases offer the most detailed information about hot atom studies, and most of the examples are chosen from this special class.

One of the first questions of concern to the chemist, upon finding that recoil tritium atoms are able to substitute for hydrogen atoms in an energetic process, is the stereochemistry of this substitution process. This question has been investigated in a series of studies with molecules containing asymmetric carbon atoms (four unlike substituents), and which therefore exist in mirror-image forms. If the reaction proceeds with retention of configuration, the substitution may well be accomplished with relatively little disturbance of the rest of the molecule; on the other hand, substitution with inversion of configuration must be accompanied by substantial motion of the other three substituent groups through angles of $(30 \div 40)^\circ$ to the new positions in the product molecule.

All experiments involving tritium replacement of hydrogen have shown, as summarized in Table I, that retention is greatly preferred, occurring in greater

TABLE I. – *Recoil tritium reactions with molecules containing asymmetric carbon atoms.*

	References
A. Experiments with crystalline materials	
1) $T^* + \text{glucose} \rightarrow \text{glucose-}t + H$ Yield 12%; retention of configuration	[19]
2) $T^* + \text{glucose} \rightarrow \text{galactose-}t + H$ Yield < 0.03%; No inversion at no. 4 carbon atom	[20]
3) $T^* + L(+)\text{-alanine} \rightarrow L(+)\text{-alanine-}t + H$ At asymmetric C—H position: Yield *L*(+) 19, *D*(—) 0 ± 1 } ⩾ 95% retention	[21]
B. Experiments with gases	
4) $T^* + \text{2-butanol} \rightarrow \text{2-butanol-}t + H$ At asymmetric C—H position: Yield ⩾ 90% retention	[22]
5) $T^* + \text{2, 3-dichlorobutane} \rightarrow \text{2, 3-dichlorobutane-}t + H$ Yield: 98.8% retention Estimated at asymmetric C—H: ⩾ 95% retention	[23]
6) $T^* + (CHFCl)_2 \rightarrow CTFClCHFCl + H$ Yield: > 99.5% retention	[23*a*]

than 95 % of the substitutions [19-23]. The gas phase experiments with 2-butanol, free from any constraints of the crystalline phase, clearly established that retention is the predominant mechanism [22]. The separation of the *meso* and *dl* forms of 2,3-dichlorobutane is illustrated in Fig. 8. Since these molecules

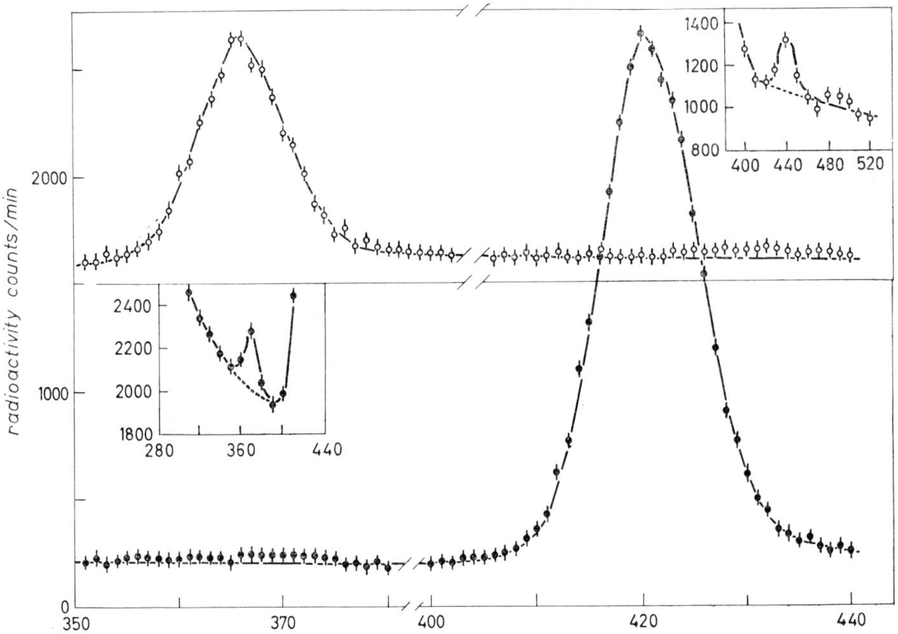

Fig. 8. – Radio gas chromatographic separation of *meso*- and *dl*-2, 3-dichlorobutane-*t* following recoil tritium reactions with 2, 3-dichlorobutane. Upper line: *meso* target; Lower line: *dl* target. Insets: same data in counts/10 min.

contain two asymmetric carbon atoms, inversion at one site does *not* produce the optical isomer, but forms the *dl* isomer from the *meso*, or vice versa. This difference permits use of the relatively simple chromatographic technique (*) for analysis [23, 23*a*].

The related stereochemical observation of a strong preference for retention of configuration during tritium substitution for hydrogen in the isomeric maleic and fumaric acids led to the suggestion [24] that « many true, 'hot' atom re-

(*) Six of the eight H atoms of 2, 3-dichlorobutane are not attached to the asymmetric carbon atoms, so the amount of inversion is estimated to be $\lessapprox (8/2)\cdot 1.2\%$ or about 5%. Even this small amount of inversion may be the result of another radical recombination mechanism not entirely suppressed by the butadiene-O_2 scavenger combination present in the irradiated system. (*Note added in proofs*: experiments with $(CHFCl)_2$ indicate that the substitution of T-for-H is accompanied by inversion in $< 0.5\%$ of the substitution reactions [23*a*].)

actions will follow mechanisms which involve, 'economy of atomic motion' »—or

$$\begin{array}{cc} \text{HOOC} & \text{COOH} \\ \diagdown & \diagup \\ \text{C}=\text{C} \\ \diagup & \diagdown \\ \text{H} & \text{H} \end{array} \qquad \begin{array}{cc} \text{HOOC} & \text{H} \\ \diagdown & \diagup \\ \text{C}=\text{C} \\ \diagup & \diagdown \\ \text{H} & \text{COOH} \end{array}$$

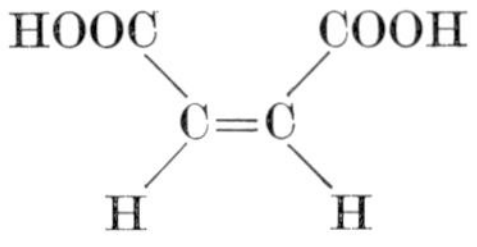

maleic acid   fumaric acid

expressed as a broad principle of the general form of the Franck-Condon principle [4, 25] « hot atom reactions requiring nuclear motions which are slow relative to the time of collision tend to be forbidden ».

Experiments with ^{38}Cl (and ^{39}Cl) also demonstrate retention of configuration in gas phase experiments with 2,3-dichlorobutane, as shown in Fig. 9 [26]. This result is especially interesting since the velocity of an energetic Cl atom is much slower than that of a tritium atom of the same energy, and one might have expected that the slower time scale for chlorine atoms would have permitted substitution with inversion of configuration.

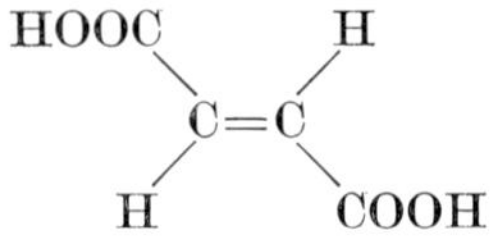

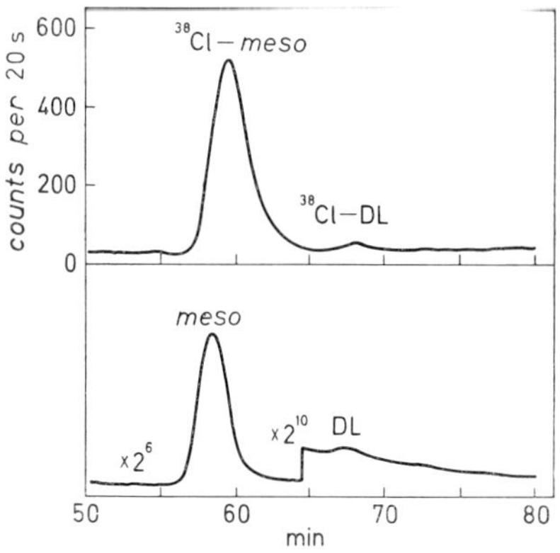

Fig. 9. – Radio gas chromatographic separation of ^{38}Cl-labeled *meso-* and *dl*-2,3-dichlorobutane following reactions of recoil ^{38}Cl with *meso* compound. Upper line: radioactivity; Lower line: macroscopic composition.

A second general type of experiment from which detailed information can be obtained involves comparisons of reaction yields from isotopic molecules. Figure 3 has already illustrated the result for one such experiment—the substitution of T for H in CH_4 is favored over T for D in CD_4 by a factor of 1.31 ± 0.03. Such a comparison includes two possible isotope effects—the *replacement* isotope effect in removing H *vs.* D, and the secondary isotope effects involved in bond formation with CH_3 *vs.* CD_3(*). Only very recently have experiments been carried out comparing hydrogen atom replacement in molecules of the type CHX_3 *vs.* CDX_3. These comparisons, involving $X = CH_3$, CD_3, and F, all demonstrate that H is more readily replaced than D by a factor of $1.25 \div 1.3$, and can have no secondary isotope effect contribution [27], since the residual CX_3 is identical in both cases.

Specific comparison of the substitution for H *vs.* D in the same molecule,

---

(*) The *substitution* isotope effect, *e.g.* H*+CD_4 *vs.* T*+CD_4, has not been measured in any system as yet.

e.g. CH_2D_2, requires the very difficult analytical separation of $CHTD_2$ *vs.* CH_2DT. This kind of experiment has not been carried out with recoil tritium, for which the isotope effect between CH_4 and CD_4 is only 1.3, but has been done with 2.8 eV tritium atoms for which the substitution isotope effect, CH_3T from CH_4 *vs.* CD_3T from CD_4, is a factor of 7. This large change is not the only variation occurring below 3 eV, for the CH_4 product ratio, HT/CH_3T, increases from 0.8 for the « infinite energy » nuclear recoil tritium to 3.6 for 2.8 eV photochemically hot tritium atoms [15]. The change in product ratios reflects the fact that 2.8 eV is quite near the energy threshold for substitution while still well above that for the thermally-observed abstraction reaction.

Competitive experiments with two parent molecules in the same system have established that the relative *total* yields for substitution (both H and D)

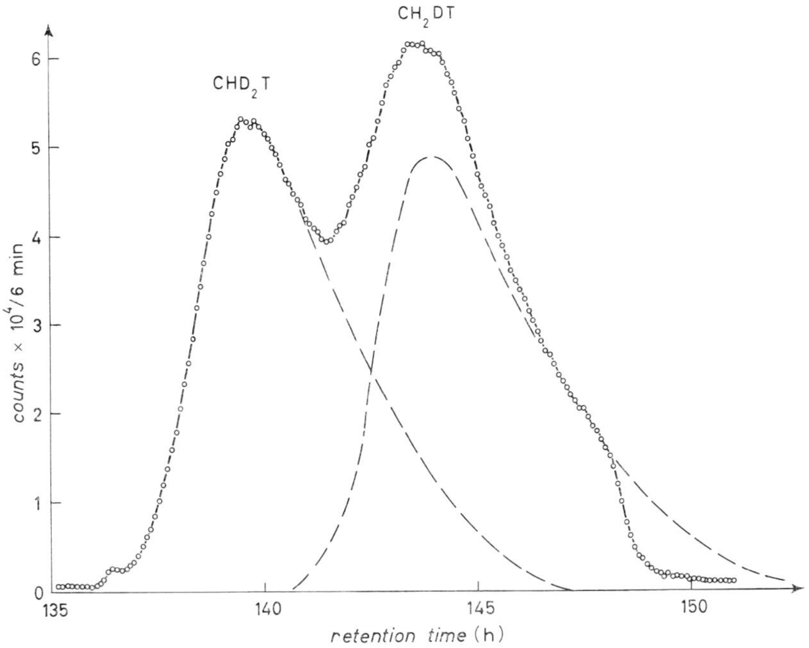

Fig. 10. – Radio gas chromatographic separation of CHD_2T and CH_2DT from reaction of 2.8 eV T with CH_2D_2. Total column length = 1200 ft. (The sudden decrease in radioactivity after 148 h is an artifact of the « column-switching » technique used in this experiment.)

by 2.8 eV tritium atoms are in the steadily decreasing ratio: CH_4, 7.2; CH_3D, 5.6; CH_2D_2, 3.1; and CD_4, 1.0. Nearly complete separation of the two products from CH_2D_2, as shown in Fig. 10, shows that the individual replacement of H is essentially equal to that of D: 1.06 ± 0.1. After taking into account that the residual methyl groups are isotopically different, CHD_2 *vs.* CH_2D, and presumably introduce a secondary isotope effect, one can conclude that the substitution of T for H or D is, at least in the energy range below 2.8 eV, not an isolated « three-particle » event ($T+RH \rightarrow RT+H$) but is very much affected by the nature and identity of both the replaced atom, and of the individual atoms in the residual radical, R. The substitution reaction yield appears to be favored by about 1.6 ± 0.2 for *each* H *vs.* D in the target methane, independent of whether the atom is the one replaced or is present in the residual radical [16*a*].

All of these isotope effects in which H atoms increase reaction yield vs. D atoms represent various cases in which the ability of the lighter atom to respond more rapidly to changing force fields facilitates the successful formation of a new chemical bond to the impacting tritium atom.

Still another type of experiment with energetic atomic species concerns the *moderator* effect—the loss of energy through elastic and inelastic nonreactive scattering from molecules in the system. Moderator experiments have generally been carried out with the inert gases, helium, neon, argon, etc., as shown in Fig. 11 and 12. Figure 11 shows the typical diminution for the hot yield from hot tritium reactions with methane *vs.* increasing mole fraction of moderator [28], while Fig. 12 shows similar results for ^{38}Cl reactions with CH_3Cl [8]. For ^{38}Cl, the diminution in yield is clearly much more rapid in Ne than in He, as is expected since the mass of Ne more closely matches that of Cl than does He. Since the experiments actually involve a comparison of the energy losses in Ne (or He) *vs.* CH_3Cl, the relative moderator efficiency of CH_3Cl can also be obtained. The fractional energy losses in nonreactive

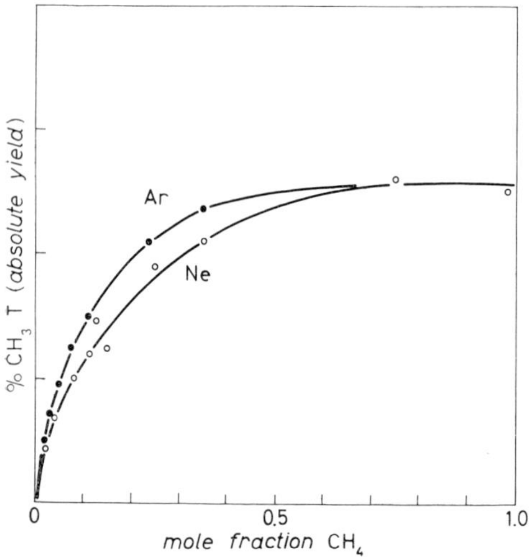

Fig. 11. – Moderator curves for the reactions of recoil tritium with CH_4, as moderated by argon and neon.

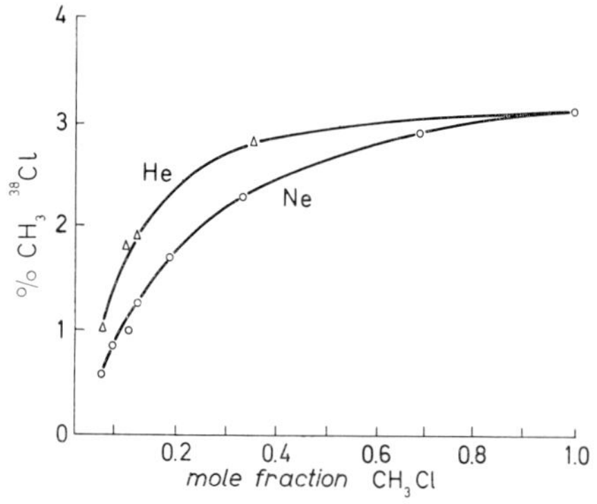

Fig. 12. – Moderator curves for the reactions of recoil ^{38}Cl with CH_3Cl, as moderated by helium and neon.

scattering from CH_3Cl are much larger than those from either inert gas ($\alpha_{CH_3Cl}=3.0$; $\alpha_{Ne}=1.0$; $\alpha_{He}=0.4$) (*). Experiments with tritium (Fig. 11) and ^{18}F also demonstrate much higher energy losses for collisions with polyatomic species than with monatomic inert gases [29]. In all cases, highly inelastic transfer of kinetic energy to the internal energy modes of the polyatomic molecules must account for the unusually large energy losses. Detailed exposition of the mathematical handling of moderator systems can be found in the references [4, 29-31].

One further type of experimental result will now be considered—the effect of the nature of the R group in RH upon the hot yield of the abstraction reaction, $T+HR \rightarrow TH+R$. Such experiments are carried out in the presence of a substantial excess of some « bulk » molecule or mixture of molecules which *a*) contains no H atoms, and hence cannot itself lead to HT; and *b*) dominates the energy loss processes and assures equivalent energy losses in systems with different RH molecules as minor components; and *c*) removes thermalized T atoms. Two examples of such molecules are $CD_2=CD_2$ [32] and *cyclo*-C_4F_6 [33]. The relative yields of HT per C—H bond from seven different saturated hydrocarbons are listed in Table II, together with the known bond dissociation energies [34] ($=\Delta H$ for the reaction $RH \rightarrow R+H$). The correlation is obviously very good for these saturated hydrocarbons.

TABLE II. – *Relative yields of* HT *from recoil tritium reaction with saturated hydrocarbons: correlation with bond dissociation energies.*

Hydrocarbon	Bond dissociation energy [34] kcal/mole	HT yield [33] (per bond)
methane, CH_4	104.0 ± 1	(1.00)
cyclopropane, C_3H_6	101 ± 3	1.21 ± 0.03
neopentane, C_5H_{12}	99.3 ± 1	1.54 ± 0.03
ethane, C_2H_6	98.0 ± 1	1.84 ± 0.05
cyclobutane, C_4H_8	95 ± 3	2.77 ± 0.10
cyclohexane, C_6H_{12}	94 ± 3	3.21 ± 0.08
cyclopentane, C_5H_{10}	93 ± 3	3.41 ± 0.11

However, we have measured the HT yields from some additional molecules, including two that show significant discrepancies from the correlation of Table II. The value for the CH_3 group in $CH_3{-}CD{=}CD_2$ is 3.4 ($\simeq 93$ kcal/mol) even though the bond dissociation energy has been well-established as about 88 kcal/mol [35]; the value for the CHF_3 molecule is 1.34 despite a bond dissociation energy of 106 kcal/mol [36].

(*) $\alpha=$ average logarithmic energy loss in nonreactive scattering.

The dynamic explanation for these differences in yield can be seen by comparing the residual radicals left behind in these two cases. The CF_3 radical left after abstraction of H from CHF_3 is already in almost the correct configuration—the F—C—F bond angles are approximately the same in CF_3 as they are in CHF_3. Consequently, no matter how rapidly an H atom is removed from CHF_3, one would expect to find the residual radical in an unexcited state. On the other hand, the residual allyl radical, CH_2—CD—CD_2, after the abstraction of H from propylene would need to change to the equilibrium allylic C—C bond distances of about 1.42 Å from the values of 1.50 and 1.34 in the original propylene molecule. In this case, if the hydrogen atom is removed from the propylene molecule before complete equilibration of the residual radical, the radical itself would be higly excited, and the HT yield correspondingly lower; we believe this to be the case. For the residual CH_3 radical left from CH_4, the equilibrium configuration is planar, and some excitation energy would be left in the radical if the abstraction reaction were completed prior to equilibration; a similar situation holds for the other hydrocarbons.

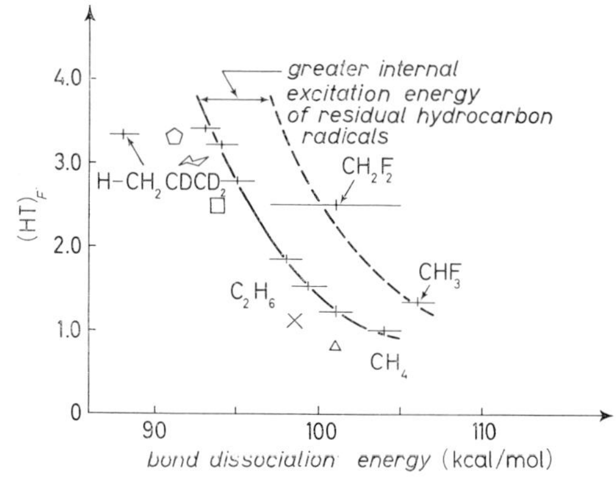

Fig. 13. – Relative HT yields from reactions of recoil tritium atoms with various molecules (methane, cyclopropane, neopentane, ethane, cyclobutane, cyclohexane, cyclopentane, CHF_3, CH_2F_2, and $CH_3CD{=}CD_2$).

We believe that the explanation for these observations is that the abstraction reaction *is* completed prior to equilibration of the residual radicals, leaving most of the hydrocarbon radicals with small amounts of excitation energy, the allyl radical with a large amount, and the CF_3 radical with almost none, as indicated in Fig. 13. The time scale for this abstraction reaction must be less than the time required for complete relaxation of these radicals, and is estimated to be a few times 10^{-14} second [35]. From this explanation, we must also conclude that one cannot successfully consider the abstraction reaction as a « three-particle » event ($T+HR \rightarrow TH+R$) because the R radicals will possess amounts of energy which are dependent upon the detailed nature and structure of the individual radical itself.

We are, of course, pleased to see the outcome of Prof. BUNKER's initial calculations of 6-atom trajectories, which show considerable residual excitation in the CH_3 radical from CH_4, since we have reached the same conclusion from the actual experimental observations.

REFERENCES

[1] *Chemical effects of nuclear transformations*, vol. **2** (I.A.E.A., Vienna, 1961).
[2] *Chemical effects of nuclear transformations*, vol. **2** (I.A.E A., Vienna, 1965).
[3] F. SCHMIDT-BLEEK and F. S. ROWLAND: *Angew. Chem.*, **76**, 901 (1964); *Angew. Chem.* (International Edition), **3**, 769 (1964).
[4] R. WOLFGANG: *Progress in Reaction Kinetics*, **3**, 97 (1965).
[5] R. WOLFGANG: *Ann. Rev. Phys. Chem.*, **16**, 15 (1965).
[6] R. WOLFGANG and F. S. ROWLAND: *Anal. Chem.*, **30**, 903 (1958).
[7] J. K. LEE, E. K. C. LEE, B. MUSGRAVE, Y.-N. TANG, J. W. ROOT and F. S. ROWLAND: *Anal. Chem.*, **34**, 741 (1962).
[8] C. M. WAI and F. S. ROWLAND: *Journ. Amer. Chem. Soc.*, **90**, 3638 (1968).
[9] E. K. C. LEE, J. W. ROOT and F. S. ROWLAND: *Chemical effects of nuclear transformations*, vol. **1** (I.A.E.A., Vienna, 1965), p. 55.
[10] A. P. WOLF: *Chemical effects of nuclear transformations*, vol. **2** (I.A.E.A., Vienna, 1965), p. 3.
[11] C. MACKAY, M. PANDOW, P. POLAK and R. WOLFGANG: *Chemical effects of nuclear transformations*, vol. **2** (I.A.E.A., Vienna, 1965), p. 17.
[12] R. J. CARTER, W. H. HAMILL and R. W. WILLIAMS: *Journ. Amer. Chem. Soc.*, **77**, 6457 (1955).
[13] R. M. MARTIN and J. E. WILLARD: *Journ. Chem. Phys.*, **40**, 2994, 3007 (1964).
[14] A. KUPPERMANN and J. M. WHITE: *Journ. Chem. Phys.*, **44**, 4532 (1966).
[15] C. C. CHOU and F. S. ROWLAND: *Journ. Amer. Chem. Soc.*, **88**, 2616 (1966).
[16] C. C. CHOU and F. S. ROWLAND: *Journ. Chem. Phys.*, **50**, 2763 (1969).
[16*a*] C. C. CHOU and F. S. ROWLAND: unpublished experiments.
[17] M. MENZIGER and R. WOLFGANG: *Journ. Amer. Chem. Soc.*, **89**, 5992 (1967).
[18] C. T. TING and F. S. ROWLAND: *Journ. Phys. Chem.*, **72**, 763 (1968).
[19] F. S. ROWLAND, C. N. TURTON and R. WOLFGANG: *Journ. Amer. Chem. Soc.*, **78**, 2354 (1956).
[20] H. KELLER and F. S. ROWLAND: *Journ. Phys. Chem.*, **62**, 1373 (1958).
[21] J. G. KAY, R. P. MALSAN and F. S. ROWLAND: *Journ. Amer. Chem. Soc.*, **81**, 5050 (1959).
[22] M. HENCHMAN and R. WOLFGANG: *Journ. Amer. Chem. Soc.*, **83**, 2991 (1961).
[23] Y.-N. TANG, C. T. TING and F. S. ROWLAND: *Journ. Phys. Chem.* (in press.)
[23*a*] G. PALINO and F. S. ROWLAND: unpublished experiments.
[24] R. M. WHITE and F. S. ROWLAND: *Journ. Amer. Chem. Soc.*, **82**, 5345 (1960).
[25] R. A. ODUM and R. WOLFGANG: *Journ. Amer. Chem. Soc.*, **85**, 1050 (1963).
[26] C. M. WAI and F. S. ROWLAND: *Journ. Phys. Chem.*, **71**, 2752 (1967).

[27] T. Smail and F. S. Rowland: *Journ. Phys. Chem.*, **72**, 1845 (1968).
[28] D. Seewald and R. Wolfgang: *Journ. Chem. Phys.*, **47**, 143 (1967).
[29] J. F. J. Todd, N. Colebourne and R. Wolfgang: *Journ. Phys. Chem.*, **71**, 2875 (1967).
[30] P. J. Estrup and R. Wolfgang: *Journ. Amer. Chem. Soc.*, **82**, 2665 (1960).
[31] R. Wolfgang: *Journ. Chem. Phys.*, **39**, 2983 (1963).
[32] J. W. Root, W. Breckenridge and F. S. Rowland: *Journ. Chem. Phys.*, **43**, 3694 (1965).
[33] E. Tachikawa and F. S. Rowland: *Journ. Amer. Chem. Soc.*, **90**, 4767 (1968).
[34] J. A. Kerr: *Chem. Rev.*, **66**, 465 (1966).
[35] E. Tachikawa, Y.-N. Tang and F. S. Rowland: *Journ. Amer. Chem. Soc.*, **90**, 3584 (1968).
[36] E. Tachikawa and F. S. Rowland: *Journ. Amer. Chem. Soc.*, **91**, 559 (1969).

Hot Atom Chemistry. - II

F. S. ROWLAND

University of California - Irvine, Cal.

1. – Unimolecular reactions.

Great attention has been given in thermal reaction systems to homogeneous, gas phase unimolecular reactions, and many of the advances in the theories of chemical kinetics have been achieved through a detailed study of these reactions. A satisfactory explanation of the apparent unimolecular behavior of many decompositions, even though initiated by bimolecular collisions, was first put forward in 1919 by LINDEMANN. This general explanation is illustrated in eqs (1) to (3):

$$\text{A} + \text{A} \xrightarrow{k_1} \text{A}^* + \text{A}\,, \tag{1}$$

$$\text{A}^* + \text{A} \xrightarrow{k_2} \text{A} + \text{A}\,, \tag{2}$$

$$\text{A}^* \xrightarrow{k_3} \text{P}\,. \tag{3}$$

The steady-state concentration of A* is determined from $\mathrm{d}A^*/\mathrm{d}t = 0 = k_1 A^2 - k_2 A A^* - k_3 A^*$ to be given by $A^* = k_1 A^2/(k_2 A + k_3)$, and the rate of formation of product is given by $\mathrm{d}P/\mathrm{d}t = k_3 A^* = k_1 k_3 A^2/(k_2 A + k_3)$.

In the high-pressure limit ($k_2 A \gg k_3$), the formation of product is first order in A, $(\mathrm{d}P/\mathrm{d}t) = (k_1 k_3/k_2) A$, and the reaction appears to be unimolecular. At lower pressures, k_3 is no longer negligible relative to $k_2 A$ and $(\mathrm{d}P/\mathrm{d}t)$ is in the « fall-off » region; at very low pressures ($k_2 A \ll k_3$), essentially all excited A* molecules react to give product, and $(\mathrm{d}P/\mathrm{d}t) = k_1 A^2$. More sophisticated treatments allow for a continuous distribution of different A* molecules, with an increasing value of k_3 for increasing excitation energy of A*. Measurements of the high pressure rates and of the « fall-off » characteristics have been made for many simple molecules; detailed discussion of both the theoretical and experimental aspects of unimolecular reactions are given in other lectures in this course.

This accumulated knowledge from thermally initiated unimolecular decompositions, and from related « chemical activation » studies, serves as background for a more detailed consideration of the energetics of hot substitution reactions. In brief, if an excited molecule from any source possesses an energy corresponding to $k_3 \sim 10^9 \div 10^{10}\ s^{-1}$, its decomposition or isomerization should be competitive with collisional stabilization in the $(0.1 \div 1.0)$ atmosphere range of gas pressures. Estimates of the energetics of hot substitution reactions can thus be obtained by a search for a pressure-dependence of the yields of hot reaction products in this pressure range.

2. – Secondary reactions of excited products from recoil substitution reactions.

Just such a pressure-dependence was first demonstrated for *cyclo*-C_3H_5T formed by recoil tritium substitution for H in *cyclo*-C_3H_6 [1]. Cyclopropane has, however, some peculiarities of behavior [1, 2], and the first detailed study of the energetics of secondary decomposition was made with *cyclo*-C_4H_8 [3]. The sequence of reactions involved for this molecule is given in (4) to (6).

$$T^* + c\text{-}C_4H_8 \rightarrow c\text{-}C_4H_7T^* + H\,, \tag{4}$$

$$c\text{-}C_4H_7T^* + c\text{-}C_4H_8 \rightarrow c\text{-}C_4H_7T + c\text{-}C_4H_8\,, \tag{5}$$

$$c\text{-}C_4H_7T^* \rightarrow CH_2{=}CHT + C_2H_4\,. \tag{6}$$

This experimental system proved to be a very well-behaved one—both $c\text{-}C_4H_7T$ and C_2H_3T were found as products, the sum of the two yields was constant, and the individual yield of $c\text{-}C_4H_7T$ increased with increasing pressure, consistent with the mechanism of reactions (4) to (6). The yields of $c\text{-}C_4H_7T$ are shown in Fig. 1, on the scale $c\text{-}C_4H_7T + C_2H_3T = 100$. As $c\text{-}C_4H_7T$ goes up in yield, that of C_2H_3T decreases

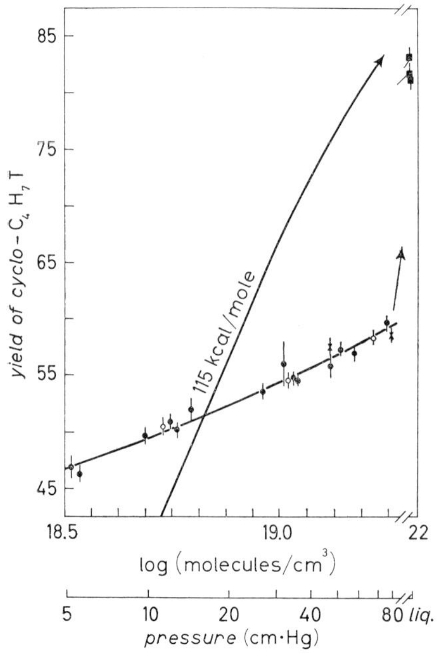

Fig. 1. – Yield of *cyclo*-C_4H_7T *vs.* pressure for reactions of recoil tritium atoms with *cyclo*-C_4H_8: •, O_2 scavenged; ○, O_2 scavenged mixture with CH_4; ■, liquid phase, without scavenger; ×, unscavenged, short radiation.

as demanded by the competitive, complementary relationship of the two in eqs. (5) and (6).

Some qualitative conclusions can be drawn from the data of Fig. 1:

a) Excited molecules of *c*-C_4H_7T are found in substantial yield after recoil tritium substitution reactions, and will decompose to C_2H_3T (+ C_2H_4) if not stabilized quickly by collision.

b) Some molecules are so excited that they decompose *even* in the rapid collision conditions of the liquid phase.

c) Measurement of primary hot yields from energetic atom reactions probably requires careful study of the extent of decomposition of these products in the particular system.

If the assumption is made that energy is rapidly equilibrated through the cyclobutane molecule, then the rate of decomposition of *cyclo*-C_4H_7T of a given excitation energy should be independent of the particular source of the original energy, and rate constants, k_6, obtained from pyrolytic or chemical activation systems should be directly applicable to the recoil tritium experiments. The values of k_6 *vs.* excitation energy are shown in Fig. 2. The pressure-dependence expected for a monoenergetic excitation energy of 5 eV (115 kcal/mole) for all *c*-$C_4H_7T^*$ is given in Fig. 1. A fourth conclusion can now be drawn from the *c*-$C_4H_7T^*$ data:

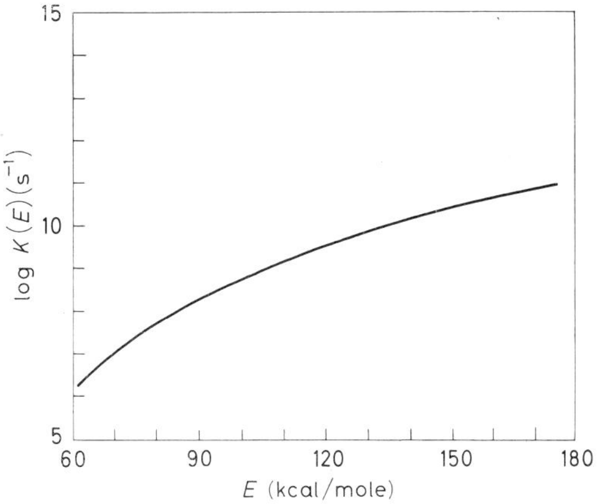

Fig. 2. – Theoretical estimates of rate constant (k_6) *vs.* excitation energy for decomposition of excited cyclobutane-*t*. Rice-Marcus model: $K(E)=A[(E-E_0+E_z^+)/(E+E_z^*)]^{n-1}$; $n=30$.

d) The excitation energies of *c*-C_4H_7T are not monoenergetic, but fall into a broad distribution with median energy around 5 eV (the 5 eV curve intersects the experimental curve near a yield of 50 %).

The kinetic energy of the reacting tritium atom is not known in these experiments. However, since the isotopic substitution of T for H is nearly thermoneutral, the high observed excitation energies imply that hot reaction can occur with appreciable cross sections to 8 or 10 eV or higher. No measure is obtained here of how much kinetic energy the departing H atom may carry away. The scale of energies for tritium atoms at reaction in the beam experiments of

MENZINGER and WOLFGANG is in reasonable consistency with these observations [4] — the reactions seem to occur chiefly in the $(5 \div 20)$ eV range.

A spectrum of excitation energies which fits the experimental data of Fig. 1 is given in Fig. 3. This spectrum is far from unique, and is generally illustrative of one kind of possible data fit. The actual distribution, as yet not established, will share several qualitative characteristics with the illustrated distribution lettered on Fig. 3: *a*) molecules with so little excitation energy ($<$ 100 kcal/mole) that they do not decompose appreciably at 6 cm pressure; *b*) molecules which survive for about 10^{-9} s and which are stabilized by 80 cm pressure, but not 6 cm; *c*) molecules which live about 10^{-11} s and which are stabilized by liquid phase collision frequencies, but not at 80 cm; and *d*) molecules so excited that they decompose even in the liquid phase.

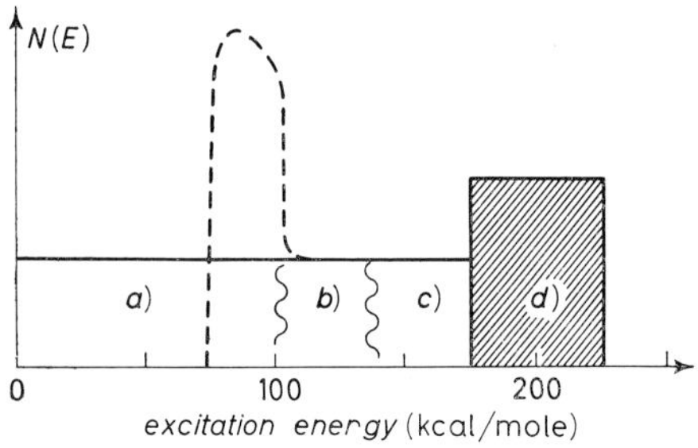

Fig. 3. – Hypothetical distribution of excitation energies of *cyclo*-C_4H_7T: ———, arbitrarily fitted to data of Fig. 1; – – –, another arbitrary fit to Fig. 1 *plus* data on CH_2TNC.

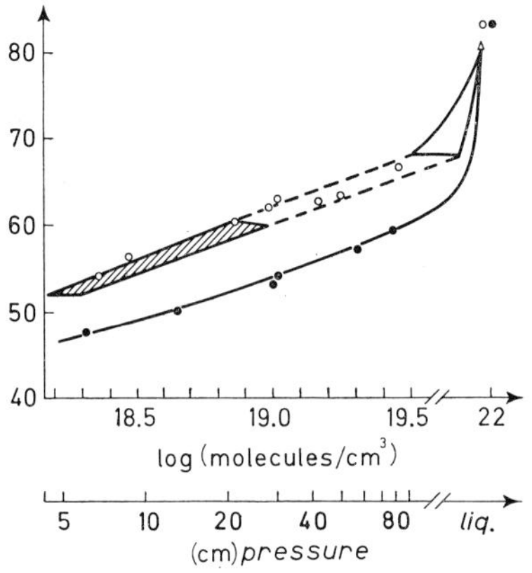

Fig. 4. – Yields of *cyclo*-C_4D_7T *vs.* pressure for reactions of recoil tritium atoms with *cyclo*-C_4D_8: ——— C_4H_8, LEE, ROWLAND (1963); •, C_4H_8; ○, C_4D_8, HOSAKA, ROWLAND (1967).

Additional information about the excitation energies of hot reaction products can be obtained through various experiments in the same style. For example, the replacement of D by T in *cyclo*-C_4D_8 has also been studied [5], as shown in Fig. 4. After correction for a lower rate of decomposition ($k_H/k_D \sim 4$) for equally excited molecules of $C_4D_7T^*$ *vs.* $C_4H_7T^*$, the data of Fig. 4 indicate roughly equivalent distributions of excitation energy for the two isotopic molecules. Measurement in solid *c*-C_4H_8 at -196 °C still shows 5% decomposition with 95% stabilization of *c*-C_4H_7T [5].

Recoil tritium experiments are generally limited by nuclear characteristics (*)

(*) The path length for 192 000 eV tritium atoms in methane is about 1 cm at 16 cm Hg pressure, and is inversely proportional to the density. At pressures below a few cm Hg, the triton range is so long that only a negligible fraction of the tritium atoms are stopped in the gas phase, and the hot reactions in the gas cannot be observed.

to gas pressures greater than a few cm Hg [6-8], prohibiting experiments at much lower cyclobutane pressures. Consequently, further information about the low energy end (segment « *a* ») of excitation energies can best be sought through choice of a target molecule with a lower activation energy for reaction. One such well-studied molecule is CH_3NC, which undergoes the isomerization reaction to CH_3CN, eq. (7), with an activation energy of only 38 kcal/mole [9]. The excitation energy corresponding to $k_7 = 10^{10}\ s^{-1}$ is only 80 kcal/mole *vs.* an excitation of 130 kcal/mole for $k_6 = 10^{10}\ s^{-1}$.

$$CH_2TNC^* \rightarrow CH_2TCN\ . \tag{7}$$

If the excitation spectrum of Fig. 3 were directly transferable to methyl isocyanide, then segments *b*), *c*) and *d*) would all be fully isomerized, or decomposed to $CH_3 + CN$, in the gas phase. The yields of CH_2TNC and CH_2TCN

TABLE I. – *Distribution of radioactivity following recoil tritium reactions with methyl isocyanide* [10]. (Yield of $HT = 100$).

Product	CH_2TNC	CH_2TCN	CH_3T	CH_2TOH	C_2H_5T
Liquid CH_3NC	19.8	14.3	7.2	<0.05	<0.1
Liquid $CH_3NC + I_2$	30.4	11.6	9.6	<0.03	<0.3
16 cm CH_3NC	<0.05	7.6	3.7	<0.01	3.7
16 cm CH_3NC, + 3 cm O_2	<0.05	8.4	3.6	1.3	<0.06
16 cm CH_3NC, 5 atm Ar, + 3 cm O_2	<0.05	15.9	4.3	1.8	<0.03

are summarized in Table I for a variety of experimental conditions [10]. Two conclusions can be reached from these data:

a) liquid-phase experiments show the formation and survival of CH_2TNC;

b) gas phase experiments show a negligible survival of CH_2TNC.

The results from *cyclo*-C_4H_8 and CH_3NC can be summed up in two successive statements: I) Recoil tritium experiments with *c*-C_4H_8 demonstrate that the T—for—H substitution products are *often* left with a high state of internal excitation. II) Recoil tritium experiments with CH_3NC demonstrate that the T—for—H substitution products are *always* left with a high state of internal excitation.

A strong temptation exists to generalize about the distribution of excitation energies following T—for—H substitution from experiments with *c*-C_4H_8 and CH_3NC. A distribution consistent with both of these experiments could be concocted merely by concentrating all of *a*) of Fig. 3 into the region between 75 and 100 kcal/mole, as shown by the dashed line of Fig. 3. However, while an assumption of a broad range of high excitation energies seems quite reasonable for T—for—H in all molecules, neither the original *cyclo*-C_4H_8 distribution, nor the new modification will fit the measured pressure-dependence for CH_3T from CH_4. The rate constant for CH_3T decomposition varies extremely rapidly with excitation energy, such that very little pressure-dependence of yield is expected or observed. Fig. 5 shows the observed data, together with the fit obtained with a broad excitation energy distribution [11, 12]. This distribution does not allow for as large a component of high excitation energies as does *c*-C_4H_8, and no single modified distribution of excitation energies can be made to fit simultaneously the data on *c*-C_4H_8, CH_3NC, and CH_4. It should further be noted that there is no experimental proof that all of the CH_2T radicals trapped by Br_2 must arise from the decomposition of excited CH_3T—the lack of appreciable pressure-dependence has also been interpreted to indicate the formation of CH_2T in a single-step, double displacement of 2 H atoms by one T atom in the initial primary step [11]. While there is as yet no experimental proof of this difficult-to-prove concept of single-step double displacement, its existence in the reactions of T* with CH_4 would only widen the gap between the *c*-C_4H_8 and CH_4 energy distributions.

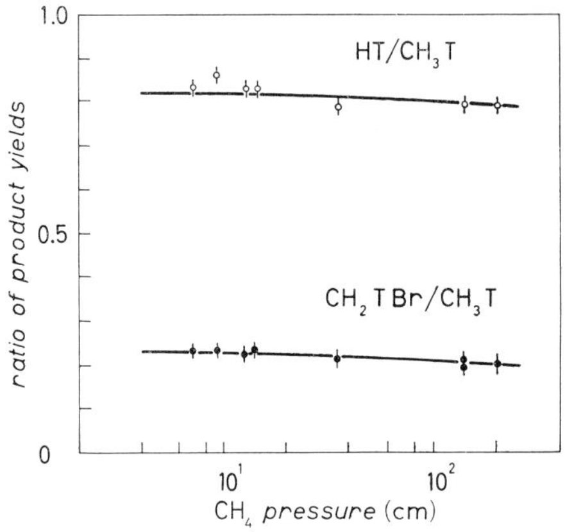

Fig 5. – Experimental pressure-dependence for reactions of recoil tritium atom. with CH_4. Solid lines are obtained from hypothetical broad energy distributions.

Nevertheless, one can generally assume that sufficient excitation energy will be deposited in T—for—H substitutions to facilitate secondary reactions of the excited products. In some instances, the point of chemical interest is reversed, and the hot atom reactions furnish a tool for the understanding of the decompositions rather than vice versa. Using recoil tritium reactions with partially-deuterated parent molecules, the mechanism of reaction (8)

$$(8) \qquad CD_3CHTF \rightarrow CD_2{=}CHT + DF$$

has been demonstrated to go as written (*), and not through $CD_3CHTF \rightarrow$ $\rightarrow CD_3CT\!: + HF$ [13]. On the other hand, the excited molecule $CHTF_2$ does

(*) The actual experiment also involves the formation of $CD_2TCH_2F^*$, but no ambiguity is introduced into the final interpretation.

decompose by loss of HF with the formation of the previously unavailable CTF (*). This reactive chemical species attacks the olefinic positions of molecules with the stereospecific retention of the substituent R group position in the resulting substituted cyclopropanes [14], as shown in reaction (9) and Fig. 6.

(9)

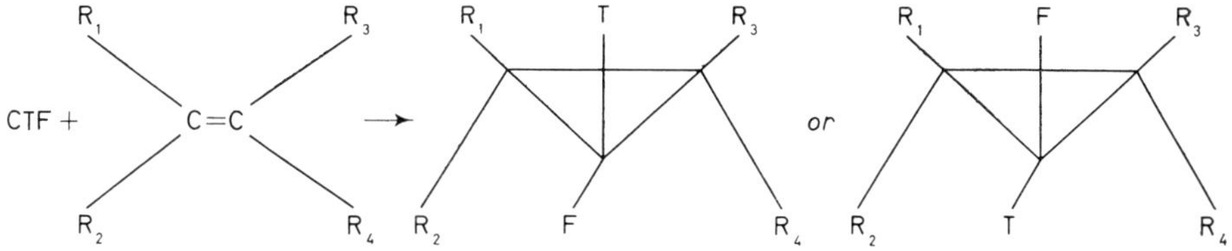

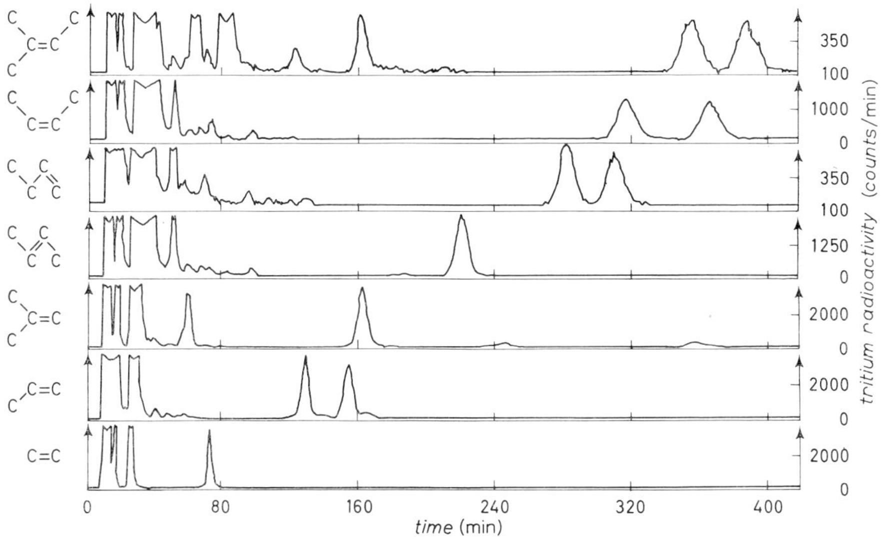

Fig. 6. – Radioactive products from recoil tritium reactions with CH_2F_2—olefin mixtures. The single or double peaks at the right of each line represent various fluorocyclopropane-*t* molecules.

The phenomenon of high product excitation is not confined to the T—for—H reaction, but is a very general phenomenon, occuring for: *a*) replacement of other groups by tritium, *e.g.* Cl, CH_3 [15, 16]; *b*) addition of tritium to π-bonds, [17] as in the sequence of (10) and (11):

(10) $$T^* + CH_3CH{=}CHCH_3 \rightarrow CH_3CHT\dot{C}HCH_3^* ,$$

(11) $$CH_3CHT\dot{C}HCH_3^* \rightarrow CH_3 + CHT{=}CHCH_3$$

(*) Decomposition also occurs to CHF and TF, but is not studied since the radioactivity is no longer attached to the C atom. (The analytical system does not directly record the yield of TF or any reaction products derived from it.)

c) substitution of ^{18}F for F or H in hydrocarbons [18] or fluorocarbons; *d*) substitution of ^{38}Cl for Cl or H [16].

The combined effects of temperature and pressure on the terminal addition of T* to 1-butene, as shown in eq. (12), and the

$$T^* + CH_3CH_2CH{=}CH_2 \rightarrow CH_3\,CH_2CHCH_2T^* \tag{12}$$

subsequent decomposition to $CH_2TCH{=}CH_2 + CH_3$, are illustrated in Fig. 7. The mere fact that the extra energy content (0.1 eV more at 124 °C than 25 °C) is experimentally significant implies that the kinetic energy of the reacting tritium atom is not in the (5÷10) eV range. The best fit to the data of Fig. 7 suggests that more than 70 % of the tritium atoms reacting by (12) have less than 3 eV kinetic energy at reaction [19].

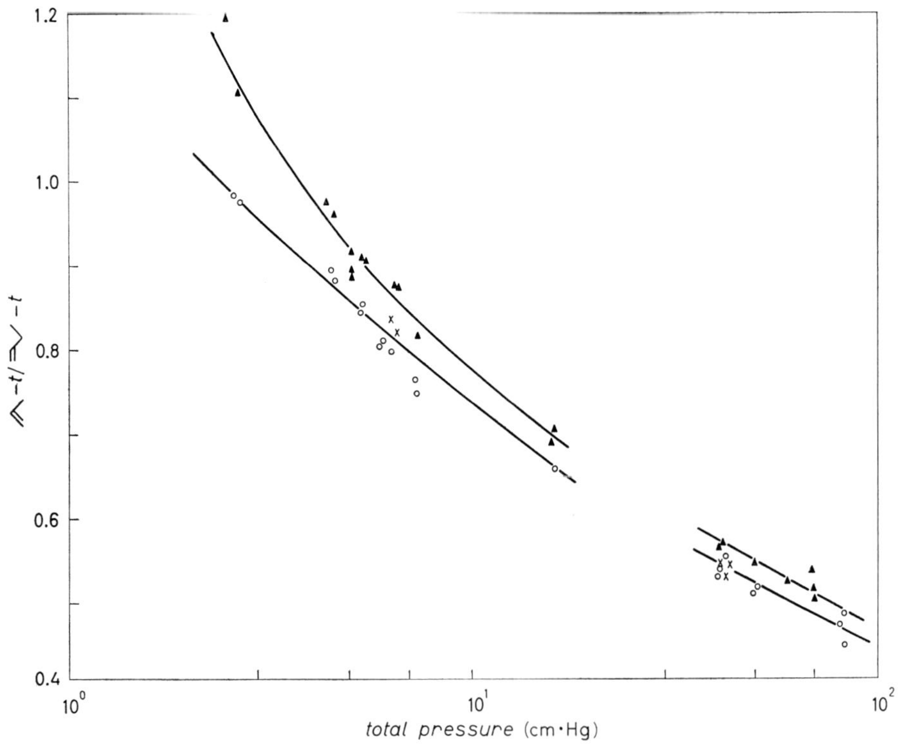

Fig. 7. – Pressure and temperature-dependence of recoil tritium reactions with 1-butene ○, 24 °C; ▲, 125 °C; ×, heated after irradiation.

Measurement of the energetics of ^{18}F (or ^{38}Cl) substitutions for H in hydrocarbons is difficult, since many of the resulting excited molecules decompose by elimination of $H^{18}F$ and leave no trace of their short-lived existence.

However, target molecules containing more than one F atom do leave reactive species, such as $CH^{18}F$ and $CF^{18}F$ from the decomposition of $CH_2F^{18}F^*$ or $CHF_2{}^{18}F^*$, and these can be identified by their characteristic reactions with olefins, as shown in Fig. 8. The measured yields of $CH^{18}F$ and $CF^{18}F$ indicate

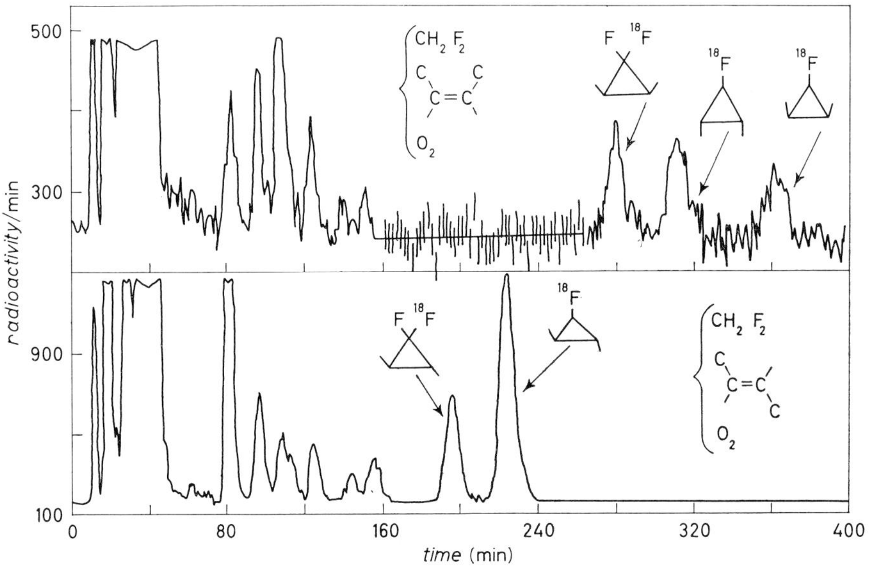

Fig. 8. – Radioactive products from recoil ^{18}F reactions with CH_2F_2—butene mixtures.

that $>80\%$ of the excited molecules $CH_2F^{18}F^*$ and $CHF_2{}^{18}F^*$ decompose at one atmosphere pressure [20]. These observations indicate that the yields of primary hot reactions will be very hard to estimate from experiments even at pressures as high as a few atmospheres.

3. – Is energy always randomized prior to reaction?

I shall now return, in concluding this discussion of the excitation energy of hot-atom products, to a consideration of the question of randomization of energy in an excited molecule.

Are we really justified in assuming, as we have tacitly done throughout this lecture, that the excited molecules formed by the energetic substitution of tritium atoms are indistinguishable in behavior from those energized by pyrolysis, chemical activation or other means?

And, if this assumption is reasonable for molecules decomposing after 10^{-9} s, is it reasonable for decomposition in 10^{-12} or 10^{-13} s?

Consider an excited molecule of isopropyl chloride, which decomposes by elimination of HCl, with the H atom coming from either end of the molecule and the Cl atom from the central carbon atom, as illustrated:

$$\text{CH}_2\text{—CH—CH}_2 \qquad \text{or} \qquad \text{CH}_2\text{—CH—CH}_2$$
$$\;\;|\qquad\;|\qquad|\qquad\qquad\qquad\;\;|\qquad\;|\qquad|$$
$$\;\;(\text{H}\quad\text{Cl})\quad\text{H}\qquad\qquad\qquad\;\;\text{H}\quad(\text{Cl}\quad\text{H})$$

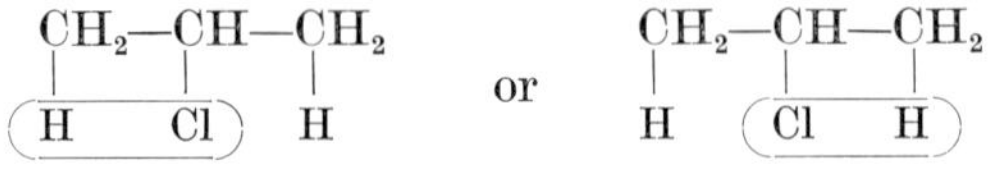

If a tritium atom substitutes for H in isopropyl chloride to form an excited

$$\overset{1}{\text{C}}\text{HT—}\overset{2}{\text{CH}}\text{—}\overset{3}{\text{CH}}_2$$
$$\;\;|\qquad\;|\qquad|$$
$$\;\;\text{H}\qquad\text{Cl}\qquad\text{H}$$

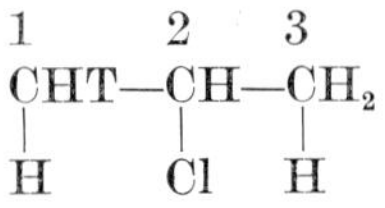

$CH_2TCH\,ClCH_3$ molecule, the initial excitation energy is introduced into the environment of C-1. Randomization over the molecule will make, excluding T/H isotope effects, loss of an individual H atom from C-1 or C-3 equally probable. If HCl is lost from C-1 and C-2, the surviving propylene is $CHT=CH—CH_3$, while loss from C-3 and C-2 leaves $CH_2TCH=CH_2$. These two molecules can be separated by radio gas chromatography, as shown in Fig. 9, and the relative yield of each determined.

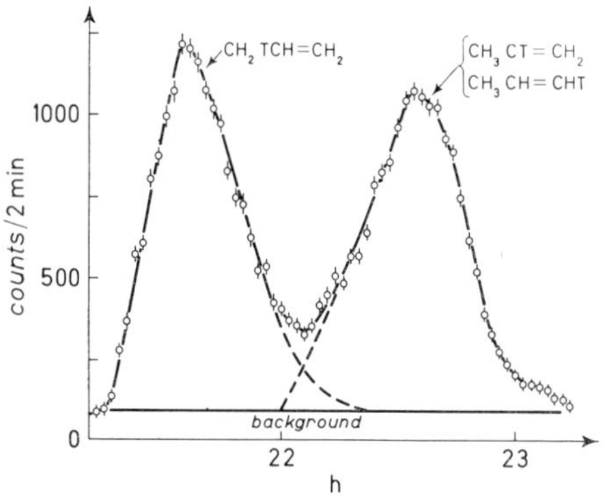

Fig. 9. – Radio gas chromatographic separation of isomers of propylene-t.

Reaction at 5 cm pressure in the gas phase results in about 4% of the tritium atoms being found as propylene-t, and $50.4\pm0.6\%$ of this is $CH_2TCH=CH_2$. On the other hand, reaction in the liquid phase gives a propylene-t yield of only 1%, of which $45.0\pm0.9\%$ is $CH_2TCH=CH_2$. (*) The chief conclusion that has been reached from these experiments is that equilibration of energy is essentially complete in the gas phase, and is very nearly so in the liquid phase [12]. Since the collision rate in liquids is about 10^{12} s^{-1}, the decomposition reaction must be proceeding on a time scale of $<10^{-12}$ s—nevertheless, energy equilibration is practically complete prior to the decomposition by loss of HCl.

(*) A value of about 50% alkyl-t is reasonable—if all 7 H's were replaced with equal probability, and then all 6 end atoms were lost with equal probability, one should find 3 parts $CH_2TCH=CH_2$, 1 part $CH_3CT=CH_2$ from substitution at C-2, and 2 parts $CH_3CH=CHT$, with 1 part lost as TCl. Isotope and position effects can cause minor deviations.

However, the second, and tentative, conclusion is that the measurably lower % alkyl in propylene-*t* (45.0 *vs.* 50.4) from the liquid phase represents a small contribution from the decomposition of propylene-*t* molecules in which the energy from substitution is still preferentially localized in the C-1 environment, with the result that loss of H from C-1 is still somewhat preferred over loss of H from C-3. This reaction *may* be an example of a secondary decomposition occurring on a time scale competitive with intramolecular equilibration of energy.

The deposition of energy in the substitution of T for H offers a wide range of time scales:

i) consecutive, multiple collisions with atoms of the same molecule, with loss of two groups, either separately or together—the single-step double-displacement reaction mentioned earlier [11, 12]; *ii*) single substitution leaving a molecule with the C-T bond near the dissociation level, followed after one or a few vibrations, by secondary decomposition with preferential loss of either the T atom, or an adjacent H atom; *iii*) single substitution, followed by randomization of energy, and secondary decomposition from an excited, internally equilibrated molecule. The evidence for iii) is present in most of the systems thus far investigated; no firm evidence exists for either i) or ii) so far, although minor anomalies such as the results with isopropyl chloride may represent such a contribution. In any event, the assumption of internal randomization of energy in hot atom systems is fully justified as a general assumption of wide, if perhaps not quite universal, validity.

REFERENCES

[1] J. K. Lee, B. Musgrave and F. S. Rowland: *Journ. Amer. Chem. Soc.*, **81**, 3803 (1959); *Canad. Journ. Chem.*, **38**, 1756 (1960).

[2] E. K. C. Lee and F. S. Rowland: unpublished experiments.

[3] E. K. C. Lee and F. S. Rowland: *Journ. Amer. Chem. Soc.*, **85**, 897 (1963).

[4] M. Menzinger and R. Wolfgang: *Journ. Amer. Chem. Soc.*, **89**, 5992 (1967).

[5] A. Hosaka and F. S. Rowland: *Fourth Informal Hot Atom Chemistry Conference, Kyoto, Japan, October* 1967.

[6] F. Schmidt-Bleek and F. S. Rowland: *Angewandte Chemie*, **76**, 901 (1964); *Angew. Chem.* (International Edition), **3**, 769 (1964).

[7] R. Wolfgang: *Progress Reaction Kinetics*, **3**, 97 (1965).

[8] R. Wolfgang: *Ann. Rev. Phys. Chem.*, **16**, 15 (1965).

[9] F. W. Schneider and B. S. Rabinovitch: *Journ. Amer. Chem. Soc.*, **84**, 4215 (1962); **85**, 2365 (1963); B. S. Rabinovitch, P. W. Gilderson and F. W. Schneider: *Journ. Amer. Chem. Soc.*, **87**, 158 (1965).

[10] C. T. Ting and F. S. Rowland: *Journ. Phys. Chem.*, **72**, 763 (1968).
[11] D. Seewald and R. Wolfgang: *Journ. Chem. Phys.*, **47**, 143 (1967).
[12] Y.-N. Tang and F. S. Rowland: *Journ. Phys. Chem.*, **72**, 707 (1968).
[13] Y.-N. Tang and F. S. Rowland: *Journ. Amer. Chem. Soc.*, **90**, 570 (1968).
[14] Y.-N. Tang and F. S. Rowland: *Journ. Amer. Chem. Soc.*, **89**, 6420 (1967).
[15] Y.-N. Tang and F. S. Rowland: *Journ. Amer. Chem. Soc.*, **87**, 3304 (1965).
[16] C. T. Ting and F. S. Rowland: *Journ. Phys. Chem.* (in press.)
[17] D. Urch and R. Wolfgang: *Chemical Effects of Nuclear Transformations*, vol. **2** (Vienna 1961), p. 99.
[18] Y.-N. Tang and F. S. Rowland: *Journ. Phys. Chem.*, **71**, 4576 (1967).
[19] R. Kushner and F. S. Rowland: *Journ. Amer. Chem. Soc.*, **91**, 1539 (1969).
[20] Y.-N. Tang, T. Smail and F. S. Rowland: *Journ. Amer. Chem. Soc.*, **91**, 2130 (1969).

Kinematics of Ion-Molecule Reactions.

A. HENGLEIN

Hahn-Meitner-Institut für Kernforschung - Berlin

1. – Introduction.

Ion-molecule reactions have been known since the early days of mass spectroscopy, *i.e.* for three or four decades [1]. For example, the H_3^+ ion has been observed in the mass spectrum of the ions produced in a hydrogen discharge. It is formed in a bimolecular reaction between H_2^+ and H_2:

$$H_2^+ + H_2 \rightarrow H_3^+ + H\,.$$

The discovery in 1952 of an ion-molecule reaction in an organic system, *i.e.* the reaction

$$CH_4^+ + CH_4 \rightarrow CH_5^+ + CH_3\,,$$

marks the beginning of more comprehensive studies of ion-molecule reactions [2]. In the following decade, hundreds of ion-molecule reactions have been found to occur in conventional mass spectrometers when these are operated at pressures of $(10^{-3} \div 10^{-4})$ Torr in the ion source. In fact, there is nearly no organic substance for which ion-molecule reactions do not occur. Their investigation yielded much information about the reactions of positive and negative ions. This information proved to be very useful for the understanding of various fields of chemistry in which gaseous ions are involved such as radiation chemistry, electric discharges, flames and the chemistry of space and the upper atmosphere. As important as these aspects were the contributions of ion-molecule studies to the field of chemical kinetics in which we are mainly interested in this course.

A common feature of these studies during the first decade after 1952 was the use of conventional mass spectrometers. The ions were formed by electron impact on gas molecules in a ionization chamber, reacted with gas molecules in the same chamber, then were extracted by a weak electric field, accelerated up to a few thousands volt and mass-analysed. In recent years, *i.e.* after about 1962, more sophisticated apparatus have been developed to study ion-molecule reactions. There are three major improvements which should be mentioned:

1) The ions were produced by electron impact in an ionization chamber but reacted with neutral molecules either in another chamber or in a molecular beam. The velocity and angular distribution of the product ions were then determined without acceleration of the product ions after their formation. These studies have given substantial information about the kinematics of ion-molecule reactions and will mainly be dealt with in this paper.

2) Sources have been developed in which the ions are produced by photo-ionization of gas molecules. By using monochromatic UV light it is possible to study reactions of ions in well defined excited states [3].

3) Apparatus have been developed where the ions react with gases under rather high pressures in the collision chamber, *i.e.* generally at about 1 Torr. In these apparatus consecutive ion-molecule reactions can be observed, *i.e.* the primary ion reacts with a molecule to form a secondary ion which in turn reacts to form a tertiary ion and so on [4]. In some devices, pressures of several 100 Torr have been reached in the collision chamber. Under these conditions, the equilibrium between the various product ions is often reached, for example, the various hydration equilibria of hydronium ions in water vapor [5]:

$$H_3O^+ + H_2O \rightleftarrows H_5O_2^+ \underset{}{\overset{H_2O}{\rightleftarrows}} H_7O_3^+ \overset{H_2O}{\rightleftarrows} H_9O_4^+ .$$

4) « Flowing afterglow » apparatus have been utilized to measure rate constants of ion-molecule reactions at thermal energies. Ions are produced in a flowing gas stream of rare-gas atoms and then mixed with a second gas. From the primary- and product-ion intensities, absolute rate constants for product formation may be determined with an accuracy of better than 30% when the flow dynamics of the system is taken into consideration [6].

This lecture will be concerned mainly with the details of the reaction mechanism for ion-molecule reactions. In this respect there are many similarities to mechanisms of molecule-molecule interactions, despite the fact that ion-molecule reactions are generally studied at energies substantially higher than neutral reactions. In both cases there exist reactions involving complexes of lifetimes longer than one period of rotation as well as those proceeding by a direct mechanism. The polarization theory will be presented and its limitations and agreement with experimental data will be shown. Several types of direct reaction theories and the energy ranges at which they are applicable will be discussed and compared with the experimental results. Finally, examples will be presented of reactions which show a change from a direct to a complex mechanism as the relative kinetic energy is decreased.

2. – Conventional studies of ion-molecule reactions.

2·1. *Experimental procedures.* – Figure 1 shows a schematic drawing of the ion source of a conventional mass spectrometer. The filament F emits electrons which are accelerated by the voltage U applied between F and the ionization chamber C. The electrons traverse the chamber as a narrow beam and are finally collected on the trap electrode T. The electron current amounts to a few microampere. The chamber C contains the gas at a pressure of about 10^{-4} Torr. Only a small percentage of the electrons collides with gas molecules. The ionization can lead to the « parent » ion of a molecule R_1R_2 or to a « fragment » ion R_1^+ or R_2^+:

$$R_1R_2 + e^- \begin{cases} \longrightarrow R_1R_2^+ + 2e^- & (2.1) \\ \longrightarrow R_1^+ + R_2 + 2e^- & (2.2) \\ \longrightarrow R_1 + R_2^+ + 2e^- . & (2.3) \end{cases}$$

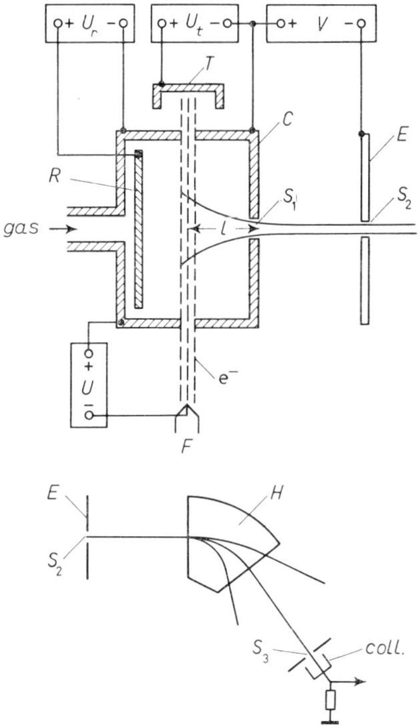

Fig. 1. – Schematic representation of the ionization chamber and of the analysing section of a conventional mass spectrometer.

In polyatomic molecules many fragment ions are formed, the relative abundances of which constitute the « mass spectrum » of the compound. These ions are called « primary » ions. The ions are pushed out of the chamber by the weak electric field across the chamber which is produced by the voltage U_r between the repeller electrode R and the chamber C. They pass through the slit S_1 of the chamber and are now accelerated by the high voltage V between C and the exit electrode E of the ion source. After having passed through the slit S_2 of the electrode, they fly as a narrow beam into the magnetic sector field H and are mass-analysed.

At the detector the primary ion current, i_p, is proportional to the concentration, $[M]$, of the gas molecules in the chamber C, *i.e.*

$$i_p \sim [M] . \tag{2.5}$$

The product ions are formed along the path l of the primary ions between the electron beam and the slit S_1 of the chamber C and are observed in the mass spectrum. The intensity of the secondary ions generally is smaller by two or more orders of magnitude than that of the primary ions, since at 10^{-4} Torr the mean free path is much higher ($\approx (10 \div 100)$ cm) than the length of l (a few mm), *i.e.* only a small percentage of the primary ions undergoes collision with gas molecules along l. The intensity of a secondary ion is proportional to the cross-section q of the ion-molecule reaction, the primary ion current, and the concentration of the gas molecules:

$$i_s \sim q \cdot l \cdot i_p \cdot [M] . \tag{2.6}$$

Therefore

$$i_s \sim [M]^2 \tag{2.7}$$

and

$$\frac{i_s}{i_p} \sim q . \tag{2.8}$$

It should be remarked that only bimolecular reactions are observed at a pressure of about 10^{-4} Torr, since the probability of a subsequent reaction of a secondary ion with a gas molecule is extremely low. Furthermore, since $i_s \ll i_p$, only those ion-molecule reactions can be traced which occur at nearly every collision, *i.e.* which have practically no activation energy. In other words, only exothermic or thermoneutral reactions will generally be detected in a conventional mass spectrometer.

The cross-section q represents a «phenomenological» cross-section, *i.e.* an average cross-section of the true cross-section, σ, at primary ion energies E between about zero eV and the energy $eF_r \cdot l$, where F_r is the strength of the repeller field across the chamber:

$$q = \frac{1}{eF_r \cdot l} \int_0^{F_r \cdot l \cdot e} \sigma_{(E)} \cdot \mathrm{d}E . \tag{2.9}$$

If the dependence of the cross-section, σ, on the energy of the incident ion is

$$\sigma = K \cdot E^a , \tag{2.10}$$

where K and a are constants, the integration of eq. (2.9) leads to

$$q = \frac{K}{1+a} \cdot (e \cdot F_r \cdot l)^a . \tag{2.11}$$

Since F_r is proportional to U_r, it follows from eqs. (2.8) and (2.11)

$$\frac{i_s}{i_p} \sim U_r^a . \tag{2.12}$$

This relationship has often been used to determine the exponent a in eq. (2.10).

In Fig. 2 it can be seen that a varies from -0.38 for the reaction $N_2^+ + N_2 \to N_3^+ + N$, to -0.5 for $Ar^+ + H_2 \to ArH^+ + H$ to 0.9 for $H_2O^+ + H_2O \to \to H_3O^+ + OH$ and to -1.2 for $C_3H_3N^+ + C_3H_3N \to C_4H_4N_2^+$ (sticky complex formation for acrylonitrile). Some other reactions which occur in acrylonitrile have exponents between -0.9 and 0.7 [7]. The highest primary ion energy was equal to about 6 eV. Measurements of this kind give only an average value of the exponent a, if the exponent changes in the energy range studied [8].

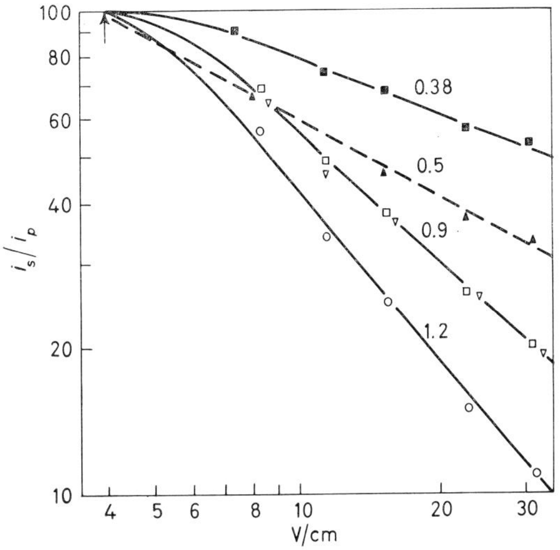

Fig. 2. – Current ratio i_s/i_p for various ion-molecule reactions as a function of the repeller field strength on a double logarithmic scale. ▲ $Ar^+ + H_2 \to ArH^+ + H$; □ $H_2O^+ + H_2O \to \to H_3O^+ + OH$; ▽ $C_3H_3N^+ + C_3H_3N \to C_6H_5N_2^+ + H$; ○ $C_3H_3N^+ + C_3H_3N \to C_6H_6N_2^+$; ■ $N_2^{+*} + N_2 \to \to N_3^+ + N$.

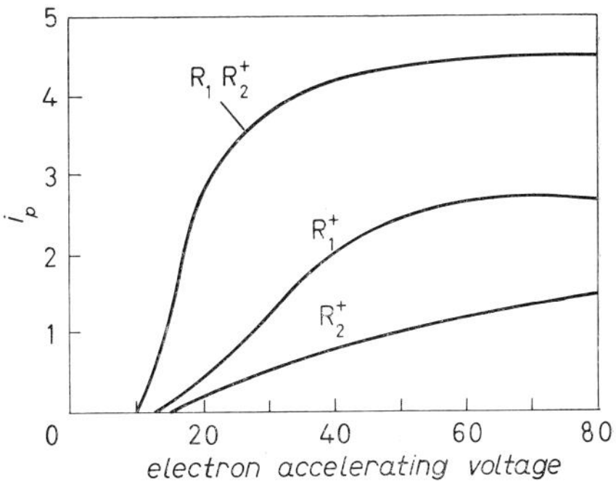

Fig. 3. – Schematic representation of ionization efficiency curves for the parent ion $R_1R_2^+$, and fragment ions R_1^+ and R_2^+.

Figure 3 shows, how the ion current depends on the electron accelerating voltage U. The parent ion $R_1R_2^+$ appears at $U = I(R_1R_2)$, where I is the vertical ionization potential of R_1R_2. At higher voltages, the cross-section for ionization increases up to a maximum at about 70 V. Fragment ions appear at higher voltages, since the molecule has to be dissociated as well as ionized by electron impact. Since the current i_s of a secondary ion is proportional to the current i_p of its precursor ion (eq. (2.6)), the ionization efficiency curve of a secondary ion has to have identical shape to that of the corresponding primary ion. By comparing the ionization efficiency curves of the primary ions with that of a secondary ion, one can often find out which primary ion was the precursor.

2·2. *Various types of reactions of positive ions.*

A) H atom or proton transfer:

(2.13) $$H_2^+ + H_2 \rightarrow H_3^+ + H\,,$$

(2.14) $$H_2O^+ + H_2O \rightarrow H_3O^+ + OH\,,$$

(2.15) $$CH_4^+ + CH_4 \rightarrow CH_5^+ + CH_3\,,$$

(2.16) $$Ar^+ + H_2 \rightarrow ArH^+ + H\,,$$

(2.17) $$H_2^+ + Ar \rightarrow ArH^+ + H\,,$$

(2.18) $$C_3H_8^+ + H_2O \rightarrow H_3O^+ + C_3H_7\,.$$

The products are « onium »-ions of high stability. The ArH^+ ion, for example, is comparable to the hydrogen chloride molecule since Ar^+ is isoelectronic with Cl. The observation of a proton transfer reaction enables one to calculate a minimum value of the proton affinity, *i.e.* the binding energy of the proton to the acceptor molecule. For example, since reaction (2.14) occurs with a large cross-section, one has to conclude that it is exothermic, *i.e.* the enthalpy level of the system $H_3O^+ + OH + e^-$ must lie below that of $H_2O^+ + H_2O + e^-$ (Fig. 4). From Fig. 4 one obtains for the proton affinity, $PA(H_2O)$, of water

(2.19) $$PA(H_2O) \geqslant D(\text{H-OH}) + I(H) - I(H_2O)\,,$$

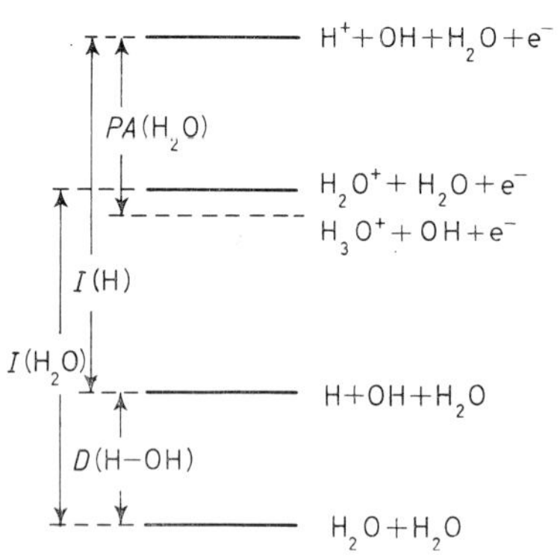

Fig. 4. – Energy level diagram for water.

where the *D* and *I* are well-known dissociation and ionization energies. In an analogous manner standard heats of the formation of gaseous ions have often been calculated from the combination of thermodynamic data and observations of ion-molecule reactions [9].

B) Hydride (H^-) transfer from the molecule to the ion:

(2.20) $$C_3H_5^+ + \text{neo-}C_5H_{12} \rightarrow C_5H_{11}^+ + C_3H_6 \,.$$

C) Condensation reactions (formation of product ions of large size):

(2.21) $$CH_3^+ + CH_4 \rightarrow C_2H_5^+ + H_2 \,,$$

(2.22) $$C_2H_4^+ + C_2H_4 \rightarrow C_3H_5^+ + CH_3 \,,$$

(2.23) $$CH_3I^+ + CH_3I \rightarrow C_2H_6I^+ + I \,,$$

(2.24) $$Xe^+ + CH_4 \rightarrow XeCH_3^+ + H \,.$$

2·3. *Reactions of excited ions.* – Certain reactions can be observed only if the ion is in an electronic [10-13] or vibrational excited state [14]. The ionization efficiency curve of the secondary ion has another shape than that of the primary ion. The appearance potential of the secondary ion is higher than that of the primary ion and corresponds to the appearance potential of the excited primary ion.

An excited state $CS_2^{+*}(^2\Pi_{\frac{3}{2}})$ is formed in CS_2 at 13.5 eV, while the ionization energy of CS_2 is 10.1 eV. Several chemical reactions of the excited CS_2^+ have been observed. For example, in the reaction with water the ion H_2OS^+ is produced (Fig. 5). It is seen that CS_2^+ in its ground state cannot be the precursor

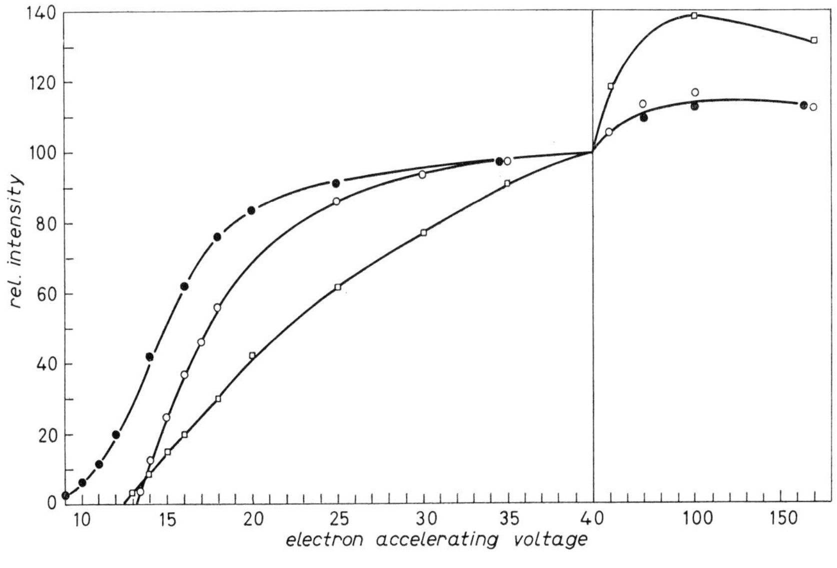

Fig. 5. – Ionization efficiency curves fo various ions formed in a CS_2-water-mixture (all curves normalized at 40 V). $CS_2^{+*} + H_2O \rightarrow H_2OS^+ + CS$; • CS_2^+, ○ H_2OS^+, □ H_2O^+.

of H_2OS^+, since the latter does not appear at the onset of the CS_2^+ curve. H_2OS^+ is first formed when the electron accelerating voltage reaches the appearance potential of CS_2^{+*} of 13.5 eV. Since water has a ionization potential of 12.6 V, the reaction $H_2O^+ + CS_2 \rightarrow H_2OS^+ + CS$ can be excluded. H_2OS^+ must be formed by the process

$$(CS_2^+)^* + H_2O \rightarrow H_2OS^+ + CS \; . \tag{2.25}$$

In some cases the excitation is in the form of vibrational instead of electronic energy. One example of such a reaction which is exothermic only if the incident ion is in a higher excited vibrational level is the reaction of the hydrogen ion with neon [14]:

$$H_2^+ + Ne \rightarrow NeH^+ + H \; . \tag{2.26}$$

2·4. *Statistical complexes in ion-molecule reactions.* – According to the statistical theory of the mass spectra of polyatomic molecules, the fragmentation does not occur simultaneously with the ionization of a molecule by electron impact. The excess vibrational energy of the ion may be distributed among the various degrees of freedom for a rather long time until energy is in a single bond to break it. In fact, some dissociation processes require more than 10^{-6} s which is comparable to the time of flight of the ions through the instrument (« metastable » ions). The decay of excited polyatomic ions is described by the usual theories of unimolecular rate processes. If a complex is formed in an ion-molecule reaction, the decomposition pattern of the intermediate complex should be similar to the mass spectrum of a molecule of the same composition as the complex. For example, the reaction of $C_2H_4^+$ with C_2H_4 leads to several product ions. If $C_2H_4^+$ and C_2H_4 at first form a complex, $C_4H_8^+$, the fragmentation pattern of this complex should be comparable to the mass spectrum of C_4H_8 (1-butene, cis-butene-2, isobutene, c-butane). Table I shows the mass

TABLE I. – *Comparison of primary and secondary mass spectra.*

Ion	1-butene	cis-butene-2	isobutene	c-butane	from $C_2H_4^+ + C_2H_4$
$C_3H_4^+$	6.5	7	11	6.5	1.5
$C_3H_5^+$	100	100	100	100	100
$C_4H_6^+$	2.5	4	2.5	3	0.2
$C_4H_7^+$	18	22	16	21	8.8
	primary ions from C_4H_8 by 70 eV electron impact				secondary ions

spectra of these C_4H_8 isomers and the relative abundances of the product ions from the reaction $C_2H_4^+ + C_2H_4$ [15]. The abundance of the most frequent ion was set equal to 100. It is seen that the spectra are qualitatively similar. Even better agreement was been found for larger molecules, for example, for octene and the products of the reaction $C_4H_8^+ + C_4H_8$ [16]. These results clearly show that «statistical» intermediate complexes exist in certain ion-molecule reactions at low primary ion energies (< 1 eV lab. s.).

2'5. *Sticky collision complexes.* – A sticky collision complex is one containing all the atoms of the reactants and having a lifetime larger than the time of flight through the mass spectrometer, about 10^{-6} s. Although many hundreds of ion-molecule reactions have been observed to date, only about a dozen sticky collision complexes have been found [13, 17, 18]. Table II shows some of these complexes and the reactions by which they are formed. It is seen that some of the complexes contain two heavy atoms such as Br or I while others are formed from ions and molecules containing several double bonds. A most interesting example is the complex HCl_2^+ which, contains only three atoms. In some cases, decomposition products of the complex have also been observed as indicated in the Table.

TABLE II. – *Sticky collision complexes in ion-molecule reactions.*

$C_2H_5I^+ + C_2H_5I$	$\to C_4H_{10}I_2^+$	$\to C_4H_{10}I^+ + I$
$C_2H_5Br^+ + C_2H_5Br$	$\to C_4H_{10}Br_2^+$	
$CH_3I^+ + C_2H_5I$	$\to C_3H_8I_2^+$	
$C_3H_3N^+ + C_3H_3N$	$\to C_6H_6N_2^+$	$\to C_6H_5N_2^+ + H$
$C_6H_5^+ + C_6H_6$	$\to C_{12}H_{11}^+$	$\to C_{12}H_{10}^+ + H$
		$\to C_{12}H_9^+ + H_2$
$C_2N_2^+ + C_2N_2$	$\to C_4N_4^+$	
$Cl^+ + HCl$	$\to HCl_2^+$	

If a sticky collision complex is considered as a type of statistical complex, its decomposition into either the reactants or products can be treated by the theory of unimolecular rate processes. The most simple formula for the unimolecular decay of a molecule has been derived by KASSEL:

$$\tau = \nu^{-1}\left(1 - \frac{D_i}{E_i}\right)^{1-\alpha}, \tag{2.27}$$

where E_i is the total internal energy, D_i the energy required to dissociate the complex in the i-th bond and τ is the lifetime. α is the number of effective degrees of vibrational freedom. (Not all degrees of freedom may sometimes participate in the energy distribution.) ν^{-1} is of the order of the frequency of a vibration.

E_i is equal to

$$E_i = E_B + E_r + E_v + 3kT \tag{2.28}$$

where E_B is the binding energy between the ion and molecule, E_r the translational energy in the c.m. system, E_v the vibrational energy of the incident ion and $3kT$ the rotational and translational energy of the molecule. With increasing energy E_r, τ will rapidly decrease because of the high exponent $(1-\alpha)$. In fact, the decrease in the current ratio i_{complex}/i_p for the reactions listed in Table II with increasing repeller voltage U_r is much more pronounced than for other reactions. For example, it can be recognized from Fig. 2 that i_s/i_p more drastically decreases for complex formation in acrylonitrile than for the formation of a product ion such as $C_4H_5N_2^+$ from the complex.

By introducing plausible values into eq. (2.28), *i.e.* $E_B = 1$ eV, $E_r = 0.1$ eV, $E_v = 0.1$ eV, and $3\,kT = 0.14$ at the temperature of 250° in the ion source, $\nu = 10^{13}\,s^{-1}$, one finds that for $\tau \geqslant 10^{-6}$ s, $\alpha \geqslant 13$. According to this picture, one should be able to detect many sticky collision complexes in ion-molecule reactions, since there are many reactions known involving polyatomic species where α should be much higher. It is still an open question why only a small number of complexes have been observed.

Finally it should be noted that the Kassel formula is correct only for some orders of magnitude in τ. Much more detailed formulae have been derived in modern theories of unimolecular rate processes. We have chosen the simplest formula just to demonstrate the problem.

3. – The polarization theory.

3·1. *Cross-section for close collisions.* – The long-range attraction between a ion and a neutral molecule results from inducing a dipole moment in the molecule. The dipole moment which is induced in a molecule of polarizability α, assumed to be constant, by an electric field of strength F is

$$\boldsymbol{m} = \alpha \cdot \boldsymbol{F} \,. \tag{3.1}$$

If the ion is considered as a point charge e, the field strength at the distance r is equal to

$$F = \frac{e}{r^2} \tag{3.2}$$

and the dipole moment m becomes

$$m = \alpha \cdot \frac{e}{r^2} \,. \tag{3.3}$$

The force between the two particles is then given by

$$K_{(r)} = \boldsymbol{m} \cdot \frac{\mathrm{d}\boldsymbol{F}}{\mathrm{d}r} = \frac{\alpha e}{r^2} \cdot \frac{-2e}{r^3} = \frac{-2\alpha e^2}{r^5}. \tag{3.4}$$

They move in the potential

$$\varphi_{(r)} = -\int_0^r K_{(r)}\,\mathrm{d}r = -\frac{e^2\alpha}{2r^4}. \tag{3.5}$$

The orbits of the ions in such a potential have been given by LANGEVIN [19] in connection with a calculation of the mobilities of ions in gases. More recent calculations have been carried out by GIOUMOUSIS and STEVENSON [20]. An orbit is determined by the initial velocity v_1 in the laboratory system of the incident ion of mass M_1 and the impact parameter b. Typical cases are shown by Fig. 6 where it is assumed that the center of mass of the system lies in the

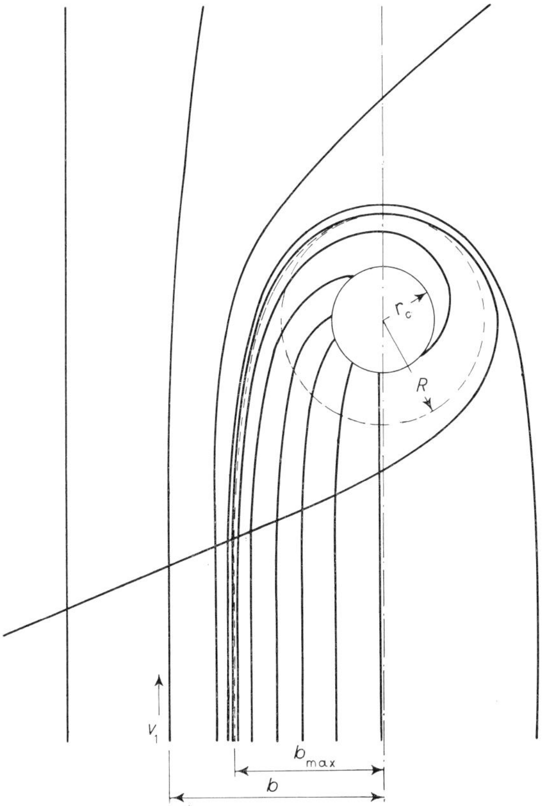

Fig. 6. – Some orbits described by the incident ion of velocity v_1 at various collision parameters b. (r_c = critical distance at which the chemical reaction takes place. b_{max}: largest impact parameter for close collisions).

molecule of mass M_2, *i.e.* $M_2 \gg M_1$ (in fact, the two particles move around the center of mass). At larger values of b (or v_1) the ion is less deflected. At a critical value b_{max} it orbits the molecule on a circle of radius R. At smaller impact parameters it continues to approach the molecule until the repulsive force between the electron shells of the particles repel it again or a chemical reaction takes place with subsequent separation of the products.

It is useful to describe the collision in terms of center-of-mass co-ordinates. The upper part of Fig. 7 shows the collision in the laboratory system. The center of mass, c.m., moves in the direction of the primary ion with the velocity

$$c = \frac{M_1}{M_1 + M_2} \cdot v_1 \, . \tag{3.6}$$

As it is shown by the middle part of Fig. 7, the ion moves in the center-of-mass system with the velocity $u_1 = v_1 - c$, while the neutral molecule moves from the right to the left with $u_2 = c$. The kinetic energy in the laboratory system is equal to

$$E_1 = \tfrac{1}{2} M_1 v_1^2 \tag{3.7}$$

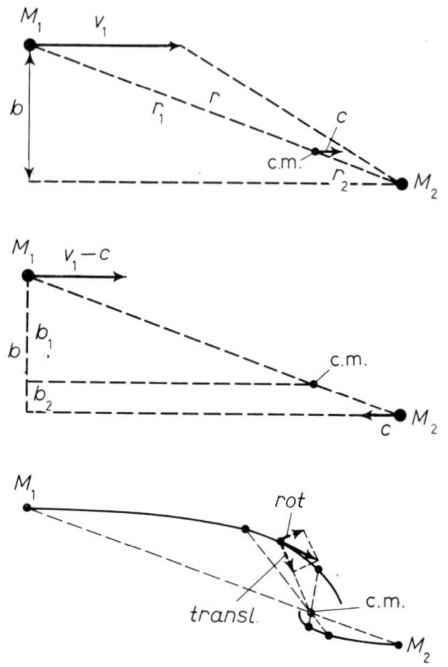

Fig. 7. – Collision between ion of mass M_1 and velocity v_1 with molecule M_2 at rest in the laboratory system (upper part) and in the center-of-mass system (middle part). The lower part shows the orbits in the c.m. system and the rotational and translational components of the velocity for the ion.

if the molecule M_2 is at rest. In the c.m. system, the sum of the kinetic energies $\frac{1}{2} M_1 u_1^2 + \frac{1}{2} M_2 u_2^2$ is equal to

$$E_r = E_1 \cdot \frac{M_2}{M_1 + M_2} = \frac{1}{2} \mu v_1^2 \, . \tag{3.8}$$

E_r is frequently called the «reative kinetic energy» of the system. μ is the reduced mass of the system

$$\mu = \frac{M_1 \cdot M_2}{M_1 + M_2} \, . \tag{3.9}$$

During the collision, the two particles are accelerated because of the attractive force given by eq. (3.4). The sum of the potential and kinetic energy is constant during the collisions:

$$\varphi + E_{\text{kin}} = E_r \,. \tag{3.10}$$

At large r, $\varphi = 0$ and $E_{\text{kin}} = E_r$. The velocity vector can always be composed of a component in the direction to the center of mass (translational component) and a component perpendicular to this direction (rotational component). This is shown in the lower part of Fig. 7. Then E_{kin} in eq. (3.10) can be written as the sum of the translational and rotational energies E_t and E_{rot}:

$$E_r = \varphi + E_{\text{rot}} + E_t \tag{3.11}$$

As long as E_t is larger than zero, the two particles are still approaching one another. E_{rot} in eq. (3.11) can be calculated in the following manner: The angular momentum in the center-of-mass system is equal to

$$L = \mu \cdot v_1 \cdot b \,. \tag{3.12}$$

The angular momentum remains constant during the collision. The rotational energy depends on the angular momentum according to

$$E_{\text{rot}} = \frac{L^2}{2\mu r^2} \,. \tag{3.13}$$

By substituting $L^2 = \mu^2 \cdot v_1^2 \cdot b^2$ and $v^2 = (2E_r/\mu)$, one obtains

$$E_{\text{rot}} = E_r \frac{b^2}{r^2} \,. \tag{3.14}$$

Equation (3.11) can therefore be written as

$$E_r = \underbrace{-\frac{e^2 \alpha}{2r^4} + E_r \cdot \frac{b^2}{r^2}}_{E_{\text{eff}}} + E_t \,. \tag{3.15}$$

The sum of the potential and rotational energies is called the « effective » potential E_{eff}. This potential is plotted *vs.* the distance r for a given collision parameter and different values of E_r (Fig. 8). It can be seen that the rotational energy produces a potential wall which increases with increasing E_r. We can easily calculate the distance R and the height $E_{\text{eff}}^{\max}$ of the potential wall by setting $\mathrm{d}E_{\text{eff}}/\mathrm{d}r$ equal to zero:

$$R = \frac{e}{b} \left(\frac{\alpha}{E_r} \right)^{\frac{1}{2}} \tag{3.16}$$

and

(3.17) $$E_{\text{eff}}^{\max} = E_r^2 \cdot \frac{b^4}{2\alpha e^2} \, .$$

With increasing kinetic energy E_r the potential wall lies at shorter distances and its maximum becomes higher.

The particles can overcome the wall, if their translation energy

(3.18) $$E_{t(R)} = E_r - E_{\text{eff}}^{\max} \, .$$

is still larger than zero at the distance R. $E_{\text{eff}}^{\max}$ is proportional to the square of E_r. Therefore, particles with an energy above a certain limit cannot cross the potential wall (case a) in Fig. 8). If $E_r = E_{\text{eff}}^{\max}$ they reach the top of the wall without closer approach. This situation corresponds to the circular orbit with radius R in Fig. 6 (case b) in Fig. 8). At lower values of E_r the collision leads to close approach (case c) in Fig. 8).

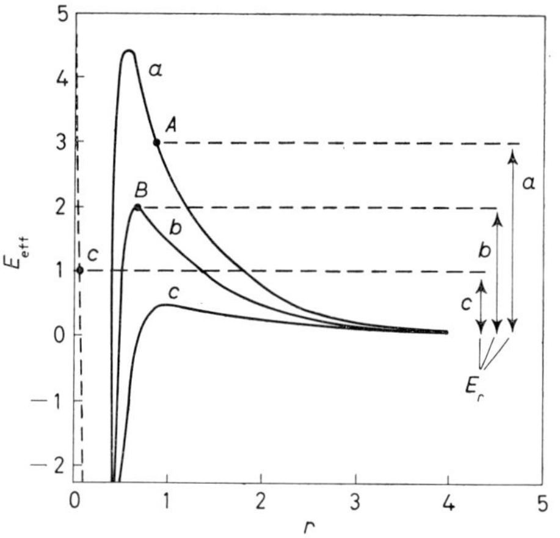

Fig. 8. – The effective potential as function of r for a given collision parameter and different relative kinetic energies. The dashed line on the left arises from the repulsive forces between the particles at very small distances.

The cross-section for collisions which lead to close approach (or to the circular orbiting as limiting case) is given by

(3.19) $$\sigma = \pi b_{\max}^2 \, ,$$

where $b_{\max}$ means the collision parameter at a given energy E_r, at which the condition $E_{\text{eff}}^{\max} = E_r$ is fulfilled. By setting $E_{\text{eff}}^{\max}$ in eq. (3.17) equal to E_r, one obtains

(3.20) $$b_{\max}^2 = e \left(\frac{2\alpha}{E_r}\right)^{\frac{1}{2}}$$

and

(3.21) $$\sigma = \pi e \left(\frac{2\alpha}{E_r}\right)^{\frac{1}{2}} = \frac{2\pi e}{v_1} \left(\frac{\alpha}{\mu}\right)^{\frac{1}{2}} = \pi e \left(\frac{2\alpha M_1}{\mu E_1}\right)^{\frac{1}{2}} .$$

From eqs. (3.16) and (3.20) one obtains

(3.22) $$R = \frac{b_{\max}}{\sqrt{2}} \, .$$

3·2. *Discussion and limitations of the polarization theory.* – The expression of eq. (3.21) gives the cross-section for close collisions. If it is assumed that the chemical reaction always occurs at a critical distance r_c which is smaller than R the reaction cross-section will be equal to the cross-section from eq. (3.21) (see Fig. 6). In most cases, however, reaction does not occur on every collision and the close collision cross-section must be multiplied by a probability factor, η, to obtain the reaction cross-section. For several simple H atom transfer reactions, the measured cross-sections agree fairly well with the calculated ones where $\eta = 1$ (see Fig. 9).

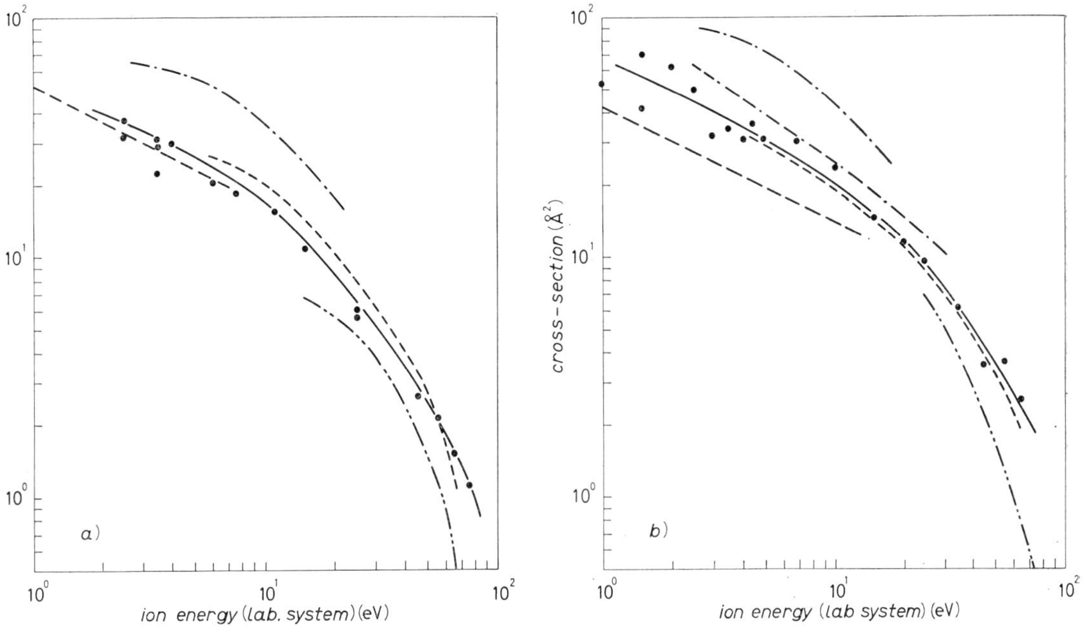

Fig. 9. – Cross-section as a function of the energy of the incident ion for two reactions of the type $X^+ + D_2 \rightarrow XD^+ + D$ where X^+ is Ar^+ (in *a*)) and N_2^+ (in *b*)) [21]. Experimental results of several authors are included [22-25, 30, 31]. *a*) $Ar^+ + D_2 \rightarrow ArD^+ + D$: —·—·— GIESE *et al.*; – – – ROBB *et al.*; —··—··— LACMANN *et al.*; ——— polarization theory. *b*) $N_2^+ + D_2 \rightarrow N_2D^+ + D$: —·—·— GIESE *et al.*; —··—··— TURNER *et al.*, KOSKI *et al.*; – – – ROBB *et al.*; —··—··— LACMANN *et al.*; ——— polarization theory.

It can be seen from Fig. 9 *a*) and 9 *b*) that the slope is equal to — 0.5 as expected from eq. (3.21) if the primary ion energy is below about 5 eV. At high primary ion energies, the slope rather suddenly becomes equal to 1.0 and at still higher energies (> 50 eV) becomes even larger.

The formula of eq. (3.21) is expected to be correct only if the cross-section is much larger than the gas kinetic cross-section σ_g of the reactants. Only under this condition can the ion be regarded as a point charge and the potential of eq. (3.5) can be used. At high energies the cross-section approaches σ_g and, therefore, the polarization theory cannot be used to calculate the cross-section of a ion-molecule reaction. Table III gives the kinetic energy E_1 of the ion at which the cross-section calculated from eq. (3.21) becomes equal to the gas kinetic cross-section for collisions between the neutral particles. One can see that eq. (3.21) generally can only be used for $E_1 < 1.5$ eV (laboratory). Only in those cases when $M_1 \gg M_2$ such as in the reactions of the type $X^+ + D_2 \rightarrow$ $\rightarrow XD^+ + D$ may the formula be applied up to several eV.

TABLE III. – *Limiting energy of the incident ion, where* σ (eq. (3.21)) *becomes equal to* σ_g [8].

Ion	Molecule	α (10^{-24} cm^3)	σ_g (10^{-16} cm^2)	Limiting ion energy (eV, laboratory)
H_2^+	H_2	0.78	21	1.0
Ar^+	H_2	0.79	24	8.3
Ar^+	D_2	0.79	24	4.3
Kr^+	H_2	0.79	26	14.0
H_2^+	Kr	2.48	26	1.6
CH_4^+	CH_4	2.6	33	1.4
$C_2H_4^+$	C_2H_4	4.3	40	1.5
O_2^+	D_2	0.79	24	6.6
D_2^+	O_2	1.60	24	0.8

Deviations from eq. (3.21) have also been found for reactions in which the neutral reactant has a high permanent dipole momentum. The potential $\varphi_{(r)}$ now contains two terms [26]:

$$\varphi_{(r)} = \frac{-e\gamma}{r^2} - \frac{e^2\alpha}{2r^4}\,. \tag{3.23}$$

γ is the permanent dipole momentum of the molecule. It is assumed here that γ is oriented in the position of minimum energy, *i.e.* it is aligned with the r-vector. This probably will only be the case at low primary-ion energies. As the velocity increases the phase angle will lag until the collision cross-section is given by eq. (3.21) (when all orientations of the dipole have equal probability during the reaction).

4. – The spectator stripping model.

4·1. *Velocity spectra of* H *atom transfer reactions.* – The « stripping » mechanism of certain ion-molecule reactions was first detected in simple discrimination experiments [27, 28] where only ions formed with small momentum transfer could be measured. Direct evidence came from the measurement of the velocity distribution of the XH^+ ions produced in reactions of the type

$$X^+ + H_2(D_2) \rightarrow XH^+(XD^+) + H(D) \,, \tag{4.1}$$

where $X^+ = Ar^+$, N_2^+, CO^+ [29-35].

Figure 10 shows the apparatus. The primary ions are produced by electron impact in a conventional ion source and accelerated by voltages between 5 and 200 V. The ion beam passes through the collision chamber which contains hydrogen or deuterium under a pressure of $(1 \div 2) \cdot 10^{-3}$ Torr.

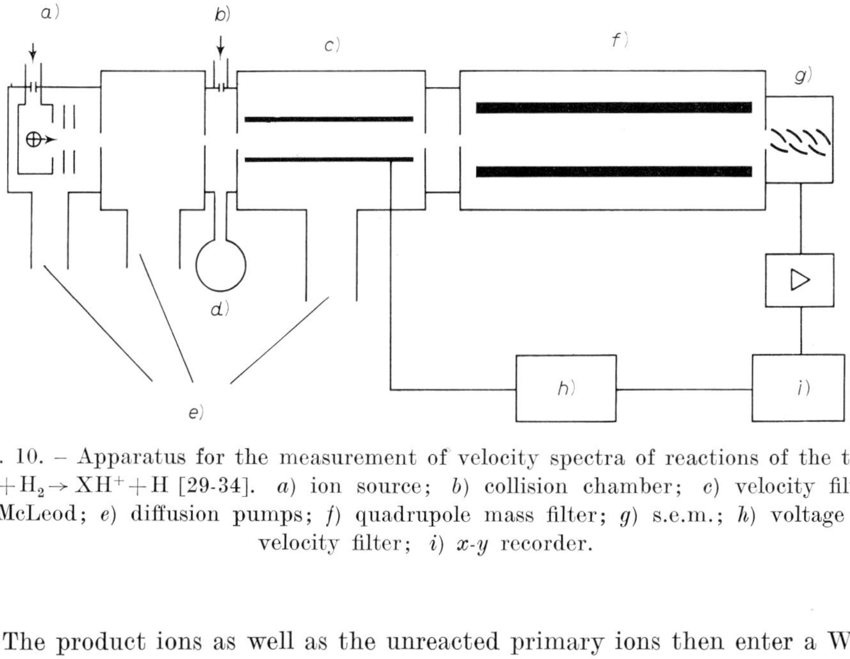

Fig. 10. – Apparatus for the measurement of velocity spectra of reactions of the type $X^+ + H_2 \rightarrow XH^+ + H$ [29-34]. *a*) ion source; *b*) collision chamber; *c*) velocity filter; *d*) McLeod; *e*) diffusion pumps; *f*) quadrupole mass filter; *g*) s.e.m.; *h*) voltage for velocity filter; *i*) *x*-*y* recorder.

The product ions as well as the unreacted primary ions then enter a Wien velocity filter (crossed electric and magnetic fields) which allows ions of equal linear velocity to pass on a straight line. The ions are then mass-analysed in a quadrupole mass filter. The velocity spectrum of the reaction is obtained by scanning the electric field of the Wien filter and automatically recording the ion intensity.

Figure 11 shows a typical spectrum for the reaction $Ar^+ + D_2 \rightarrow ArD^+ + D$. The ion intensity is plotted *vs.* the velocity in units of the velocity v_1 of the incident ion in the laboratory system. The velocity distribution of the Ar^+ ion, mass 40, was measured with and without D_2 in the collision chamber. As can be seen from the Figure, appreciable attenuation of the primary ion beam

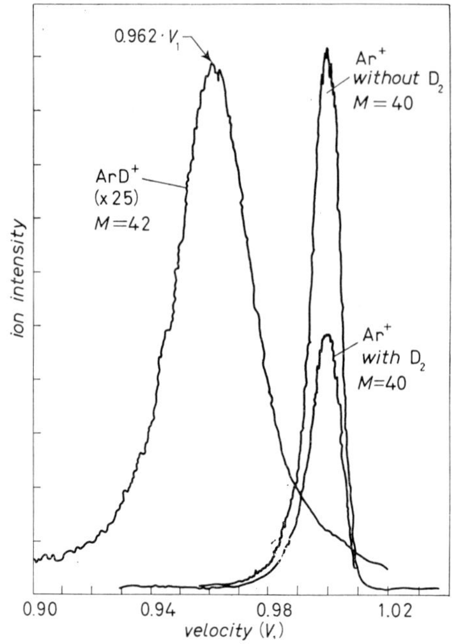

Fig. 11. – Velocity spectrum of the reaction $Ar^+ + D_2 \rightarrow ArD^+ + D$ at 26 eV (lab. s.) of the primary ion.

has taken place. The band obtained at mass 42 is produced by the product ion. The center of this band lies at $0.962v_1$. The band is symmetric and broader than the primary ion band.

The position of the center of the product ion band will now be discussed. As a first approximation, the D_2 molecule may be regarded at rest, since its thermal velocity is much smaller than the velocity of the incident Ar^+ ion. The thermal motion of the D_2 molecule does not shift the band center, since movements of D_2 in all directions are equally probable, but a broadening of the band will occur. In the following treatment, the mass and velocity of the various reactants and products are characterized by the following indices: 1) primary ion, 2) neutral reactant, 3) product ion, 4) neutral product.

Figure 12 shows the Newton diagrams and velocity spectra in the laboratory system for intermediate complex formation and stripping in the reaction $Ar^+ + D_2 \rightarrow ArD^+ + D$. An intermediate ArD_2^+ contains the relative kinetic energy E_r of the system as internal energy and moves with the velocity

$$c = \frac{M_1}{M_1 + M_2} v_1 \tag{4.2}$$

in the forward direction, *i.e.* the direction of the incident ion. All or a part of the internal energy

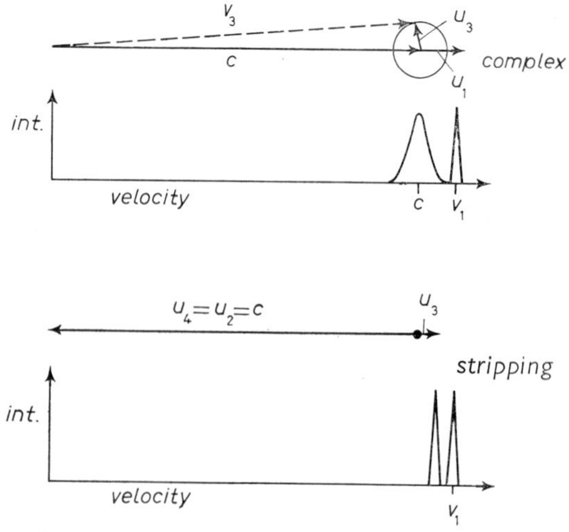

Fig. 12. – Newton diagrams and velocity spectra for the complex and stripping mechanisms for reactions of the type $X^+ + H_2 \rightarrow XH^+ + H$.

as well as a part of the heat of reaction may appear as kinetic energy of the products. If the complex dissociates isotropically in the center-of-mass system, the tip of the velocity vector u_3 of the product ion will be found with equal probability at any point on the surface of a circle around the c.m. point with radius u_3. Depending on the amount of kinetic energy which is freed in the decomposition of the complex, various circles with different radii may be drawn. Conservation of momentum requires that

$$M_1 u_1 = M_2 u_2 \,. \tag{4.3}$$

Since $u_2 = c$,

$$u_1 = \frac{M_2}{M_1} c \,. \tag{4.4}$$

The spectrum is expected to contain a broad band of the production centered at the velocity $v_3 = c$. By introducing the masses of $Ar^+(M_1)$ and $D_2(M_2)$ into eq. (4.2), one obtains a velocity v_3 of $0.909 v_1$ for the band center which does not agree with that of Fig. 11.

In the stripping process, the D atom is picked up by the Ar^+ ion without transfer of momentum to the other D atom, the spectator atom. This means that the product ion moves strictly in the forward direction. In the center-of-mass system, the spectator D atom continues to move with the velocity $u_4 = u_2$ of the D_2 molecule which is equal to c, and the product ion moves with the velocity u_3. Conservation of momentum requires that

$$u_3 = \frac{M_4}{M_3} \cdot c \,. \tag{4.5}$$

Combining eqs. (4.4) and (4.5) one obtains

$$u_3 = \frac{M_4}{M_3} \cdot \frac{M_1}{M_2} \cdot u_1 \,. \tag{4.6}$$

Since $M_4/M_2 = 0.5$ and $M_1/M_3 \approx 1$, u_3 is equal to about $0.5 u_1$ for the reactions of the type $X^+ + H_2 \rightarrow XH^+ + H$. The production band therefore appears about half-way between v_1 and c. The velocity of the production in the laboratory system is equal to

$$v_3 = c + u_3 = c\left(1 + \frac{M_4}{M_3}\right) . \tag{4.7}$$

Substitution of c from eq. (4.2) leads to

$$v_3 = \frac{M_1}{M_3} \cdot v_1 \tag{4.8}$$

since $M_3 + M_4 = M_1 + M_2$. For $Ar^+ + D_2$ one obtains $v_3 = 40\,v_1/42 = 0.953\,v_1$ which is in good agreement with Fig. 11. It has therefore been concluded that this reaction occurs via a stripping mechanism.

The spectator stripping model is only a limiting case of a collision model. All kinds of intermediate situations between this model and the complex model are conceivable. Even for a given reaction and given kinetic energy not all of the reactive collisions will obey the stripping mechanism. It must always be expected that the velocity spectrum contains some contribution resulting from central collisions, *i.e.* reactions with small impact parameters which result in velocities backwards in the centre of mass system. In the case of the $X^+ + D_2$ reaction at several 10 eV of the incident ion, this contribution is only of the order of a few percent.

Similar spectra have been obtained for $X^+ = N_2^+$ and CO^+ in eq. (4.1). The spectra have been investigated at various primary ion energies. In the « intermediate » energy range from about 20 to 50 eV the band of the production is always centered at the position predicted by the spectator stripping model. Deviations at higher and lower energies will be discussed later in Subsect. **5**˙1 and **5**˙4.

4˙2. *Isotope effects and heavy spectator groups.* – The stripping mechanism of simple H atom transfer reactions can also be investigated by using HD as target gas in the collision chamber:

$$Ar^+ + HD \begin{cases} \nearrow ArH^+ + D\,, & (4.9) \\ \searrow ArD^+ + H\,. & (4.10) \end{cases}$$

The product ions ArH^+ and ArD^+ are expected at the same position, *i.e.* at $v_3 = c$, in the velocity spectrum, if they are produced from an intermediate complex $ArHD^+$.

However, if stripping occurs they should produce two separate bands having the same positions as the ArH^+ band from the reaction $Ar^+ + H_2$ and the ArD^+ band from $Ar^+ + D_2$, respectively. A two-band velocity spectrum of the products has been found in agreement with the stripping model. Furthermore, it has been shown that the cross-sections for the reactions according to eqs. (4.9) and (4.10) are equal to half the cross-sections of the reactions of Ar^+ with H_2 and D_2, respectively, at the same energy of the incident ion since at the same pressure in the collision chamber, HD contains half as many H or D atoms per cm^3 as H_2 or D_2, respectively [32, 33]. Another isotope effect exists with respect to the intensity ratio ArH^+/ArD^+ which will be dealt with in Subsect. **5**˙4.

Stripping has also been shown to occur in the reactions

$$X^+ + CH_4 \rightarrow XH^+ + CH_3\,, \qquad (4.11)$$

$$X^+ + HBr \rightarrow XH^+ + Br\,, \qquad (4.12)$$

which have about 100 times smaller cross-sections than the reactions with H_2 [35]. The band of the XH^+ ion always had the same position as that of XH^+ from $X^+ + H_2$ if the energy of X^+ was in the intermediate range of a few 10's eV. The mass of the spectator does not influence the velocity of the product as is expected from eq. (4.8).

4'3. *Angular distribution of product ions.* – The full kinematic analysis of a chemical reaction requires the measurement of the velocity or energy distribution of the products at various scattering angles. A number of investigations of this kind has been carried out by American authors [36-39]. The stripping mechanism of H atom transfer reactions $X^+ + H_2 \rightarrow XH^+ + H$ has been confirmed this way.

A schematic diagram of the apparatus used by HERMAN *et al.* [38] is shown by Fig. 13.

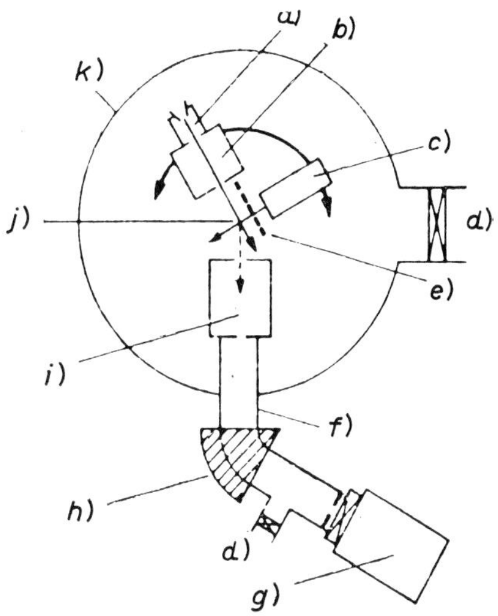

Fig. 13. – Apparatus for the measurement of angular and energy distributions [38]: *a*) ion source; *b*) focusing and decelerating lenses; *c*) molecular-beam source; *d*) pumps; *e*) chopper wheel; *f*) mass spectrometer; *g*) electron multiplier, *h*) magnet; *i*) energy analyser; *j*) collision region; *k*) scattering chamber.

The ion beam intersects the molecular beam at a fixed 90° angle. Ions from the collision zone pass through an energy analyser, a 60° sector mass spectrometer and are detected by an electron multiplier. The beam sources are mounted on a rotable lid of the scattering chamber and can thus pivot around the collision center. The ions are formed by electron impact, accelerated to 70 eV and then decelerated and focused by a series of 14 circular and semicircular elements.

This lens assembly is a modification of a design of GUSTAFSSON and LINDHOLM [40]. The molecular beam is pulsed by a rotating chopper wheel to distinguish the N_2D^+ ions formed in the reaction zone from those ions produced in collisions of N_2^+ ions with D_2 molecules outside the zone. The energy analyser consists of a fine metal grid to which a retarding potential is applied. Only ions of higher energy can pass and are detected. By changing the retarding potential and recording the ion intensity, an integral energy distribution is obtained.

The Newton diagram for the reaction $N_2^+ + D_2 \rightarrow N_2D^+ + D$ in perpendicular reactant beams is shown by Fig. 14. In the laboratory system, the center

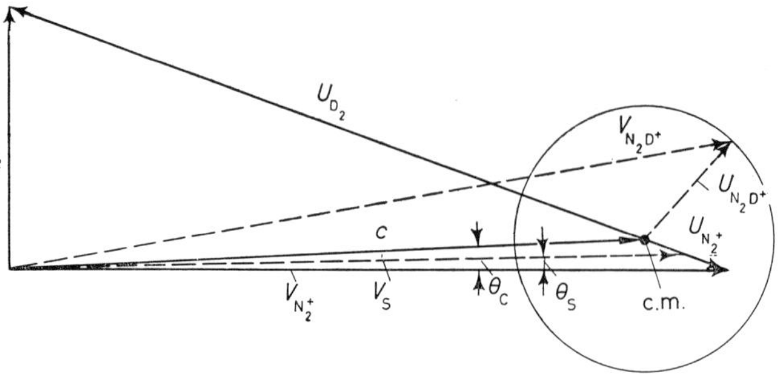

Fig. 14. – Newton diagram for the reaction $N_2^+ + D_2 \rightarrow N_2D^+ + D$ in crossed beams. v_{D_2} corresponds to 0.025 eV, thermal energy at room temperature, and the length of the $v_{N_2^+}$ vector corresponds to an incident ion energy of 1.5 eV.

of mass moves with the velocity c at an angle θ_c with respect to the direction of the incident N_2^+ ion. The velocities $u_{N_2D^+}$ and $v_{N_2D^+}$ describe the movement of the product ion in the center-of-mass and laboratory system, respectively. In the special case of stripping, the N_2D^+ ion moves in the direction of the vector $u_{N_2^+}$ in the c.m. system with a veloctiy of about $\frac{1}{2} \cdot u_{N_2^+}$ (see eq. (4.6)). In the laboratory system, its velocity is characterized by the vector v_s in Fig. 14, which forms the angle θ_s with the forward direction.

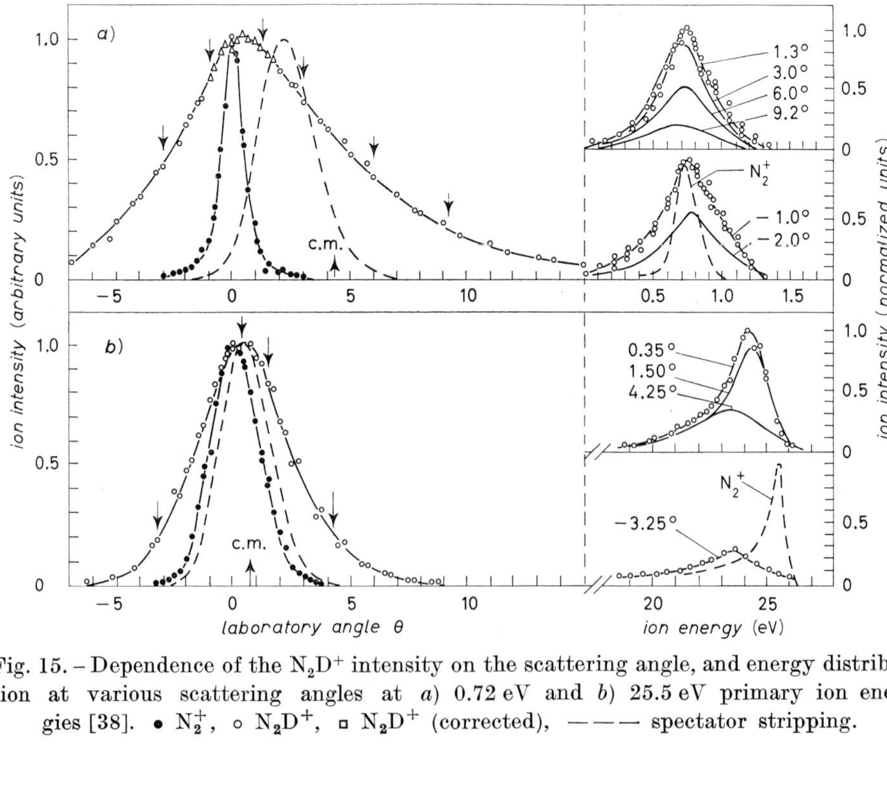

Fig. 15. – Dependence of the N_2D^+ intensity on the scattering angle, and energy distribution at various scattering angles at *a*) 0.72 eV and *b*) 25.5 eV primary ion energies [38]. • N_2^+, ○ N_2D^+, □ N_2D^+ (corrected), ——— spectator stripping.

Figure 15 *b*) shows how the ion intensity depends on the scattering angle in the laboratory system at an energy of 25.5 eV of the N_2^+ ion. The dashed line indicates the angular distribution of the product ion as expected from the spectator stripping model taking into account the thermal-velocity distribution of D_2. The peak of the intensity occurs at the expected angle θ_s and the band is a little broader than the primary ion band. The right-hand side of Fig. 15 shows energy distributions of the N_2D^+ ion at various scattering angles. « Contour maps » [28] of the reaction obtained from these measurements are shown in Fig. 16 *b*).

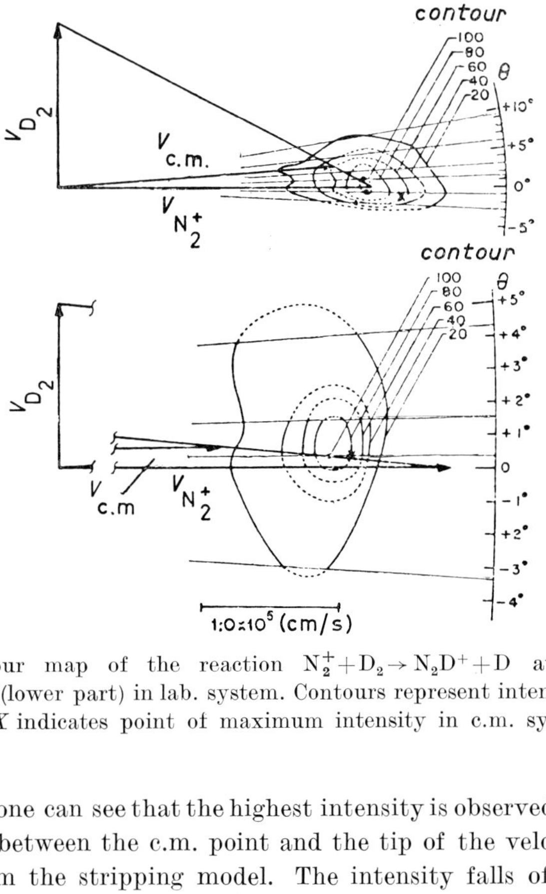

Fig. 16. – Contour map of the reaction $N_2^+ + D_2 \rightarrow N_2D^+ + D$ at 0.72 eV (upper part) and 25.5 eV (lower part) in lab. system. Contours represent intensity in laboratory system. *X* indicates point of maximum intensity in c.m. system [38].

Once again, one can see that the highest intensity is observed at a point lying about half-way between the c.m. point and the tip of the velocity vector $u_{N_2^+}$ as expected from the stripping model. The intensity falls off rather rapidly around this point. Similar contour maps have been obtained by GENTRY *et al.* [39]. A map for the reaction of N with H_2 at a relative kinetic energy of 8.1 eV is given by Fig. 17.

These maps give some idea about the deviations from the spectator stripping model mentioned in Sect. 4.1. For example, one sees from Fig. 17 that a second intensity maximum lying in the backward hemisphere of the c.m. system can

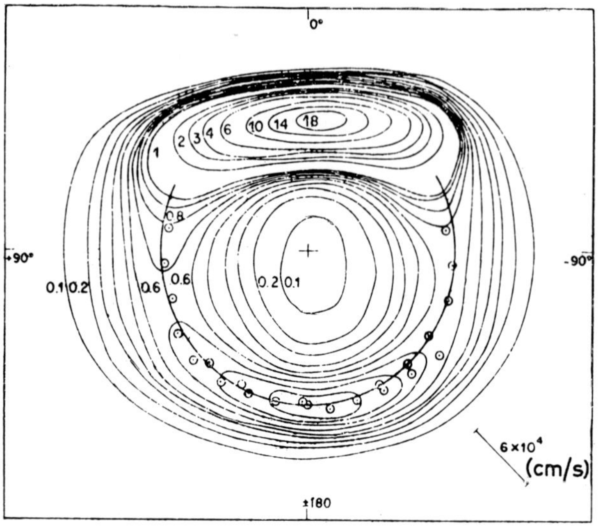

Fig. 17. – Contour map of the reaction $N_2^+ + H_2 \rightarrow N_2H^+ + H$ at 8.1 eV [39] in c.m. system.

be observed which may be attributed to backward scattering from central collisions. The maximum amounts to only about 3 % of the maximum due to the stripping collisions. The low-intensity maximum was first found by DOVERSPIKE *et al.* [36] at low energies. They observed that it was less pronounced at higher energies of the incident ion.

4·4. *Energy considerations.* – One of the most interesting questions in chemical kinematics is the transformation of translational and internal energy of the reactants into internal or translational energy of the products. If E and E' are the sums of the translational energies of the reactants and products in the laboratory system, E_r and E_r' the corresponding relative kinetic energies, W the heat of the reaction and U the internal energy of the products, conservation of energy requires that

$$E + W = E' + U \,, \tag{4.13}$$

or

$$E_r + W = E_r' + U \,, \tag{4.14}$$

assuming that the reactants are not internally excited. The difference in the kinetic energies of the products and reactants is called the translational exoergicity of the reaction:

$$Q = E' - E = E_r' - E_r = W - U \,, \tag{4.15}$$

or

$$U = W - Q \,. \tag{4.16}$$

If spectator stripping occurs in a reaction like $X^+ + RH \rightarrow XH^+ + R$ and the molecule RH is at rest

$$Q = \tfrac{1}{2} M_3 v_3^2 - \tfrac{1}{2} M_1 v_1^2 \,. \tag{4.17}$$

Since $\frac{1}{2} M_1 v_1^2 = E_1$ and $v_3 = (M_1/M_3) v_1$ (eq. (4.8)),

$$Q = -E_1 \cdot \frac{M_3 - M_1}{M_3} = -E_1 \frac{m}{M_1 + m} \,, \tag{4.18}$$

where m is the mass of the transferred atom ($m = M_3 - M_1$). $E_1(m/(M_1 + m))$ represents the relative kinetic energy of the system consisting of the incident ion M_1 and the transferred atom. The internal energy of the product XH^+ is equal to

$$U = W + E_1 \cdot \frac{m}{M_1 + m} \,. \tag{4.19}$$

According to eq. (4.19) the internal energy of the product (vibrational and rotational energy) increases with increasing energy of the incident ion. If U exceeds the dissociation energy D, the product is not stable and the cross-section of the reaction becomes zero. The corresponding « critical » energy of the incident ion is equal to

$$E_{1\text{crit}} = (D - W) \frac{M_1 + m}{m} \,. \tag{4.20}$$

The critical energy can be calculated if D and W are known. Since $m \ll M_1$, there should be an isotope effect of a factor of about 2 in the critical energies for H and D atom transfer. Table IV contains the calculated critical energy for various reactions of the type $X^+ + H_2(D_2) \rightarrow XH^+(XD^+) + H(D)$; it is assumed here that $D(X^+ - H)$ is about 5 eV and $W = 0.5$ eV [30, 31].

The critical energy calculated from eq. (4.20) is only correct for collisions described by the spectator stripping model. It has already been mentioned that this model will only be true to a certain degree. If the mechanism changes the reaction may still be observed at energies above the critical energy. In order to compare the calculated critical energies with the experimental results, the cross-section *vs.* energy curve may be extrapolated to zero cross-section. Only that range of the curve should be used for extrapolation over which the velocity spectra and angular distributions correspond to the stripping model.

TABLE IV. – *Critical energies in the laboratory and c.m. systems calculated from eq.* (4.20) *for various reactions* [30, 31].

Reaction	$E_{1\,\text{crit}}$ (eV)	$E_{r\,\text{crit}}$ (eV) (*)
$Ar^+ + H_2 \rightarrow ArH^+ + H$	184	8.8
$Ar^+ + D_2 \rightarrow ArD^+ + D$	94	8.5
$N_2^+ + H_2 \rightarrow N_2H^+ + H$	131	7.7
$N_2^+ + D_2 \rightarrow N_2D^+ + D$	68	8.5
$CO^+ + H_2 \rightarrow COH^+ + H$	131	7.7
$CO^+ + D_2 \rightarrow COD^+ + D$	68	8.5

(*) $E_{r\,\text{crit}} = E_{1\,\text{crit}} \cdot \dfrac{M_2}{M_1 + M_2}$.

The cross-sections of the reactions of Ar^+ with H_2 and D_2 are plotted *vs.* the relative kinetic energy in Fig. 18 *a*). The upper scales give the kinetic energy

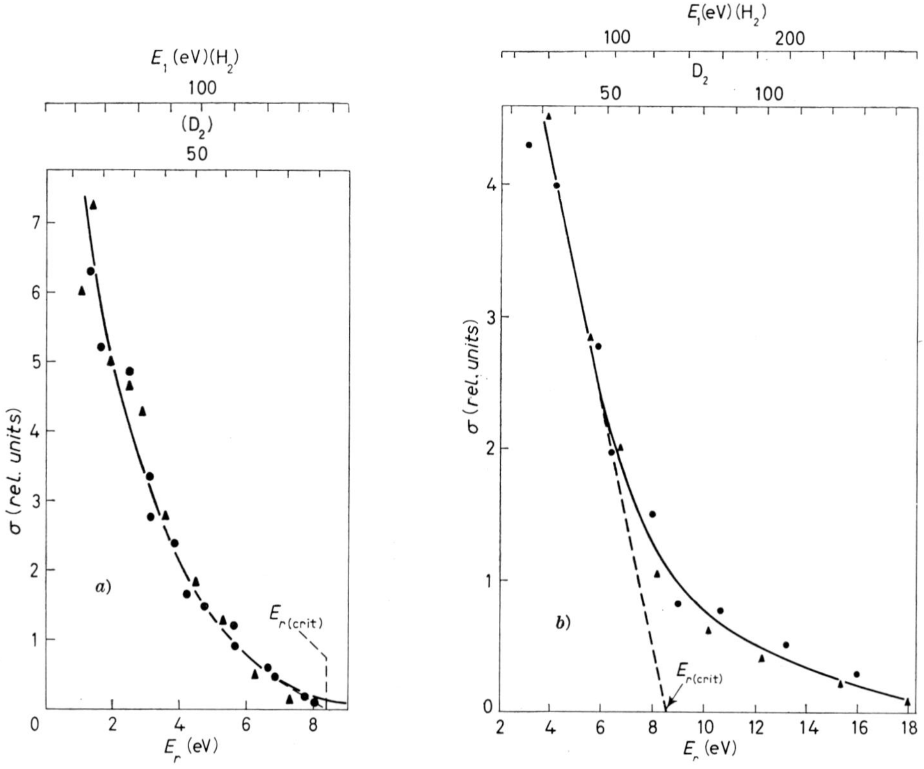

Fig. 18. – Dependence of the cross-section on the relative kinetic energy for the reactions of Ar^+ (*a*)) and CO^+ (*b*)) with H_2 and D_2 and extrapolation for the critical energy. *a*) ▲ $Ar^+ + H_2 \rightarrow ArH^+ + H$, ● $Ar^+ + D_2 \rightarrow ArD^+ + D$; *b*) ▲ $CO^+ + H_2 \rightarrow COH^+ + H$, ● $CO^+ + D_2 \rightarrow COD^+ + D$.

E_1 in the laboratory system of the incident primary ion. As expected the cross-sections for the two reactions are equal when they are compared at the same relative kinetic energy. The velocity spectra indicated stripping up to 8 eV of relative kinetic energy. At higher energies, the velocity spectra show band shifts to higher velocities which will be described in Sect. 5'1. The extrapolation of the curve leads to critical energies in the center-of-mass system which are slightly higher than 8 eV and are in good agreement with the values listed in Table IV.

The agreement is not so good for the reactions of CO^+ with H_2 and D_2 as can be seen from Fig. 18 *b*). The shifts of the product's bands to higher velocities are much more pronounced in these reactions at higher kinetic energies (see Subsect. 5'1). The shifts become noticeable at about 6 eV relative kinetic energy. Extrapolation of the curve at this energy leads to critical energies close to the calculated values in Table IV.

5. – Modifications of the stripping model.

5'1. *Deviations from spectator stripping at high energies.* – As described in Subsect. 4'1, in the velocity spectra of the reactions

$$X^+ + H_2(D_2) \rightarrow XH^+(XD^+) + H(D)$$

the product-ion band is at the position expected from the spectator stripping model at intermediate energies of the primary ion. However, at higher energies the band shifts to higher velocities with respect to the velocity of the incident ion than is expected from the spectator stripping model [31].

The effect is more pronounced for $X^+ = N_2^+$ or CO^+ than Ar^+, and becomes noticeable at energies slightly below the extrapolated critical energy (eq. (4.20)). The effect has first been found in the Wien filter experiments [31] (Fig. 10) and later confirmed by DOVERSPIKE *et al.* [37] and GENTRY *et al.* [39] who used quite different apparatus. The band shift is accompanied by a decrease in the band width. The band width is largest for $X^+ = CO^+$ and smallest for $X^+ = Ar^+$. This decrease is only in part due to the fact that the thermal movement of the hydrogen molecules becomes less important as the energy of the incident ion increases.

Figure 19 shows the velocity of the center of the product-ion band as a function of the kinetic energy for $X^+ + D_2 \rightarrow XD^+ + D$. The lower dashed lines give the velocity in units of v_1 calculated for spectator stripping (eq. (4.8)). v_{max} in the Figure is the kinetic energy of the product-ion in the forward direction, if Q is equal to zero. It can be seen that the position of the band center does not reach v_{max} but approaches it in the reaction of CO^+ with D_2.

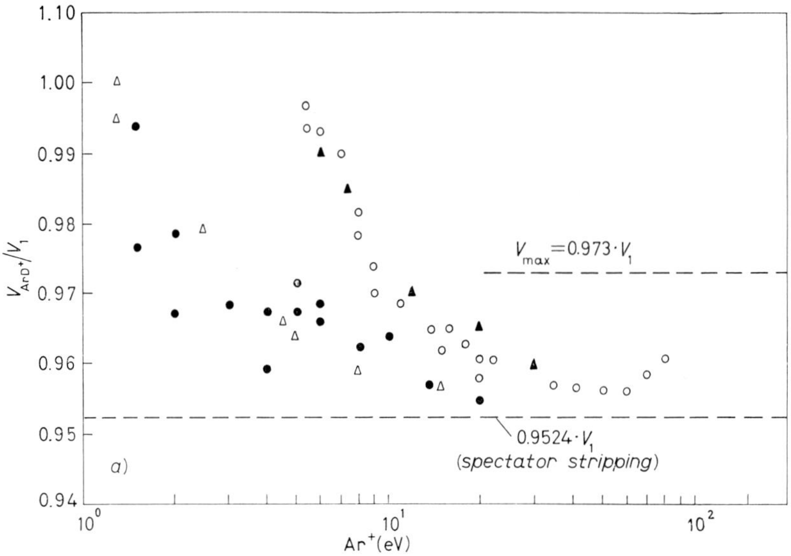

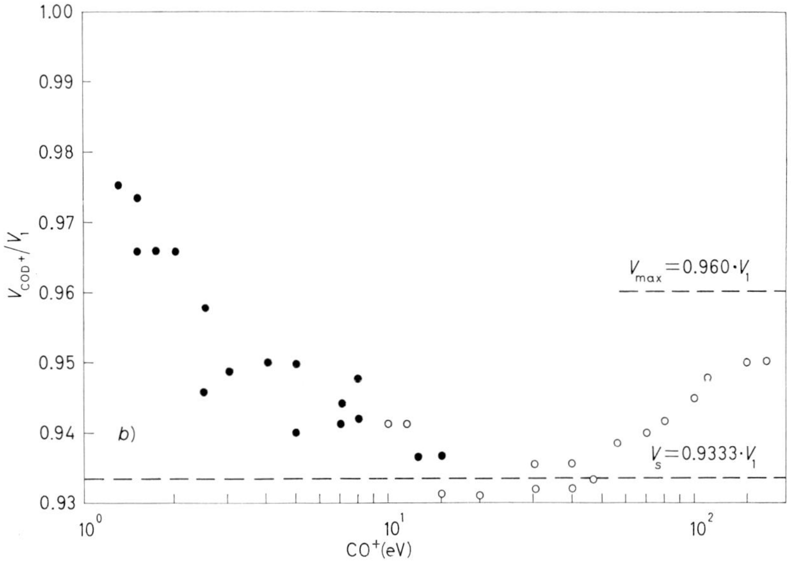

Fig. 19. – Velocity of the maximum of the product ion band of the reaction Ar^+ *a*) and CO^+ *b*) with D_2 over a wide range of Ar^+ energies [35, 48]. Results of other authors are included [38, 45]. (Remark: the results of various authors fairly agree for intermediate and high ion energies. At low energies, the measurements carried out with the Wien filter show an increase at higher primary ion energies as the measurements with the retarding-potential method. DING has proposed that stray fields from the Wien filter may result in less accurate measurements at low ion energies [35]). *a*) $Ar^+ + D_2 \to ArD^+ + D$: Wien filter: ○ DING *et al.*, 1967; ▲ FINK-KING, 1967; retarding-potential method: ● DING *et al.*, 1968; △ HERMAN *et al.*, 1967; *b*) $CO^+ + D_2 \to$ $\to COD^+ + D$: ○ Wien filter; ● retarding-potential method.

Some measurements have also been carried out with the apparatus shown by Fig. 20. The ions were produced by electron impact, accelerated to 70 eV, mass-selected by a Wien filter, decelerated by a « Lindholm » system of electrodes to the desired energy, shot through the collision chamber, energy-analysed with a retarding potential and finally mass-analysed and re-

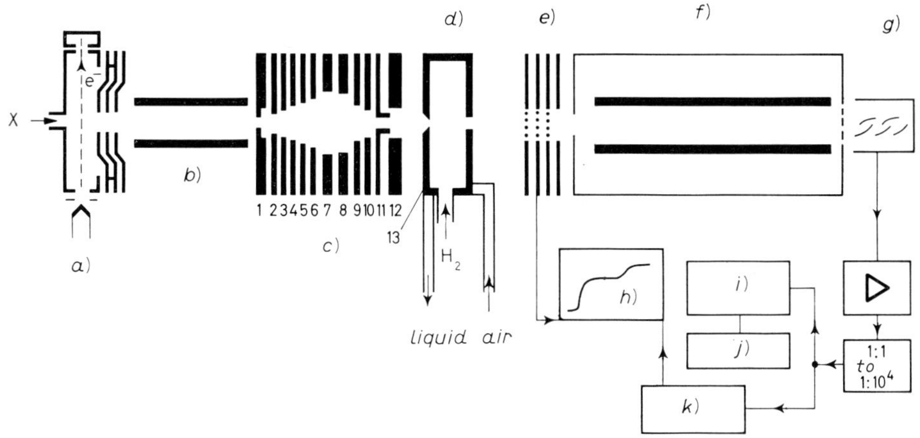

Fig. 20. – Apparatus for measuring the distribution of kinetic energy in the forward direction of a ion-molecule reaction: *a*) ion source; *b*) Wien filter; *c*) decelerator; *d*) collision chamber; *e*) energy filter; *f*) mass filter; *g*) electron multiplier; *h*) *x*-*y* recorder; *i*) counter; *j*) printer; *k*) integrator.

corded [35, 48]. The collision chamber of this apparatus could be cooled by liquid air. The velocity spectra of the reactions of the type $X^+ + D_2 \rightarrow XD^+ + D$ did not change when the reactions were carried out at — 196 °C. Furthermore, the cross-section did not change between — 196° and 20° C over the entire energy range of the incident ions.

5'2. *Recoil stripping.* – The shift of the product-ion band mentioned above has been attributed to the thermal motion of the hydrogen molecules in the collision chamber [39]. Because of the isotropic velocity distribution of the target molecules, the relative kinetic energy may be larger or smaller than the nominal value calculated by assuming a stationary target. Thus, molecules moving sufficiently rapidly in the direction of the primary ion beam may still react to form a stable product even if the energy of the incident ion exceeds the critical energy. The product-ion band, of course, appears at a higher velocity than calculated for the stripping process with an hydrogen molecule at rest. The temperature of the target gas should have a strong influence on the cross-section at high energies since the number of hydrogen molecules having a sufficiently high-velocity component in the forward direction decreases sharply as the temperature goes down. As already mentioned in Subsect. **5˙1**, no temperature dependence has been observed. The band shift cannot be explained, therefore, by the thermal motion of the target gas molecules.

On the other hand, it has been postulated that the two hydrogen atoms involved in the reaction experience a repulsion, *i.e.* the spectator atom is not « idle » but receives some momentum [31, 33]. Such a recoil of the spectator atom must be postulated since the product-ion band is always broader than the primary-ion band as has already been mentioned in Subsect. **4˙1**. The angular distribution of the product ion is also broader than expected from the spectator model as can be recognized from the lower part of Fig. 15. As long as the momentum of the spectator atom is isotropically distributed, the center of the product band still appears where it is expected from the spectator stripping model. Let us assume that the repulsion between the two hydrogen atoms occurs rather early in the approach of the X^+ ion (*) and that the new X^+-H bond is finally formed in a collinear configuration X^+-H-H indicated in Fig. 21. Various orientations of the H_2 molecule with respect to the direction of the incident X^+ ion are shown in this Figure. Depending on the orientation of the H-H axis with respect to the direction of the incident ion, the product ion receives an additional recoil momentum into the forward or backward hemisphere of the c.m. system (cases *c*) and *a*) in Fig. 21). At low or intermediate primary ion energies, there will be practically no dependence of the reaction probability on the angle θ, *i.e.* a symmetric velocity distribution around the velocity given by eq. (4.8) will be obtained. However, at energies near the critical energy only those collisions will lead to a stable product in which there is strong backward recoil of the spectator. In these cases the relative kinetic energy between the incident ion and picked-up hydrogen atom is smaller than that calculated for a stationary atom. At higher energies, the repulsion between the two H atoms may be regarded as being the result of a binary collision, as described in the following paragraph.

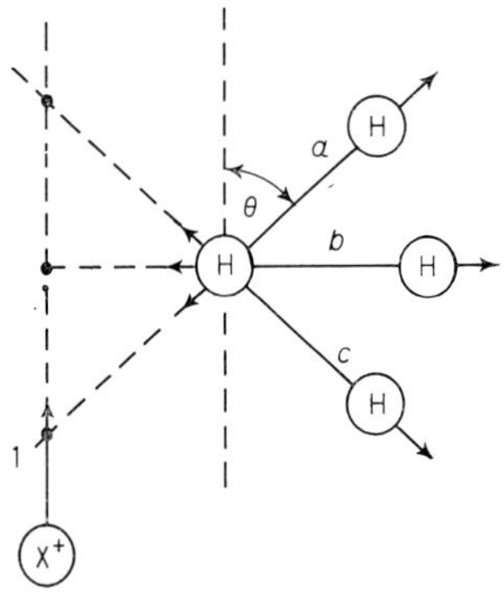

Fig. 21. – Collision of the incident X^+ ion with H_2 molecules of different orientations.

5˙3. *Impulse approximation at high energies.* – At energies of the incident ion above the critical energy the polarization forces as well as chemical binding

(*) A similar assumption is made in the Dipr-model (« direct interaction with product repulsion ») recently developed by POLANYI and coworkers [57].

of the colliding particles can be ignored. BATES *et al.* [46] have described reactions of the type

$$X^+ + {}^1H—{}^2H \rightarrow X^{+1}H + {}^2H$$

by a classical impulse approximation in which the H_2 molecule is regarded as a loose cluster of two H atoms. We label them with the indices 1 and 2. According to this model the ^1H atom which is initially bound to ^2H first suffers a binary collision with X^+ and then suffers another binary collision with ^2H, the two binary collisions being such as to reduce its energy of motion relative to X^+ below that required for separation. Figure 22 shows the main features

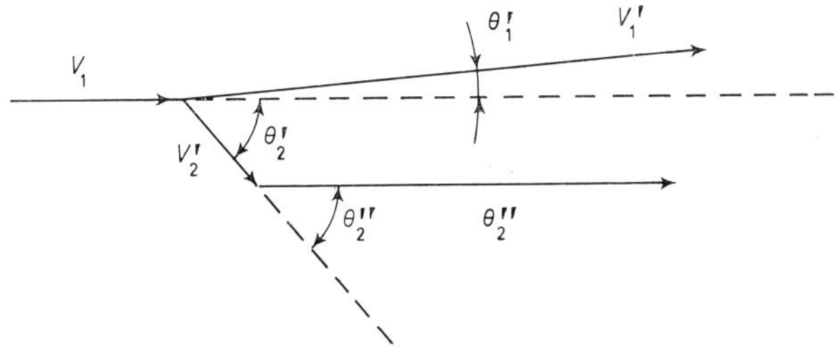

Fig. 22. – The capture mechanism (impulse approximation) at very high energies.

of this $X^+—{}^1H—{}^2H$ sequence. Before the collision the projectile X^+ has the velocity v_1 and the target H_2 is at rest. In the first binary collision, X^+ is scattered by the ^1H atom with velocity v_1', into an angle θ_1', while ^1H is scattered with velocity v_2' and angle θ_2'. In the second binary collision, ^1H is scattered with velocity v_2'' into an angle θ_2'' while ^2H is scattered in a way that need not be specified. The result of these two collisions is a reduction of the relative kinetic energy between X^+ and ^1H. Capture of ^1H by X^+ may occur if

$$v_2'' \approx v_1' \,. \tag{5.1}$$

It can be shown by classical mechanics that $\theta_2' \approx \theta_2'' \approx 45°$ in order to satisfy the condition of eq. (5.1). The cross-section of the capture process is given by the product of the probabilities for scattering into a cone with a half-angle between θ_2' and $\theta_2' + d\theta_2'$ with velocities between v_2' and $v_2' + dv_2'$ in the first collision and for scattering into a cone with a half-angle θ_2'' to $\theta_2'' + d\theta_2''$ under the azimutal angle, $d\psi/4\pi r^2$ where r is the equilibrium distance of the H_2 molecule. In order for capture to occur the final relative velocity of X^+ and ^1H, $v_1' - v_2''$, must be contained in a volume of velocity space determined by the dissociation energy D of XH^+. In the center-of-mass system, the cross-section for capture is

$$q = \gamma \frac{4\cdot\sqrt{2}\,\pi\sigma_1(90^\circ)\,\sigma_2(90^\circ)}{3r^2}\left(\frac{2D}{m_2 v_1^2}\right)^{\frac{3}{2}}, \tag{5.2}$$

where σ_1 and σ_2 are the differential cross-sections of the two binary collisions and γ is a factor close to unity. At very high energies, the electron shells of the particles penetrate each other and the scattering occurs under the influence of the Coulomb repulsion of the atomic nuclei. The differential cross-section of such « Rutherford » scattering processes varies as v^{-4}. Since both σ_1 and σ_2 are proportional to v^{-4}, q will be proportional to v_1^{-11} at very high energies E_1 of the incident ion. Absolute cross-sections for the reaction $Ar^+ + D_2 \rightarrow ArD^+ + D$ calculated from eq. (5.2) are plotted *vs.* E_1 in Fig. 23.

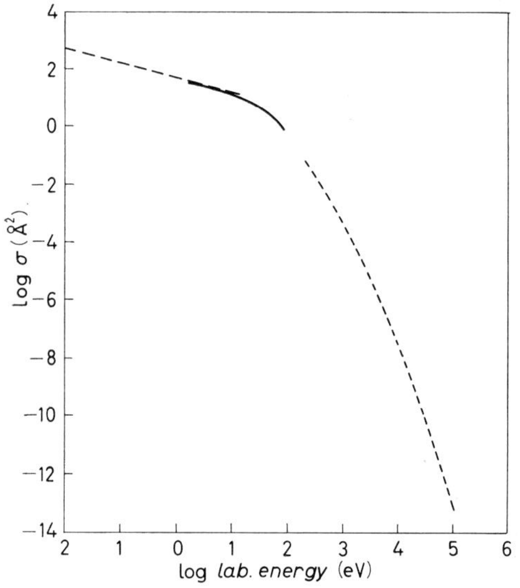

Fig. 23. – Reaction $Ar^+ + D_2 \rightarrow ArD^+ + D$. Broken line: cross-section *vs.* energy calculated from the polarization theory (eq. (3.21)). Dashed line: cross-section *vs.* energy calculated from the impulse approximation (eq. (5.2)). Full line: experimental values. — — — Polarization theory; - - - impulse approximation.

The required differential scattering cross-sections σ_1 and σ_2 were computed using classical mechanics and potentials calculated with eigenfunctions derived by the Hartree method [46].

Quantum-mechanical modifications of the impulse approximation do not alter the results significantly [47]. The cross-sections calculated from eq. (3.21) (polarization theory for low energies) are also given in the Figure. The solid line shows the experimentally determined cross-sections which were measured in an energy range where the assumptions of neither the polarization theory nor the impulse approximation are fulfilled. In this range the spectator stripping model, a special case of the impulse approximation in which the second binary collision does not occur, fits the experimental results.

5.4. *Deviations at low energies.* – At kinetic energies of a few electron volts of the incident ion, the ratio of the product-ion velocity to the initial velocity increases for the reactions $X^+ + H_2 \rightarrow XH^+ + H$ (see Fig. 19 *a*), *b*)). Furthermore, the angular distribution of the product-ion peaks at an angle closer to the primary-ion peak than is expected from the spectator stripping model (Fig. 15 *a*) and 16 *a*)).

These results are also inconsistent with the formation of a long-lived complex during the reaction. In this case, a symmetrical angular distribution around the c.m. would be expected and the maximum in the velocity spectra would lie at lower energies than are observed. For these simple hydrogen-transfer reactions, therefore, there is no transition from a direct to complex formation mechanism even at c.m. energies of the order of 0.02 eV. Thus the common supposition is shown to be unjustified that a complex will be formed in a reaction at low energy if there is a strong attraction between the reactants [38]. HERMAN *et al.* [38] have constructed a model to explain the experimental results for the $X^+ + D_2$ reaction at low kinetic energies. A schematic representation is shown in Fig. 24. This « modified stripping » model takes into account the long range polarization forces which have been shown to determine the cross-section for close approach at low energies (see Subsect. 3·1).

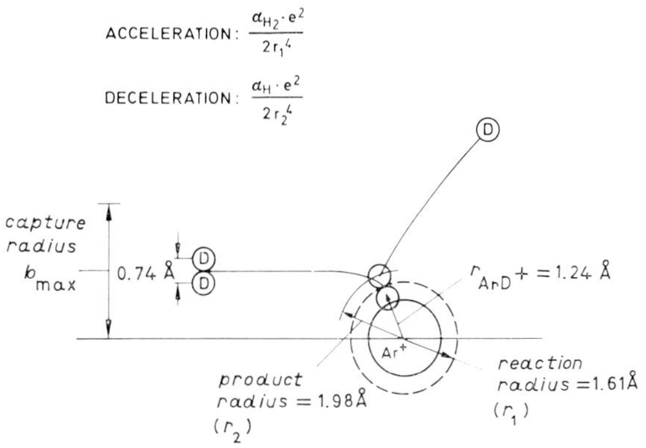

Fig. 24. – Schematic representation of the modified stripping model [38].

The D_2 approaching is accelerated towards the ion by ion-induced dipole forces. When the « reaction radius » r_1 is reached, the initial kinetic energy of the two particles has been increased by the amount $\alpha_{D_2} \cdot e^2/2r_1^4$, where α_{D_2} is the polarizibility of the D_2 molecule. One D atom is detached in a strippinglike process, and the other D proceeds with the same speed but a different direction from that of D_2 at the moment of reaction. In receding, it is decelerated by ion-induced dipole forces acting between ArD^+ and D. The deceleration potential is equal to $\alpha_D \cdot e/2r_2^4$, where α_D is the polarizability of the D atom and r_2 is the « product radius ». r_1 and r_2 are related by one half the bond distance of D_2, and are determined from the measurements of the energy of the product ion at the lowest relative kinetic energy. For the $Ar^+ + D_2$ reaction the increase in energy due to the acceleration of the reactants is greater than the decrease in energy due to the deceleration of the products. The net result, therefore, is an increase in the velocity of the ArD^+ ion over that expected from spectator stripping. Although several assumptions are necessary for the quantitative application of the model, it gives plausible values for r_1 and r_2 (Fig. 24) and agrees with the experimentally obtained dependence of Q on E [38] (see below). The exoergicity Q (eq. (4.15)) can be calculated for the maximum of the product ion band (see Fig. (19)). Q is equal to

$$Q = E_3 + E_4 - E_1 \, . \tag{5.3}$$

By expressing E_3 and E_4, the translational energies of the products in the laboratory system, as functions of E_1 and the velocity ratio $f = v_3/v_1$ which is plotted on the scales of Fig. 19, one obtains

$$Q = E_1 \left(\frac{M_3}{M_1} f^2 + \frac{(fM_3 - M_1)^2}{M_4 \cdot M_1} - 1 \right) . \tag{5.4}$$

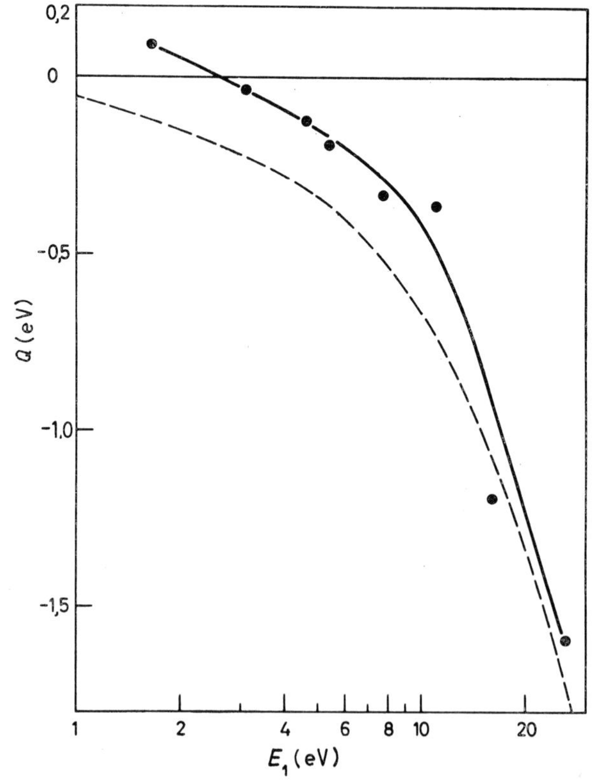

Fig. 25. – Reaction $N_2^+ + D_2 \rightarrow N_2D^+ + D$. Translational exoergicity as function of the incident ion energy (lab. s.). Solid line: from the data of Fig. 19. Dashed line: from eq. (4.19) for spectator stripping.

Q is plotted for the reaction $N_2^+ + D_2 \rightarrow N_2D^+ + D$ in Fig. 25 (solid line). The Figure also shows $Q(E_1)$ as a calculated from eq. (4.18) for spectator stripping (dashed line). It is seen that the two curves approach each other in the intermediate energy range of 20 eV. As the initial energy decreases the Q value becomes less negative and reaches a value of about $+0.1$ V at the lowest energy. A positive Q value is only possible if the reaction is exothermic and part of the heat of the reaction appears as translational energy of the products. Since the heat of the reaction is about 1 eV, only a small amount of the heat appears as translational energy of the products. Most of the heat of reaction is apparently contained in the product ion as vibrational or rotational excitation. At high relative energies the contribution to changes in internal energy from the heat of reaction is negligible.

Finally, isotope effects in reactions of the type

$$X^+ + HD \begin{cases} \nearrow XH^+ + D \\ \searrow XD^+ + H \end{cases}$$

may be used to discuss the collision mechanism.

Figure 26 shows the dependence of the intensity ratio ArH^+/ArD^+ on the energy E_1 of the incident ion for the reactions of Ar^+ with HD. The ratio is smaller than unity at very low energies [44]. This effect can be explained by formation of a complex $ArHD^+$ according to the theory of unimolecular reactions. Since the ArD^+ bond is a little stronger than the ArH^+ bond and since the frequency factor for the splitting-off H is a little larger than for D, the ratio ArH^+/ArD^+ should be slightly smaller than unity.

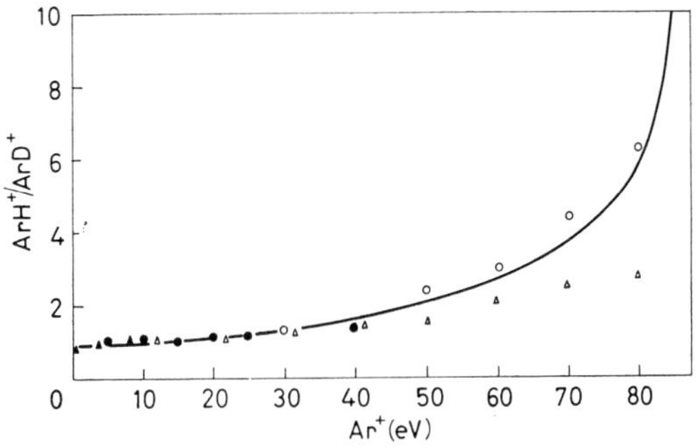

Fig. 26. – ArH^+/ArD^+ intensity ratio in the reaction $Ar^+ + HD$ as a function of initial energy in the laboratory system [32]. Results of other authors are included [41-44]. △, FUTRELL and ABRAMSON, 1966; ●, BERTA, ELLIS and KOSKI, 1966; ▲, KLEIN and FRIEDMANN, 1964; ○: LACMANN and HENGLEIN, 1965.

Conclusions about the kinematics of a reaction which are based on the interpretation of isotope effects are less convincing than direct measurements of the angular and velocity distribution of the products. Since at low energies these experiments lead to different conclusions, the first interpretation of the curve in Fig. 26, *i.e.* stripping at high energies and complex formation at low energies [32, 33] may have to be revised. However, a satisfying explanation of the basis of the « modified stripping » model has not yet been given. It might be noted that the term « complex » in beam studies must not always have the same meaning as the term « complex » in the mechanical-statistical treatment, statistical distribution of energy among the degrees of freedom. The results of the beam experiments and isotope effects in the reaction of Ar^+ with hydrogen could perhaps be understood in terms of an intermediate which lives long enough to allow statistical distribution of part of the excess energy but short enough the decay in an anisotropic manner in the c.m. system.

6. – Complex → stripping transitions.

6·1. *The reaction* $CH_4^+ + CH_4 \rightarrow CH_5^+ + CH_3$. – Kinematic evidence for the formation of intermediate complexes which isotropically decay and transitions between the complex and stripping mechanism have first been obtained in the simple discrimination experiments mentioned in **4·1** [27, 28]. For example, the proton transfer

$$CH_4^+ + CH_4 \rightarrow CH_3 + CH_5^+ \tag{6.1}$$

has been found to proceed via a complex at low and as stripping process at high energies. HENCHMAN and his co-workers [49] have developed a pulse technique for studying the kinematics of ion-molecule reactions below $E_1 = 1.5$ eV. It involves the pulsing of various electrodes in a mass spectrometer ion source and achieves velocity analysis of the ions by measuring their residence times in the ionization chamber. They could definitely show that the H atom transfer in methane

$$CH_4^+ + CH_4 \rightarrow CH_5^+ + CH_3 \tag{6.2}$$

proceeds via a complex at low ion energies.

The spectra of the translational energy in the forward direction for both primary CD_4^+ ion and the product-ion CD_5^+ of the corresponding D transfer are shown by Fig. 27 for two energies of the incident ion. The spectra have been obtained by using the apparatus of Fig. 20 in which CD_5^+ ions are collected that experienced scattering up to 10°. The positions expected for the center of the CD_5^+ band according to the complex and stripping models are marked by « K » and « S », respectively. Two bands appear in the spectrum at $E_1 = 7$ eV (laboratory system) which has been explained by simultaneous occurrence of complex and stripping processes. At lower energies, complex formation is predominant, while stripping is more important at higher energies [50].

It seems noteworthy to emphasize that appreciable complex formation occurs in this reaction even at a c.m. energy of 3.5 eV ($E_1 = 7$ eV). At this c.m. energy, reactions of the type $X^+ + D_2 \rightarrow XD^+ + D$ ($X^+ = Ar^+$, N_2^+, CO^+)

were found to obey the stripping model as mentioned in Subsect. **4**˙1. It has been proposed that the large number of internal degrees of freedom in the reactants (or complex) of the methane reaction favours complex formation [50].

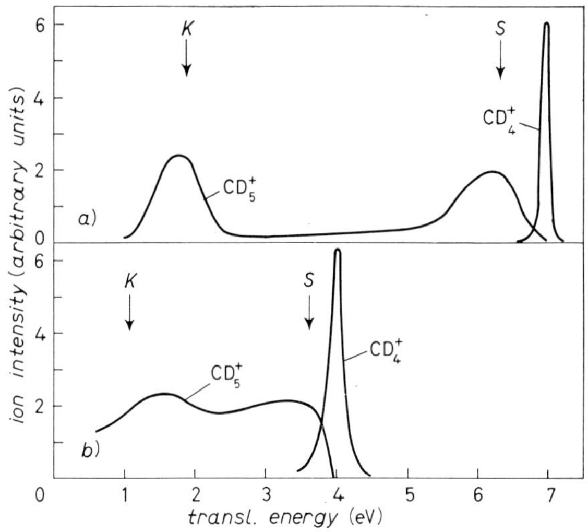

Fig. 27. – Energy spectra of primary and secondary ion for the reaction $CD_4^+ + CD_4 \rightarrow$ $\rightarrow CD_5^+ + CD_3$ at two different laboratory energies of the primary ion. *S* and *K* indicate the energy positions expected for the product ion according to the spectator stripping and complex models, respectively. *a*) $E_1 = 7$ eV, *b*) $E_1 = 4$ eV.

6˙2. *Reactions of the type* $X^+ + CD_4 \rightarrow XD^+ + CD_3$. – Reactions of this type where $X^+ = Ar^+$ or N_2^+ have already been mentioned in Subsect. **4**˙2. They proceed as stripping processes at high energies. Figure 28 shows the velocity of the product ion band (in units of v_1) as a function of the energy of the N_2^+ ion. The upper and lower curves give the band width (80% width), the middle curve shows the velocity of the band center. It can be seen that the band is

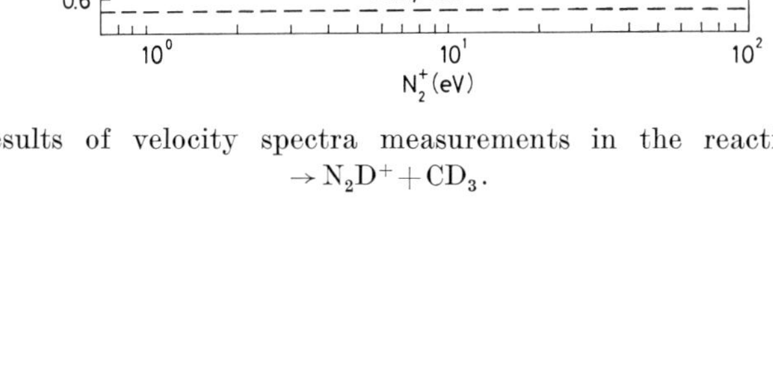

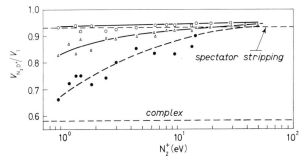

Fig. 28. – Results of velocity spectra measurements in the reaction $N_2^+ + CD_4 \rightarrow$ $\rightarrow N_2D^+ + CD_3$.

strongly broadened at low energies, the bulk of the intensity moving to lower velocities. This effect is explained by a stronger interaction of the incident N_2^+ ion with the CD_3 spectator at low energies. Within the range of available primary ion energies, the band of the product ion never reaches a symmetric distribution around the c.m. velocity of $0.581\,v_1$. It may be expected from a rough extrapolation of the curve that a complex is formed at primary ion energies of a few tenths of an eV. A similar effect has been observed for the reaction of Ar^+ with CD_4 [48].

6·3. *Reactions of the methanol ion. Kinematics of an endothermic process.* – The results obtained for the reaction

$$CH_3OH^+ + CD_4 \rightarrow CH_3OHD^+ + CD_3\,, \tag{6.3}$$

are shown by Fig. 29. The energy in the forward direction of the product ion in units of the incident ion energy E_1 is plotted here *vs.* E_1 or the c.m. energy. The vertical lines indicate the 80% band width, the points give the position

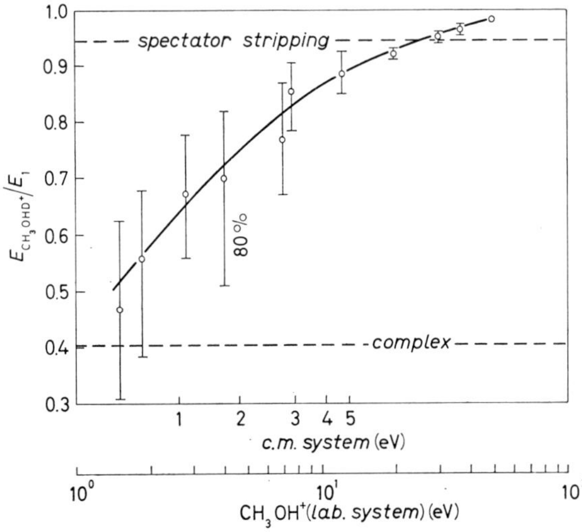

Fig. 29. – Results of translational energy distribution measurements in the reaction $CH_3OH^+ + CD_4 \rightarrow CH_3OHD^+ + CD_3$. The vertical lines indicate the energy limits which include 80% of the product-ion intensity. The dashed lines are the calculated energies.

of the center of the band. As can be recognized, stripping occurs at high energies. With decreasing E_1, the band becomes broader and shifts towards lower energies. At $E_1 \sim 1$ eV (laboratory system), the product band has reached a symmetric distribution around the energy $E_3 = (M_1 \cdot M_3/(M_1 + M_2)^2)E_1 = 0.403 \cdot E_1$ expected for intermediate complex formation.

A « Complex → Stripping-Transition-Energy » may be defined as that energy (c.m. system) of the reactants where the bulk of the intensity of the product-ion appears at a velocity half-way between the velocity for stripping (eq. (4.8)) and the velocity of the center of mass. Inspection of Fig. 29 shows that the transition energy amounts to 1.3 eV for the reaction of CH_3OH^+ with CD_4 [51].

The reaction

$$CH_3OH^+ + D_2 \rightarrow CH_3OHD^+ + D \tag{6.4}$$

has a threshold energy as can be recognized from the dependence of the cross-section on energy in Fig. 30. The current ratio of the secondary and primary ion is used here as a measure of the relative cross-section. A rough extrapolation to zero cross-section at low energies leads to a threshold of 0.2 eV. It should be emphasized that the CH_3OH^+ ion will be formed in various vibrationally excited states by electron impact, each of which probably has a distinct threshold energy for its reaction with CD_4. The curve in Fig. 30 therefore represents a complicated superposition of excitation functions for various states of CH_3OH^+. The threshold probably is due to the endothermicity of the reaction.

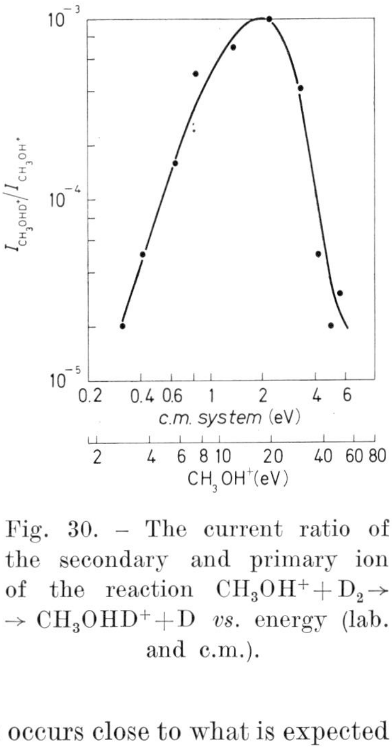

Fig. 30. – The current ratio of the secondary and primary ion of the reaction $CH_3OH^+ + D_2 \rightarrow$ $\rightarrow CH_3OHD^+ + D$ *vs.* energy (lab. and c.m.).

In Fig. 31, the energy of the product ion in units of E_1 is plotted *vs.* E_1 or the c.m. energy, the band width indicated by the length of the vertical lines has been corrected for the width of the band of the primary ion. The lines therefore give the increase in bandwidth. The points show the position of the band center. At high energies, the process occurs close to what is expected from the spectator stripping model. The band shifts to lower energies and reaches symmetric distribution around the energy expected for intermediate complex formation: $E_3/E_1 = M_1 M_3/(M_1 + M_2)^2 = 0.840$. During this shift, the band at first becomes broader and then narrows again slightly above the threshold. Apparently, the complex can decompose into the products just above the threshold without providing them with translational energy (in the c.m. system). At higher energies more « exploding » complexes are produced until the impulsive mechanism becomes predominant. As can also be seen from the Figure, there exists a narrow intermediate energy range around 4 eV (c.m system) where two peaks appear in the spectrum of the product ion, as has already been observed from the reaction $CD_4^+ + CD_4$ (Fig. 27, upper part). The complex → transition energy amounts to about 4 eV (c.m. energy) [51].

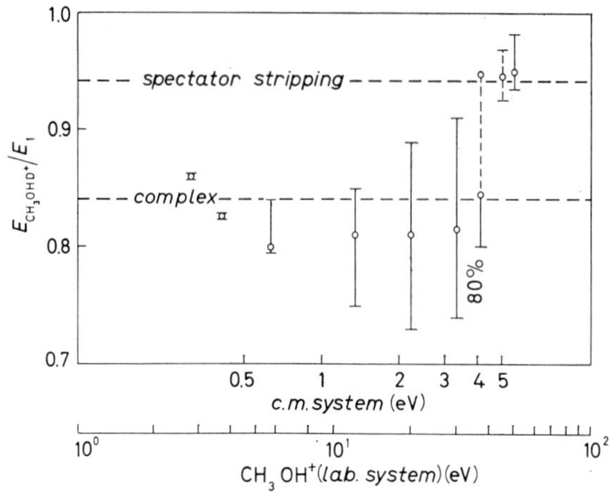

Fig. 31. – Energy measurements of the product ion from $CH_3OH^+ + D_2 \rightarrow CH_3OD^+ + D$. The vertical lines give the increase in bandwidth as compared to the band of the primary ion.

6'4. *Internal-energy relaxation during a chemical reaction.* – The complex → → stripping transition energies of various reactions are plotted in Fig. 32 *vs.* the number N of atoms in the incident ion plus neutral molecule. In each of these

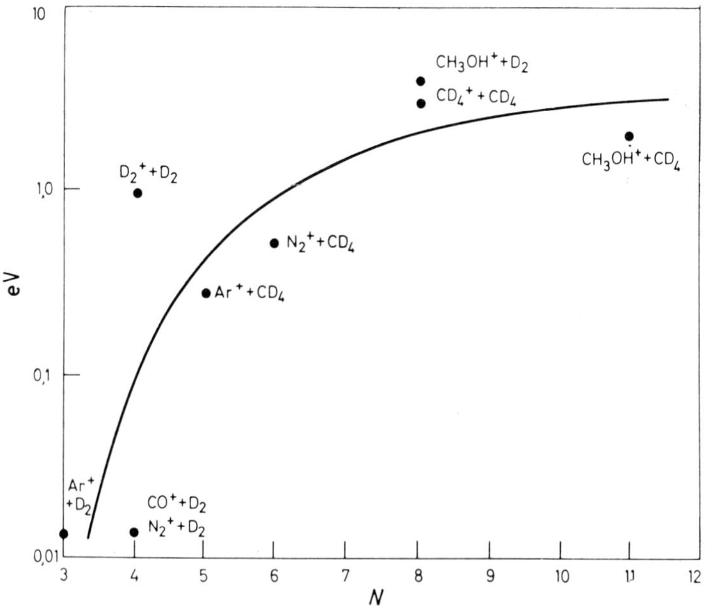

Fig. 32. – The complex→stripping transition energy *vs.* the number of atoms in the reactants for various D atom transfer reactions.

reactions, a D atom is transferred from the molecule to the ion. The transition energies for the reactions $X^+ + D_2$ ($X^+ = Ar^+, N_2^+, CO^+$) could not be determined kinematically because of the strong forward scattering of the product-ion at low energies due to the polarization force (Fig. 19). They were roughly estimated from the isotope effects described by Fig. 26. The data from DURUP and Durup's [52] work were taken for the reaction $D_2^+ + D_2$.

The transition energy (c.m. system) increases with the number of atoms in the reactants. In the logarithmic scale on the ordinate axis in Fig. 32, the transition energy approaches a « limiting » value of about 3 eV for systems consisting of many atoms, *i.e.* the strengh of a chemical bond.

During the approach of the ion X^+ to the molecule D-R, a new bond $X^+ \dots$ D—R is being formed. The critical degree of freedom (which with respect to the $X^+ \dots$ D bond is at first of translational, later of vibrational nature) contains the relative kinetic energy of the system $X^+ + D$ after the collision, if spectator stripping occurs. *I.e.*, there is not time enough to allow the « flowing » of energy into the other internal degrees of freedom of either the molecule or ion. At lower energies, *i.e.* at longer collision times, the energy distribution among the various degrees of freedom becomes more and more perfect until a complex is formed that lives longer than half a period of rotation and has completely lost its « memory » of the forward direction. In the ideal case, a « statistical » complex as described in Subsect. **2**.4 is formed [51).

Apparently, the formation of a complex is not only a question of the interaction time, but is strongly influenced by the number of internal degrees of freedom available for the distribution of the excess energy. It is concluded from Fig. 32 that at about $N = 8$ (corresponding to 18 degrees) sufficiently rapid energy relaxation occurs in order to make complex formation still possible at 3 eV (c.m. system) in polyatomic systems. The collision time at this energy can readily be calculated from the relative velocity and an assumed interaction distance of 6 Å. A time of about 10^{-13} s was obtained this way. It may be taken as a rough estimate of the relaxation time for internal energy distribution in polyatomic systems (including the critical degree of freedom). Internal energy distribution is a very rapid process, provided that a large number of degrees of freedom is available [51].

This fact is the prerequisite for the application of the quasi-equilibrium theory of mass spectra of polyatomic molecules [53]. The transition state of a chemical reaction can only be regarded as a species of well-defined thermodynamic properties, if internal energy distribution is established during the approach of the reactants. If a complex is formed in kinematic studies (life time larger than half a period of rotation), it can be assumed that the energy is well distributed. Since the transition energy in Fig. 32 is far above 1 eV for polyatomic systems ($N > 8$), which corresponds to « temperatures » far higher than generally used in chemistry, it can be concluded that the transition state theory is applicable. Only in the case of a few simple systems, such as $Ar^+ + D_2$, where an impulsive mechanism is observed even at room temperature, the transition state theory may not always be used.

7. – Comparisons with crossed molecular beam studies.

Kinematic studies have frequently been carried out with crossed molecular beams of alkali atoms and various alkali, organic and molecular halides [54, 55]. It seems suitable to make a few comparisons between these investigations and the kinematic studies of ion-molecule reactions.

Since the energy of the incident ion can easily be changed over a wide range, ion-molecule reactions are particularly suitable for studying the changes in the collision mechanism with energy. The observed complex $\to$ stripping transitions with increasing energy and finally the description of D atom transfer reactions by the impulse approximation at very high energies are typical examples for changes in the kinematics of a chemical reaction. It also seems noteworthy that similar mechanisms are well-known in nuclear physics, such as the formation of a compound nucleus at low energies, stripping in deuteron reactions and spallation processes at higher energies. Changes in the kinematics of chemical reactions between neutral reactants have not yet been studied since it is still difficult to produce beams of high-energy particles with sufficient intensity.

On the other hand, a lot of information has been obtained for reactions in crossed molecular beams at temperatures of a few 100 °K. The distribution of the heat of reaction among the various degrees of freedom of the reactants has frequently been determined by either kinematic studies or by direct investigation of the states of the products such as by the method of infrared chemiluminescence [56]. Impulsive mechanisms (such as «rebound» in $K + CH_3I \to \to KI + CH_3$ or «stripping like» in $K + Br_2 \to KBr + Br$) as well as complex formation (such as $M_1 + M_2X \to M_2 + M_1X_1 \cdot M =$ alkali, $X =$ halide) have been observed and a lot of knowledge about the behaviour of such complexes has been obtained from the statistical treatment by the RRK method. Furthermore, comparisons of reactions with nonreactive scattering in these systems have given valuable information about intermolecular potentials. Chemical potential surface calculations which can be compared with the experimental results gave additional information about the dynamics of these reactions. Detailed studies of this kind have not yet been carried out on ion-molecule reactions. The work with ion beams of low energy is still in the beginning. The determination of a few Q-values (translational exoergicity) and of the scattering pattern of a few D atom transfer reactions down to about 0.1 eV c.m. energy shows at least that studies in this energy range are possible and promising. The opinion has sometimes been expressed that kinematic studies of ion-molecule reactions at low energies are not so fruitful, since the particles are moving in the boring r^{-4} potential of the ion-induced dipole interaction which prevents one from recognizing the more interesting potential of the chemical forces. We do not quite share this opinion since after all the reactions of ions constitute an important part of chemistry.

A few words should also be said about the question of internal energy-relaxation and the impulsive character of a reaction. The observation of complex $\rightarrow$ stripping transitions shows that all kinds of intermediate situations between these extreme cases are possible. For example a nonisotropic scattering pattern of the products from a simple atom transfer reaction must not necessarily mean that a strictly « impulsive » mechanism is operative, *i.e.* that one end of a polyatomic molecule is attacked while the rest of the molecule acts as a hard sphere. In polyatomic systems, the time for energy distribution among the various degrees of freedom (including the critical degree) of a « complex » that is decaying or just being formed has been found to be so short (some 10^{-14} to 10^{-13} s) that a certain degree of statistical distribution of the excess energy is conceivable even if the « complex » lives shorter than half a period of rotation. Phenomena related to this problem have not only been met in ion-molecule studies but seem also to have been observed in crossed molecular beam studies. For example, in reactions of the type $K+RI \rightarrow KI+R$ where R represents an organic group, the scattering pattern indicates an impulsive mechanism of the rebound type although the organic rest internally retains a significant fraction of the excess energy, this fraction being the larger the more internal degrees of freedom are present in R [54].

* * *

I am indebted to Dr. LACMANN, Dr. DING, Dr. ROSE, and Mr. MITTMANN for their kind help in revising the manuscript.

REFERENCES

[1] H. D. SMYTH: *Rev. Mod. Phys.*, **3**, 347 (1931); T. R. HOGNESS and R. W. HARKNESS: *Phys. Rev.*, **32**, 784 (1928).
[2] V. L. TALROZE and A. K. LYUBIMOVA: *Dokl. Akad. Nauk SSSR*, **86**, 909 (1952).
[3] W. A. CHUPKA, M. E. RUSSELL and K. RAEFEY: *Journ. Chem. Phys.*, **48**, 1518 (1968).
[4] F. H. FIELD: *Journ. Amer. Chem. Soc.*, **83**, 1523 (1961).
[5] P. KEBARLE and A. M. HOGG: *Journ. Chem. Phys.*, **42**, 798 (1965).
[6] F. C. FEHSENFELD, A. L. SCHMELTEKOPF, P. D. GOLDAN, H. I. SCHIFF and E. E. FERGUSON: *Journ. Chem. Phys.*, **44**, 4087 (1966).

[7] A. Henglein: *Zeits. Naturforsch.*, **17** a, 44 (1962).
[8] N. Boelrijk and W. H. Hamill: *Journ. Amer. Chem. Soc.*, **84**, 730 (1962).
[9] F. W. Lampe, J. L. Franklin and F. H. Field: *Progress in Reaction Kinetics*, vol. **1** (Oxford, 1961), p. 69.
[10] A. Henglein and G. A. Muccini: *Zeits. Naturforsch.*, **15** a, 584 (1960).
[11] V. Cermak and Z. Herman: *Journ. Chim. Phys.*, **57**, 717 (1960).
[12] A. Henglein: *Zeits. Naturforsch.*, **17** a, 37 (1962).
[13] A. Henglein, G. Jacobs and G. A. Muccini: *Zeits. Naturforsch.*, **18** a, 98 (1963).
[14] W. Kaul, U. Lauterbach and R. Taubert: *Zeits. Naturforsch.*, **16** a, 624 (1961); T. F. Moran and L. Friedman: *Journ. Chem. Phys.*, **39**, 2491 (1963).
[15] F. H. Field, J. L. Franklin and F. W. Lampe: *Journ. Amer. Chem. Soc.*, **79**, 2419, 2665 (1957).
[16] R. Fuchs: *Zeits. Naturforsch.*, **16** a, 1026 (1961).
[17] R. F. Pottie and W. H. Hamill: *Journ. Phys. Chem.*, **63**, 877 (1959).
[18] A. Henglein: *Zeits. Naturforsch.*, **17** a, 44 (1962).
[19] P. Langevin: *Ann. Chim. Phys.*, **5**, 245 (1905).
[20] G. Gioumousis and D. P. Stevenson: *Journ. Chem. Phys.*, **29**, 294 (1958).
[21] D. Hyatt and K. Lacmann: *Zeits. Naturforsch.*, **23** a, 2080 (1968).
[22] C. F. Giese and W. B. Maier II: *Journ. Chem. Phys.*, **39**, 739 (1963).
[23] J. B. Homer, R. S. Lehrle, S. C. Robb and D. W. Thomas: *Nature*, **202**, 795 (1964).
[24] B. R. Turner, M. A. Fineman and R. F. Stebbings: *Journ. Chem. Phys.*, **42**, 4088 (1965).
[25] E. R. Weinar, G. R. Hertel and W. S. Koski: *Journ. Amer. Chem. Soc.*, **86**, 788 (1964).
[26] T. F. Moran and W. H. Hamill: *Journ. Chem. Phys.*, **39**, 1413 (1963).
[27] A. Henglein and G. A. Muccini: *Zeits. Naturforsch.*, **17** a, 452 (1962).
[28] A. Henglein and G. A. Muccini: *Zeits. Naturforsch.*, **18** a, 753 (1963).
[29] A. Henglein, K. Lacmann and G. Jacobs: *Ber. Bunsenges. Phys. Chem.*, **69**, 279 (1965).
[30] K. Lacmann and A. Henglein: *Ber. Bunsenges. Phys. Chem.*, **69**, 286 (1965).
[31] K. Lacmann and A. Henglein: *Ber. Bunsenges. Phys. Chem.*, **69**, 292 (1965).
[32] A. Henglein, K. Lacmann and B. Knoll: *Journ. Chem. Phys.*, **43**, 1048 (1965).
[33] A. Henglein: *Adv. Chem. Ser.*, **58**, 63 (1966).
[34] A. Ding, K. Lacmann and A. Henglein: *Ber. Bunsenges. Phys. Chem.*, **71**, 596 (1967).
[35] A. Ding: Dissertation, Technische Universität Berlin, 1968.
[36] L. D. Doverspike and R. L. Champion: *Journ. Chem. Phys.*, **46**, 4718 (1967).
[37] L. D. Doverspike, R. L. Champion and T. L. Bailey: *Journ. Chem. Phys.*, **45**, 4385 (1966).
[38] Z. Herman, J. Kerstetter, T. Rose and R. Wolfgang: *Disc. Faraday Soc.*, **44**, 123 (1967).
[39] W. R. Gentry, E. A. Gislason, Y. T. Lee, B. H. Mahan and Ch.-W. Tsao: *Disc. Faraday Soc.*, **44**, 137 (1967).
[40] E. Gustafsson and E. Lindholm: in J. B. Hasted: *Physics of Atomic Collisions* (London, 1964), p. 128.
[41] J. H. Futrell and F. P. Abramson: *Adv. Chem. Ser.*, **58**, 107 (1966).
[42] M. A. Berta, B. Y. Ellis and W. S. Koski: *Adv. Chem. Ser.*, **58**, 80 (1966).
[43] F. S. Klein and L. Friedmann: *Journ. Chem. Phys.*, **41**, 1789 (1964).
[44] D. P. Stevenson and D. O. Schissler: *Journ. Chem. Phys.*, **23**, 1353 (1955).

[45] R. D. Fink and J. S. King: *Journ. Chem. Phys.*, **47**, 1857 (1967).
[46] D. R. Bates, C. J. Cook and F. J. Smith: *Proc. Phys. Soc.*, **83**, 49 (1964).
[47] G. K. Ivanov and Yu. S. Sayasov: *V Intern. Conference, Physics of Electronic and Atomic Collisions* (Leningrad, 1967), p. 238.
[48] A. Ding, A. Henglein, D. Hyatt and K. Lacmann: *Zeits. Naturforsch.*, **23** a, 2084 (1968).
[49] L. Matus, D. J. Hyatt and M. J. Henchman: *Journ. Chem. Phys.*, **46**, 2439 (1967).
[50] A. Ding, A. Henglein and K. Lacmann: *Zeits. Naturforsch.*, **23** a, 779 (1968).
[51] A. Ding, A. Henglein, D. Hyatt and K. Lacmann: *Zeits. Naturforsch.*, **23** a, 2090 (1968).
[52] J. Durup and M. Durup: *Journ. Chim. Phys.*, **64**, 386 (1967).
[53] H. M. Rosenstock, M. B. Wallenstein, A. L. Wahrhaftig and H. Eyring: *Proc. Nat. Acad. Sci.*, **38**, 667 (1952).
[54] D. R. Herschbach: *Adv. Chem. Phys.*, **10**, 319 (1966).
[55] W. B. Miller, S. A. Safron and D. R. Herschbach: *Disc. Faraday Soc.*, **44**, 108 (1967).
[56] J. C. Polanyi: *Trans. Roy. Soc. Canada*, **5**, 105 (1967).
[57] P. J. Kuntz, M. H. Mok and J. C. Polanyi: *Journ. Chem. Phys.*, **50**, 4623 (1969).

Gas-Phase Proton-Transfer Reactions.

G. G. Volpi

Istituto di Chimica dell'Università - Perugia

1. – Introduction.

This lecture is focussed on some aspects of proton-transfer reactions, as investigated in the gas phase by high-pressure mass spectrometry. The importance of such reactions in solution chemistry has long been recognized in a large variety of instances; and around them is founded one of the most venerable and useful concepts of chemistry, that of acids and bases. In the gas phase, however, the occurrence of proton transfer from an ion to a molecule has been discovered only in the last decade and our knowledge about these processes is still scanty [1]. This in spite of the fact that protonated species may play an important role in radiation chemistry, [2] and in particular in the polymerization of unsaturated hydrocarbons [3]. Proton-transfer processes belong to the general area of ion-molecule reactions, and therefore their study is carried out along two lines: first the establishment of the occurrence and mechanism involved in these processes in ionized gases, and the extraction of information concerning chemical properties of protonated ions in the absence of solvent effects; second, the detailed investigation of selected reactions, in order to extract a knowledge of their dynamics [4]. From this second point of view, proton transfer processes are extremely attractive for the kineticist, because of their unique simplicity, at least in principle. However, the techniques which have been employed so far for the investigations of these reactions (basically mass spectrometry at moderate ($10^{-4} \div 10^{-2}$) Torr or high ($> 10^{-2}$ Torr) pressure in the ion source,) give information concerning the first one of the aforementioned lines, namely only about the reaction mechanism and, possibly, rates. But, as far as the dynamics are concerned, only a limited amount of information can be inferred, and therefore one must presumably await the exploitation of molecular beam methods in this field too. In the meantime, we think it worthwhile to review some aspects of proton transfer processes which have emerged out of high-pressure mass spectrometric studies, stressing the results which we believe to be more interesting from the viewpoint of chemical physicists and more inviting in the sense of a full exploitation of the reaction dynamics.

2. – The apparatus. H_3^+ as a proton donor.

The apparatus which has been employed in these experiments [5] is shown, schematically, in Fig. 1. It consists basically of a cylindrical box, where ionization is produced by the beta-rays of tritium, adsorbed on the walls of the chamber itself. The use of beta-rays as the ionizing medium is a departure from the electron bombardment method commonly used in mass spectrometry,

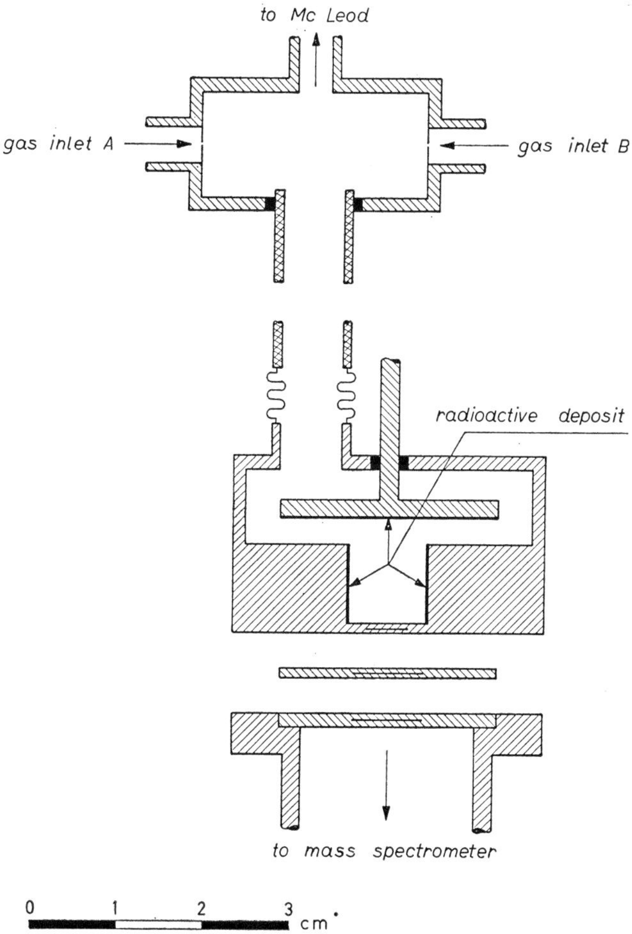

Fig. 1. – A schematic drawing of the apparatus.

and was chosen because: *a*) it avoids the presence of heated filaments, which may cause unwanted pyrolysis effects, *b*) it allows the construction of a « closed » box, where the pressure can be raised up to ~ 1 Torr without the danger of burning filaments (this simplifies the differential pumping requirements), *c*) it produces homogeneous ionization in the bulk of the gas even at the higher pressures. An obvious drawback of this set-up is that ions are produced everywhere in the box, and when a field is applied to the repeller plate they are pushed toward the slit of the mass spectrometer; therefore one does not have a definite residence time or energy for the ions, but rather broad distributions.

When hydrogen is admitted into the chamber at a pressure in the range 0.1 to 1.0 Torr (as is typical of these experiments) protonated hydrogen, H_3^+, is produced, through the reaction

$$H_2^+ + H_2 \rightarrow H_3^+ + H \tag{1}$$

and H_3^+ is practically the only ion present in the mass spectrum. As long ago as 1936, reaction (1) and the recombination of H_3^+ with electrons were suggested as the source of H atoms responsible for the radiation-induced exchange between hydrogen and deuterium. [6] Experimental evidence, however, came in the late fifties [7] that hydrogen atoms are not responsible for the radiation-induced exchange, and only recently the newly proposed mechanism was verified. [5]

When a mixture of hydrogen and deuterium is irradiated, H_2^+ and D_2^+ ions are produced and react through reactions of type (1), to give isotopically mixed ions, for example H_2D^+. These latter initiate a proton-transfer chain, such as

$$H_2D^+ + D_2 \rightarrow HD + D_2H^+ ,$$

which leads to an isotopic exchange.

Protonated hydrogen, H_3^+, is probably an equilateral triangle, and its heat of formation is about 280 kcal/mole. The proton affinity of hydrogen is thus about 85 kcal/mole. [8] Hydrogen, therefore, behaves [5] as a base in the Brønsted sense with protonated argon:

$$ArH^+ + H_2 \rightarrow Ar + H_3^+ ,$$

but H_3^+ is a Brønsted acid with xenon:

$$H_3^+ + Xe \rightarrow XeH^+ + H_2 .$$

Likewise, H_3^+ behaves as a proton donor with all the molecules which have a higher proton affinity than hydrogen, and these include hydrocarbons, alcohols etc. This property, together with its ease of production and the possibility of using D_3^+ for deuterium, to decide some mechanistic questions, makes it particularly suitable for the investigation of gas-phase proton-transfer processes, even with a very simple apparatus.

3. – Proton transfer rates and mechanisms.

When a gas, with a proton affinity higher than that of H_2, is allowed into the chamber at a pressure much smaller than that of hydrogen, the intensity of H_3^+ ions, as recorded by the mass spectrometer, decreases, because of the occurrence of proton transfer between H_3^+ and the gas. From such a decrease it is possible to extract the rate constants for the proton transfer reaction, provided the residence time and the energy of the reacting ions in the chamber are known. In practice, however, as indicated above, the ions have a broad velocity distribution, so that to carry out the kinetic analysis it is necessary to postulate the energy-dependence of the rate constants themselves. The simplest way of doing that is to assume that the rate constants do not depend on the energy; it will be recognized that this is equivalent to invoking the Langevin model for these reactions. [4] This model assumes that the interaction between the ion and the molecule is of the charge-induced dipole type, and therefore is probably well suited for the reactions between H_3^+ and molecules without a permanent dipole, *e.g.* the saturated hydrocarbons. This analysis gives for these reactions between H_3^+ and saturated hydrocarbons rate constants of the order of 10^{-9} cm^3 $molecule^{-1}$ s^{-1}. Rate constants of the same order of magnitude were also obtained carrying on the same analysis with the data of the reactions with unsaturated and aromatic hydrocarbons, and alcohols. Since, however, some of these molecules have a permanent dipole, the assumed independence of the rate constants on the ion energy cannot be justified further. Therefore such values must be regarded only as phenomenological, *i.e.* related to the present apparatus.

Moreover, deuterium substitution methods show that, in all the investigated reactions, the net effect is passage of a proton from H_3^+ to the molecule. This fact, and the extremely high values of the rate constants, would suggest that the proton jumps when the ion and the molecule are still at large distances. This model, which resembles closely those which apply to charge transfer [10] and stripping reactions [11], imposes well-defined restrictions on the momentum transfer modes. Therefore it can be proved or disproved only on the basis of a detailed kinematic analysis of the products.

Proton transfer from H_3^+ to the systems investigated, namely the hydrocarbons and the alcohols, is highly exothermic, and the protonated intermediates have enough energy to decompose into fragments. These fragmentation processes were indeed observed, and are the general mode through which these reactions proceed: moreover, for most of these systems more than one fragmentation pattern takes place. A typical example is the reaction between H_3^+ and ethane, [8, 12] which we now discuss in detail to show the kind of information that can be obtained with the present technique.

In the mass spectrum obtained by adding a small partial pressure of ethane to hydrogen, the decrease in H_3^+ ions is balanced by the appearance of protonated ethane, $C_2H_7^+$ and ethyl ion, $C_2H_5^+$. Their relative abundance is generally a function of the pressure of H_2, P_{H_2}, the pressure of C_2H_6, $P_{C_2H_6}$, and the strength of the electric field, E. Since we are not interested now in the reactions of the above ions with ethane, we will consider here only the relative « yields » of $C_2H_7^+$ and $C_2H_5^+$ as a function of P_{H_2} and E, extrapolating them to the limit $P_{C_2H_6} \to 0$. At a fixed P_{H_2} and E, these « yields » represent the distribution of the products of the reaction between H_3^+ and C_2H_6, when the H_3^+ ions have the energy distribution typical of the field E and the reaction takes place at a total pressure of hydrogen P_{H_2}. Scans of the mass spectrum as a function of this latter parameter give information on the role of H_2 as a « third body » in the reaction studied. Fig. 2 *a*) shows an example of the results obtained. It can be seen that P_{H_2} favors a higher « yield » of the protonated ion, $C_2H_7^+$, *i.e.* prevents its fragmentation into $C_2H_5^+$ and H_2: We can therefore write

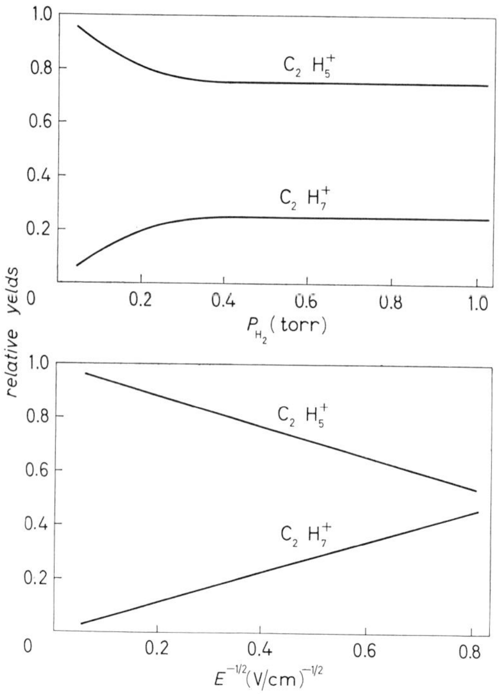

Fig. 2. – « Yields » of product ions for the reaction of H_3^+ with ethane: *a*) as a function of hydrogen pressure at field strength $E=4$ V/cm; *b*) as a function of $E^{-\frac{1}{2}}$ at $P_{H_2}=0.6$ Torr.

(2)
$$
\begin{aligned}
&H_3^+ + C_2H_6 \to C_2H_7^{+*} + H_2 \\
&\qquad\qquad\qquad \hookrightarrow C_2H_5^+ + H_2 \\
&\qquad\qquad + M \\
&\qquad\qquad\qquad \hookrightarrow C_2H_7^+ ,
\end{aligned}
$$

where the asterisk denotes internal excitation and M is the « third body ».

The lifetime of $C_2H_7^{+*}$ can be estimated from these results to be of the order of 10^{-7} s.

On the other hand, the Figure shows also that some of the $C_2H_5^+$ ions are formed through a process which is not affected by total pressure. We can write, therefore

$$H_3^+ + C_2H_6 \rightarrow (C_2H_7^+)^* + H_2 \qquad (3)$$
$$\quad\quad\quad\quad\quad\quad\quad \llcorner\!\rightarrow C_2H_5^+ + H_2\,,$$

where the protonated intermediate in parenthesis has a much shorter lifetime than the one in (2), so that it cannot be collisionally deactivated, at least in this range of total pressures. If we operate now at a P_{H_2} high enough that the intermediate in (2) is completely deactivated, we can get some hint about the effect of the kinetic energy of H_3^+ ions in bringing about the reaction along the path (2) or the path (3), simply by varying the field E. It can be seen from Fig. 2 *b*) that this effect is dramatic. The path (2) is dominant over path (3) at the lowest energies, but its relative importance decreases drastically as the energy increases.

This situation, which we have outlined in the particular case of ethane, is probably quite general, since we have observed it also in the reactions of H_3^+ with both the hydrocarbons and the alcohols. We have, therefore, that with these series of compounds proton transfer from H_3^+ leads to two different modes of reaction: the first one dominates at the lowest energies, and leads to a protonated intermediate with a lifetime long enough to be stabilized by collision. We call this process « *proton capture* », and in the following Section shall describe some of its features, such as the length of these lifetimes and their trends along the series. The second mode of reaction is favored by the higher energies and may proceed through more than one reaction path. For reasons which will become clear in Sect. **5** it will be called « *bond fission* ».

4. – Proton capture.

In the reactions with the lower members of the alkane series, the above described case of ethane is intermediate between that of methane, and that of the higher alkanes. For methane, only proton capture is observed: protonated methane, CH_5^+, when not stabilized by collision, decomposes into methyl ions, CH_3^+, and H_2. Its lifetime is of the order of 10^{-6} s, while that of protonated ethane, $C_2H_7^+$, is about 10^{-7} s. For propane and the butanes, on the other hand, no proton capture takes place; but only bond fission, to give an alkyl ion and a lower alkane or hydrogen. This striking difference along the alkane series in the formation of a protonated species is probably related to structural differences. For protonated methane, CH_5^+, trigonal bipyramidal structure can be thought of, with the five C—H bonds perfectly equivalent. For ethane, an analogous structure would be somewhat strained and therefore with a lower lifetime for fragmentation. For the higher members of the series strains would be such that the lifetime would be extremely low and make the protonated species unobservable.

Proton capture occurs in the reactions of H_3^+ with the saturated alcohols to give ROH_2^+ ions, which lose hydrogen with lifetimes of the order of $(10^{-6} \div 10^{-7})$ s. In this series, too, the importance of proton capture as compared to bond fission decreases sharply as the molecular complexity increases.

Particularly interesting is proton capture in the reactions of H_3^+ with simple aromatic hydrocarbons. [13] With benzenes, the main loss process of the protonated molecule is a hydrogen atom. One could ask whether these proton capture products are the common $C_6H_7^+$ ions (namely whether the proton is σ-bonded to the molecule), or else they are π-complexes, with the proton bonded to the π electrons. Use of D_3^+ allowed us to establish that the incoming proton does not lose its identity in the protonated molecule, as far as the decomposition is concerned. Therefore, the protonated species we form in these reactions is a true π-complex, with lifetimes in the range $(10^{-6} \div 10^{-7})$ s. Collisional stabilization probably rearranges them into a stable σ-structure. This conclusion seems to be supported by radiolysis experiments. [14]

5. – Bond fission.

Proton transfer leads to bond fission in most of the investigated H_3^+ reactions, with the necessary requirement that the overall process be exothermic. Moreover, in several of the reactions studied, more than one mode of fission is possible, and in some instances it proved possible to estimate the relative probability of breaking a particular bond of the reacting molecule.

In the case of propane and the butanes [15], this was accomplished using suitably deuterated compounds. In these hydrocarbons, RH, proton transfer leads to C—H bond rupture according to the scheme:

$$RH \xrightarrow{H_3^+} R^+ + H_2\,, \tag{4}$$

where R^+ ions are *n*- and *sec*-propyl ions for propane, *iso*- and *tert*-butyl ions for isobutane, and *n*- and *sec*-butyl ions for *n*-butane. For each hydrocarbon, both structures of the product alkyl ions R^+ are formed in comparable yields. However, in propane and in *n*-butane, the rupture of a C—H bond at a secondary carbon atom is more probable than at a primary one, and in isobutane such a rupture at a tertiary carbon is still more probable than at a primary one. This effect of positional discrimination correlates with the energetics of these processes. As a « rule of thumb », we can state that the probability of fission of a C—H bond depends on whether it is primary, secondary or tertiary according to the ratio

$$\text{primary} : \text{secondary} : \text{tertiary} \approx 1 : 2 : 3 \, .$$

Other bond fission processes involve rupture of C—C bonds. For propane and the butanes, we can write

$$R'R'' \xrightarrow{H^+} R''^+ + R'H \, . \tag{5}$$

When R′H is methane, R''^+ is an ethyl ion in propane, a *sec*-propyl ion in isobutane, and an *n*-propyl ion in *n*-butane. A reaction leading to ethyl ions and ethane occurs in *n*-butane, but not in isobutane, where the process, although exothermic, would require a rearrangement in the carbon skeleton. However, no methyl ions are produced in any of the cases studied, showing that when alternative exothermic paths are possible for a given C—C rupture, reaction (5) proceeds along the more exothermic one. This rule applies also to the other series of compounds investigated. The reactions above described were shown to occur in radiolyses of H_2-hydrocarbon mixtures [16]; the mechanisms and the reaction rates derived have been used in order to explain several processes [17, 18] in radiation chemistry.

Other typical examples of proton transfer leading to bond rupture are those occurring in the reactions between H_3^+ and the C_1—C_4 saturated alcohols [19]. The dissociative channels available after H^+ transfer to the alcohols follow two general exothermic patterns: i) rupture of a C—O bond, with formation of a water molecule and an alkyl ion ii) rupture of a C—H or a C—C bond with formation of H_2 or a saturated hydrocarbon and a protonated carbonyl compound. For all the C_1—C_4 alcohols, C—H and C—C bond fissions are all approximately equally probable, but C—O bond fission is constantly dominant; this shows that the interaction of the proton is more probable on that side of the molecule which contains the more electronegative element.

REFERENCES

[1] F. W. Lampe, J. L. Franklin and F. H. Field: *Progr. Reaction Kinetics*, **1**, 67 (1961).

[2] P. Ausloos and S. G. Lias: *Disc. Faraday Soc.*, **39**, 36 (1965).

[3] V. Aquilanti, A. Galli, A. Giardini-Guidoni and G. G. Volpi: *Trans. Faraday Soc.*, **63**, 926 (1967); **64**, 124 (1968).

[4] A. Henglein: This volume, p. 139.
[5] V. Aquilanti, A. Galli, A. Giardini-Guidoni and G. G. Volpi: *Journ. Chem. Phys.*, **43**, 1969 (1965).
[6] H. Eyring, J. O. Hirschfelder and H. S. Taylor: *Journ. Chem. Phys.*, **4**, 479 (1936).
[7] S. O. Thompson and O. A. Schaeffer: *Journ. Am. Chem. Soc.*, **80**, 553 (1958).
[8] V. Aquilanti and G. G. Volpi: *Journ. Chem. Phys.*, **44**, 2307, 3574 (1966).
[9] G. Gioumousis and D. P. Stevenson: *Journ. Chem. Phys.*, **29**, 282 (1958).
[10] J. B. Hasted: *Physics of Atomic Collisions* (London, 1964).
[11] D. R. Herschbach: This volume, p. 77.
[12] V. Aquilanti, A. Galli and G. G. Volpi: *Journ. Chem. Phys.*, **47**, 831 (1967).
[13] V. Aquilanti, A. Giardini-Guidoni and G. G. Volpi: *Trans. Faraday Soc.*, **64**, 3282 (1968).
[14] F. Cacace and S. Caronna: *Journ. Am. Chem. Soc.*, **89**, 6848 (1967).
[15] V. Aquilanti, A. Galli, A. Giardini-Guidoni and G. G. Volpi: *Journ. Chem. Phys.*, **48**, 4310 (1968).
[16] P. Ausloos and S. G. Lias: *Journ. Chem. Phys.*, **40**, 3599 (1964).
[17] F. Cacace, M. Caroselli and A. Guarino: *Journ. Am. Chem. Soc.*, **89**, 4584 (1967).
[18] R. H. Lawrence jr. and R. F. Firestone: *Advan. Chem. Ser.*, **58**, 278 (1966).
[19] V. Aquilanti, A. Galli and G. G. Volpi: *Ric. Sci.*, **36**, 359 (1966); **37**, 244 (1967).

Chemiionization.

R. S. Berry

University of Chicago - Chicago, Ill.

I. - Chemiionization Processes and Experimental Information.

1. - Introduction and classification.

Chemiionization includes all the processes that result in the formation of free charges, electrons or ions, under the conditions of chemical reactions. The definition is perhaps best given if we list the processes we include, and name some of the processes that are normally not considered as chemiionization processes. The discussion will treat

1) Autoionization (or preionization): $AB^* \to AB^+ + e$;

2) Associative ionization, also called the Hornbeck-Molnar process: $A^* + B \to AB^+ + e$;

3) Penning ionization: $A^* + B \to A + B^+ + e$;

4) Collisional ionization: $A^* + B \to A^+ + B + e$;

5) Thermal dissociation to ions: $AB + M \to A^+ + B^- + M$.

Except perhaps for autoionization, all these processes may occur as thermal collision processes in systems whose temperatures are low enough to permit molecules to exist. In other words we restrict chemiionization processes to mean ionization processes occurring in systems whose effective temperatures are low enough for molecules or neutral atoms to exist. Autoionization is included because, as we shall see, it is sometimes intimately related to the other processes, especially associative ionization and Penning ionization. Other processes which are related to chemiionization but cannot in any sense be called chemical ionization are

6) Ionization from collisions of heavy particles in the medium energy range of $(25 \div 100)$ eV;

7) Excitation transfer: $A^* + B \to A + B^*$ or $A^{**} + B \to A^* + B$;

8) Vibrational or rotational relaxation of ion-molecules in collision with electrons: $AB^+(v, J) + e \to AB^+(v', J') + e$;

9) Vibronic or rotational (rovibronic) perturbations of Rydberg states, (*e..g* *l*-uncoupling).

The most important ionization processes with which we shall *not* deal are the direct processes of photoionization, electron impact ionization and high energy collisional ionization. By a « high enery » collision, we imply that the time-dependent representation of the collision Hamiltonian contains large

TABLE I. – *A collection of related low-energy inelastic processes of a molecule-ion* AB^+ *and an electron.*

	Initial state		Final state		Process
	Electronic	Vibrational	Electronic	Vibrational	
1)	bound	bound	free	bound	Autoionization $AB^* \to AB^+ + e$
2*a*)	bound	free	free	bound	Associative ionization $A + B^* \to AB^+ + e$ (Hornbeck-Molnar process)
2*b*)	free	bound	bound	free	Dissociative recombination $AB^+ + e \to A + B^*$
3)	bound	free	free	free	Penning ionization $A + B^* \to A^+ + B + e$
4)	bound	free	free	free	Collisional ionization $A + B^* \to A + B^+ + e$
5)	bound	bound	bound	free	Thermal dissociation $AB + M \to A^+ + B^- + M$
6)	bound	bound	free	bound	Medium-energy collisional ionization $A + B \to A^+ + B$
7)	bound	free	bound	free	Excitation transfer
8)	free	bound	free	bound	Vibrational or rotational relaxation
9)	bound	bound	bound	bound	Rovibronic perturbations in Rydberg states

Fourier components in the energy range of the ionization threshold of the target molecule.

The chemiionization processes 1)-5) and the other related processes 6)-9) are easily classified and their close relationships are particularly evident if they are tabulated in terms of their initial and final states. More strictly, we may represent the initial and final state approximately as products of electronic functions $\psi_{i,f}(\boldsymbol{r}, \boldsymbol{R})$ and vibrational-rotational functions $\chi_{i,f}(\boldsymbol{R})$ in the usual context of the Born-Oppenheimer approximation. The operator that couples initial and final states may be a part of the adiabatic Hamiltonian (as it is in the cases of configuration interaction or spin-orbit coupling) or it may be the nuclear kinetic energy. In any event, the processes 1)-9) are classified according to the free or bound nature of the initial and final electronic states ψ_i and ψ_f, and of the corresponding vibrational functions χ_i and χ_f. Table I shows these processes together. Only entry 5) is not among the processes that can be treated as coupling phenomena among the states of $AB^+ + e$. In fact if one has a means for obtaining the bound and free electronic wave functions ψ and the nuclear functions χ which give bound and continuum states with approximately equal ease, then it becomes possible to relate the processes quite quantitatively, as we shall see.

2. – Autoionization.

Autoionization or preionization has long been recognized as an important mechanism for loss of atomic or molecular electrons. However its full exploration and elucidation has only occurred relatively recently, with the observation of atomic autoionization resonances in atomic spectra [1] and with the use of high-resolution photionization and mass spectrometry to analyze molecular autoionization. The atomic situation deserves only passing mention in this context. In the first known cases [2, 3] rare gas atoms were shown to have two kinds of transient states above their first ionization limits. Each of the two sets of state gives rise to a Rydberg series converging to the $^2P_{\frac{1}{2}}$ first excited state of the inert gas ion. The broader series corresponds to excitation of a bound *np* electron to an *nd* orbital with the core taking on the $^2P_{\frac{1}{2}}$ state rather than the $^2P_{\frac{3}{2}}$ ground state. The sharper series corresponds to excitation of the *p* electron to an *ns* orbital, also moving in the field of an excited core. In either case, the Rydberg electron is capable of autoionizing by taking rotational energy from the $^2P_{\frac{1}{2}}$ core and leaving it in its $^2P_{\frac{3}{2}}$ ground state.

The distinctive features first reported for helium by MADDEN and CODLING [1] are the primitive examples of « long-lived » doubly-excited states. The simplest of these states are those of helium. We may usefully characterize an atomic state by its principal configuration; *e.g.* $1s^2$ for the ground state of

He or $1s^2 2p$ for the first excited state of Li. In He, we may excite the electrons to the configuration $2s2p$, for example. The energy of this state (relative to $He^{+2}+2e$) is roughly that of two L-electrons in the field of a nucleus with charge $+2$, or approximately -2 Rydberg units. The energy of $He^{+}+e$ is -4 Rydberg units, so that the $2s2p$ configuration has far more than enough energy to ionize. The process itself requires that one electron return to the $1s$ orbital, giving the other its energy and causing the latter to go free. The process is due, in physical terms, to the electron-electron interaction and more specifically to the fact that their motions are correlated enough that neither electron can be strictly characterized by its own quantum numbers n, l, and m. Correlation acts as the coupling mechanism in this case. Formulated mathematically, the coupling operator is the electron-electron interaction (or the deviation of the true electronic potentials from the Hartree-Fock mean fields) and the formal coupling is essentially configuration interaction. In contrast to conventional configuration interaction, here one must deal with a bound state interacting with a continuum in which it is imbedded. A parametric characterization and formal solution for a single quasi-bound state in a continuum was developed by FANO [4], who showed that the unusual line shapes in this type of process were due to the interference in phase between the transition amplitudes connecting ground state with quasi-bound and with continuum excited states. Microscopic treatments of the parameters in Fano's theory have been carried out by several people, including FANO and COOPER [5] and COMES and SÄLZER [6, 7]. The problem of overlapping resonances has been treated by MIES and KRAUSS [8].

The particular process of atomic autoionization via correlation coupling or configuration interaction is not really a good example of chemiionization because the relevant quasi-bound states are normally at rather high energies. However in some important molecular examples, notably NO, one may expect some autoionization to occur through configuration interaction. In NO, the coupling involves two or three manifolds. The NO^+ ion has the configuration

$$KK(\sigma 2s)^2(\sigma' 2s)^2(\pi 2p)^4(\sigma 2p)^2 \,.$$

The lowest empty orbital of NO^+, the orbital from which the electron has come, is an antibonding $\pi 2p$ orbital. The other two manifolds are based on a Rydberg electron in the field of an excited NO^+ core. The two excited core configuration and lowest corresponding states are

$$KK(\sigma 2s)^2(\sigma' 2s)^2(\pi 2p^4)(\sigma 2p)(\pi 2p)\,, \qquad {}^3\Pi$$

and

$$\ldots\,(\pi 2p)^3(\sigma 2p)^2(\pi' 2p)\,, \qquad {}^3\Sigma^+\,, \qquad {}^3\Delta$$

in the separated-atom notation for the orbitals. Thus the autoionization coupling involves a transition of one electron from the Rydberg orbital into the continuum and of another electron from the $\pi' 2p$ orbital into either the $\sigma 2p$ or the $\pi 2p$ orbital. The photoionization spectrum of NO shows no sharp structure in the region $(0 \div 2)$ eV above threshold [9]. It is not known whether the lack or breadth of structure is due to rapid autoionization or to the dominance of direct ionization.

In contrast to NO, the H_2 molecule exhibits a very pronounced structure in its photoionization spectrum. The technology of this field has changed dramatically in the past three years. Figure 1 *a*) shows the photoionization spectrum of H_2 as reported by DIBELER, REESE and KRAUSE [10] in 1965; Fig. 1 *b*) is that of CHUPKA and BERKOWITZ [11], taken three years later. COMES and WELLERN [12] also have photoionized H_2 with light from a monochromator and measured the current of H_2^+ so generated; their resolving power has been distincly lower than that of CHUPKA and BERKOWITZ, but is significantly better than that of the earlier work of DIBELER, REESE and KRAUSE. The recent work has given a new value of 15.425 eV to the thermodynamic ionization energy of H_2. This value is based on upper and lower limits set by identification of the lowest-energy optically-allowed state of H_2^* that autoionizes spontaneously and the highest-energy optically-allowed state that can be ionized only by application of an external electric field [13]. (Strictly, the value of the ionization energy is given as $124.417 > IP(H_2) > 124.393$ cm^{-1}.)

The most striking characteristic of the autoionizing states of H_2 is their narrowness. By using cold (— 78 °C) para-H_2, CHUPKA and BERKOWITZ were able to resolve a large number of individual rotational lines. The apparent widths of most of these lines were limited by the apparatus; in some cases, lines were found with widths as great as 0.2 Å, ($\tau \approx 10^{-13}$ s), somewhat greater than the apparatus width. In other words the lifetimes of the most rapidly autoionizing states of H_2 are many times the vibrational period of the H_2^+ core. (The vibrational frequency of H_2^+ is 2297 cm^{-1}, corresponding to a period of $1.45 \cdot 10^{-14}$ s) [14]. These lifetimes are in sharp contrast to the peaks in the helium spectrum, whose widths are of order 10^{15} cps.

The intensities in the H_2 photionization spectrum are of interest. The background continuum due to direct photoionization rises only very slowly; in the region $(0 \div 1.5)$ or $(0 \div 2)$ eV above threshold, the major contribution to to ionization comes from autoionization and not from the direct process. As the energy increases still further, the proportion of intensity in the continuum becomes predominant.

Qualitatively, the photoionization spectra of at least some other molecules show the same sharp features and predominance of autoionization at low energies. Recent unpublished data of BERKOWITZ and CHUPKA on photoionization of N_2 shows peaks much like those in H_2; the background direct ionization continuum in this case is relatively stronger than in H_2 [15, 16] but a large part of the ionization is clearly due to autoionizing excited states.

Autoionization as a decay mechanism sometimes competes with spontaneous

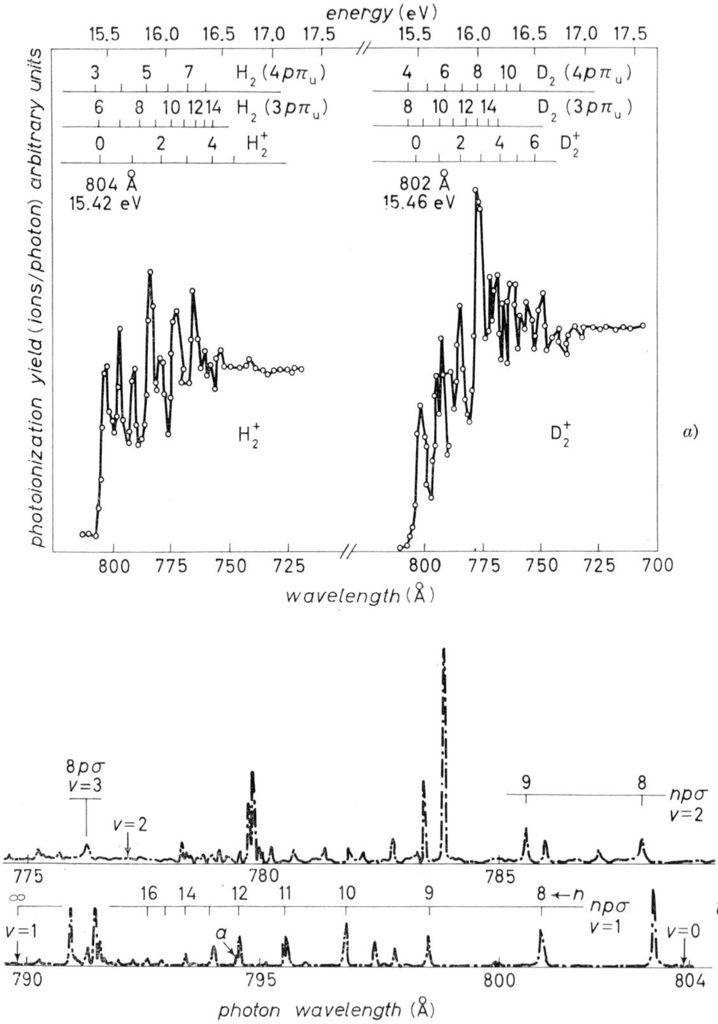

Fig. 1. – *a*) Photoionization spectrum of H_2, 1965 (ref. [10]); *b*) Photoionization spectrum of H_2, 1968 (ref. [11]). (Resolution: 0.04 Å).

emission and sometimes with predissociation. The competition with emission arises only for the slowest sorts of autoionization, with characteristic times of $(10^{-9} \div 10^{-7})$ s. Predissociation at its fastest has a characteristic time of order 10^{-13} s, so that only the very fastest autoionization processes can compete with predissociation. However, rapidly predissociating states are relatively rare. Moreover the relative intensities of photoion peaks associated with autoionizing states are frequently in the same ratio as the intensities of the same peaks in the absorption spectrum. This observation is strongly suggestive that for such states, the principal decay mechanism is autoionization and that the loss of an electron occurs so fast that other processes such as predissociation do not offer readily available decay channels in most excited states of small molecules. The theory of competition between autoionization and predissociation has been discussed recently [17] and is reviewed in the second part of this article.

3. – Associative ionization, Penning ionization, collisional ionization and dissociative recombination.

The processes described in this Section are those in which an excited electron gains energy from the internal energy or from the kinetic energy of relative motion of two colliding free particles and, in so doing, makes a transition from a bound state to a free state. We let A^* collide with B; whether the process is called associative, Penning or collisional ionization depends on whether the final state leaves the positive charge on the bound $(AB)^+$, on B^+ or on A^+. The longest-known of these is apparently the associative ionization process which was suggested by Franck as an explanation for experimental results obtained by Mohler and Boeckner [18] and previously by Mohler, Foote and Chenault [19]. The experiments showed that cesium vapor irradiated with light of wavelengths corresponding to cesium principal series lines (and therefore of insuffcient energy to cause atomic ionization) did in fact give rise to ionization. It was Franck's suggestion that the extra energy required to remove an electron comes from formation of a molecule of Cs_2^+. The process is sometimes called the Mohler-Franck process.

Associative ionization is also identified with the names of Hornbeck and Molnar. Their study [20] of the appearance potentials of He_2^+, Ne_2^+ and Ar_2^+, relative to those for He^+, Ne^+ and Ar^+, demonstrated clearly that the diatomic molecule-ion is generated by collision of an atom in its ground state with an excited atom and not with a ionized atom. As Hornbeck and Molnar point out, this mechanism had been suggested previously by Arnot and M'Ewen on the basis of the pressure dependence of the intensities of M_2^+ ions [21], in a much earlier mass spectrometric study of He_2^+.

It is an interesting historical note that the data of Hornbeck and Molnar, especially their Fig. 7, may well be the first experimental indication of the e-He resonance now known to exist at 19.30 eV and to be due to the formation of a transient He^- ion [22,23]. We quote from footnote 8 of ref. [21]: « The dip shown in Fig. 7 is a « ghost » which we do not explain. It was found only in helium, and its position on the voltage scale varied with gas presssure, repeller voltage, and other instrument settings. By making several « runs » under different conditions the entire excitation curve ... could be plotted ». The actual resonance is narrower than the resolution of HORNBECK and MOLNAR.

The Penning process is another mechanism for chemiionization that has been recognized for many years, since it was first reported by KRUITHOFF and PENNING [24]. A process found more recently is not precisely either Penning ionization or associative ionization. This is the rearrangement ionization process reported by KUPRIANOV [25] and also investigated by HOTOP and NIEHAUS [26]: *e.g.*, $A^* + H_2 \to AH^+ + H + e$, in contrast to the Penning process which would give $Ar + H_2^+ + e$ or the associative process, giving ArH_2^+.

We also include some discussion of the quite important inverse to associative ionization, dissociative recombination.

3'1. *Dissociative recombination.* – Dissociative recombination rates or cross-sections were, for many years, accessible only through estimates of the rates for the increase of the measured associative ionization rates. The cross-sections for dissociative recombination of alkali diatomic molecule-ions with electrons are of order $(10^{-12} \div 10^{-14})$ cm^2, quite enormous as atomic cross-sections go.

Interest in dissociative recombination was stimulated sharply by a proposal of Bates and Massey to explain an atmospheric phenomenon. In about 1940, it was found that the degree of ionization in the E-layer of the atmosphere (ca. 110 km high) drops at sunset at a rate much greater than existing models could explain; BATES and MASSEY [27] proposed the dissociative recombination as a general mechanism to account for the rapid recombination rate. BATES [27] soon thereafter proposed the specific process

$$NO^+ + e \to N + O^*$$

as a mechanism for the enhanced production of oxygen atomic lines at twilight [29, 30]. The $NO^+ + e$ thermal recombination process was studied recently by GUNTON and SHAW [31] and by YOUNG and ST. JOHN [32]. GUNTON and SHAW prepared ions by pulsed photoionization and followed the recombination by observing the microwave reflectivity. YOUNG and ST. JOHN produced ions continuously by chemiionization in mixtures containing N and O atoms and monitored the recombination by application of pulsed potentials on their collecting system. The two very different methods gave approximately the same very large values for the rate coefficient, ca. $4 \cdot 10^{-7}$ cm^3/s from the photoionization and $(5 \pm 2) \cdot 10^{-7}$ cm^3/s from the chemiionization. The corresponding cross-section is of order 10^{-13} cm^2, comparable to that for $Cs_2^+ + e$. Other dissociative recombination cross-sections are also of this magnitude; only

$He_2^+ + e$ seems to have an anomalously low cross-section for low-energy collisions. That this is truly the case is the conclusion of the analysis of the existing data by FERGUSON, FEHSENFELD SCHMELTEKOPF [33] and of the potential curves of He_2^+ by MULLIKEN [34, 35] and GINTER [36]. In essence, the repulsive He*-He curves which give impetus to the dissociating neutrals probably cross the attractive He_2^+ potential well at relatively high energy, *i.e.* at high He_2^+ vibrational states. This suggests that dissociative recombination cross-sections for $He_2^+ + e$ are only large for highly excited vibrational states of He_2^+.

3·2. *Associative and Penning ionization.* – The relationship between dissociative recombination and associative ionization rate coefficients, k_{dr} and k_{ai} respectively, is a direct consequence of microscopic reversibility and the precise relation between the coefficients or cross-sections should be expressed in terms of the microstates and specific collision velocities. However the experimental values are gross values taken from systems in or near thermal equilibrium. Hence we can derive an approximate relationship between the rate coefficients and make an estimate of the cross-sections for associative ionization, from the relationships at equilibrium:

$$\frac{dN(e)}{dt} = 0 = k_{ai} N(A^*) N(B) - k_{dr} N(AB^*) N(e) \tag{1}$$

so that

$$\frac{k_{ai}}{k_{dr}} = \left[\frac{\sigma_{ai}\, v(A^*\text{-}B)}{\sigma_{dr}\, v(AB^*\text{-}e)}\right]_{\text{effective}} = \frac{N(AB^+)N(e)}{N(A^*)N(B)} = \frac{Z_{trans}(AB^+\text{-}e)Z(AB_+)Z(e)}{Z_{trans}(A^*\text{-}B)Z(A^*)Z(B)} \tag{2}$$

where $Z_{trans}(X - Y)$ is the translational partition function for relative motion of X and Y, and the $Z(Y)$'s without subscripts are partition functions for Y's internal degrees of freedom. The translational factor is $[m_e/m_{A+B}]^{\frac{3}{2}}$, typically of order 10^{-6}. We may assume that the rotational partition function of AB^+ is essentially $2kTI\hbar^{-2}$ (I = moment of inertia) and is about 10^2, and the vibrational partition function $[1 - \exp[\hbar\omega/kT]]^{-1}$ of AB^+ and the electronic partition function of all species are of order unity. Consequently $(k_{ai}/k_{dr}) \simeq 10^{-4}$ and the corresponding effective cross-section $\sigma_{ai} = 10^{-2}\sigma_{dr}$. Thus the cross-sections for associative ionization are reasonably close to gas kinetic cross-sections, of order $10^{-15} \div 10^{-16}$, sometimes reaching $10^{-14}\ cm^2$, in keeping with naive preconceptions one might be expected to have in view of the relatively short range of atom-atom interactions. As an example, the cross-section for $Hg + Ar^*$ (presumably ^3P) $\rightarrow HgAr^+ + e$ is $2.5 \cdot 10^{-15}\ cm^2$ [37, 38]

Associative ionization and Penning ionisation are the most important mechanisms for ion production in flames and other chemical systems, and may play roles in the upper atmosphere, in many shock-heated gas systems and possibly even in stellar atmospheres. For example the *primary* chemi-ion in most hydrocarbon-oxygen flames is HCO^+, even though H_3O^+ is the predominant ionic species. The HCO^+ is produced by the reaction [39, 40]

$$CH + O \rightarrow HCO^+ + e\,,$$

where CH may be in its ground state or in an excited state. The H_3O^+ ions are produced by proton transfer:

$$HCO^+ + H_2O \rightarrow H_3O^+ + CO.$$

In the acetylene-oxygen flame, a system still not well understood at all, the primary chemi-ion may be $C_3H_3^+$ [39, 41], which was thought to be due to

$$CH + C_2H_2 \rightarrow C_3H_3^+ + e\,,$$

although the reaction might involve an HCO^+ precursor [42, 43]. The $C_3H_3^+$ ion is chemically interesting because of the presumed stability of the cyclopropenium ion,

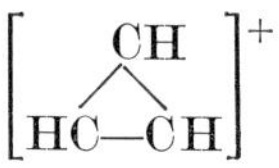

with a pseudo-aromatic system of two π-electrons. Such a structure is the most obvious to propose as the $C_3H_3^+$ species; its properties have been studied in solution [44].

The study of associative and Penning ionization was relatively dormant for several years (except for the work of HORNBECK and MOLNAR) but has now become a very lively field. One has a number of options available when one tries to study the problem experimentally. For example, one may ask what are the consequences of specific initial states of the excited partner. Does it matter whether He is excited to a metastable or an optically allowed state? One may also ask about the competition between the various possible mechanisms or gross outgoing channels: does a given reaction, *e.g.* $He^* + H_2$, give $He + H_2^+$, or $HeH^+ + H$? Knowing which, one may ask why. (Alternatively, one may first try to decide what the results might be on the basis of a theoretical model for the reaction. In fact, the experimental results were known first in this case.) One may go further in analysing the exit channels by asking what the electron energy is and inferring from the results what molecular vibrational states are produced in associative ionization, or what the kinetic energy of the heavy particles is in the final state of the Penning process.

All of these aspects of chemiionization have now been examined in at least a preliminary way. With regard to initial states, it now seems that three kinds of excited states of A^* may be involved when A^* collides with M to give $AM^+ + e$ or $A + M^+ + e$. One type is the normal optically-allowed, short-lived state like the $2\,^1P$ state of He^*. Another is the relatively low-lying metastable state, like the 2^3S and 2^1S states of He; because of their long lifetimes and well-known properties, these states were the most commonly invoked and studied. Finally there are relatively long-lived highly-excited states whose existence in electron-impact-excited beams was established by CERMAK and HERMAN [45]. These states are presumably high Rydberg states that owe their long radiative lifetimes ($> 10^{-5}$ s) to the small oscillator strength of the highly spread-out states. As HOTOP and NIEHAUS and KUPRIANOV have pointed out [25, 46], the radiative lifetime of a Rydberg state increases with principal quantum number n as $n^{4.5}$.

The comparison of « allowed » and long-lived states of A^* has been made for $He^* + He$ [47] and for $Ar + Ar^*$ [47*a*]. KAUL, SEYFRIED and TAUBERT [47] used fast pulsed excitation to allow them to separate the He_2^+ ions produced by short-lived He^* from those produced by metastables. The method does not distinguish which states are involved, but does show that there are long-lived and short-lived progenitors, and that they contribute approximately equally to the total He_2^+ production. TETER, NILES and ROBERTSON have more recently examined the pressure-dependence of specific levels by applying intense chopped monochromatic irradiation [48]. The reaction $He^* + He \rightarrow He_2^+ + e$ is only energetically possible (in a system near room temperature) if He^* is in a state higher than the $3\,^3S$. Transitions from the states $3\,^3P$, $3\,^3D$, $3\,^1P$ and $3\,^1D$ all showed a decrease in intensity with pressure, which was interpreted as the result of associative ionization. Excitation transfer was ruled very unlikely in this system because large irradiation-induced changes in the population of one state did affect the intensities of lines from other states. The authors derived cross-sections for the various states; the smallest where the $3\,^3P$ and $3\,^1P$, ca. $2 \cdot 10^{-16}$ cm^2, while the largest, the $3\,^3D$, is approximately $2 \cdot 10^{-15}$ cm^2. The emission lines from states with principal quantum numbers higher than 3 showed negligible pressure-dependence, even though their radiative lifetimes are longer. For example KAUL, SEYFRIED and TAUBERT give theoretical radiative lifetimes of $9.7 \cdot 10^{-8}$, $1.4 \cdot 10^{-7}$ and $2.2 \cdot 10^{-7}$ s for $3\,^3P$, $4\,^3P$ and $5\,^3P$, and $1.4 \cdot 10^{-8}$, $3.2 \cdot 10^{-8}$, $6 \cdot 10^{-8}$ and $1.1 \cdot 10^{-7}$ s for the $n\,^3D$ ($n = 3$-6) series [47]. The implication of the experimental results is that the size of the orbital containing the excited electron is critically important, and that in simple systems, one can expect associative ionization to be important only for a very restricted bundle of incoming channels. It could be that the probability is also a sensitive function of which incoming potential curve the colliding pair follows. The range of cross-sections found by TETER, NILES and ROBERTSON could be interpreted to indicate either that a few specific potential curves are important or that the essential dependence on initial state is only a sensitive function of the state of the isolated excited atom, and not of the particular potential curve. For $Ar^* + Ar$, it appears that three states of Ar^* with radiative lifetimes varying by about a factor of 30 all have relatively similar cross-sections, about $3 \cdot 10^{-14}$ cm^2, and are the primary contributors to associative ionization in this gas, at least when argon is excited by impact of electrons whose energy falls between 15 and 100 eV [47*a*].

Further information on specific state dependences comes from the work of HERCE, PENTON, CROSS and MUSCHLITZ on $He^* + CH_4$ (and CD_4) [49] and of MUSCHLITZ, PENTON and HERCE on $He^* + H_2$ [50]. The He^* source was designed specifically to eliminate all low-lying optically-allowed states (by making the passage time sufficiently long) and also highly excited long-lived states, by passing the excited atoms through a quenching and sweeping field. The differences between $2\,^1S$ (at 20.6 eV) and $2\,^3S$ He^* (at 19.8 eV) could be determined by varying the accelerating potential on the exciting electron beam in the He^* source and thereby varying the $n(2\,^1S)/n(2\,^3S)$ ratio (because the two excitation processes have quite different dependences on electron energy). Although the differences are real and far outside experimental uncertainties, they are not grossly different. For example the ratios of intensities of CH_4^+, CH_3^+ and CH_2^+ (as measured in a mass spectrometer) are 38:57:5 for He^* ($2\,^1S$) and 47:49:4 for He^* ($2\,^3S$). The inference to be drawn here is the not-so-surprising supposition that the spin of the excited electron does not affect the ease with which it can gain enough energy to ionize. Still more information about the role of initial states comes from the recent analysis by CERMAK and HERMAN [51] of the reaction $He^*(2\,^3S, 2\,^1S) + Hg \rightarrow Hg^+ + He + e$. In this work the emitted electron intensity was monitored as a function of the electrons' kinetic energy. very close to the difference between the excitation energy $E(\mathrm{He}, 2\,^3P) - E(\mathrm{He}, 1\,^1S)$ and the various ionization energies of Hg. The triplet He therefore gives up its energy to the Hg atom without changing its own momentum very much. In the $2\,^1S$ channel, the electrons carried significantly less energy than the maximum allowed on the basis of the exothermicity of the reaction. The implication here is immediate, that the ionization occurs so as to give relative kinetic energy to the He and Hg^+ nuclei in a rather effective way, *e.g.* in a transition to a repulsive curve.

Associative ionization occurs with molecular excited precursors as well as with atoms. For example HERMAN and CERMAK have shown that associative ionization of alkali atoms occurs in collisions with $N_2(A\,^3\Sigma_u^-)$ [38] or with $CO(a\,^3\Pi)$ [52]. That the excited state is a low-lying metastable and not a highly excited species can be demonstrated by the shape of the ion production curve as a function of the energy of the electron used to excite the CO, for example. The ion production shows the sharp peak near threshold characteristic of excitation associated with a change of multiplicity [35]. The associative ionization of $N_2^* + NO$ to give N_3O^+ (in very small yield) seems to be due to a long-lived N_2^* state having the same multiplicity as the ground state, *e.g.* the $a\,^1\Pi_g$, $a'\,^1\Sigma_u^-$ or $a\,^1\Delta$ state [54].

The role of highly excited states in ionizing collisions can also be examined by following ion intensities as functions of excitation. HOTOP and NIEHAUS have carried out collision studies with excited He, Ne, Ar, Kr and Xe on H_2 and HD [26] and with He, Ne and Ar on a variety of diatomic and polyatomic species [46] including H_2O, NH_3, SO_2, C_2H_5OH and SF_6. In the work with H_2, HOTOP and NIEHAUS found that the processes involving He and Ne followed curves indicating that He* and Ne* were in low-lying metastable states (*e.g.* $2\,^3S$ or $2\,^1S$ for He); the heavier rare gases appeared to be in highly excited Rydberg states. In the work on polyatomics, all the excited rare-gas species were of the high-energy variety.

In contrast to the various low-lying states, whose behavior is relatively similar, the high-lying states seem to differ considerably from the low-lying metastables. First, and not surprisingly, the highly excited states are capable of collisional ionization. Strikingly, the collisional ionization process (with the highly excited X denoted as X**),

$$X^{**} + M \rightarrow X^+ + M + e\,,$$

was only observed when M was at least a triatomic, and was not found for $M = H_2$, N_2, O_2, NO or CH_4. It may be that the energy for thermal collisional ionization of X** must come from vibrational energy within M, rather than from the relative kinetic energy of X and M.

A second characteristic of the highly excited species is their ability to undergo rearrangement ionization. HOTOP and NIEHAUS found that the reaction

$$Ar^{**} + H_2 \rightarrow ArH^+ + H$$

is the most probable process, with a cross-section an order of magnitude greater than that for Penning ionization, and associative ionization is unobservable for this system. By contrast, Penning ionization of H_2 by He* is the most probable process almost a factor of 10 more probable than rearrangement ionization. Again, associative ionization is less important than the other mechanisms but with He* or Ne* on H_2, it is an observable process. Earlier, KUPRIANOV [25], working with highly excited helium, had found relatively large ion current of HeH^+ and a much smaller current of HeH_2^+.

The cross-sections for collisional detachment are apparently quite large; HOTOP and NIEHAUS report values of order 10^{-12} cm^2 for the species they observed [46]. The Bohr radius for a state with $n \sim 10$ is ca. 50 Å, corresponding to a geometric area of about $0.8 \cdot 10^{-12}$, so that the cross-sections might be interpretable, at least partially, in terms of the attractive field of the passing molecule. In the case of SF_6, HOTOP and NIEHAUS observed SF_6^-, produced by direct capture from Ar**, and also with a cross-section of about 10^{-12} cm². Similar observations of electron capture from Rydberg states were recently observed by BERKOWITZ and CHUPKA [55].

Cross-sections for the rearrangement process are apparently smaller. KUPRIANOV [25] gives an estimate of about $4\cdot10^{-11}$ to 10^{-13} cm² for the total cross-section including that for ArH^+ production.

Associative ionization and Penning ionization frequently have comparable cross-sections [52], *e.g.* in the case of Ar^*+Hg. These cross-sections, and more specifically their ratio, are apparently sensitive functions of the temperature of the colliding species [51, 56]. We shall return to this point when we discuss the mechanisms of the various processes.

A very simple system, chemiionization of H_2 by H^* and H_2^*, gives us virtually the only experimental information about the role of the kinetic energy of relative motion in a ionizing collision. The system was studied by CHUPKA, RUSSELL and REFAEY [57] as part of an examination of ion production and ionic reactions in photoexcited hydrogens. The chemiionization reactions whose occurrence they inferred from peaks in the H_3^+ intensity are

$$H^*+H_2\rightarrow H_3^++e\,, \qquad H_2^*+H_2\rightarrow H_3^++H+e\,,$$

and possibly

$$H_2^*+H_2\rightarrow H_3^++H^-\,,$$

although no search for H^- has yet been reported. Cross-sections for the electron-producing reaction were estimated as about 10^{-16} cm². The excited hydrogen atoms are produced by irradiation with monochromatic light ($\Delta\lambda\sim 2.5$ Å). The photoexcited H_2 undergoes dissociation into $H(1s)+H(2s, 2p)$ and also, at specific energies, predissociation via the $D\,{}^1\Pi_u(1s3p\pi)$ state ($v=3, 4$) into $H(1s)+H(2s)$, probably along the $B'\,{}^1\Sigma_u(1s3p\sigma)$ potential [58]. When predissociation occurs from a known state to a known dissociation limit, the relative kinetic energy of the product fragments is known. In the H_2 case, the dissociation threshold is 14.676 eV and the ($v=3, 4$) bands 0.05 and 0.16 eV above this. The system's thermal energy adds about 0.02 eV more. The relative intensities and, by inference, quantum yields of $H(2s)$ are known for the $v=3$ and $v=4$ bands, so that a comparison of the peak intensities in the H_3^+ product curve allows an inference of the relative cross-sections for H_3^+ production from $H(2s)+H_2$ collisions at 0.07 and 0.18 eV. The inference is that $\sigma(0.07)/\sigma(0.18)\simeq 1.8$. The occurrence of reaction between H_2^* and H_2 was presumed to be responsible for a weak peak and a broad « rising feature » of the H_3^+ curve which could not be attributed to predissociation. The excited H_2^* states are not identified.

It is worth noting that the energy analysis of the electrons from Penning ionization can be used as a spectroscopic probe to locate states of the M^+ ion in the process $A^* + M \rightarrow A + M^+ + e$, if metastables of known energy are used, or to locate metastable states of A^* if known detector molecules M are used. CERMAK has used this approach to study the states of some forty molecule ions, and to determine various ionization potentials for formation of these ions [59, 60]. The ionization potentials could, in many cases, be compared with values from photoelectron spectroscopy; the two methods give results in quite good agreement. The intensities and peak shapes indicate that Penning ionization is a Franck-Condon process, in the sense that the distribution of vibrational states of the M^+ ion corresponds to vertical excitation from the initial vibrational distribution of M.

3'3. *Ionization in thermal dissociation.* – If a hot gas contains a substance having a large electron affinity, then conditions may exist in which a substantial fraction of the negative charge exists in the form of heavy negative ions rather than as electrons. This is the situation, for example when a flame is seeded with a halogen; the conductivity of the flame drops because the electrons become bound to halogen atoms, via dissociative attachment to halogen molecules.

The most favorable systems for production of large concentrations of negative ions are of course those in which the gas contains both a donor, *e.g.* a metal with a low ionization prtential, and an acceptor which forms the negative ion. The system is thus the same as the one based on the metal-acceptor salt, *e.g.* an alkali halide. Molecules such as alkali halides have a ionic dissociation limit only slightly above the dissociation limit corresponding to separated atoms in their ground states. This separation is

$$Q(MX) = I.P.(M) - E.A.(X) ,$$

where I.P.(M) and E.A.(X) denote ionization potential and electron affinity of the donor and acceptor, respectively.

The potential curve based on the ionic dissociation limit is a Coulombic attractive potential. The other low-lying potential curves exhibit only short-range interactions. Hence, as R grows smaller, the ionic curve would cross some of the neutral-atom curves with lower dissociation limits. In cases like the alkali halides, the ionic curve would cross all the lower potential curves, and become the molecular ground state curve for $R \simeq R_e$. The adiabatic non-crossing rule may prevent some of the crossing from occurring; this rule would make all crossings into avoided crossings (for states of the same symmetry and multiplicity) if the molecular electronic wave function were to follow the nuclear motion adiabatically. However some of the crossings occur at very large internuclear separations; in CsBr, the crossing of the ionic and atomic ground state curves is at 51.5 a.u. It is not likely that the electronic wave function can undergo an adiabatic avoided-crossing transition when the internuclear separation is so large, particularly when the transition involves an electron transfer between acceptor and donor. Consequently, the likelihood is large that at least some « ionic » molecules have states which, in effect, may follow the ionic potential curve all the way to complete dissociation. In cases for which the crossing distance R_x is of order 5 a.u., one might expect the adiabatic curves, with their avoided crossings, to be followed.

The likelihood of non-adiabatic passage to the dissociation limit opens the possibility for production of ionization directly by thermal collisions, a process we might for convenience call thermal-ionic (or thermionic) dissociation. The thermal collisions, leading to dissociation, carry the molecule all the way to its ionic dissociation limit before the electronic wave function responds by recognizing that lower dissociation limits, including the adiabatic limit, are available. Such a situation has been observed. Gaseous CsBr molecules, dissociated by collision with shock-heated argon, gives as primary dissociation products the ions Cs^+ and Br^- [61]. The concentrations of Cs^0 and Br^- are monitored by time-resolved absorption spectroscopy. In the temperature range $(2500 \div 4500)$ °K, the Br^- concentration rises well before the Cs^0 concentration, and at higher temperatures shows a distinct maximum. The time lag corresponds to about 10^5 collisions. The system then relaxes back to distribution of atoms, ions and electrons in thermal equilibrium. This relaxation time corresponds to about $5 \cdot 10^5 \div 10^6$ collisions for an atom or molecule. The same overpopulation effect is observed with several other alkali halides, including KBr, CsF, KF and CsI. The species NaI, NaBr and LiBr dissociate to atoms [61*b*], and apparently NaCl and KCl do also [61*a*], whereas KI and RbI seem to be intermediate cases.

The experiments demonstrate, among other things, that overpopulation of the state corresponding to the ionic dissociation limit can be achieved by thermal collisions. More pertinent to the context of chemiionization is the demonstration that thermal dissociation to ions is possible at all. We may suspect, for example, that some of the ionization in rocket exhausts or seeded flames occurs by thermal-ionic dissociation.

II. – Interpretation and Theory.

To survey the theoretical and interpretive aspects of chemi-ionization we shall first summarize some of the generalizations and inferences from the experiments described in Part. I. Next we examine the interpretations of mechanisms in qualitative terms, based on potential curve and correlation diagram arguments. Finally we survey existing quantitative theoretical models.

1. – A summary of observation.

Nuclear energy changes: In autoionization of molecules, the most probable transitions seem normally to be those involving the smallest change in vibrational kinetic energy [11]. Penning and associative ionization appear in general to be governed by the Frank-Condon principle [59, 60, 62, 63]. In the Penning ionization of a molecule M, in particular, the vibrational states of M^+ correspond to vertical transitions from the position-momentum distribution of the initial state of the free M molecule, and not from an M molecule very strongly perturbed by a nearby A^*. Penning ionization can sometimes give heavy products with a significant amount of net relative kinetic energy [51]. Associative ionization cross-sections seem to drop with increasing relative kinetic energy[57]. Collisional ionization of highly excited Rydberg states may be effected better by transfer of energy from vibration than translation [46].

Electronic transitions: Low-lying metastable and short-lived states of A^* give rather similar patterns of products in associative, Penning and rearrangement ionization [45, 47, 50]. Highly excited states of A^* seem to be quite different from low-lying states with regard to their patterns of ionization products [25, 46]. Associative ionization in a homonuclear atomic system (for which Penning ionization is apparently very improbable and rearrangement ionization is meaningless) may be restricted to a narrow band of Rydberg states [48].

2. – Interpretation in terms of potential curves.

2·1. – The easiest processes to interpret in terms of adiabatic potential curves are the autoionization processes. So long as autoionization is sufficiently slow, it is useful to describe the process as a transition of the molecule from one potential curve or surface to another. Figure 2 *a*) and 2 *b*) illustrate two such situations. In Fig. 2 *a*), potential curves are shown for a Rydberg state of H_2^* and for $H_2^+ + e$ (with zero kinetic energy). The autoionization is presumably induced by vibrational coupling [10, 64-66]. The most probable kind of transition is that shown by the curved arrow. Figure 2 *b*) indicates some potential curves of NO^* and the lowest curve for NO^+; this is a system for which autoionization by configuration mixing or electron-electron interaction is possible.

In either sort of autoionization, the implication of the molecule's tendency to conserve nuclear kinetic energy of the transition to exhibit Frank-Condon character is that a specific initially excited state will decay preferentially into one or a small number of vibrationally excited final states of the molecule-ion. The electron energy analysis of DOOLITTLE and SCHOEN [67] shows that this is the case for at least one autoionizing state and suggests that the rule may be general.

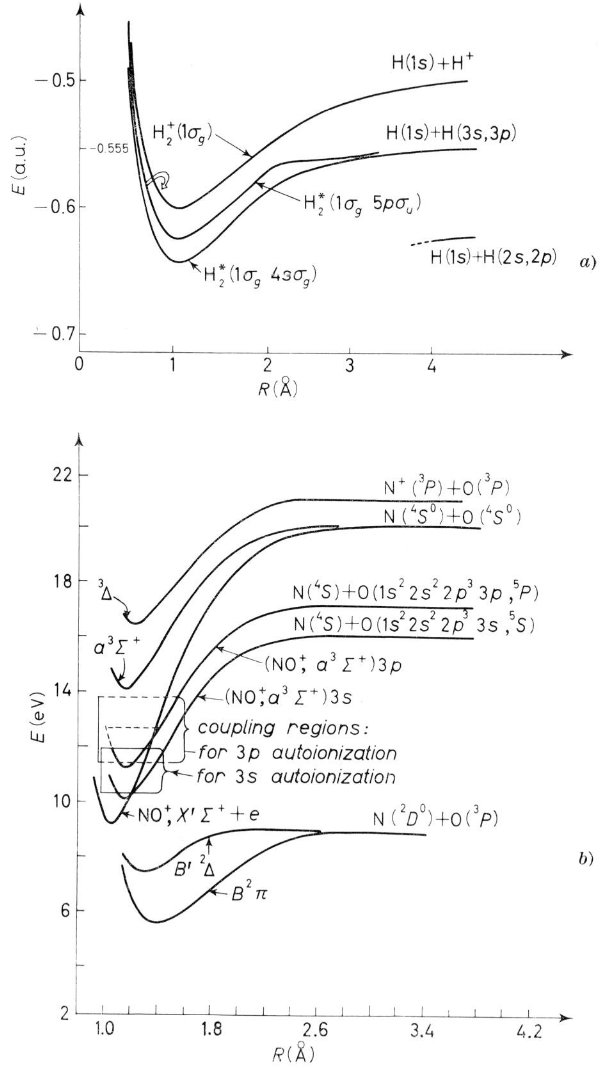

Fig. 2. – Potential curves associated with molecular autoionization. *a*) Close potential curves, vibronic coupling, no crossing point necessary: attractive potentials of H_2; no account is taken of Mulliken's « obligatory humps » (ref. [34]). *b*) Electron-electron interaction (correlation) coupling; crossing point may or may not be a requirement: NO. (Vibronic coupling may also occur in this system.)

The question arises as to whether a crossing point must be invoked to explain autoionization or, for that matter, any of the collisional chemiionization processes. It has been rather common in the literature to base arguments on the existence and position of crossing points. The basis of the crossing point arguments is in the δ-function approximation to the Franck-Condon principle, namely that both the position and momentum must be identical in order for a « curve-crossing » transition to occur between two states. Any real system deviates from this strictest of all approximate statements of the Franck-Condon principle. The degree to which it deviates depends on de Broglie wavelengths of the relative nuclear motion in the two states, or more specifically on the rate at which these wavelengths diverge from each other. If the de Broglie wavelengths diverge very rapidly as the particles move away from the crossing distance, then the only contribution to the total transition probability will come from the vicinity of the crossing point, R_x. If the de Broglie wavelengths are rather similar for a wide range of separations R, then the transition probability will come not only from contributions near R_x but from probability amplitude in a wide range of R. Two rather extreme situations are shown in Fig. 3. In the first case, the vicinity of the crossing point is clearly the only region, both in energy and in position, from which an integral will accumulate a nonvanishing value, if it involves the bound slowly-oscillating function and the rapidly-oscillating continuum function. In the second case, the wavelengths of the two vibrational states are similar enough that one can expect contributions to a transition integral from the entire range of the vibrational amplitude.

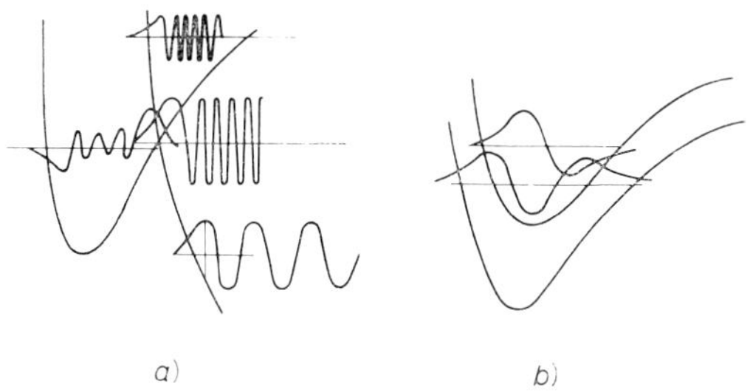

Fig. 3. – Schematic diagram of wave functions for: *a*) Case where crossing point condition is necessary; *b*) case where crossing point is irrelevant.

To which situation does autoionization correspond? In the case of vibrationally-coupled autoionization of H_2, the potential curves of the bound excited states are very close and similar to that of H_2^+. The result in this case is that the crossing point argument is entirely inapplicable and in fact whether or not a crossing occurs has no significant effect on the transition probability [7, 68]. Other cases of electron-light molecule collisions such as the vibrational excitation of H_2 or N_2 by formation of a transient negative ion state in e—H_2 or e—N_2 scattering [70] also are best described in terms of interactions over a range of R. In the case of autoionization of NO^* by electron-electron interaction, the applicability of crossing point arguments has not been questioned yet.

From the curves of Fig. 2 *b*), it appears that perhaps the crossing point arguments are applicable for autoionization of all the Rydberg states built on an NO^+, $a'\,^3\Sigma^+$ core, except the lowest, the $3s$ Rydberg states.

2·2. *Potential curve arguments for associative, collisional and Penning ionization.* – The interpretation of ionization processes in thermal collisions, at the level of potential curve arguments, has been started by MULLIKEN [34], by HERMAN and CERMAK [51, 52, 60] and by HOTOP and NIEHAUS [26]. We shall try to describe the problem in a slightly more general way to include processes like the production of $H+H^*$ from H_2^+ + electrons with energies in the 50 eV region, as observed by DUNN; this was part of a general study of e-H_2^+ collisions [71, 72].

First, we may ordinarily assume that in the collision of A^* with M, the ground-state potential fo the ion AM^+ is attractive and exhibits a minimum. There may be a rather shallow well, as is perhaps the case for $LiHe^+$, or a reasonably deep one, as in the case of H_2^+. The entire range of well depths for diatomic molecule ions probably lies between about 0.2 and 3.5 eV, because the long-range interaction is always an attractive r^{-4} charge-induced dipole interaction, and the repulsive forces in a positive ion must set in at least as rapidily as in the corresponding neutral molecule because the ion has one less electron to contribute to the screening of the nuclear-nuclear repulsion. We shall not consider excited ion-molecule states whose potential curves are repulsive, on the basis that they are associated with energies too high to contribute to chemiionization. Nevertheless these states do exist and we must take into account excited states of neutral molecules whose cores are essentially molecule-ions in repulsive states. One example is the $H_2^+(1\sigma_u)$ or any higher excited state of H_2^+ [72].

The next problem is that of constructing the potential curve or curves of the excited $(AM)^*$ or A^*-M interaction. Let us distinguish three cases:

1) $(AM)^*$ corresponds to an electron outside AM^+ in its ground state; in this case the potential curve of AM^* ordinarily lies everywhere below the curve of the ground state of AM^+, and is thus attractive.

2) $(AM)^*$ corresponds to an electron bound to an excited AM^+, and has an attractive potential.

3) $(AM)^*$ corresponds to an electron bound to AM^+ in a repulsive state.

Other cases exist, but will not be examined in detail here. For example we might consider the case of an $(AM)^*$ in which two electrons are excited, and which does not correspond in any simple way to one electron outside an ion-molecule core. Cases of this kind will ultimately have to be included in the interpretation of chemiionization in terms of the details of the correlation of separated atom and united atom orbitals and states.

Even among the three cases just listed, we should take into account the binding properties of the excited electron. These properties must ordinarily have a relatively small effect on the potential, compared with the state of the AM^+ core, because the excited electron normally is sufficiently far from the nuclei that its bonding force is small [73]. The bonding or antibonding force is greater for an excited electron if its orbital φ^* has a symmetry different from any of the occupied orbitals of AM^+, than if there is an occupied orbital of AM^+ of the same symmetry as φ^*. MULLIKEN [35] has referred to the former as « penetrating » and the latter as « nonpenetrating »; the nonpenetrating orbitals are kept out of the core by the effective repulsion of the exclusion principle.

Suppose the excited electron of A* is in an orbital which correlates with a bonding orbital of AM*. Then the potential curve of AM* will lie everywhere below the curve for the *corresponding* state of the AM^+ core, whether the core state is the ground state of AM^+ or some higher state, attractive or repulsive. The same is true if the excited orbital φ^* is only weakly antibonding. If φ^* is strongly antibonding, then the potential curve of AM* may in principle rise above that of AM^+. The outer parts of the low He_2^* Σ states are repulsive because the φ^* orbitals correlate adiabatically with highly promoted orbitals in the united atom limit and thereby become antibonding. In this case and in most cases, it is unlikely that the AM* curve rises above that of the corresponding AM^+ curve because the repulsive energy must be greater than the ionization energy (term value) of φ^* for this to occur. It does occur in the H_2^- ion; here the curve for H_2^- lies below that of H_2 for distances greater than about 3 a.u. but appears to cross above it as R becomes smaller than 3 a.u. [74]. Naturally an H_2^- state in the continuum of H_2+e is not a stable state and in fact corresponds to a transient negative ion with a lifetime of about 10^{-15} s when $R=R_e(H_2)$. Similarly, the AM* state just described will autoionize.

The curves for the lower He_2^* states exhibit minima at moderately small R for a reason that has a rather general applicability [34]. The nature of the Σ states changes as R grows smaller. The changes are from configurational bases composed of orbitals that correlate with high-energy united atom orbitals, into configurational bases that correlate with low-lying orbitals and states of the united atom. In other words these He_2^* states can be thought of in terms of avoided curve crossings, where the approximate « crossing » curves would be restricted to specific configurations.

The easiest systems to analyse are the homonuclear diatomics, *e.g.* H*+H or He*+He. MULLIKEN has shown how some of the low-lying excited states of He_2^*, for example, must have potential maxima at large internuclear distances [34] which would make it difficult for $He(1s^2)+He(1s2s, 2\,^{1,3}S)$ to reach sufficiently small internuclear distances in a thermal collision to « find » the potential for He_2^++e. The high excited states, by contrast, can be expected to have potential curves nearly parallel to and slightly below the curve for He_2^+. The charge density in the vicinity of the core decreases as n^{-3}, and the bonding force or contribution to bonding (or antibonding) energy, correspondingly. Since the Rydberg term value is proportional to n^{-2}, the antibonding contributions to the repulsive energy are overridden by the attraction of the core for its Rydberg electron, at least in the limit of high n.

A simple model system is $H^* + H \rightarrow H_2^+ + e$. This system has been studied theoretically in some detail by Nielsen and this author [17, 18]. The pertinent curves of H_2^* are the attractive potential curves based on $H(1s) + H^*(3s, p, d)$ and the repulsive curves based on $H(1s) + H^*(2s, 2p)$. The attractive curves correspond to situations like that of Fig. 3 *b*), while the repulsive curves probably fit the crossing point model. The attractive curves are very similar to those of H_2^+, and transitions occur over the entire range of R from ca. 1 to 4 a.u. The repulsive states arising from $H(1s) + H^*(2s)$ have a core of $1\sigma_u$ (antibonding) rather than $1\sigma_g$, and are very steep at small enough R values to approach the H_2^+ curve.

Heteronuclear cases have been considered by HERMAN and CERMAK [51, 52] and by HOTOP and NIEHAUS [26]. The rare gas-metal atom systems studied by HERMAN and CERMAK have potentials corresponding to case 2), because the lowest excited states of the rare gases are all well above the ionization potentials of the alkalis or mercury. The electrons released in Penning or associative ionization must carry away a large amount of energy (which is what is observed) or, in principle, could leave the ionized system with considerable excitation. That the electron carries away *all* the exoergicity of the $Hg + He(2\,^3S) \rightarrow \rightarrow Hg^+ + He + e$ reaction was taken to mean that the potential curves for the initial and final states are essentially parallel, whereas the observed energy deficiencies of electrons from $Hg + He(2\,^1S)$ were interpreted as implying that in the range of internuclear separations where the transition occurs, points on the potential diagram representing the initial state lie lower, relative to the asymptote of the initial state, than do the corresponding points of the final ionized state, relative to their own asymptote. Thus a downward vertical transition from $(HgHe)^*$ to $(HgHe)^+ + e$ gives the electron less than the total exoergicity of the reaction, leaving some as kinetic energy of relative nuclear motion. Associative ionization in the $(HeHg)^*$ system was attributed to an excited attractive state of $HgHe^+$ which dissociates to $Hg^+(^2P) + He$.

It may be that one could also look at this process in terms of adiabatic curves and an avoided crossing. From the $He(^1S) + H^+(^2P)$, one can construct a $^2\Sigma$ state which may bind a Rydberg σ-electron, making the state of the system $^1\Sigma$ or $^3\Sigma$. A state of the same symmetry is generated by the $He^*(2\,^1S$ or $2\,^3S) + Hg(\ S)$; the dissociation limits of these states are all close together, as shown in Fig. 4. If the noncrossing rule is applied, one obtains adiabatic curves that lie rather close together; the situation now is an example of our case 1). In such a case, one expects associative ionization to give all possible vibrational states of AM^+. As the internuclear distance grows smaller during the collision, the 2*s* electron of He* goes into a Rydberg orbital of (HgHe)* as the Hg^+ core goes from its lowest state to a ^2P state. Then, during the remainder of the collision, the Rydberg electron may be lost by ordinary vibronic coupling. The adiabatic transition through a Rydberg intermediate would not be the correct description if the collision were to carry the He*+Hg pair to distances shorter than R_x of Fig. 4 *a*) while the He 2*s* remained the best

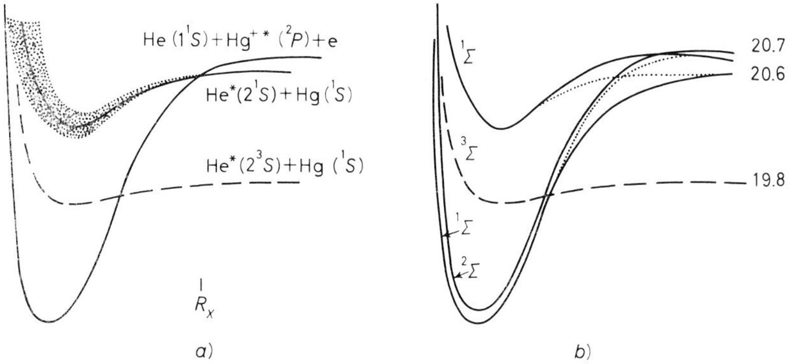

Fig. 4. – Potential curves for He*+Hg. *a*) Curve-crossing model, with He(2 ^1S)+Hg giving rise to a transient resonance (shaded area) for $R < R_x$; *b*) adiabatic model with one of the Rydberg states mixing adiabatically with He(2 ^1S)+Hg curve to give the lowest curve shown; autoionization proceeds into the curve indicated as $^2\Sigma$.

description of the excited electron in the dynamic system. In this case, one should properly consider the HgHe* « molecule » as a transient bound state of $HgHe^+ + e$, analogous to the H_2^- state mentioned previously. The potential curve is shaded to indicate the transient nature of the state. Coupling, in this event, is probably through electron-electron interaction, rather than through vibronic interaction.

The interpretation given by Hotop and Niehaus for $He^* + H_2$ (or $Ne^* + H_2$) is essentially this [26]. The system goes rapidly into a transient state best described as $(He + H_2^+)$, with vibrational excitation in the H_2^+. The most probable decay made for this compound system is directly into $He + H_2^+$. If the H_2^+ is sufficiently excited vibrationally, then the system may go to $HeH^+ + H$. If the H_2^+ has essentially no vibrational energy, a HeH_2^+ molecule may be formed.

Highly excited species like Ar** were described differently by Hotop and Niehaus, in light of the totally dominant role of rearrangement ionization of H_2, at the expense of Penning or associative ionization. This situation was ascribed to direct collisional ionization of the Ar** to give a compound system

$[Ar^+ + H_2]$. A crossing of the $Ar^{**} + H_2$ and $Ar + H_2$ curves was postulated by Hotop and Niehaus, but their conclusions follow even if no crossing takes place. The system is then in a state that one could call the associative ionization product state. However this state lies well above the dissociation limit for the H—H or Ar—H bond in ArH_2^+, so that decay eventually occurs to give $ArH^+ + H$. It would be interesting to try to find ArH_2^+ by removing vibrational energy of the compound state through collisions with inert third bodies.

As a final analysis in terms of potential curves, we may examine the relationship between Penning and associative ionization in simple systems, like $He^* + Ar$ or, even simpler, $H^* + H$. In the latter case, Penning ionization can only occur if the collision is fast enough to provide the kinetic energy needed to ionize H^*. The most probable transitions in the H_2 system seem to be at thresholds [68]. Thus if a colliding pair $H(1s) + H(nl)$ has just enough energy to form $H + H^+ + e$ with the H^+ and H moving apart slowly, and with n of 3 or 4, then the Penning process is probably competitive with associative ionization. This in turn suggests that the ratio of H_2^+ to H^+ will be a sensitive function of the temperature of the system. In the He*—Ar system, the electron must carry away a large amount of energy. Here we must go one step further and inquire what the relation is between the initial nuclear kinetic energy and the fraction of the energy carried away by the electron. Herman and Cermak [52] considered a similar situation and interpreted the AM^+/M^+ ratio in terms simply of whether or not the A*-M system made its downward transition to a point to the left of the point R_0, the classical turning point for a pair of free particles $A + M^+$ with zero kinetic energy. In Fig. 5 we show a situation for which associative and Penning ionization differ only insofar as the Franck-Condon transition from the upper curve's turning point reaches above or below the dissociation limit of AM^+. The requirement for such a system is that R_0 for AM^+ falls to the left of R_0 for $A^* + M$, so that both free and bound states of AM^+ are accessible via vertical transitions from the turning points of the potential for free $A^* + M$. The ratio of AM^+ to M^+ is thus determined by the fraction of colliding pairs with relative kinetic energy above E_T, the transition energy E_T is defined as the value of $E(A^* + M)$ at the distance $R_0(AM^+)$. Hence E_T is the energy above which Frank-Condon transitions give Penning ionization and below which the transitions give the associated product. To account for noncentral collisions in the simplest way for this system, one merely constructs a new curve for each orbital rotational state of $A^* + M$ and applies the same reasoning. (In practice, of course, one needs examine only a few widely spaced rotational states.) The principal effect of rotation is to raise more of the AM^+ curve above the dissociation limit and thus to move $R_0(AM^+)$ to the right. This, in turn, lowers E_T and therefore produces more Penning ionization and less associative ionization than one would infer from head-on collisions only.

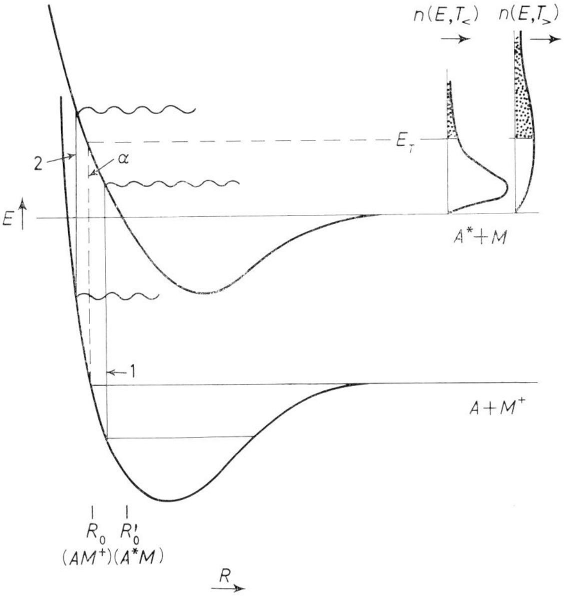

Fig. 5. – Model for associative ionization and Penning ionization when $E(A^*+M) > E(A+M^++e)$, as for example in He^*+Ar or Hg. The assumption is made that the two potential curves are sufficiently different that most transitions occur in the vicinity of the classical turning point. The energy E_T and the thermal distribution of collision energies $n(E, T)$ govern the ratio of associative to Penning ionization. If the energy of relative motion of A^*+M, E, is less than E_T, the Franck-Condon transition at the turning point corresponds to a vertical line, like line 1, to the right of line α. Line α corresponds to a transition leaving $A+M^+$ free but with zero kinetic energy of relative motion. Transitions of A^*+M with $E > E_R$ correspond to lines such as line 2, into the vibrational continuum of $A+M^+$, and therefore to Penning ionization. Lines such as line 1 leave $A+M^+$ in a bound vibrational state and therefore to associative ionization. Two thermal distributions of colliding atoms, A^*+M, are shown; $n(E, T_<)$ represents a relatively low temperature and $n(E, T_>)$, a relatively high temperature. The portion of the distribution giving rise to Penning ionization is shaded. Note that the mechanism proposed in this figure requires (at least in the δ-function approximation for Frank-Condon transitions at the turning point) that the turning point R_0, for which $A+M^+$ has zero kinetic energy, be less than R_0, the corresponding internuclear distance for A^*+M.

Note that if $R_0(A^*M)$ is less than $R_0(AM^+)$, then associative ionization is very improbable. This may be the reason that He^*+H_2 and particularly Ar^{**} or $Kr^{**}+H_2$ produce very little associative ionization [26].

3. – Quantitative theories.

3·1. *The Demkov-Komarov theory.* – This theory [75] is not strictly a chemi-ionization theory but is closely enough related to our subject to warrant mention. We consider a colliding A+B pair with sufficient kinetic energy to ionize, but with A and B in their ground or low-lying states. The model supposes that the effective repulsions in the A-B system are great enough to carry its potential up through the infinite set of $AB^+ + e$ Rydberg states and into the continuum. This model is reasonable insofar as there are generally, some strongly repulsive potential curves available to an A-B colliding pair regardless of whether there are also attractive curves. The assumption is made that at each curve crossing, the interaction of adiabatic states takes place over a length L. If W_0 is the probability of reaching the ionization energy E_0 in a collision (without going out of the collision in some lower state), and if v is the relative velocity, then the total probability of ionization is shown to be

$$W = W_0 \exp[-2LE_0/\hbar v] .$$

The quantity W_0 cannot be obtained in any simple way, but the dependence of the total probability on E_0, L and v is clear. The theory is probably applicable for collisions in the (20÷100) eV region. It is not clear whether any neutral-neutral systems are yet known which are described by this theory. Ionizing collisions of two-atom systems at still higher energies are interpretable in terms of the correlation diagrams for the molecular orbitals, according to a diabatic theoretical model developed by FANO and LICHTEN [76], and further by LICHTEN [77].

3˙2. *Weak interaction theories.* – Two models have been used to treat Penning ionization at the level of a semiquantitative theory. One is the simple « critical radius » model; the other is based on the dipolar exchange of a quantum from one collisional partner to the other. In the « critical radius » model, one constructs a characteristic distance R_c for the particular system and assumes that all collisions occurring within the characteristic distance give rise to reaction. The cross-section σ_{eff} is taken to be the geometric one, πR_c^2, and the rate coefficient R as $\bar{v}\sigma_{\text{eff}}$. This approach is the basis, for example, of the Gioumousis-Stevenson [78] treatment of ion-molecule reactions, where the critical radius is defined as the Langevin orbiting radius. The rate coefficient is

$$k(\text{ion-molecule}) = 2\pi(2C_4/m)^{\frac{1}{2}},$$

where C_4 is the potential parameter in the expression for the long-range interaction

$$V(R) = -C_4 R^{-4} ,$$

and m is the reduced mass for the collision. FERGUSON developed a model for collisions of atoms with He* based on the effective radius associated with the momentum transfer cross-section [79]. Here, the interaction varies as R^{-6}, rather than as R^{-4}. This gives a cross-section for a collision with relative kinetic energy E of

$$\sigma(\text{momentum}) = 1.39\cdot 10^{-16}(C_6/E)^{\frac{1}{3}}\ \text{cm}^2\ .$$

Bell, Dalgarno and Kingston [80] have recently extended the use of the Langevin orbiting radius to potentials of the form

$$V(R) = C_s R^{-s}\ .$$

They assume that all collisions in which the particles pass the centrifugal barrier must lead to possible reactions and then use experimental data to give rough estimates of the probability of electron loss when the system is in close collision. The cross-section for close collisions is

$$\sigma(\text{orbit}) = (1.66\cdot 10^{-6}(C_6/E)^{\frac{1}{3}}\ ,$$

and the probability for ionization is taken to be 0.4 for $He^* + O$, and unity for alkalis. The ionization rate coefficients so estimated fall in the range $(4\div 10)\cdot 10^{-10}$ cm³/s, except for methane which gives only $0.97\cdot 10^{-10}$.

A more elaborate and fundamental treatment is that of Katsuura, Watanabe and Mori [81-84]. In essence the theory is an impact parameter treatment (See Lecture on Electronic Excitation Transfer) in which the interaction occurs by simultaneous optical transitions $A^* \to A$ and $M \to M^+ + e$. The coupling operator is thus the product of two electric dipole operators, μ_A and μ_M for the system, and the transition probability, cross-section and rate coefficient are functions of the product of the squares of the transition dipole matrix elements. In a sense the problem is closely related to excitonic energy transfer. The wave function is expressed as a superposition of states corresponding to $A^* + B$ and $A + B^+ + e$. The time-dependent Schrödinger equation for the coefficient can be integrated for straight-path collisions with a given impact parameter b, and then over impact parameters. The assumption is made that the interparticle interaction potential can be represented as

$$V(R) = -2R^{-3}(A^*|\mu_A|A)(M|\mu_M|M^+ + e)\,e^2/3$$

analogous to a dipole-dipole R^{-3} interaction of static systems. With this assumption the equation for the cross-section can be integrated explicitly to give Penning ionization cross-section $\sigma_p(r)$ for collisions of a given velocity v:

$$\sigma_v(v) = 2\pi D(3\pi^2|\mu_A|^2\cdot|\mu_M|^2/\hbar v)^{\frac{2}{5}}\ ,$$

where we let μ_A and μ_D stand for the transition dipole matrix elements and D is the definite integral

$$D=\int_0^\infty (1-\exp[-x^{-5}])\,x\,dx \simeq 0.74460 .$$

Averaging over temperatures in a Boltzmann distribution, one obtains a rate coefficient k_p

$$k_p(T)=19.07(|\mu_A|^2|\mu_M|^2/\hbar)^{\frac{2}{5}}(2kT/M)^{\frac{3}{10}} .$$

The refined method developed by WATANABE and KATSUURA [82] takes into account the imperfect coupling of angular momentum to the internuclear axis, with the result that cross-sections are 23 % smaller than in the earlier form of the theory. It was also possible to show that the assumption of perfect angular momentum coupling during the collision sets an upper bound on the cross-section.

The theory of KATSUURA has the limitation that it can only be applied when A* is connected optically to a lower state. It cannot be used, for example, to treat the helium metastables. The cross-sections predicted by this method are rather large. Typically, for $He(2\,^1P)+CH_4 \to He(1\,^1S)+CH_4^+ + e$, σ_p is about $(7.6 \div 7.7)\cdot 10^{-15}$ cm² at 300 °K and $(5.5 \div 5.6)\cdot 10^{-15}$ at 1500 °K, and for $He(2\,^1P)$ + Ar, about the same size. From the work of HERCE and MUSCHLITZ [85] and of HOTOP and NIEHAUS [56], the cross-sections for Penning ionization of Ar by $He(2\,^3S)$ and by $He(2\,^1S)$ are both one order of magnitude smaller. If the optical transition model were generally valid, one would expect the latter two processes to have much smaller cross-sections. Furthermore one would expect the $He(2^1S)$ to be much more effective than the $He(2^3S)$, simply because of the added spin-forbidden character for the transition $2^3S \to 1^1S$. Still another reservation derives from the results of KAUL, SEYFRIED and TAUBERT [47] on associative ionization in helium, where long- and short-lived excited states contributed roughly equally to ion production. The theory of KATSUURA remains to be tested. Calculations of all the alkali-rare gas Penning cross-sections were carried out by SHELDON [86]. Instead of deriving bound-free optical transition probabilities as KATSUURA did, directly from experimental data, SHELDON used a quantum defect method and found cross-sections between $0.9\cdot 10^{-15}$ and $2.8\cdot 10^{-15}$.

A model has been developed by WARKE [87], and elaborated by CHAN [88], which falls somewhat between weak and close-coupled limits. The model has been applied to e—O_2^+ recombination by both authors and to e—N_2^+ and e—He_2^+ by WARKE. The essence of the model is this. The incoming electron is represented as being in an *atomic* state, free in the initial state ψ_i and bound in the final state ψ_f. The chosen states are the distorted Coulomb wave around the atomic ion, *e.g.* O^+, for ψ_i and the lowest unfilled bound state of the same atomic ion for ψ_f. The coupling is presumed to be due to the nuclear vibration which gives rise to a fluctuating potential, specifically the fluctuating potential of the second (and neutral) atom. Warke's semiclassical model assumes that all the allowed final states of $O(^1D) + O(^3P)$ or of $O(^3P) + O(^3P)$ contribute in proportion to their statistical weights, and that the bound-free vibrational overlaps are unity. The results so obtained for O_2^+, $1.14 \cdot 10^{-7}$ cm³/s, are in rather good agreement with the experimental value for the recombination coefficient [89], $(2.2 \pm 0.4) \cdot 10^{-7}$ cm³/s. CHAN has shown that Warke's assumption about the overlap integrals overestimates them very much. CHAN calculates both electronic and nuclear terms quantum-mechanically within the perturbed single atom picture (and with the assumption that electronic contributions may be computed at a single internuclear distance); he obtains a value of $(2.8 \pm 0.2) \cdot 10^{-7}$ cm³/s based on the assumption that a $^1\Sigma_u^+$ curve for $O(^1D) + O(^3P)$ crosses the O_2^+ curve near $v = 0$. With this assumption, the value for the recombination coefficient is in quite reasonable agreement with the experimental value.

3'3. *Close-coupled near-adiabatic theories.* – To go beyond a weak interaction theory, one must ask about the way A* and M interact during their collision. At one extreme, the collision may constitute a fast perturbation with Fourier components whose frequencies are comparable to or higher than the characteristic frequencies of the electronic transitions in the free A or M particles. This is not the situation in chemiionization. The processes we are considering have collision times so long that the Born-Oppenheimer approximation or the adiabatic model for the collision is the most suitable starting point.

Near-adiabatic theoretical treatments now exist at various stages of refinement for molecular autoionization [17, 64-66, 68] and some types of predissociation [17], for dissociative recombination [90] and for associative ionization [68, 91]. In the case of autoionization and predissociation in H_2, HD and D_2, quantitative nonparametric calculations of lifetimes are available [17, 64-68]; cross-sections for associative ionization and dissociative recombination have been computed for $H(1s) + H(3s, p, d)$ [68].

In all the near-adiabatic treatments, the initial and final states Ψ_i, Ψ_f, are represented as products of electronic functions $\psi_{i,f}(\boldsymbol{r}, \boldsymbol{R})$ and vibrational functions $\chi_{i,f}(\boldsymbol{R})$ or vibrational and rotational functions. (We let $\boldsymbol{r}$ represent all electronic coordinates and $\boldsymbol{R}$, all nuclear co-ordinates.) The states may fall into any of the categories indicated in Table I; the properties of the corresponding process are contained in the properties of the appropriate element of the T-matrix,

$$(\Psi_f|T|\Psi_i) = (\Psi_f|\mathscr{V}(1 + \mathscr{G}^+ \mathscr{V})|\Psi_i)$$

where the Green's function

$$\mathscr{G}^{+} = (E - \mathscr{H} + i\varepsilon)^{-1}$$

and $\mathscr{V}$ is the part of the total Hamiltonian $\mathscr{H}$ which couples Ψ_f and Ψ_i. Often one can neglect the propagator term, the term $\mathscr{G}^{+}\mathscr{V}$, and obtain transition probabilities according to the Wentzel-Fermi golden rule [92]:

$$P(i \to f) = 2\pi\hbar^{-1}\left|(\Psi_f|\mathscr{V}|\Psi_i)\right|^2 \varrho(f)$$

where $\varrho(f)$ is the density of final states. As described in the Lecture on Electronic Energy Transfer, the physics of the problem enters when we choose the representation for Ψ_f and Ψ_i and the coupling operator $\mathscr{V}$.

In some problems, particularly in rearrangement collisions, basis functions Ψ_f and Ψ_i are sometimes chosen which are eigenfunctions of different Hamiltonians. If such bases are selected, the left- and right-hand ψ's in the T matrix must be different, to correspond to the two different Hamiltonians. Furthermore one must exercise care to keep initial and final states orthogonal, if they represent different channels. Fortunately, for some systems and our ionization processes are among them, one can avoid this difficulty by taking Ψ_f and Ψ_i to be eigenstates of a single adiabatic Hamiltonian for the *entire* system, and thereby ensure that they represent mutually exclusive states.

In the treatment of ionization processes in H_2 and its isotopes, the initial approximations were based on atomic hydrogen Rydberg and continuum functions [64, 66]. It has proved feasible to carry the calculations to a more quantitative level by deriving the functions $\psi_{i,f}$ and $\chi_{i,f}$ from the effective potential of the H_2^+ core [93]. This adiabatic potential, with both Coulomb and approximate exchange contributions, can be expressed in terms of its spherical harmonic components. The non-adiabatic coupling operator $\mathscr{V}$ in this system is the nuclear kinetic energy operator or, in physical terms, the coupling is the result of the imperfect ability of a Rydberg electron to keep up with the motion of the nuclei and the core electron, particularly with their vibrational motion. That this is so can be seen from some characteristic times. The Bohr period for an electron with principal quantum number n is approximately $10^{-16} n^3$ s. The vibrational period for H_2 is approximately 10^{-14} s, and the rotational period is $10^{-12}/J(J+1)$ s. Hence for $n \sim 5$, the Bohr period and vibrational period are comparable and one can expect them to couple. The same is true of states with $J \sim 10$, but for low J states, coupling is only expected at high n.

The essential characteristics of the transition probability for autoionization are these:

a) The autoionization rate decreases as n^{-3} ($n =$ initial principal quantum number) for a given initial vibrational quantum number v and vibrational state change Δv [65, 66];

b) the autoionization rate is much greater for $\Delta v = -1$ than for $\Delta v = -2$, etc., for a given initial v and n; this is called the vibrational propensity rule;

c) the autoionization rate increases with v; actually it increases approximately as the root-mean-square vibrational energy, for a given initial n.

These results have their counterparts for predissociation into attractive Rydberg states and for associative ionization. There, the transition probabilities are greatest when the nuclear kinetic energy suffers the smallest change. Hence associative ionization transitions from the vibrational continuum of $H^* + H$ are most likely if they leave the H_2^+ in a vibrational state in the energy range of the H_2^+ well close to the dissociation limit of H_2^*. In the predissociation case, transitions are most probable from an excited vibrational state of the n-th Rydberg state into the vibrational continuum of the $(n-1)$-st Rydberg state, if it is energetically possible.

Certain things can be said about the microscopic mechanism of the coupling in vibronically-induced autoionization, predissociation, associative ionization and dissociative recombination. In contrast to the suppositions of the earliest treatments, the dominant part of the coupling comes from the monopole, not from higher spherical harmonics. This can only be recognized when the theory takes into account the finite size of the core, a characteristic which the original treatments neglected. This has important consequences for the selection rules; in particular it means that $J = 0 \rightarrow J = 0$ transitions (and other $\Delta J = 0$ transitions also) are allowed.

A second and perhaps more important characteristic is the complete lack of dependence for these processes on crossing points. As was discussed earlier, when the deBroglie wavelengths of initial and final states are similar, the transition amplitude is accumulated over a range of R, and not just in the vicinity of a crossing point.

Some attention has also been given to higher-order coupling in H_2; *e.g.* one Rydberg vibronic state may be coupled with another which, in turn, is coupled to the ionization continuum. Fortunately for the sake of simplicity, only a few states are coupled strongly even in the largest clusters. The effects of this coupling can be large; a state which would be expected to autoionize with a rate of order $10^8\ s^{-1}$ in a first-order approximation may, in a favorable case, « steal enough evanescence » (to coin a term analogous to « stealing intensity ») to autoionize at a rate of order $10^{10}\ s^{-1}$. Such effects are the results of accidental near-degeneracies and cannot be expected to appear with regularity.

One quantitative comparison can be made between theory and experiment for a specific state, the $8p\sigma$, $v = 3$ state. The autoionization peak for this state was found by CHUPKA and BERKOWITZ [11] to have a width of approximately $10^{11}\ s^{-1}$, distinctly greater than the instrumental width. The width calculated by NIELSEN and this author is $7.9 \cdot 10^{11}$. The discrepancy presumably lies in the inaccuracy of the electronic wave functions in the core region, and can be removed with more accurate functions. RUSSEK, PATTERSON and BECKER do not give a specific result for this state, but for autoionization of the $8\pi p$ state (no dependence on vibrational quantum number), the rate is $1.0 \cdot 10^{10}\ \mathrm{s}^{-1}$. The value is presumably low partially because of the neglect of the monopole contribution.

For molecules more complicated than H_2, the theory is still not a in quantitative state. BARDSLEY [90] has treated dissociative recombination, with particular reference to the interesting (and, in the atmosphere, important) case of $NO^+ + e$. In NO, both electron-electron interaction and vibronic coupling may contribute to dissociative recombination and the other processes. BARDSLEY has attributed the recombination to two processes:

a) The direct process,

$$NO^+ + e \rightarrow NO^*[\dots (\pi 2p)^4(\sigma 2p)(\pi' 2p)^2,\ B'\,{}^2\Delta \text{ or } I\,{}^2\Sigma]$$

or

$$NO^*[\dots (\pi 2p)^3(\sigma 2p)^2(\pi' 2p)^2,\ B\,{}^2\Pi \text{ or } L'\,{}^2\Phi]$$

where the vibrational energy in the final state is greater than the dissociation energy of that state;

b) an indirect process, with $NO^+ + e$ passing first into an excited Rydberg state having a core in its own ground state, and then into one of the final states just given for the direct process.

The presumption is that the transition into the final state is induced by electron-electron coupling while the initial step of the indirect processes is induced by vibron c coupling.

Bardsley's method is set up to be used as a parametrized or semiempirical treatment, and makes use of the crossing point model. The basic expression for the cross-section is

$$\sigma_{\mathrm{dr}}(E) = \sigma_c \cdot S_f(E)\ ,$$

where σ_c is a cross-section for capture into a transient or resonance state and $S_f(t)$ is a survival factor for the resonance state. The cross-section

$$\sigma_c(E) = \pi^2 (2m\varepsilon)^{-1} r \Gamma_c f(\varepsilon)\ ,$$

where r is the ratio of degeneracies of the resonant and initial molecule-ion state, ε is the electron kinetic energy, Γ_c is the capture width, essentially the configuration interaction matrix element giving the resonance its existence, and $f(\varepsilon)$ is the probability that the nuclei lie in the vicinity of the crossing point R_x:

$$f(\varepsilon) = |\chi(R_x)|^2 |dE/dR|^{-1}_{R=R_x} ,$$

where the numerator is simply the square of the nuclear wave function and the denominator is the slope of the potential for the neutral (dissociating) state at the crossing point. The survival factor can be determined in principle from the electronic and vibrational functions ψ_i, ψ_f, χ_i and χ_f but was taken as unity by BARDSLEY, since no reliable means was available to evaluate it. The width Γ_c also is not evaluated, but if it is essentially constant for low energies, the temperature-dependence of the rate coefficient can be found by integrating $\sigma(E)$ over the electron energy. The result is

$$k_{dr} = (2\pi^2\hbar^2 r)(2\pi m^3 kT)^{\frac{1}{2}} \Gamma_c f(0) .$$

From spectroscopic data, BARDSLEY is able to estimate Γ_c and $f(0)$ to get contributions

$$k_{dr}(B\,{}^2\Pi + B'\,{}^2\Delta) \simeq 2.6 \cdot 10^{-7}\ cm^3/s .$$

The experimental values [31, 32] are between 3.3 and $7 \cdot 10^{-7}$ cm³/s.

The indirect process is likely to make a significant contribution to the total capture rate for gas temperatures up to about 1000 °K, with the capture energy going into excitation of vibration and the electron, into Rydberg states with $n \simeq 7$. The contribution of the indirect process decreases with temperature, and may fall off as fast as $T^{-\frac{3}{2}}$. Thus in the upper atmosphere, the indirect process probably plays an important role.

* * *

The author would like to thank Drs. HOTOP, LANDMAN and NIEHAUS for recording the notes of his lectures. He would also like to express his gratitude to Drs. J. BERKOWITZ, V. CERMAK, Y.-T. LEE and J. C. PERSON for their helpful comments during the preparation of this manuscript.

REFERENCES

[1] R. P. MADDEN and K. CODLING: *Journ. Opt. Soc., Am.*, **54**, 268 (1964); *Phys. Rev. Lett.*, **12**, 106 (1964); **10**, 516 (1963); *Astrophys. Journ.*, **141**, 364 (1965).

[2] H. BEUTLER: *Zeits. Phys.*, **93**, 177 (1935).

[3] R. E. HUFFMAN, Y. TANAKA and I. C. LARRABEE: *Journ. Chem. Phys.*, **39**, 902 (1963).

[4] U. Fano: *Phys. Rev.*, **124**, 1866 (1961).
[5] U. Fano and J. W. Cooper: *Phys. Rev.*, **137**, A 1364 (1965).
[6] F. J. Comes and H. G. Sälzer: *Phys. Rev.*, **152**, 24 (1966).
[7] F. J. Comes, H. G. Sälzer and G. Schumpe: *Zeits. Naturforsch.*, **23** a, 137 (1968).
[8] F. H. Mies and M. Krauss: *Journ. Chem. Phys.*, **45**, 4455 (1966); F. H. Mies: *Phys. Rev.*, **175**, 164 (1968).
[9] G. L. Weissler, J. A. R. Samson, M. Ogawa and G. R. Cook: *Journ. Opt. Soc. Am.*, **49**, 338 (1959); A. J. C. Nicholson: *Journ. Chem. Phys.*, **39**, 954 (1963), and references therein; K. Watanabe: *Journ. Chem. Phys.*, **22**, 1564 (1954).
[10] V. H. Dibeler, R. M. Reese and M. Krause: *Journ. Chem. Phys.*, **42**, 2045 (1965).
[11] W. A. Chupka and J. Berkowitz: *Journ. Chem. Phys.*, **48**, 5726 (1968).
[12] F. J. Comes and H. O. Wellern: *Zeits. Naturforsch.*, **23** a, 881 (1968).
[13] H. Beutler and H.-O. Jünger: *Zeits. Phys.*, **100**, 81 (1936). It is of historical interest that the original value of 15.428 eV $((124.429 \pm 13)\ \text{cm}^{-1})$ for the ionization potential of H_2^+ was based on an autoionization argument quite analogous to those of ref. [11, 12]. H. Beutler and H.-O. Jünger (this reference) assigned lines at $\lambda < 804$ Å as having extra width due to autoionization. Y. Tanaka showed (*Tokyo Scientific Research Institute Journal*, **42**, Physics Section, 49 (1944)) that the apparent widths of these « lines » were due to several overlapping lines. However the photoionization studies demonstrated unequivocally that these lines, as well as a few at still lower energies, are associated with autoionization but that the apparatus of Beutler and Jünger must have been unable to resolve their true widths. It may now be possible to determine many widths from new spectral data. (Y. Tanaka: private communication).
[14] G. Herzberg: *Molecular Spectra and Molecular Structure*. - I: *Spectra of Diatomic Molecules* (New York, 1950).
[15] G. R. Cook and P. H. Metzger: *Journ. Chem. Phys.*, **41**, 321 (1963).
[16] J. Berkowitz and W. A. Chupka: private communication.
[17] S. E. Nielsen and R. S. Berry: *Chem. Phys. Lett.*, **2**, 503 (1968).
[18] F. L. Mohler and C. Boeckner: *Journ. Res. Nat. Bur. Stand.*, **5**, 51, 399, 831 (1930). (See also B. Gudden: *Lichtelektrische Erscheinungen* (Berlin, 1928), p. 225.)
[19] F. L. Mohler, P. D. Foote and R. L. Chenault: *Phys. Rev.*, **27**, 37 (1926).
[20] J. A. Hornbeck and J. P. Molnar: *Phys. Rev.*, **84**, 621 (1951).
[21] F. L. Arnot and M. B. M'Ewen: *Proc. Roy. Soc.*, A **166**, 543 (1938); A **171**, 106 (1939).
[22] G. E. Chamberlain: *Phys. Rev.*, **155**, 46 (1967), and references therein.
[23] D. Andrick, H. Ehrhardt and M. Eyb: *Zeits. Phys.*, **214**, 388 (1968).
[24] A. A. Kruithoff and F. M. Penning: *Physica*, **4**, 430 (1937).
[25] S. E. Kuprianov: *Žurn. Ėksp. Teor. Fiz.*, **48**, 467 (1965); **51**, 1011 (1966); *Sov. Phys. JETP*, **21**, 311 (1965); **24**, 674 (1967).
[26] H. Hotop and A. Niehaus: *Proc. Conf. on Heavy Particle Collisions, Queen's University of Belfast* (April, 1968), p. 194; *Zeits. Phys.*, **215**, 395 (1968); V. Fuchs and A. Niehaus: *Phys. Rev. Lett.*, **21**, 1136 (1968).
[27] D. R. Bates and H. S. W. Massey: *Proc. Roy. Soc.*, A **187**, 261 (1946); A **192**, 1 (1947).
[28] D. R. Bates: in *Emission Spectra of Night Sky and Auroral* (Reports of the Gassiot Committee, The Physical Society) (London, 1948), p. 21.
[29] J. W. Chamberlain: *Physics of the Aurora and Airglow* (New York, 1961).

[30] E. W. McDaniel: *Collision Phenomena in Ionized Gases* (New York, 1964).
[31] *a*) R. C. Gunton and T. M. Shaw: *Phys. Rev.*, **140**, A 748, A 756 (1965); *b*) R. P. Stein, M. Scheibe, M. W. Syverson, T. M. Shaw and R. C. Gunton: *Phys. Fluids*, **7**, 1641 (1964).
[32] R. A. Young and G. St. John: *Phys. Rev.*, **152**, 25 (1966).
[33] E. E. Ferguson, F. C. Fehsenfeld and A. L. Schmeltekopf: *Phys. Rev.*, **138**, A 381 (1965).
[34] R. S. Mulliken: *Phys. Rev.*, **136**, A 962 (1964).
[35] R. S. Mulliken: *Journ. Am. Chem. Soc.*, **86**, 3183 (1964).
[36] M. L. Ginter: *Journ. Chem. Phys.*, **42**, 561 (1965).
[37] Z. Herman and V. Čermák: *Coll. Czech. Chem. Comm.*, **28**, 799 (1963).
[38] Z. Herman and V. Čermák: *Nature*, **199**, 588 (1963).
[39] A. Fontijn, W. J. Miller and J. M. Hogan: *Proc. X Symposium (International) on Combustion* (Pittsburgh, Pa., 1965), p. 545.
[40] H. F. Calcote: in *Progress in Astronautics and Aeronautics*, vol. 12, *Ionization in High-Temperature Gases*, edited by K. E. Shuler (New York, 1963), p. 107.
[41] G. B. Kistiakowsky and J. V. Michael: *Journ. Chem. Phys.*, **40**, 1447 (1964).
[42] G. P. Glass, G. B. Kistiakowsky, J. V. Michael and H. Niki: *Journ. Chem. Phys.*, **42**, 608 (1965).
[43] I. D. Gay, G. P. Glass, R. D. Kern and G. B. Kistiakowsky: *Journ. Chem. Phys.*, **47**, 313 (1967).
[44] R. Breslow, J. T. Groves and G. Ryan: *Journ. Am. Chem. Soc.*, **89**, 5048 (1967) and references therein for elaborate studies of substituted species.
[45] V. Čermák and Z. Herman: *Coll. Czech. Chem. Comm.*, **29**, 953 (1964).
[46] H. Hotop and A. Niehaus: *Journ. Chem. Phys.*, **47**, 2506 (1967).
[47] W. Kaul, P. Seyfried and R. Taubert: *Zeits. Naturforsch.*, **18** a, 432, 884 (1963).
[47*a*] P. M. Becker and F. W. Lampe: *Journ. Chem. Phys.*, **42**, 3857 (1965).
[48] M. P. Teter, F. E. Niles and W. W. Robertson: *Journ. Chem. Phys.*, **44**, 3018 (1966); C. H. Turner and W. W. Robertson: *Proc. XXI Annual Gaseous Electronics Conference, Boulder, Colo., October 1968.*
[49] J. A. Herce, J. R. Penton, R. J. Cross and E. E. Muschlitz jr.: *Journ. Chem. Phys.*, **49**, 958 (1968).
[50] E. E. Muschlitz, J. R. Penton and J. A. Herce: *Proc. V Int. Conf. on Phys. of Electronic and Atomic Collisions (ICPEAC), Leningrad, 1967* (Leningrad, 1967).
[51] V. Čermák and Z. Herman: *Chem. Phys. Lett.*, **2**, 359 (1968). See also [26].
[52] Z. Herman and V. Čermák: *Coll. Czech. Chem. Comm.*, **31**, 649 (1966).
[53] Cf. N. F. Mott and H. S. W. Massey: *Theory of Atomic Collisions*, III Ed. (London, 1965), p. 591.
[54] V. Čermák: *Journ. Chem. Phys.*, **43**, 4527 (1965).
[55] W. Chupka and J. Berkowitz: private communication.
[56] H. Hotop and A. Niehaus: private communication.
[57] W. Chupka, M. E. Russell and K. Refaey: *Journ. Chem. Phys.*, **48**, 1518 (1968).
[58] T. Namioka: *Journ. Chem. Phys.*, **40**, 3154 (1964); **41**, 2141 (1964); **43**, 1636 (1965).
[59] V. Čermák: *Journ. Chem. Phys.*, **44**, 1318, 3771, 3781 (1966).
[60] V. Čermák: *Coll. Czech. Chem. Comm.*, **33**, 2739 (1968).
[61] *a*) R. S. Berry, T. Cernoch, M. Coplan and J. J. Ewing: *Journ. Chem. Phys.*, **49**, 127 (1968); *b*) J. J. Ewing, R. Milstein and R. S. Berry: *Proc. VII Int. Shock Tube Symposium, Toronto, June 1969*; *c*) Similar experiments, based on molecular absorption and alkali emission, were recently reported:

R. Hartig, H. A. Olschewski, J. Troe and H. G. Wagner: *Ber. Bunrenges. Phys. Chem.*, **72**, 1016 (1968). The inference was made that NaCl and KCl dissociate principally to atoms.

[62] A. L. Schmeltekopf, F. C. Fehsenfeld, G. I. Gilman and E. E. Ferguson: *Planet. Space Sci.*, **15**, 401 (1967).

[63] W. W. Robertson: *Journ. Chem. Phys.*, **44**, 2456 (1966).

[64] R. S. Berry: *Journ. Chem. Phys.*, **45**, 1228 (1966).

[65] J. N. Bardsley: *Chem. Phys. Lett.*, **1**, 229 (1967).

[66] A. Russek, M. R. Patterson and R. L. Becker: *Phys. Rev.*, **167**, 17 (1968).

[67] P. H. Doolittle and R. J. Schoen: *Phys. Rev. Lett.*, **14**, 348 (1965).

[68] S. E. Nielsen and R. S. Berry: submitted for publication.

[69] J. N. Bardsley, A. Herzenberg and F. Mandl: *Proc. Phys. Soc.*, **89**, 321 (1966).

[70] A. Herzenberg: *Journ. Phys.* B (*Proc. Phys. Soc.*), **1**, 548 (1968).

[71] G. H. Dunn and B. Van Zyl: *Phys. Rev.*, **154**, 40 (1967); G. H. Dunn: private communication.

[72] J. M. Peek: *Phys. Rev.*, **154**, 52 (1967).

[73] T. Berlin: *Journ. Chem. Phys.*, **19**, 208 (1951).

[74] H. S. Taylor and F. E. Harris: *Journ. Chem. Phys.*, **39**, 1012 (1963).

[75] Yu. N. Demkov and I. V. Komarov: *Žurn. Ėksp. Teor. Fiz.*, **50**, 286 (1966); *Sov. Phys. JETP*, **23**, 189 (1966).

[76] U. Fano and W. Lichten: *Phys. Rev. Lett.*, **14**, 627 (1965).

[77] W. Lichten: *Phys. Rev.*, **164**, 131 (1967).

[78] G. Gioumousis and D. P. Stevenson: *Journ. Chem. Phys.*, **29**, 294 (1958).

[79] E. E. Ferguson: *Phys. Rev.*, **128**, 210 (1962).

[80] K. L. Bell, A. Dalgarno and A. E. Kingston: *Journ. Phys.* B (*Proc. Phys. Soc.*), **1**, 18 (1968).

[81] K. Katsuura: *Journ. Chem. Phys.*, **42**, 3771 (1965).

[82] T. Watanabe and K. Katsuura: *Journ. Chem. Phys.*, **47**, 800 (1967).

[83] M. Mori, T. Watanabe and K. Katsuura: *Journ. Phys. Soc. Japan*, **19**, 380 (1964); see M. Mori: *Journ. Phys. Soc. Japan*, **26**, 773 (1969); H. Nakamura: *Journ. Phys. Soc. Japan*, **26**, 1473 (1969), for more general extensions.

[84] T. Watanabe: *Journ. Chem. Phys.*, **46**, 3741 (1967).

[85] J. R. Herce and E. E. Muschlitz jr.: *Proc. XX Annual Gaseous Electronics Conf., San Francisco, October 1967.*

[86] J. W. Sheldon: *Journ. Appl. Phys.*, **37**, 2928 (1966).

[87] C. S. Warke: *Phys. Rev.*, **144**, 120 (1966).

[88] F. T. Chan: *Journ. Chem. Phys.*, **49**, 2533 (1968).

[89] M. A. Biondi: *Ann. Geophys.*, **20**, 34 (1964); W. H. Kasner and M. A. Biondi: *XIX Ann. Gaseos Electronic Conf., Georgia Inst. of Tech., Atlanta, Ga., October, 1966.*

[90] J. N. Bardsley: *Journ. Phys.* B (*Proc. Phys. Soc.*), **1**, 348, 365 (1968).

[91] S. E. Nielsen and J. S. Dahler: *Journ. Chem. Phys.*, **45**, 4060 (1966).

[92] G. Wentzel: *Zeits. Phys.*, **43**, 524 (1927).

[93] R. S. Berry and S. E. Nielsen: *Journ. Chem. Phys.*, **49**, 116 (1968).

Transfer of Electronic Excitation [1].

R. S. Berry

University of Chicago - Chicago, Ill.

1. – A survey of some systems.

The study of electronic-energy transfer in gases is one of the oldest in the field of atomic-molecular processes. Moreover the principal methods by which electronic energy transfer was first studied are still some of the most important experimental approaches in the area. These are the study of sensitized fluorescence and of fluorescence quenching in gases.

The process is demonstrated with a classic example, the fluorescence of a mixture of mercury and thallium vapor irradiated with light from a mercury lamp. This process, called sensitized fluorescence, was first studied in 1922 by Cario and Franck [2] and has been studied periodically since then; an example of the most recent work on the subject is that of Hudson and Curnutte [3, 4] in 1966. The processes are these:

$$
(1)\quad \left\{
\begin{aligned}
&\mathrm{Hg}\,(6\,^1S_0) \underset{10^{-7}\,\mathrm{s}}{\overset{2537\,\text{Å}}{\rightleftarrows}} \mathrm{Hg}\,(6\,^3P_1) \rightarrow \underset{\text{(metastable)}}{\mathrm{Hg}\,(6\,^3P_0)}\,,\\
&\mathrm{Hg}\,(6\,^3P_{0,1}) + \mathrm{Tl}\,(6\,^2P_{1/2}) \rightarrow \mathrm{Hg}\,(6\,^1S_0) + \mathrm{Tl}\,(8\,^2S_{1/2'},\,6\,^2D_{5/2'},\,6\,^2D_{3/2'},\,7\,^2P_{3/2'},\,7\,^2P_{1/2})\\
&\qquad\qquad\qquad\qquad\qquad\qquad\quad \Big\downarrow \ \text{principally 3519-29, 3230, 2918, 2826, 2768, 2704 and 2580 Å}\\
&\qquad\qquad\qquad\qquad\qquad\qquad\quad \mathrm{Tl}\,(7\,^2S_{1/2'},\,6\,^2P_{3/2'},\,6\,^2P_{1/2})\,.
\end{aligned}
\right.
$$

People customarily monitored their experiments by observing the intensities of fluorescence of the mercury at 2537 Å, of the thallium lines, *e.g.*, at 3519-29 and 3230 Å, or of the thallium visible line at 5350 Å, due to the $7\,^2S_{1/2} \rightarrow 6\,^2P_{3/2}$ transition. The direct proportionality of the thallium line intensity to thallium pressure and to the illumination intensity, and of the decrease in mercury fluorescence with thallium pressure indicate clearly that the thallium atoms gain their excitation by collision with optically excited mercury atoms. (Establishing the conditions where wall collisions and unwanted atom-atom collisions are eliminated is not entirely trivial, but has been done; cf. ref. [3, 4]).

With the mercury-thallium system as a prototype, we can go on to classify types of electronic energy exchange in a general way. These types may be described in terms of the way they are observed. Let us write a reference system

$$\begin{array}{ccc} A & & \\ \big\downarrow \text{initial excitation} & & \\ A^* + B & \rightarrow A + & B^* \\ & & \downarrow \\ & & B\,. \end{array} \tag{2}$$

Strictly, A and B may differ chemically from A* and B*; we simply let A* be the electronically excited reactant and B*, the primary product of the elementary transfer reaction. Then we have the following classification:

1) Monitoring property of A*:
   - *A*) Fluorescence (A* short-lived)
   - *B*) Phosphorescence } for A* metastable
   - *C*) Absorption spectrum } for A* metastable
   - *D*) Electron emission from surface detector } for A* metastable

2) Monitoring property of B*:
   - *A*) Fluorescence (electronic or vibrational)
   - *B*) Phosphorescence
   - *C*) Ionization
   - *D*) Dissociation and rearrangement.

Before examining more specific systems and the microscopic theory of electronic-energy exchange, we may review the simplest phenomenological formulation for dealing with electronic-energy transfer, namely the Stern-Volmer treatment of fluorescence quenching [1*a*]. Suppose species A has a spontaneous emission rate k_1, and that the quenching rate is a bimolecular process whose rate is given in terms of the concentrations n_{A*} and n_B, with the constant k_2. (We define an effective cross-section $\tilde{\sigma}_q$ and mean velocity $\bar{v}$, so $k_q = \tilde{\sigma}_q \bar{v}$; often $\tilde{\sigma}$ is defined so that $k_q = 2\tilde{\sigma}_q \bar{v}$.) Let the excitation rate be $k_0 n$. Then in a steady-state condition, the fluorescence rate for A in the absence of B is

$$I_0 = k_1 n_{A*} = k_0 n\,. \tag{3}$$

If B is present, then the intensity

$$I = k_1 n_{A*} = k_0 k_1 n/(k_1 + k_2 n_B)\,, \tag{4}$$

so that the experimentally measurable intensity ratio

$$\frac{I_0}{I} = 1 + (k_2/k_1)\,n_B\,, \tag{5}$$

affording a laboratory measure of k_2/k_1. Since k_1's and $\bar{v}$'s can be determined, one has a direct measure of k_2 and of $\tilde{\sigma}_q$. The effective quenching cross-section $\tilde{\sigma}_q$ for the Hg($6\,^3P_1$) - Tl system was only recently measured [3]. With $\tilde{\sigma}_q$ defined as $k_2/\bar{v}$ (not $k_2/2\bar{v}$), it has values of about 200 Å, corresponding to k_2's of ca. $5\cdot10^{-9}$ cm^3/s at temperatures of (800 ÷ 850) °C. In other words the cross-sections for this kind of process are large, corresponding to interaction distances of order 8 Å. This in turn implies that electronic-energy exchange can occur even in rather distant collisions, or that the exchange can occur through the outer part of the electronic wave functions of the colliding atoms. Not all electronic-excitation processes have such high probabilities per collision, but the fact that some do clearly has a strong influence on the kind of mechanism to which we attribute the transition.

When the fluorescence of A is quenched by collision of A* with B, the exchanged energy may appear in the kinetic energy of relative motion, or it may appear in internal degrees of freedom of B. If an atom of Hg excited to the $6\,^3P_1$ state collides with an argon atom, the only degree of freedom that might receive the excitation energy is the kinetic energy of relative motion. Such a transfer is very inefficient and almost all the relaxation in an Hg($6\,^3P_1$)-Ar system occurs by emission of radiation from the mercury [1a]. If the collision partner is N_2, CO or NO, the situation is very different. The Hg($6\,^3P_1$) atom may give up 0.218 eV to its molecular collision partner and reach the radiatively metastable $6\,^3P_0$ state, or it may give up 4.18 eV (corresponding to $\lambda = 2537$ Å) and reach the ground $6\,^1S_0$ state. Most of the energy so transferred may in principle go into vibrational energy of the molecule. In fact, it seems to go only partially into vibration.

Scheer and Fine [5] monitored the rate of production of metastable Hg($6\,^3P_0$) by collision of Hg($6\,^3P_1$) with CO and N_2. Both molecules are moderately efficient producers of the metastable; CO has a cross-section of 0.57 Å^2 and N_2, 0.42 Å^2. The energy deficit that must go into translation is slightly greater with N_2. This fact might account for the difference in cross sections, but a more detailed examination of the system shows that there are qualitative differences between the interaction of CO and of N_2 with Hg($6\,^3P_1$). Specifically, the CO molecule is capable of de-exciting the photo-excited mercury all the way to its ground electronic state, whereas such de-excitation is virtually impossible with N_2.

Polanyi and his coworkers [6-8] have measured the infrared emission from specific vibrational states of (vibrationally) excited CO and NO in order to determine how the energy is partitioned between vibration and kinetic energy of relative motion and to try to infer the mechanism of the energy transfer process. Neither CO nor NO absorbs all the excitation energy of $Hg(6\,^3P_1$ or $6\,^3P_0)$ in vibration. Carbon monoxide absorbs up to 50% of the energy in vibration, but the next most probable processes leave CO with only a few vibrational quanta. The transition probability is a slowly decreasing function of the vibrational quantum number of CO in the final state, dropping to an indetectably small figure for $v > 9$. Transfer of all the energy to vibration would leave CO in the state $v = 20$. Moreover $Hg(6\,^3P_1)$ transfers its energy to CO about 14 times more efficiently than does $Hg(6\,^3P_0)$, in returning to the $6\,^1S_0$ state. The NO molecule is qualitatively similar, but has the added complication that an electronically excited metastable $^4\Pi$ state can be generated by collision of unexcited NO with $Hg(6\,^3P_1)$. To interpret the results, one must assume a strong and specific interaction between Hg* and CO or NO, and a mechanism that operates during the collision to convert electronic energy to vibration and relative kinetic energy. In the Hg-CO case, the partition can be interpreted in terms of the dynamics of repulsion between Hg and CO in the latter part of the collision; in the Hg-NO case, where up to $\frac{2}{3}$ of the energy may appear in vibration, it seems necessary to invoke some vibrational relaxation of the Hg-NO compound system during the collision [7, 8].

An extreme situation, although not at all uncommon, is the process of photosensitized dissociation. Here, A* transfers its energy to B and leaves B* in a *dissociating* state. The situation is extreme in the sense that a drastic chemical reaction is the result of the excitation transfer. One example is in the work on the photosensitized decomposition of ethylene [9-11]. In essence, the following two situations are found: with mercury,

$$Hg \xrightarrow{2537\,\text{Å}} Hg^*, \tag{6}$$

$$Hg^* + C_2H_4 \rightarrow Hg + C_2H_2 + H_2, \tag{7}$$

while with the higher-energy cadmium,

$$Cd \xrightarrow{2288\,\text{Å}} Cd^*, \tag{8}$$

$$Cd^* + C_2H_4 \rightarrow Cd + C_2H_3 + H. \tag{9}$$

The second dissociation process requires about 4.5 eV, while the first requires only about 1.8. Although the 4.9 eV mercury line has enough energy to induce the higher-energy process, it seems that either the cross section for the high-energy process is small near the energy threshold, or the detailed interactions of mercury and cadmium with ethylene are rather different.

Other systems leading to « chemistry » are rearrangement reactions, *e.g.*,

(10) $$Na^* + X_2 \ (X = \text{halogen}) \rightarrow NaX + X$$

and

(11) $$Na^* + H_2 \rightarrow NaH + H\,,$$

which were among the earliest photochemical reactions to be studied and also among the first to be interpreted on a microscopic level [1*a*, 12, 13].

In some cases, such as the $Hg(^3P_1) + NO$ system previously mentioned, the transferred excitation may create a state both vibrationally and electronically excited. For example the process

(12) $$Xe \xrightarrow{1470\,\text{Å}} Xe^*\,,$$

(13) $$Xe^* + CO \rightarrow Xe + CO^*$$

gives rise to a collection of electronically excited CO states which can only be explained on the basis of a complex XeCO species [14]. Whether direct excitation of vibrations is frequent with excited inert gas-diatomic interactions is not clear. The metastable argon energy transfer to N_2 produces both electronic and vibrational excitations. The process was interpreted, not altogether satisfactorily, without reference to an ArN_2 intermediate [15]. The argon, with approximately 11.6 eV of energy, excites the $C^3\Pi$ state of N_2 with a somewhat surprising vibrational distribution. Of course energy transfer can occur also from one excited molecule to another. An example is $N_2^* + NO \rightarrow N_2 + NO^*$ [16]. It is not clear whether compound intermediates are formed in the N_2-NO system. Compound intermediates are known to occur with a large variety of polyatomics, particularly aromatic molecules, for which the exothermicity of the reaction can « get lost » in the many vibrational degrees of freedom of the system. Such dimers are called excimers or excidimers, and can be quite long-lived [17]. Excimers are frequently long-lived enough to exhibit emission spectra. These spectra are generally broad and at energies well below those of the separated molecular species. The reason for both the breadth and the low energy of excimer emission is generally this: the excimer transition involves an upper potential surface with a stable potential minimum, and a lower surface that, in most cases, has no minimum corresponding to a stable bound state. Thus the Franck-Condon conditions on the emission process in the excimer and on the separate components tells us that the potential surfaces are closer together near the excimer geometry than for well-separated species, meaning that the excimer emission is displaced, relative to that of the separated species, to long wavelengths. We also infer that the lower surface is normally rather steep in the region of the excimer geometry, which implies that the emission spectrum of the excimer must be broad.

Molecular excitation transfer in dense systems takes on a character very different from that in rarefied gases. In molecular solids and even in liquids, to a lesser degree, the excitation can propagate as a quasi-particle, an exciton [18]. The subjects of excitons and excimers are beyond the scope of our present discussion, but it is worth noting that many of the interpretations of exciton mobility and excimer formation have direct analogues in the field of collisional energy transfer.

The formation of excimers from an excited molecule and a normal molecule is a particularly cogent example of the way attractive forces are generally stronger when at least one collision partner is electronically excited than when both partners are in their ground states. It is possible to make a somewhat more specific statement about the chemical behavior of an electronically excited species, *i.e.* that it is *electrophilic* in nature [19]. An atom or molecule with one or more excited electrons has a positive core that is less shielded than it is in the ground state. Thus the excited species exhibits an attraction for electrons of other molecules greater than that of the same species in its ground state. This has the result that reaction and quenching cross-sections of excited species are larger for electron-rich or polarizable collision partners like olefins than for tightly-bound systems like saturated hydrocarbons [19]. Thus the chemical character of an excited atom or molecule is somewhat like that of the molecule-ion to which it would ionize.

One final category of electronic energy transfer is the *intra*-molecular transfer of energy from an electronic degree of freedom to vibration. In small molecules, this sometimes takes the form of predissociation, or of one of the varieties of indirect ionization, which are discussed in the chapter on chemi-ionization. The coupling may also make itself apparent in the form of « anomalously long » radiative lifetimes [20] or in « anomalously low » luminescence yields [21]. The coupling mechanism is now reasonably well understood to be a consequence of a breakdown of the Born-Oppenheimer approximation [22-26]. Some microscopic treatments have been developed for diatomics, but as yet the interpretation of polyatomic molecules intramolecular energy transfer has been at a phenomenological level.

2. – Theories of electronic-energy transfer.

Up to this point, we have been discussing the macroscopic manifestations of electronic-excitation transfer. Now we turn to the problem of interpreting the data. Before discussing the theories, mechanisms, and generalizations, we must point out that until quite recently there was essentially no theory of the electronic-energy transfer process. At best, we were able to say a little about general rules that sometimes apply, and a modest amount about what kinds of theories one can write, what are the implications of each and how one would go about setting up a given problem. Several recent theories [27-30] are still restricted to moderately specific and very limited classes of systems and are relatively untested. For one set of processes, the transitions between the fine structure components of ^2P terms of excited alkali atoms in alkali-inert gas collisions, a theory has been developed by NIKITIN [31-33] which seems to promise a quantitatively accurate analysis of the physical situation. In cases where larger amounts of energy are transferred, the theories remain at a relatively more primitive state.

2˙1. *General rules.* – There are two general rules that have often been applied to interpret the process of electronic-energy transfer. They have in common that they both can be stated as separability rules for applying conservation laws to individual subsystems of the complete colliding system. The two generalizations are the Wigner spin selection rule and the energy resonance rule. The Wigner spin rule states that the total spin of the system is conserved throughout the inelastic collision. It implies that spin and orbital angular momenta are separable, so that the conservation of total angular momentum can be stated as two conservation laws, for spin and orbital angular momenta separately. The energy resonance rule states that excitation transfer is most efficient when the *internal* energy given up by A is precisely equal to the *internal* energy absorbed in the process $B \to B^*$. Stated as a conservation rule or a propensity rule, we can restate it thus: excitation transfer is most efficient when the relative kinetic energy of the colliding species is most nearly conserved. Two other slightly different but closely related generalizations would state that those exchanges are most favored that require the least transfer of *a*) electronic to nuclear kinetic energy, or *b*) electronic to nuclear angular momentum.

It now seems that none of these rules can be treated as a general law for electronic excitation transfer. There are cases to support the applicability of both Wigner spin conservation rule and the energy resonance rule, but there are also fatal counterexamples for both. The transfer of excitation from Ar* to N_2, for example, was shown by FISHBURNE [15] to conform nicely to the Wigner rule, but the simple He*+He processes clearly violate this rule. More specifically, for $n > 4$, the $n\,^1P \to n\,^3F$ « forbidden » process is at least 9 times more important than the corresponding transfer to $n\,^1F$. For $n \leqslant 4$, the Wigner rule is obeyed. The He*+He system was most recently studied by KAY and HUGHES [34], and previously by several others, including St. JOHN and NEE [35], and PENDELTON and HUGHES [36]; the work of MAURER and WOLF [37] first indicated an apparent violation of the Wigner rule. The transfer of excitation from Se* was shown by CALLEAR and his coworkers [1c] to conform reasonably well to the energy resonance rule. However their work was undertaken because the data for transfer of energy from Hg* to simple molecules could in no way be interpreted with this rule, and the question naturally arose as to whether the rule was ever applicable [1c]. The best one can do with the rules, it seems, is to see whether either applies to a given process and then use this information to help determine what kind of theory or mechanism is applicable to that particular process.

2·2. *Microscopic theories.* – The possible sorts of microscopic theories of electronic-energy transfer can be put in three classes, based on the mechanism, or more specifically on how we may describe the interaction responsible for the transfer. These are:

1) Theories based on interactions describable in terms of radiationlike interactions. These include true radiative-transfer processes and interactions that can be written as multipole-multipole interactions;

2) Theories requiring explicit description of the states of the donor A* or of the acceptor B during the course of the collision, but in which A and B essentially retain their identity,

and

3) theories in which A and B become so strongly coupled that a description of the transfer process requires explicit consideration of the compound system.

The problem, like all scattering problems, can be formulated so that the probability per collision for a specific process is (aside from a constant) given by the appropriate matrix of the S- or T-operator [1b]: $P(\alpha \to \beta) \propto |(\beta|T|\alpha)|^2$. What is this matrix and how is our formulation of it related to the physics of the problem? The matrix is defined so that if $|\beta)$ represents the final state of interest for an isolated system, *e.g.*, A+B* long after the collision, and $|\alpha)$ represents the system as it is initially prepared, then $(\beta|T|\alpha)$ is the projection $(\beta|\Psi_\alpha^{(+)})$ of the final state on the fully time-developed system $|\Psi_\alpha^{(+)})$ which began as $|\alpha)$ and experienced the collision. The formal relationship between the T-operator and the Hamiltonian of the full system, $\mathscr{H}$, and the interaction potential $\mathscr{U}$, the part of $\mathscr{H}$ that keeps states $|\alpha)$ and $|\beta)$ from being stationary states throughout the collision is well known [1b] and can be written as

$$T = \mathscr{U} + \lim_{\varepsilon \to 0} \mathscr{U}(E - \mathscr{H} + i\varepsilon)^{-1}\mathscr{U}\,, \tag{14}$$

where E is the energy of the system and ε serves as a convergence parameter. If one neglects the second term in the above expression and supplies the constants, one obtains the Fermi golden rule for transition probabilities,

$$P(\alpha \to \beta) = \frac{2\pi}{\hbar} |(\beta|\mathcal{U}|\alpha)|^2 \varrho(\beta)\,, \tag{15}$$

where $\varrho(\beta)$ is the density of final states $|\beta)$ (unity if $|\beta)$ is energy-normalized). Including the second term of T takes into account the propagating changes in $|\alpha)$ and $|\beta)$ that occur during the course of the collision.

We put physics into our problem when we decide what representation to use for $|\alpha)$ and $|\beta)$ and when we decide how to evaluate

$$(\beta|T|\alpha) = (\beta|\mathcal{U}|\alpha) + \left(\beta \left| \mathcal{U} \frac{1}{E - \mathcal{H} + i\varepsilon} \mathcal{U} \right| \alpha\right). \tag{16}$$

These decisions come at several levels. One is at the level of determining how much of the system should be described classically, and how much should be said in terms of quantum mechanics. This decision, as applied to the relative motion of the nuclei (or centers of mass of A and B), tells us whether to describe this motion in terms of classical trajectories or in terms of waves of translational motion [38]. It generally takes the concrete form of an impact parameter formulation in the classical case, or of a partial-wave expansion in the quantum case. The impact-parameter method requires that we evaluate the T-matrices of many collisions, each with its specified impact parameter, and average over all these impact parameters. If the collisions are completely classical, we can average the squares of the T-elements. (To see effects of interference, we must first sum the T-elements and then square.) Within this context, we may use straight-path trajectories or select a suitable interaction potential from which we can determine the trajectories for various impact parameters and initial velocities.

The impact-parameter method is used in the most recent theoretical treatments of energy exchange, *e.g.*, those of CALLAWAY and coworkers [28, 29] and of WATANABE [30]. These methods in fact used straight trajectories to describe the collisions.

The partial-wave method requires that the wave functions of the system be expanded in partial waves, states of definite angular momentum about the center of mass of the system. In this case one evaluates the T-matrix in a representation of basis functions that are eigenfunctions of the operator K^2 for the (squared) angular momentum of the centers of mass of A and B about their common center of mass. The total transition probability is given from the sum over all final angular-momentum states and the average over all initial states. If one wants only total or partial cross-sections, but no angular information, then one can sum the squares of the T-matrix elements (because of the orthogonality of the angular-momentum eigenfunctions). For angular distributions, we must sum the T's over all contributions to a given final direction and then square.

For light particles, *e.g.*, electrons or sometimes hydrogen or helium atoms, and for low velocities, the partial-wave method is more suitable, as we can see from the following order-of-magnitude calculation. Let us calculate the value of the angular-momentum quantum number K for a collision with an impact parameter b of 3 Å, an effective mass of 10^{-24} g and a velocity of $3\cdot 10^5$ cm/s, roughly the thermal velocity of a proton.

$$(17) \qquad mvb = 3\cdot 10^{-8}\cdot 10^{-24}\cdot 3\cdot 10^{5} \simeq K\hbar \simeq K\cdot 10^{-27}, \qquad \text{so } K \simeq 9 .$$

In practice, this is low enough that we can often carry out full partial-wave calculations for angular momenta as high as 9 or even much higher. However K of 10 is also about the point where impact-parameter calculations are as appropriate as partial-wave expansions. It is occasionally said that the partial-wave method is more accurate. In fact, put on the same footing with regard to computational elaboration, the methods are equally accurate for high K. In the impact parameter method, one sums the contributions from a set of selected trajectories specified by their velocities and values of the impact parameters, b. In the partial-wave method, we may calculate not every partial wave, but only a selection, using the knowledge that cross-sections are smoothly-varying functions of K for high K; we must choose our partial waves and impact parameters to give the same density in phase space and choose them to give roughly equal weight to the important ranges for our particular collision. In this way we assure that both calculations are on the same footing and our only criterion of choice is computational convenience. For systems where only the low-K collisions are important, smoothness of $\sigma(K)$ cannot be supposed, and when a collision cross-section is dominated by low-K collisions, we use the partial-wave representation to great advantage.

The method of partial waves was used, for example, by Buckingham and Dalgarno, to describe resonant excitation transfer [27]. They applied the method to $He^* + He$; in so doing, they used one of the most generally helpful approximation procedures for treating the partial waves, the Jeffrey's approximation. This method, in turn, is a modification of the WKB approximation. Its power for atomic problems lies in the fact that many atomic collisions occur in states of high enough quantum number that the semi-classical character of the approximation is valid [1*a*, 39].

In many atomic inelastic-collision processes, the internal degrees of freedom and particularly the electronic degrees are customarily described quantum-mechanically for electronic-energy exchange. For vibration-vibration and translation-vibration energy exchange, semi-classical and classical methods like the

WKBJ have been applied. One essentially classical method for treating electronic processes was introduced by GRYZINSKI [40], and pursued by several people including GERJUOY [41], VRIENS [42] and ABRINES and PERCIVAL [43] for electron collisions. At this time, we cannot be altogether optimistic about its applicability to electronic-energy exchange because of the complexity of the requisite classical description of collisions of bound electrons in perturbed systems.

Having disposed of the problem of the relative motion, we arrive at the second level of decision. We must decide what representation to use for the internal parts of the states $|\alpha)$ and $|\beta)$. We may describe the compound system in terms of a simple product of functions for free A* and B in the incoming channel and for free A and B* in the outgoing channel; this is the simplest approximation and would be the most reasonable first approach to a theory of our first type, the weak or multipole interaction form of theory.

This is the kind of theory used by CALLAWAY and BAUER [28] for treating alkali $^2P_{\frac{3}{2}} \rightarrow {}^2P_{\frac{1}{2}}$ processes due to collisions of excited alkali atoms with inert gas molecules, and (by WATANABE [30] and others as well) for obtaining phenomenological expressions for excitation transfer. Clearly the method is useful for such cases; it is unlikely to be a successful one (cross-sections ≈ 1.5 times experimental values) for collisions involving large energy or momentum transfer and is certainly inapplicable to rearrangement collisions.

The next approximation would take account of one colliding partner acting on the other at the lowest level of perturbation theory, *e.g.*, allowing for splittings of states that are degenerate in the lowest approximation, or allowing the wave function of B to be perturbed by the incoming A*, etc. This level of treatment has been very successful for the treatment of several problems: LICHTEN [44], F. J. SMITH [1*d*, 45] and F. T. SMITH [1*d*, 46] have applied it very successfully to the resonant-charge transfer problem, especially to interpret the oscillatory dependence of the cross-section on angle and energy; NIKITIN has treated the theories of vibration-vibration [1*d*] as well as collisional spin-orbit relaxation [31-33] at this level. In all these situations, it seems likely that the degeneracy splitting is a very important and sometimes a dominant aspect for understanding the experimental data of the kind now obtainable [1*d*, 47, 48]. The reason is clear when we examine the expressions for the partial cross-sections σ_K(tot), σ_K(transf) and σ_K(elastic) for total, reactive and elastic scattering for a two-channel problem; let the phase shifts for the split-degeneracy states by η_K^+ and η_K^-; then

$$\sigma_K(\text{tot}) = (2\pi/k^2)(2K+1)[\sin^2\eta_K^+ + \sin^2\eta_K^-]\,, \tag{18a}$$

$$\sigma_K(\text{transf}) = (\pi/k^2)(2K+1)[\sin^2(\eta_K^+ - \eta_K^-)]\,, \tag{18b}$$

and

$$\sigma_K(\text{elastic}) = \sigma_K(\text{tot}) - \sigma_K(\text{transf})\,. \tag{18c}$$

Hence both σ_K(transf) and σ_K(elastic) reflect interference effects due to any differences in the energy dependences of the phase shifts for the two states; moreover if η_K^+ and η_K^- vary slowly with K for reasonably high K, then the interferences from many individual angular-momentum states will tend to fall at the same energies and angles, giving an oscillatory pattern in the intensity as a function of energy (or of angle, for a fixed energy). Such patterns were observed by EVERHART [48] and by ABERTH and LORENTS [47].

The next more elaborate level of theory is really the first capable of dealing with the formation of a compound system. It is possible to treat the electronic (and vibrational) parts of the system in a low order of perturbation theory and at the same time to use a potential for A-B interaction that provides for strong attractions and long-lived collisions. However such a treatment is likely to be inconsistent and is often not much easier than one which is somewhat more consistent and accurate. This method depends on the fact that the electrons of the system adjust almost instantaneously as A and B progress throughout their collision. We must calculate the electronic part of the T-matrix elements for each A*-B and A-B* distance and, if A or B is more than just a single atom, we must calculate the electronic part of T for each orientation of the system as well. We make this calculation using the stationary state wave functions of the full (AB)* compound system but with the nuclear kinetic energy omitted. With this method, A* + B and A + B* correspond to the large-R adiabatic limits of different specific states of the total system, so long as A and B are not identical. If A and B are identical (or, *e.g.*, if A is He and B is He^+), and if we call the particles A_1 and A_2 to emphasize their identity, then the asymptotic states are superpositions of the symmetric and antisymmetric stationary states $A_1A_2^* \pm A_1^*A_2$. The method is called the perturbed-stationary-state method or adiabatic strong-coupling approximation. The term « adiabatic » implies that the electrons are represented at every internuclear distance by a function as near to a stationary state function as the mathematical representation allows. The term « strong-coupling » implies that the selected representation for the system treats (AB)* as a compound system and does not necessarily restrict the representation to a weak-perturbation picture of A + B* or A* + B.

The most commonly used criterion for the application of the adiabatic method, known as the Massey criterion, can be said in this way: the interacting particles pass through the region of interaction in a time long compared with $\hbar/\Delta E_{\text{min}}$, where ΔE_{min} is the minimum separation of the adiabatic stationary-state energies in the interaction region. Thus, if R_1 is an approximate maximum interaction distance, we require

$$\frac{R_1}{v_{\text{rel}}}\cdot\frac{\Delta E}{\hbar} > 1 \tag{19}$$

to apply the adiabatic method. We shall see its application in the context of chemi-ionization and predissociation processes later. Let us make an order-of-magnitude estimate for the adiabatic criterion. Again, let $R_1 = 3$ Å and $v_{rel} = 3\cdot 10^5$ cm/s, roughly twice the mean thermal velocity of a hydrogen atom.

$$\Delta E > \frac{3\cdot 10^5}{3\cdot 10^{-8}} = 10^{13}\ \text{s}^{-1}\ \text{or}\ \approx 300\ \text{cm}^{-1} \tag{20}$$

is the criterion for the minimum separation of the adiabatic-state energies, if the adiabatic model of the collision is applicable for thermal systems of light atoms. This is not always satisfied at thermal energies, or even at energies of several eV. The criterion becomes more severe as one gets into the 100 eV region. The most difficult situation in energy exchange or charge-exchange arises when one must use a strongly coupled representation, which is far from satisfying the Massey criterion. So few cases of energy transfer have been examined at the level of a strong-coupling adiabatic theory that it is not clear whether this situation of nonadiabatic strong coupling has ever been encountered or specifically recognized in any real system. A number of treatments have been developed for the case in which deviations from adiabatic character are small and can be treated as perturbations. In fact it may well be that this is the level at which most energy transfer processes in the $(0 \div 100)$ eV range will have to be described.

This brings us to the third type of judgement we must make in constructing a theory of electronic energy transfer. This is the decision about what force we think is responsible for the interaction. Mathematically, this means we must decide what to use for the coupling operator $\mathcal{U}$. In the case of the long-range theory, type 1, $\mathcal{U}$ is presumably given as a multipole-interaction operator, *e.g.*, as a dipole-dipole, quadrupole-induced dipole or induced dipole-induced dipole interaction. This is the longest-range and most weakly coupled form in which we can express the perturbations of A and B on each other; it requires that the charge distributions of A and B are essentially nonpenetrating. In a theory of type 2 or type 3, we expect to have to be much more specific about the details of interaction between A and B.

Two types of operators may play the dominant role in the coupling mechanism, the electrostatic potential terms or the nuclear kinetic energy. In the limit of weak interactions, both can be expressed as multipole interactions. For strong coupling, we must consider using the potential or kinetic energy explicitly. The former is sometimes sufficient, particularly if A and B are identical; in this case the state $A_1 + A_2$, representing the incoming channel, is a nonstationary state of the system (being neither symmetrical nor antisymmetrical) so that the adiabatic or Born-Oppenheimer Hamiltonian is not diagonalized in the representation based on the initial condition of the problem. In this case, the adiabatic Hamiltonian, and particularly the electron-electron terms e^2/r_{ij}, may be large enough that we can say

$$\mathscr{H}_{\text{adiabatic}} \simeq \mathscr{U} . \tag{21}$$

This has proved satisfactory, for example, in the treatment of He+He* given by BUCKINGHAM and DALGARNO [27]. One only needs the matrix elements $(A_1^* A_2 | \mathscr{H}_{\text{adiabatic}} | A_1 A_2^*)$ to get the first approximation to the exchange probability. Note that the exchange of energy between identical particles has precisely the form that arises in the related context of exciton transfer in molecular solids. Strictly, in the molecular solids problem, $A_1^* A_2$ is replaced by $A_1^* A_2 \dots A_n$, the wave function for the entire crystal in which a quantum of excitation is localized on A_1, and correspondingly $A_1 A_2^*$ becomes $A_1 A_2^* \dots A_n$. The integration over co-ordinates of all particles in $A_3, \dots, A_n$ reduces the problem to the form of a pair interaction. (This is not *strictly* accurate because of the influence of A_j^* on its environment. If one deals with charge exchange rather than excitation exchange, the effects can be important, particularly in a dense system [49]).

If A and B are not identical, then the matrix elements of $\mathscr{H}_{\text{adiabatic}}$ may still appear to be responsible for the coupling; this is the case in a type-2 theory in which one mathematical representation is based on approximate functions and not stationary states of $\mathscr{H}_{\text{adiabatic}}$. It may also be the case in the situation (overlapping the previous one) in which the final channel A + B* has some dissociation or ionization process associated with it, and if we use an approximate representation for A + B* which contains new-particle motion. This situation takes the form of a collision analogue of the Beutler-Fano autoionization process [50] in which electron-electron interactions induce configuration mixing between free and bound configurations and the real system exhibits a finite lifetime as a bound species.

In Nikitin's description of the rare gas-excited alkali collisions, [31-33] the adiabatic electrostatic terms are collected into two parts. The first is a polarization part, which expresses the long-range behavior of the interaction (and is the principal operator of CALLAWAY and BAUER [28]). The other part is the exchange interaction which has an exponential dependence on the distance between the atoms, and can be represented as $A \exp[-\alpha R]$ where A is of order 10^{-1} a.u. (the results are not terribly sensitive to its value) and $\alpha^2 = 8I^*$ ($I^* =$ = ionization energy of the excited alkali atom, in a.u.). The value of α would be exact if the interaction could be described entirely in terms of the asymptotic long-distance part of the wave function of the excited electron.

If one assumes that the unsymmetrical collision problem can be solved to arbitrary accuracy in the adiabatic or Born-Oppenheimer approximation, then one must use the nuclear-kinetic-energy operator to develop the coupling.

In certain cases, one can transform the nuclear-kinetic-energy matrix elements into derivatives of the potential: if ψ_i, ψ_f are eigenfunctions of the adiabatic Hamiltonian $\mathscr{H}_0 = \mathscr{T}_{\text{electron}} + \mathscr{V}(r, R)$, then

$$(22) \qquad (\psi_i|[\mathscr{H}_0, \partial/\partial R]|\psi_f) = (E_i - E_f)(\psi_i|\partial/\partial R|\psi_f) = -(\psi_i|\partial\mathscr{V}/\partial R|\psi_f)\,.$$

Thus if the interaction coupling ψ_i with ψ_f occurs in a narrow range $R_0 \leqslant R \leqslant R_0 + \delta R$, one may write for the matrix element of nuclear (vibrational) momentum

$$(23) \qquad (\psi_i|\partial/\partial R|\psi_f) = \begin{cases} (E_f - E)^{-1}(\delta R)^{-1}(\psi_i|V(R_0 + \delta R) - V(R_0)|\psi_f)\,, & R_0 < R < R_0 + \delta R\,, \\ 0 \quad \text{otherwise}\,. \end{cases}$$

When one can justify the assumption that the transition probability of interest has this localized « crossing point » behavior, then (23) is useful. However in some cases it is now known that the crossing-point model cannot be used and one must use (22).

The kinetic energy of relative motion of A and B, $(-\hbar^2/2M)\nabla_R^2$, can be separated into angular and « vibrational » or radial parts. The perturbations due to the radial part are apparently the most important ones for processes like predissociation, autoionization and perhaps for some pure collisional processes like associative ionization. The angular-momentum parts, however, do certainly play a role, and play it in two ways. First, there are direct couplings between electronic and nuclear angular momentum, in which rotational energy is transferred and the total angular momentum of the nuclei changes. In the second type of coupling, the nuclear angular momentum is constant but its direction changes during the collision in a way that produces a reorientation of the electronic angular momentum and thereby, leads to a change of electronic state and sometimes, of transfer of electronic energy from one molecule to another [30]. This last process has been called a perturbation due to axis-switching, and has been discussed by BATES [38] and by HOUGEN and WATSON [51] for example.

An enlightening comparison between angular and radial contributions to inelastic collisions is made in the problem of the alkali line structure relaxation [31-33]. The separate-particle states are characterized by the J, M_J values of the isolated ^2P alkali: $\frac{3}{2} \pm \frac{3}{2}$; $\frac{3}{2} \pm \frac{1}{2}$; and $\frac{1}{2} \pm \frac{1}{2}$. In the collision region, the adiabatic states are characterized only by their components of spin, orbital and total angular momentum *along* the internuclear axis: $^2\Pi_{\frac{1}{2}}$, $^2\Pi_{\frac{3}{2}}$ and $^2\Sigma_{\frac{1}{2}}$. The quantum number J is spoiled by the collision even for the adiabatic states. In the adiabatic approximation and with the internuclear axis defining the axis of quantization, the axial component of angular momentum is left unspoiled by the radial component of the nuclear kinetic energy. Thus the radial component only mixes atomic alkali states with the same axial component; in other words the radial component is principally effective in coupling the $J=\frac{3}{2}$, $M_J=\frac{1}{2}$ and $J=\frac{1}{2}$, $M_J=\frac{1}{2}$ states (and similarly for $M_J=-\frac{1}{2}$). The angular component of nuclear motion couples three states: $\frac{3}{2}$, $\frac{1}{2}$; $\frac{1}{2}$, $\frac{1}{2}$ and $\frac{3}{2}$, $-\frac{3}{2}$ (and similarly for the remaining three with M_J-signs reversed). Hence rotational motion induces transitions from atomic states with $J=\frac{1}{2}$ to states with $J=\frac{3}{2}$ and with M_J values different from that in an initial $J=\frac{1}{2}$ state. Naturally the radial part dominates in slow collisions, and the angular contribution appears to be less important than the radial for K-Ar or Cs-Ar collisions. Very fast collisions, for which the Massey parameter is much less than unity, can be treated by a sudden axis-switching approximation. An intermediate or near-resonant situation, rather than the fast limit, seems to be appropriate for cases like Na-Ar.

If the systems A and B are complex, then of course we must allow in our model for vibrational and rotational state changes in the transfer process, and the problem rapidly becomes discouragingly complex if we want to describe it in detail.

Despite all the complexity we have just gone through, we must recognize that people have been making qualitative microscopic models to interpret electronic-excitation transfer for some time. These models, beginning with that of KALLMANN and LONDON [52], and followed by many others [12, 13, 53, 54], all focused on the question of how a system crossed from one potential curve to another, with emphasis on the relation between the cross-section and the distance at which curve crossing (or, strictly, avoided crossing) occurred, as well as on other variables in the neighborhood of crossing. The most explored cases are those of the alkali, mercury and cadmium quenching reactions. For all of these, the essential explanation of fluorescence quenching by molecules has been this. The molecule Y and excited metal atom M* approach on a curve lying between the ionic M^+-Y^- state's Coulomb curve and the curve for ground state M-Y. The ionic curve cuts the M*-Y, the system makes a transition to the attractive Coulomb curve and accelerates toward collapse. Then the Coulomb curve crosses the ground state MY curve, the compound system finds itself in a compound system and either the entire system flies apart, as in $Hg+H_2 \rightarrow Hg+2H$, or a reaction occurs to dissociate the original molecule, as in $Na+H_2 \rightarrow NaH+H$. In either case, the system cannot find its way back to the curve giving M* + molecule and a quenching reaction occurs. The basic crossing probability per collision was formulated by LANDAU [55] and ZENER [56] as

$$P=\exp[-2\pi\varepsilon_{12}/hqv]\,,$$

where ε_{12} is twice the separation of the adiabatic curves at the « crossing » (with the adiabatic noncrossing rule applied), q is the difference in slope of the zero-order potential curves and v is the relative velocity. Other expressions have been developed also [12].

To close, we may make one further point about the complexity of high-energy processes. The A-B relative kinetic energy operator is still the coupling operator, but in the high-energy region, we must take into account the fact that the electrons on A move relative to those on B. This means the *electronic* wave function of the system must contain a factor modulating the wave function of the electrons on A relative to those on B, a factor which can be put int he form $\exp\left[i\left(KR - \sum_j k_j r_j\right)\right]$ where K is the relative momentum of A and B in units of $\hbar$ and the sum is taken over the electrons on A. This factor makes even simple systems rather more nasty than they are in the low-energy limit, and leads to significant changes, usually reductions, in cross-sections, as LEVY and THORSON [57] have recently shown.

APPENDIX

We conclude this discussion of electronic excitation transfer by giving a brief resume of the Magnus method, the method used in one form by CALLAWAY and BAUER [28] to treat the coupled-state problem in the weak coupling limit.

For detailed derivations and discussions, we refer the reader to the work of MAGNUS [58], of HEFFNER and LOUISELL [59], and of PECHUKAS and LIGHT [60]. In essence, one begins with the expression for the time development of the system,

$$\Psi^{(+)}(t) = U(t, t_0)\Psi^{+}(t_0)\,, \qquad t > t_0,$$

where $\Psi^{(+)}(t)$ has the usual meaning, namely the wave function for the entire system which develops from an early time t_0 to any later time t.

Presumably we can extrapolate to arbitrarily early times to obtain

$$\Psi^{(+)}(-\infty) = \Phi_{\text{init}}\,,$$

the function corresponding to the initial, uncoupled state of the system.

The time development operator $U(t, t_0)$ is related to the Hamiltonian operator by the differential equation that is the operator-equivalent of the Schrödinger equation:

$$\text{(A.1)} \qquad \frac{\mathrm{d}U}{\mathrm{d}t}(t, t_0) = \mathscr{H}(t)\,U(t, t_0)\,.$$

The problem now is to integrate this equation, given a particular representation and set of initial conditions. There are two basic ways of doing this.

The first is the direct iteration

$$U(t,t_0)=1-(i/\hbar)\int_{t_0}^{t}\mathscr{H}(t_1)\,\mathrm{d}t_1+(i/\hbar)^2\int_{t_0}^{t}\mathrm{d}t_1\int_{t_0}^{t_1}\mathrm{d}t_2\mathscr{H}(t_1)\mathscr{H}(t_2)+\dots \tag{A.2}$$

known as the Born expansion. In any truncated form, the approximate U operator is not unitary and therefore does not conserve particles. It is also a slowly convergent series for all but the most weakly coupled systems. Another alternative representation is based on retaining the exponential form of the solution of (A.1). We write

$$U=\exp[A],$$

where A is a pure imaginary operator because of the unitarity of U. By expressing (A.1) as an equation for A, and solving for A in an iterated form $A=\sum_j A_j$, one obtains

$$U=\exp[A_1+A_2+\dots], \tag{A.3}$$

where

$$\begin{cases} A_1=(-i/\hbar)\displaystyle\int_{t_0}^{t}\mathscr{H}(t_1)\,\mathrm{d}t_1, \\ A_2=-\dfrac{1}{2}(-i/\hbar)^2\displaystyle\int_{t_0}^{t}\mathrm{d}t_1\int_{t_0}^{t_1}\mathrm{d}t_2[\mathscr{H}(t_2),\mathscr{H}(t_1)], \qquad \text{etc.}\end{cases} \tag{A.4}$$

The crucial point of the derivation, which we shall not repeat here, is that $\exp[A]\exp[B]\neq\exp[A+B]$ for noncommuting A and B; rather, the Baker-Hausdorff expansion must be used:

$$\exp[A]\exp[B]=\exp[A+B+\tfrac{1}{2}[A,B]+\dots].$$

For the solution of collision problems, two sorts of approximations have been made. In the Callaway-Bauer treatment, the Hamiltonians at different times were assumed to commute so $U=\exp[A_1]$. This approximation is exactly true in the classical and sudden limits. In other cases, one can evaluate $[\mathscr{H}(t_2),\mathscr{H}(t_1)]$ and carry the expansion to second order.

If, for example, one writes the Hamiltonian for a collision problem as

$$\mathscr{H}(t)=(2M)^{-1}p_R^2+V[R(t)], \tag{A.5}$$

then

$$[\mathscr{H}(t_2),\mathscr{H}(t_1)]=(2M)^{-1}[p_R^2,V(R_1)-V(R_2)], \tag{A.6}$$

where $R_i=R(t_i)$. The expression (A.6) gives the first-order quantum correction to the classical (commuting Hamiltonian) expression for an impact-parameter representation of the collision.

REFERENCES

[1] For recent reviews and other useful references, see: *a*) A. C. G. MITCHELL and M. W. ZEMANSKY: *Resonance Radiation and Excited Atoms* (London, 1934; reprinted, 1961); *b*) N. F. MOTT and H. S. W. MASSEY: *Theory of Atomic Collisions*, III Ed. (London, 1965), p. 645; *c*) A. B. CALLEAR: *Energy transfer in molecular Collisions*, chap. 7, in *Photochemistry and Reaction Kinetics*, edited by P. G. ASHMORE, F. S. DAINTON and T. M. SUGDEN (London, 1967); *d*) *Proc. V Int. Conf. on Physics of Electronic and Atomic Collisions, Leningrad, 1967* (Leningrad, 1967).

[2] G. CARIO and J. FRANCK: *Zeits. Phys.*, **11**, 161 (1922); **17**, 202 (1923).

[3] B. C. HUDSON and B. CURNUTTE jr.: *Phys. Rev.*, **148**, 60 (1966).

[4] B. C. HUDSON and B. CURNUTTE jr.: *Phys. Rev.*, **152**, 56 (1966).

[5] M. V. SCHEER and J. FINE: *Journ. Chem. Phys.*, **36**, 1264 (1962).

[6] G. KARL and J. C. POLANYI: *Journ. Chem. Phys.*, **38**, 271 (1963).

[7] G. KARL, P. KRUUS and J. C. POLANYI: *Journ. Chem. Phys.*, **46**, 224 (1967).

[8] G. KARL, P. KRUUS, J. C. POLANYI and W. M. SMITH: *Journ. Chem. Phys.*, **46**, 244 (1967).

[9] D. J. LE ROY and E. W. R. STEACIE: *Journ. Chem. Phys.*, **9**, 829 (1941); **10**, 676 (1942).

[10] A. B. CALLEAR and R. C. ROBB: *Disc. Farad. Soc.*, **17**, 21 (1954); A. B. CALLEAR and R. J. CVENTANOVIC: *Journ. Chem. Phys.*, **23**, 1182 (1955); **24**, 873 (1956).

[11] S. ARAI and S. SHIDA: *Journ. Chem. Phys.*, **38**, 694 (1963).

[12] J. L. MAGEE and T. RI: *Journ. Chem. Phys.*, **8**, 687 (1940); **9**, 638 (1941).

[13] H. J. LAIDLER: *Journ. Chem. Phys.*, **10**, 34, 43 (1942).

[14] T. G. SLANGER: *Journ. Chem. Phys.*, **48**, 586 (1968).

[15] E. S. FISHBURNE: *Journ. Chem. Phys.*, **47**, 58 (1967).

[16] R. A. YOUNG and G. A. ST. JOHN: *Journ. Chem. Phys.*, **48**, 898 (1968).

[17] *a*) J. B. BIRKS: *Nature*, **214**, 1187 (1967); *b*) E. A. CHANDROSS, J. FERGUSON and E. G. MCCREA: *Journ. Chem. Phys.*, **45**, 3546 (1966); E. A. CHANDROSS and J. FERGUSON: *Journ. Chem. Phys.*, **45**, 3554, 3564 (1966); *c*) L. GLASS, I. H. HILLIER and S. A. RICE: *Journ. Chem. Phys.*, **45**, 3886 (1966); *d*) T. AZUMI and H. AZUMI: *Bull. Chem. Soc. Jap.*, **39**, 1829, 2317 (1966); *e*) J. LANGELAAR, R. P. H. RETTSCHNICK, A. M. F. LAMBOOY and G. J. HOYTINK: *Chem. Phys. Lett.*, **2**, 609 (1968); *f*) J. B. BIRKS: *Chem. Phys. Lett.*, **2**, 625 (1968).

[18] See, for example, the recent review: S. A. RICE and J. JORTNER: in *Physics and Chemistry of the Organic Solid State*, vol. **3**, edited by D. FOX and M. LABES (New York, 1967), p. 199.

[19] M. BELLAS, Y. ROUSSEAU, O. P. STRAUSZ and H. E. GUNNING: *Journ. Chem. Phys.*, **41**, 768 (1964); A. YARWOOD, O. P. STRAUSZ and H. E. GUNNING: *Journ. Chem. Phys.*, **41**, 1705 (1964).

[20] A. E. DOUGLAS: *Journ. Chem. Phys.*, **45**, 1007 (1966) and references therein.

[21] G. B. KISTIAKOWSKY and C. S. PARMENTER: *Journ. Chem. Phys.*, **42**, 2942 (1965).

[22] G. W. ROBINSON and R. P. FROSCH: *Journ. Chem. Phys.*, **37**, 1962 (1962); **38**, 1187 (1963).

[23] *a*) M. BIXON and J. JORTNER: *Journ. Chem. Phys.*, **48**, 715 (1968); *b*) J. JORTNER and R. S. BERRY: *Journ. Chem. Phys.*, **48**, 2757 (1968); *c*) D. P. CHOCK, J. JORTNER and S. A. RICE: *Journ. Chem. Phys.*, **49**, 610 (1968); *d*) K. F. FREED and J. JORTNER: *Journ. Chem. Phys.*, **50**, 2916 (1969).

[24] *a*) W. SIEBRAND and D. F. WILLIAMS: *Journ. Chem. Phys.*, **49**, 1860 (1968); *b*) W. SIEBRAND: *Journ. Chem. Phys.*, **47**, 2411 (1967); *c*) W. SIEBRAND: *Journ. Chem. Phys.*, **46**, 440 (1967).
[25] S. H. LIN: *Journ. Chem. Phys.*, **46**, 440 (1967).
[26] *a*) W. RHODES, B. R. HENRY and M. KASHA: *A stationary state approach to radiationless transitions. Radiation bandwidth effect on excitation processes in polyatomic molecules.* in *Proc. Natl. Acad. Sci.* (to be published); *b*) W. RHODES: *Journ. Chem. Phys.*, **50**, 2885 (1969); *c*) J. H. YOUNG: *Nonradiative electronic-vibrational transitions in molecules*, in *Journ. Chem. Phys* (to be published); *d*) P. J. GARDNER and M. KASHA: *Journ. Chem. Phys.*, **50**, 1543 (1969); B. R. HENRY and M. KASHA: *Ann. Rev. Phys. Chem.*, **19**, 161 (1968).
[27] R. A. BUCKINGHAM and A. DALGARNO: *Proc. Roy. Soc.*, A **213**, 506 (1952).
[28] J. CALLAWAY and E. BAUER: *Phys. Rev.*, **140**, A 1072 (1965).
[29] H. J. CALLAWAY and J. Q. BARTLING: *Phys. Rev.*, **150**, 69 (1966).
[30] T. WATANABE: *Journ. Chem. Phys.*, **46**, 3741 (1967). See also other references therein.
[31] E. E. NIKITIN: *Journ. Chem. Phys.*, **43**, 744 (1965).
[32] E. E. NIKITIN: *Opt. and Spectr.*, **22**, 379 (1967).
[33] E. I. DASHEVSKAYA and E. E. NIKITIN: *Opt. and Spectr.*, **22**, 866 (1967).
[34] R. B. KAY and R. H. HUGHES: *Phys. Rev.*, **154**, 61 (1967).
[35] R. M. ST. JOHN and T.-W. NEE: *Journ. Opt. Soc. Am.*, **55**, 426 (1965).
[36] W. R. PENDLETON and R. H. HUGHES: *Phys. Rev.*, **138**, A 683 (1965).
[37] W. MAURER and R. WOLF: *Zeits. Phys.*, **115**, 410 (1940).
[38] D. R. BATES: *Atomic and Molecular Processes*, chap. 14 (New York, 1962).
[39] H. S. W. MASSEY and C. B. O. MOHR: *Proc. Roy. Soc.*, A **144**, 188 (1934).
[40] M. GRYZINSKI: *Phys. Rev.*, **115**, 374 (1959); **138**, A 305, A 332, A 336 (1965).
[41] E. GERJUOY: *Phys. Rev.*, **148**, 54 (1966); J. D. GARCIA and E. GERJUOY: *Phys. Rev. Lett.*, **18**, 944 (1967); E. GERJUOY: invited papers from *V Int. Conf. on Physics of Electronic and Atomic Collisions* (Boulder, Colo., 1968).
[42] L. VRIENS: *Phys. Rev.*, **141**, 88 (1966).
[43] R. ABRINES and I. C. PERCIVAL: *Proc. Phys. Soc. London*, **88**, 861, 873 (1966).
[44] W. LICHTEN: *Phys. Rev.*, **139**, A 27 (1965).
[45] F. J. SMITH: *Phys. Lett.*, **20**, 271 (1966).
[46] F. T. SMITH, R. P. MARCHI and K. G. DEDRICK: *Phys. Rev.*, **150**, 79 (1966).
[47] W. ABERTH and D. C. LORENTS: *Phys. Rev.*, **144**, 109 (1966).
[48] F. P. ZIEMBA, G. J. LOCKWOOD, G. H. MORGAN and E. EVERHART: *Phys. Rev.*, **118**, 1552 (1960).
[49] R. GLAESER and R. S. BERRY: *Journ. Chem. Phys.*, **44**, 3797 (1966).
[50] U. FANO: *Phys. Rev.*, **124**, 1866 (1961); U. FANO and J. W. COOPER: *Phys. Rev.*, **137**, A 1364 (1965).
[51] J. HOUGEN and J. K. G. WATSON: *Canad. Journ. Phys.*, **43**, 298 (1965).
[52] H. KALLMAN and F. LONDON: *Zeits. Phys. Chem.*, **132**, 207 (1929).
[53] O. K. RICE: *Phys. Rev.*, **37**, 1187, 1551 (1931).
[54] E. TELLER: *Journ. Phys. Chem.*, **41**, 109 (1937).
[55] L. LANDAU: *Zeits. Phys. Sov.*, **2**, 46 (1932).
[56] C. ZENER: *Proc. Roy. Soc.*, A **137**, 696 (1932).
[57] H. LEVY II and W. THORSON: *Phys. Rev.*, **181**, 230, 244, 252 (1969).
[58] W. MAGNUS: *Comm. Pure Appl. Math.*, **7**, 649 (1954).
[59] H. HEFFNER and W. H. LOUISELL: *Journ. Math. Phys.*, **6**, 474 (1965).
[60] P. PECHUKAS and J. C. LIGHT: *Journ. Chem. Phys.*, **44**, 3897 (1966).

Nonequilibrium Effects in Chemical Kinetics.

J. Ross

MIT - Cambridge, Mass.

In the lecture by Wagner [1] we have seen the relation of the rate coefficient to the reactive scattering cross-section, a relation which involved translational and internal distribution functions. If the rate of the reaction is very slow, it seems reasonable (although it may not necessarily be so) that these distribution functions are close to their equilibrium form, even though the system is not at *chemical* equilibrium. If, however, the rate of the chemical reaction approaches the rate of relaxation of energy distribution in either internal degrees of freedom or translation, then a coupling between the various relaxations may be expected. Thus we may say that the process of a chemical reaction perturbs the various distributions (to some degree) and the rate of the reaction is influenced by this perturbation.

The analysis of nonequilibrium effects in chemical rates requires solutions of appropriate transport equations. We proceed with a simple way of obtaining the Boltzmann equation of transport, including bimolecular reactions in a gas at low density, and then discuss briefly Chapman-Enskog-type solutions for this equation and their consequences for chemical kinetics [18].

1. – Microscopic reversibility.

Consider the reaction $A(i)+B(j)=C(l)+D(m)$, where i, j, l, m denote the internal quantum state of each respective reactant. If $W(lm|ij)$ is the transition probability per unit time from i, j to l, m, then the statement of microscopic reversibility [2] is

$$W(lm|ij)=W(i^*j^*|l^*m^*)\,. \tag{1}$$

The internal state i^* differs from i only in the sign of the z-component of the total angular momentum. In the absence of external fields we average over all possible z components and hence need not distinguish between i and i^* [3].

The differential scattering cross-section $\sigma(lm|ij; v\Omega)$, a function of the initial relative speed v and the solid angle Ω, is given by the number of scattered particles per unit time and unit flux, (v/V), in a given volume V

$$\sigma(lm|ij; v\Omega)\,\mathrm{d}\Omega = \frac{1}{(v/V)}\, W(lm|ij; v\Omega)\, \frac{Vp'^2\,\mathrm{d}p'\,\mathrm{d}\Omega}{h^3}\,. \tag{2}$$

The number of scattered particles per unit time is the product of the transition probability per unit time and the number of final states in the region considered, which is $(V/h^3)p'^2\,\mathrm{d}p'\,\mathrm{d}\Omega$, with p'/μ' being the final relative speed and μ' the reduced mass of the products. For the reverse reaction we have a similar equation

$$\sigma(ij|lm; v'\,\Omega)\,\mathrm{d}\Omega = \frac{1}{(v'/V)}\, W(ij|lm; v'\,\Omega)\, \frac{Vp^2\mathrm{d}p\,\mathrm{d}\Omega}{h^3}\,. \tag{3}$$

For given internal states i, j, l, m the requirement of conservation of energy yields

$$\mu v\,\mathrm{d}v = \mu' v'\,\mathrm{d}v'\,, \tag{4}$$

so that from a combination of eqs. (1-4) we have

$$p^2\sigma(lm|ij; v\Omega) = p'^2\sigma(ij|lm); v'\,\Omega)\,. \tag{5}$$

Let $\boldsymbol{p}_A$, $\boldsymbol{p}_B$, $\boldsymbol{p}_C$, $\boldsymbol{p}_D$ be the momentum vectors of the reactants in the laboratory co-ordinates, and $\boldsymbol{P}$ the momentum vector of the center of mass. Then we may write

$$\begin{cases} \mathrm{d}\boldsymbol{p}_A\,\mathrm{d}\boldsymbol{p}_B = \mathrm{d}\boldsymbol{p}\,\mathrm{d}\boldsymbol{P} = \mathrm{d}\boldsymbol{P}p^2\,\mathrm{d}p\,\mathrm{d}\Omega\,, \\ \mathrm{d}\boldsymbol{p}_C\,\mathrm{d}\boldsymbol{p}_D = \mathrm{d}\boldsymbol{p}'\,\mathrm{d}\boldsymbol{P} = \mathrm{d}\boldsymbol{P}p'^2\,\mathrm{d}p'\,\mathrm{d}\Omega\,, \end{cases} \tag{6}$$

and now obtain

$$\frac{p'}{\mu'}\sigma(ij|lm; v'\,\Omega)\,\mathrm{d}\boldsymbol{p}_C\,\mathrm{d}\boldsymbol{p}_D = \frac{p}{\mu}\sigma(lm|ij; v\Omega)\,\mathrm{d}\boldsymbol{p}_A\,\mathrm{d}\boldsymbol{p}_B\,. \tag{7}$$

2. – Boltzmann equation.

We wish to obtain the Boltzmann equation of transport by a simple argument [4, 5]. Consider the number of $C(l) \equiv C_l$ produced from A_i in the chemical reaction written above; that number produced per unit time and unit volume with momentum $\boldsymbol{p}_C$ from A_i with momentum $\boldsymbol{p}_A$ and from B_j with momentum $\boldsymbol{p}_B$ is

$$\frac{p}{\mu}\sigma_R\,(lm|ij;\,v\Omega)\,n_{A_i}f_{A_i}\mathrm{d}\boldsymbol{p}_A n_{B_j}f_{B_j}\mathrm{d}\boldsymbol{p}_B\,\mathrm{d}\Omega\,, \tag{8}$$

where n_{A_i} is the number density of A_i and f_{A_i} is the normalized momentum distribution function. The subscript R on the cross-section denotes reaction. Hence the total number of C produced per unit time and unit volume from A_i with momentum $\boldsymbol{p}_A$ in the forward direction of the reaction as written is

$$\mathrm{d}\boldsymbol{p}_A\sum_{jlm}\iint\frac{p}{\mu}\sigma_R\,(lm|ij;\,v\Omega)\,n_{A_i}f_{A_i}n_{B_j}f_{B_j}\mathrm{d}\boldsymbol{p}_B\,\mathrm{d}\Omega\,, \tag{9}$$

which must also equal the rate of disappearance of A_i per unit volume in the range $\mathrm{d}\boldsymbol{p}_A$. The rate of appearance of A_i per unit volume in the range $\mathrm{d}\boldsymbol{p}_A$ is obtained by first writing an expression analogous to (8), then making use of eq. (7), and then integrating over the remaining variables $\boldsymbol{p}_B$, Ω:

$$\mathrm{d}\boldsymbol{p}_A\sum_{jlm}\iint\frac{p'}{\mu'}\sigma_R\,(ij|lm;\,v'\Omega)\,n_{C_l}f_{C_l}n_{D_m}f_{D_m}\,\mathrm{d}\boldsymbol{p}_B\,\mathrm{d}\Omega\,. \tag{10}$$

The difference between the expressions (9) and (10) is the rate of change of $n_{A_i}f_{A_i}\mathrm{d}\boldsymbol{p}_A$ due to reaction.

Similar arguments may be used to obtain the rate of change of the number density of A_i with momenta in the range $\mathrm{d}\boldsymbol{p}_A$ due to elastic and inelastic collisions. On addition of the terms due to all three effects we have the Boltzmann equation for a dilute gas in which a chemical reaction occurs:

$$\begin{aligned}\frac{\partial(n_{A_i}f_{A_i})}{\partial t} &= \sum_{j,S}n_{A_i}n_{S_j}\iint\frac{p_{AS}}{\mu_{AS}}\sigma_{AS}\,(ij;\,p_{AS}\Omega)[f'_{A_i}f'_{S_j}-f_{A_i}f_{S_j}]\,\mathrm{d}\Omega\,\mathrm{d}\boldsymbol{p}_S\,+\\ &+\sum_{j,l,m,S}\iint\frac{p_{AS}}{\mu_{AS}}\sigma_{AS}(lm|ij;\,p_{AS}\Omega)(n_{A_l}n_{S_m}f'_{A_l}f'_{S_m}-n_{A_i}n_{S_j}f_{A_i}f_{S_j})\,\mathrm{d}\Omega\,\mathrm{d}\boldsymbol{p}_S\,+\\ &+\sum_{jlm}\iint\frac{p_{AB}}{\mu_{AB}}\sigma_R(lm|ij;\,p_{AB}\Omega)(n_{C_l}n_{D_m}f'_{C_l}f'_{D_m}-n_{A_i}n_{B_j}f_{A_i}f_{B_j})\,\mathrm{d}\Omega\,\mathrm{d}\boldsymbol{p}_B\,.\end{aligned} \tag{11}$$

The new symbols are: S, any of the four species A, B, C, D; $\sigma_{AS}(ij;\,p_{AS}\Omega)$, the differential elastic scattering cross-section for collisions between A_i and S_j; $\sigma_{AS}(lm|ij;\,p_{AS}\Omega)$, the differential inelastic scattering cross-section for collisions between A_i and S_j; and primed distribution functions, indicating « post-collision » values of momenta of unprimed distribution functions within the same bracket.

Integration of eq. (11) over $\boldsymbol{p}_A$ makes the elastic integral term vanish, since elastic collisions cannot change the number density n_{A_i}. Summation over the internal states of A (summation over i) make the inelastic integral term vanish, since inelastic collisions cannot change the number density of A,

$$n_A = \sum_i n_{A_i} .$$

Hence, after the integration and summation we are left with the theoretical rate equation [5-7]

$$-\frac{\mathrm{d}n_A}{\mathrm{d}t} = \varkappa_{f,t} n_A n_B - \varkappa_{r,t} n_C n_D , \tag{12}$$

$$\varkappa_{f,t} \equiv \sum_{lmij} x_{A_i} x_{B_j} \iint \frac{p_{AB}}{\mu_{AB}} \sigma_B (lm|ij;\, p_{AB}\,\Omega) f_{A_i} f_{B_j} \mathrm{d}\boldsymbol{p}_A \mathrm{d}\boldsymbol{p}_B \mathrm{d}\Omega , \tag{13}$$

with x_{A_i}, x_{B_j} being the mole fractions of A, B in states i, j.

The distribution functions f_{A_i} etc. and the mole fractions x_{A_i} etc. must be obtained from solutions of the Boltzmann equation and may be functions of time, implicit or explicit, and may be explicit functions of concentrations. If such is the case, the ratio of rate coefficients does not equal the equilibrium constant, in general

$$K \neq \frac{\varkappa_{f,t}}{\varkappa_{r,t}} . \tag{14}$$

However, if the distribution functions and mole fractions have their equilibrium values,

$$\begin{cases} \varkappa_{f,t} \to k_{f,t}^{(0)} , \\ \varkappa_{r,t} \to k_{r,t}^{(0)} , \end{cases} \tag{15}$$

then of course the rate coefficients $k_{f,t}^{(0)}$, $k_{r,t}^{(0)}$ are independent of time and concentrations and their ratio equals the equilibrium constant,

$$\begin{cases} K = \dfrac{k_{f,t}^{(0)}}{k_{r,t}^{(0)}} , \\ K \equiv e^{-\beta\Delta\mu^0}, \qquad \Delta\mu^0 = (\mu_C^0 + \mu_D^0) - (\mu_A^0 + \mu_B^0) . \end{cases} \tag{16}$$

3. – Chapman-Enskog solution of the Boltzmann equation.

The Chapman-Enskog method of solution of the Boltzmann equation is based on the premise that the several relaxation processes, momentum (translation), internal energy, chemical reaction, have characteristic relaxation times which are not too close in value to each other. For simplicity, consider the idealized system in which reaction may occur, but in which the internal distributions need not be considered. The Boltzmann equation may be written now

$$\partial(n_A f_A)/\partial t = \sum_S E_{AS} + R_A \,, \tag{17}$$

$$E_{AS} \equiv n_A n_S \int \dots \int (p/\mu)\sigma_{AS}(p, \Omega)[f_A(p_A')f_S(p_S') - f_A(p_A)f_S(p_S)]\, \mathrm{d}p_S \,\mathrm{d}\Omega \,, \tag{18}$$

$$R_A \equiv \int \dots \int (p/\mu)\sigma_R(p, \Omega)[n_C n_D f_C(p_C')f_D(p_D') - n_A n_B f_A(p_A)f_B(p_B)]\, \mathrm{d}\Omega \,\mathrm{d}p_B \,. \tag{19}$$

The presence of the reaction (R_A) perturbs the momentum distributions from their equilibrium value, and elastic collisions (E_{AS}) tend to restore translational equilibrium. If the momentum relaxation time is faster than the chemical relaxation time, then the deviation of the translational distribution from its equilibrium value will consist of a minor perturbation. We proceed by introducing a perturbation parameter ε into the Boltzmann equation

$$\varepsilon\{\partial(n_A f_A)/\partial t - R_A\} = \sum_S E_{AS} \,, \tag{20}$$

and into the distribution function as well as the time derivative [8, 6]

$$f_A = f_A^{(0)}[1 + \varepsilon\varphi_A^{(1)} \dots] \,, \tag{21}$$

$$f_A^{(0)} = \exp[-\beta[(p_A^2/2m_A) + \varepsilon_A - \mu_A^0]] \,, \tag{22}$$

$$\frac{\partial}{\partial t} = \frac{\partial}{\partial t}\Big)_0 + \varepsilon \frac{\partial}{\partial t}\Big)_1 + \varepsilon^2 \frac{\partial}{\partial t}\Big)_2 + \dots \,. \tag{23}$$

The function $\varphi_A^{(1)}$ is the first-order perturbation. In the equilibrium distribution $f_A^{(0}$ the new symbols are: ε_A, the one-level internal energy of A; μ_A^0, the standard chemical potential of A; $\beta \equiv (kT)^{-1}$.

The lowest (zeroth) order solution of eq. (20) is obtained from equating the coefficients of ε^0 to zero

$$E_{AS}^{(0)}(n_A n_S f_A^{(0)} f_S^{(0)}) = 0 \,, \tag{24}$$

and is simply $n_S f_S^{(0)}$. The densities n_S have arbitrary values; no relation exists among the corresponding chemical potentials

$$\mu_S = kT \ln n_S + \mu_S^0 \,, \tag{25}$$

but the affinity $\mathscr{A}$ is defined to be

$$\mathscr{A} = -[(\mu_C + \mu_D) - (\mu_A + \mu_B)] , \tag{26}$$

and so vanishes at chemical equilibrium. In the lowest order the rate of the reaction is (from eq. (12))

$$-\left.\frac{\mathrm{d}n_A}{\mathrm{d}t}\right)_0 = k_f^{(0)} n_A n_B - k_r^{(0)} n_C n_D , \tag{27}$$

$$k_f^{(0)}(T) = \int \dots \int (p/\mu)\sigma_R f_A^{(0)} f_B^{(0)} \,\mathrm{d}\Omega \,\mathrm{d}\boldsymbol{p}_A \,\mathrm{d}\boldsymbol{p}_B , \tag{28}$$

$$k_r^{(0)}(T) = \int \dots \int (p/\mu)\sigma_R f'^{(0)}_C f'^{(0)}_D \,\mathrm{d}\Omega \,\mathrm{d}\boldsymbol{p}_A \,\mathrm{d}\boldsymbol{p}_B , \tag{29}$$

from which the corollary of the law of mass action, eq. (16), is readily obtained. The rate coefficients $k_f^{(0)}$, $k_r^{(0)}$ are independent of time, concentration, and affinity.

In the next order of approximation (ε^1) we have the set of equations

$$f_A^{(0)} \left.\frac{\mathrm{d}n_A}{\mathrm{d}t}\right)_0 - R_A^{(0)} = \sum_S E_{AS}^{(1)} , \tag{30}$$

$$E_{AS}^{(1)} = n_A n_S \int \dots \int (p/\mu)\sigma_{AS} f_A^{(0)} f_S^{(0)} [\varphi_A^{(1)}(p_A') + \varphi_S^{(1)}(p_S') - \varphi_A^{(1)}(p_A) - \varphi_S^{(1)}(p_S)] \,\mathrm{d}\Omega \,\mathrm{d}\boldsymbol{p}_S , \tag{31}$$

$$R_A^{(0)} = \int \dots \int (p/\mu)\sigma_R [n_C n_D f'^{(0)}_C f'^{(0)}_D - n_A n_B f_A^{(0)} f_B^{(0)}] \,\mathrm{d}\Omega \,\mathrm{d}\boldsymbol{p}_B , \tag{32}$$

which, by use of eq. (27), may be altered to

$$f_A^{(0)} [k_r^{(0)} n_C n_D - k_f^{(0)} n_A n_B] - R_A^{(0)} = \sum_S E_{AS}^{(1)} , \tag{33}$$

or

$$\left\{ -f_A^{(0)} k_f^{(0)} + \int \dots \int (p/\mu)\sigma_R f_A^{(0)} f_B^{(0)} \,\mathrm{d}\Omega \,\mathrm{d}\boldsymbol{p}_B \right\} n_A n_B [1 - \exp[-\beta\mathscr{A}]] = \tag{34}$$

$$= \sum_S n_A n_S \int \dots \int (p/\mu)\sigma_{AS} f_A^{(0)} f_S^{(0)} [\varphi_A^{(1)}(p_A') + \varphi_S^{(1)}(p_S') - \varphi_A^{(1)}(p_A) - \varphi_S^{(1)}(p_S)] \,\mathrm{d}\Omega \,\mathrm{d}\boldsymbol{p}_S ,$$

with similar equations for the other species. Since eq. (34) is linear in $\varphi_A^{(1)}$, the solution of that equation may be written in the form

$$\varphi_A^{(1)} = [1 - \exp[-\beta\mathscr{A}]] \varphi_A'^{(1)}(\boldsymbol{p}_A, n_S) , \tag{35}$$

and hence the theoretical rate coefficients may be written as

$$\varkappa_{f,t} = k_{f,t}^{(0)} - (1 - \exp[-\beta \mathscr{A}_f])\, \zeta(n_s)\,, \tag{36}$$

$$\varkappa_{r,t} = k_{r,t}^{0} - (1 - \exp[-\beta \mathscr{A}_r])\, \eta(n_s)\,. \tag{37}$$

The explicit form of the function $\varphi_A^{\prime(1)}(p_A, n_s)$ has to be obtained from a solution of the first-order (ε^1) Boltzmann equation, eq. (25). This may be done by further expansions in complete sets (Sonine polynomials); for details see refs. [4, 7, 9-11].

The functions $\zeta(n_s)$, $\eta(n_s)$ depend in general on the concentrations of the species, and their magnitude is determined in part by the various scattering cross sections which appear in the Boltzmann equation. The general form of the theoretical rate coefficients, eqs. (36, 37) remains the same on inclusion of internal degrees of freedom for all reacting species [12-15], and the magnitude of the functions ζ, η then depends in part on all elastic, inelastic, and reactive cross sections of the system. Moreover, the form of the theoretical rate coefficients eqs. (36, 37) can be obtained without the use of the Chapman-Enskog perturbation method [7].

4. – Discussion.

Substitution of the rate coefficients eqs. (36, 37) into the rate equation, eq. (12), allows a rearrangement [16, 13, 17, 7] of terms. Since the brackets containing the affinities $\mathscr{A}_f$, $\mathscr{A}_r$ are

$$(1 - \exp[-\beta \mathscr{A}_f]) = 1 - \frac{n_C n_D}{K\, n_A\, n_B}\,; \qquad (1 - \exp[-\beta \mathscr{A}_r]) = 1 - \frac{n_A n_B K}{n_C n_D}\,, \tag{38}$$

the rate equation may be written

$$-\frac{\mathrm{d}n_A}{\mathrm{d}t} = [k_{f,t}^{(0)} - \zeta - \eta K] n_A n_B - [k_{r,t}^{(0)} - \frac{\zeta}{K} - \eta]\, n_C n_D\,, \tag{39}$$

where we define new rate coefficients $\varkappa_{f,c}$, $\varkappa_{r,c}$ by the expressions in the two brackets respectively. The ratio of the theoretical rate coefficients does not in general equal the equilibrium constant

$$K \neq \frac{\varkappa_{f,t}}{\varkappa_{r,t}} \tag{40}$$

because the terms ζ and η are functions of concentration of species. The rearranged rate coefficients $\varkappa_{f,c}$, $\varkappa_{r,c}$ do however satisfy the relation

$$K = \frac{\varkappa_{f,c}}{\varkappa_{r,c}} \tag{41}$$

even though each rate coefficient is a function of concentration.

The rate equation, eq. (39) reduces to the usual phenomenological rate equation

$$-\frac{dn_A}{dt} = k_f n_A n_B - k_r n_C n_D$$

if the rate coefficients $\varkappa_{f,c}$, $\varkappa_{r,c}$ are independent of concentrations, since the phenomenological rate coefficients k_f, k_r are defined to be independent of concentrations. This reduction has been shown to hold for three classes of reactions [7]:

1) The reaction $A + B = C + D$ proceeding sufficiently slowly so that the coupling to other relaxation processes (elastic, inelastic) is negligible ($\zeta, \eta = 0$).

2) The reaction $A + B = C + D$ occurring in the presence of large excess of *one* component (the *same* component for both the forward and reverse direction).

3) The reaction $A + M = B + M$ occurring in presence of large excess of M.

In the first case $\varkappa_{f,c}$, $\varkappa_{r,c}$ reduce to $k^{(0)}_{f,t}$, $k^{(0)}_{r,t}$ respectively. For all three cases we have of course the relation $K = k_f/k_r$ by definition. The coefficients $\varkappa_{f,c}$, $\varkappa_{r,c}$ do not reduce to k_f, k_r for arbitrary concentration, even if the system is very close to chemical equilibrium.

The calculation of the rate coefficients $k^{(0)}_{f,t}$, $k^{(0)}_{r,t}$ requires knowledge about reactive cross sections, eqs. (28, 29), but the calculation of either the coefficients $\varkappa_{f,t}$, $\varkappa_{r,t}$ or $\varkappa_{f,c}$, $\varkappa_{r,c}$ requires in addition knowledge about inelastic and (possibly) elastic scattering cross sections, as well as solutions of transport equations which describe the coupling of the various relaxation processes. This statement holds whether or not $\varkappa_{f,c}$, $\varkappa_{r,t}$ reduce to the phenomenological coefficients k_f, k_r. Conversely, the measurement of rate coefficients as a function of temperature, whether the coefficients are concentration-dependent or not, is not necessarily related only to the threshold energy of the reactive cross-section, but may also be affected by the threshold energies of inelastic cross-section.

* * *

I am grateful to Drs. J. GROSSER, V. KEMPTER, and A. SCHULTZ for preparing a preliminary manuscript from the lecture.

REFERENCES

[1] H. WAGNER: This volume, p. 258.
[2] L. D. LANDAU and E. M. LIFSHITZ: *Quantum Mechanics* (Reading, Mass., 1958).
[3] For a fuller discussion of this point see J. C. LIGHT, J. ROSS and K. SHULER in *Kinetic Processes in Gases and Plasmas* (New York, 1969) p. 281.
[4] I. PRIGOGINE and E. XHROUET: *Physica*, **15**, 913 (1949).
[5] M. E. ELIASON and J. O. HIRSCHFELDER: *Journ. Chem. Phys.*, **30**, 1426 (1959).
[6] J. ROSS and P. MAZUR: *Journ. Chem. Phys.*, **35**, 19 (1961).
[7] N. S. SNIDER and J. ROSS: *Journ. Chem. Phys.*, **44**, 1087 (1966).
[8] S. CHAPMAN and T. G. COWLING: *The Mathematical Theory of Nonuniform Gases* (London, 1960).
[9] R. D. PRESENT: *Journ. Chem. Phys.*, **31**, 747 (1959).
[10] C. W. PYUN and J. ROSS: *Journ. Chem. Phys.*, **40**, 2572 (1961).
[11] R. D. PRESENT: *Journ. Chem. Phys.*, **48**, 4875 (1968).
[12] B. WIDOM: *Journ. Chem. Phys.*, **34**, 2050 (1961); *Advan. Chem. Phys.*, **5**, 353 (1963).
[13] N. S. SNIDER: *Journ. Chem. Phys.*, **42**, 548 (1965).
[14] H. O. PRITCHARD: *Journ. Phys. Chem.*, **66**, 2111 (1962).
[15] H. J. KOLKER: *Journ. Chem. Phys.*, **44**, 582 (1966).
[16] O. K. RICE: *Journ. Phys. Chem.*, **65**, 1972 (1961); **67**, 1733 (1963).
[17] B. WIDOM: *Science*, **148**, 1555 (1965).
[18] For some additional referenes see: *a*) E. W. MONTROLL and K. E. SHULER: *Advan. Chem. Phys.*, **1**, 361 (1958); *b*) B. WIDOM: *Journ. Chem. Phys.*, **34**, 2050 (1961); *c*) K. E. SHULER and G. E. WEISS: *Journ. Chem. Phys.*, **38**, 305 (1963); *d*) J. KECK and G. CARRIER: *Journ. Chem. Phys.*, **43**, 2284 (1965); *e*) J. KECK: *Disc. Faraday Soc.*, **33**, 173 (1962); *f*) For systems with internal degrees of freedom additional problems, not discussed in this lecture, may arise; see L. WALDMANN: *Proc. Internat. Symp. Transport Properties of Gases*, Brown University, Providence, R. I.; L. WALDMANN: *Handbuch der Physik*, vol. **12** (Ed. S. FLÜGGE), (Berlin, Göttingen, Heidelberg, 1958), p. 295.

Rate Constants and Reaction Cross-Sections.

H. Gg. Wagner

University of Bochum - Bochum

A detailed understanding of chemistry, of chemical processes, chemical equilibria and chemical structure started to grow, when quantitative methods were applied to the investigation of chemical problems. The progress in the various fields, however, was quite different and closely related to the development of new theoretical concepts (*e.g.* thermodynamics, quantum theory) and of new physical methods (*e.g.* X-rays, IR-, Raman- and NMR-spectroscopy). Compared to our knowledge and understanding of chemical equilibria and the structure of matter, *i.e.* the static part of chemistry, our understanding of the chemical change itself, of rates of chemical reactions, is much less developed and largely empirical.

The first quantitative study of the rate of a chemical reaction was Wilhelmy's investigation of the inversion of sugar around 1850. It was established that the rate of a chemical reaction, *e.g.* the change of concentration with time of two components A and B in a solvent forming products P, can be formally described by the rate law

$$\frac{\mathrm{d}[P]}{\mathrm{d}t} = k[A]^m[B]^n,$$

where $[P]$, $[A]$, $[B]$ are the concentrations of the reactants and products, k is called rate constant, and the sum $n+m=\nu$ the order of the reaction. (n and m need not be whole numbers.) This type of a rate law represents one way to empirically describe the velocity of a chemical reaction, in terms of the concentrations of reactants, to which a measured rate is proportional.

This type of a rate law is often quite useful. For reactions in gases the rate constant obtained from such a rate law has no meaning which can easily be interpreted. (The situation in solutions can be different. In a more adequate description of the rate of a chemical reaction the driving force is proportional to the chemical potential and in solutions the chemical potential can sometimes be approximated by an expression $[A]^n$ with some number n.)

In his famous paper of 1887 ARRHENIUS postulated a mechanism for a chemical reaction, assuming the existence of an equilibrium between the reactant and an isomer form of it

$$A \rightleftarrows A^* \rightarrow P$$

and showed, that the temperature-dependence of the corresponding rate constant can be represented by

$$k = -A \exp\left[-\frac{E}{RT}\right].$$

It was, however, not possible at that time to draw firm conclusions with respect to more specific mechanisms.

In the meantime the kinetic theory of gases had reached a state which would have allowed quantitative conclusions about the mechanism of homogeneous gas phase reactions. However, the experimental results available were rather unreliable. When they were improved by eliminating the influence of such things as wall reactions, impurities a.o. (BODENSTEIN), it was realized that in many cases the reactions observed were not single-step reactions but consisted of various reaction steps.

This is demonstrated, *e.g.*, by the rate law for the formation of HBr in the H_2-Br_2 reaction. At temperatures around 350 °C BODENSTEIN and LIND (1906) found

$$\frac{\mathrm{d}[\mathrm{HBr}]}{\mathrm{d}t} = \frac{\mathrm{k}'[\mathrm{H_2}][\mathrm{Br_2}]^{\frac{1}{2}}}{1 + k''[\mathrm{HBr}]/[\mathrm{Br_2}]}.$$

It was evident that this rate expression does not describe a single reaction step and the constants cannot be simply related to molecular encounters. Further experiments showed, that such a rate law could be due to the following mechanism:

(1) $$\mathrm{Br_2 + M \xrightarrow{k_1} 2Br + M},$$

(2) $$\mathrm{Br + H_2 \xrightarrow{k_2} HBr + H},$$

(3) $$\mathrm{H + Br_2 \xrightarrow{k_3} HBr + Br},$$

(4) $$\mathrm{H + HBr \xrightarrow{k_4} H_2 + Br},$$

(5) $$\mathrm{Br + Br + M \xrightarrow{k_5} Br_2 + M}.$$

Writing down the rates of formation of the various species and applying Bodenstein's hypothesis of quasi-stationarity (the rate of change of the atom or radical concentration is small compared to that of main reactants),

$$\frac{d[H]}{dt} = k_2[Br][H_2] - k_3[Br_2][H] - k_4[HBr][H] \approx 0$$

one obtains the rate of formation of HBr as written above. The constants in that expression are related to the rate constants of the various reaction steps: $k' = 2k_2 \cdot (k_1/k_5)^{\frac{1}{2}}$ and $k'' = k_4/k_3$. In addition $k_1/k_5 = K$ where K is the well known equilibrium constant for $Br_2 \rightleftarrows 2Br$. If the mechanism is correct, one can obtain with some accuracy from that formula the rate constant for reaction step (2), k_2, and also the ratio k_4/k_3 in a limited range of conditions. One sees immediately that these expressions do not hold at the very beginning of the reaction, where the Br_2 dissociation is rate determining. Furthermore the rate equation does not lead to equilibrium. Another important property of that mechanism is, that it contains closed reaction sequences or chain reactions.

For the evaluation of the rate expression for the H_2-Br_2 reaction we used the rates of elementary reaction steps. An elementary process is called unimolecular, bimolecular, trimolecular according to the number of molecules reacting.

Having a reaction $\sum \nu_i B_i = 0$, the rate of reaction is $(1/\nu_i)\,d[B_i]/dt$. For the reaction $H + Br_2 \rightarrow HBr + Br$, the rate of consumption of H in that reaction is $d[H]/dt = -k[H][Br_2]$. It is a bimolecular reaction which follows a second-order rate law. During the dissociation of

$$(1) \qquad Br_2 + M \rightarrow 2Br + M\,,$$

$$(5) \qquad 2Br + M \rightarrow Br_2 + M\,,$$

the rate of formation of Br is given by

$$\frac{1}{2}\frac{d[Br]}{dt} = k_1[Br_2][M] - k_5[Br]^2[M] = -\frac{d[Br_2]}{dt}\,.$$

Here the first step (1) is a unimolecular reaction which follows a second-order rate law, while the recombination is a bimolecular reaction following a third-order rate law. The collision partner **M** does not undergo chemical change.

If all elementary steps taking part in a reaction are known, one can derive an overall rate law. The reverse is not true however. One has to be very careful in interpreting an overall rate law on the basis of a proposed mechanism and in deriving rate constants of elementary reactions from such an expression. This is shown by the H_2-I_2-reaction

$$H_2+I_2 \rightarrow 2\,HI\,.$$

Up to about 600 °K this reaction follows a rate law

$$\frac{d[HI]}{dt} = k(T)[H_2][I_2]\,,$$

which is second order and seems to describe a bimolecular process. New experimental results however make it probable, that the reaction proceeds via two *I* atoms. This would mean, that the rate expression in this case does not describe an elementary process but is an overall rate law.

For a comparison with the theory of reactive interaction of particles only rate constants of elementary steps can be used. Up to now it is only a small number of elementary reaction steps in the gas phase for which the rate constants and the products are known reasonably well.

Looking at the rate constant of an elementary reaction of the type $A+B \rightarrow$ $\rightarrow C+D$ one usually realizes two things

1) the number of molecules reacting in unit time is in most cases small compared with the number of A-B collisions per unit time,

2) while the number of non-reactive collisions varies as $T^{\frac{1}{2}}$ the number of reactive collisions very often shows a much stronger temperature-dependence, which can be approximated by an Arrhenius expression $k \sim \exp[-E_a/RT]$, where the apparent energy of activation is related to some energy barrier, which has to be overcome.

The energy content of reactants and products can be described by certain distribution functions. In many cases one can use as a good approximation the equilibrium distribution functions and apply thermodynamic relations. The apparent energy of activation can be defined as

$$E = RT^2 \frac{d \ln k}{dT}\,.$$

One can also relate the rate constants k_1 for the forward reaction and k_5 for the backward reaction by

$$\ln k_1 - \ln k_5 = \ln K = -\frac{\Delta G}{RT}\,,$$

where K is the equilibrium constant and G the Gibbs free energy. Tolman derived from statistical arguments the relation

$$E = \bar{\bar{\varepsilon}} - \bar{\varepsilon} ,$$

where $\bar{\varepsilon}$ is the average energy of all molecules and $\bar{\bar{\varepsilon}}$ the average energy of molecules respectively molecular clusters reacting per unit time.

All these quantities have the character of thermodynamic properties and therefore give very limited information about the microscopic reaction mechanics.

Therefore we may ask:

1) what conditions, with respect to energy and collision geometry, must be satisfied by molecules to be able to react,

2) how often are these conditions fulfilled?

In order to discuss these questions it is not sufficient to consider the concentration of one species as a whole. One has to take into account that the total concentration of species A consists of particles A in various states of electronic excitation, vibration, rotation, and translation. In addition a certain particle A(i), where i stands for a set of inner quantum numbers, can approach a particle B(j) under various angles and with various relative velocities. The reaction products may be C(k) and D(l) where j, k and l again stand for sets of inner quantum numbers. Therefore

$$A(i) + B(j) \rightarrow C(k) + D(l) .$$

As in the case of elastic or inelastic collisions such a reactive collision can be characterized by a cross-section. A detailed treatment of the relation between cross-sections for chemical reactions and rate constants has been given by Fowler and Guggenheim, Eliason and Hirschfelder, and by Light, Ross and Shuler (see references).

Before going into details, it should be mentioned that in the case of unimolecular chemical reactions the essential collision processes are inelastic collisions characterized by inelastic collision cross-sections (see contribution of Beck), while the reactive process itself, *e.g.* an isomerization of an « energized » molecule, is not characterized by a cross-section but by a probability of reaction, which in this case is an *intra*molecular process.

In order to simplify the derivation of the relation between cross-section and rate constant we assume that the elementary process is described by binary collisions. In addition, the density shall be low enough and the forces of short range, so that no correlation between collisions occurs. Further, the reaction shall proceed so slowly that the distribution functions are practically not changed during the reaction. The case where this assumption does not hold, and therefore elastic and inelastic collision processes have to be taken into account, is discussed in the contribution of Ross.

In order to define a cross-section for reaction in a general way one proceeds as in the case of a cross-section for elastic or inelastic processes.

Two molecular beams of particles $A(i)$ and $B(j)$ with normalized velocity distributions $f_{A(i)}(\boldsymbol{v}_{A(i)})\,d\boldsymbol{v}_{A(i)}$ and $f_{B(j)}(\boldsymbol{v}_{B(j)})\,d\boldsymbol{v}_{B(j)}$ are supposed to cross in a volume V. Their particle number densities shall be $n_{A(i)}$ and $n_{B(j)}$, the relative velocity $g_{ij}=|\boldsymbol{v}_{A(i)}-\boldsymbol{v}_{B(j)}|$. We look for the number $dN_{A(i),B(j),C(k)}(\Omega)$ of product molecules arriving per unit time at a solid angle Ω' in the angle element $d\Omega'$ as a result of the reaction $A(i)+B(j)\rightarrow C(k)+D(l)$. The differential cross-section I_R for that reaction is defined by

$$dN_{A(i),B(j),C(k)}(\Omega')=I_R\left(\begin{matrix}ij\\kl\end{matrix},\,\Omega',\,g_{ij}\right)\boldsymbol{g}_{ij}n_{A(i)}f_{A(i)}(\boldsymbol{v}_{A(i)})n_{B(j)}f_{B(j)}(\boldsymbol{v}_{B(j)})\,d\boldsymbol{v}_{A(i)}d\boldsymbol{v}_{B(j)}V\,d\Omega'\,.$$

The total cross-section σ_R for reactive collision is obtained simply as $\sigma_R\left(\begin{matrix}ij\\kl\end{matrix},\,g_{ij}\right)=\int I_R(\ldots,\Omega)\,d\Omega$. It can also be given as $\sigma_R\left(\begin{matrix}ij\\kl\end{matrix},\,E\right)$ where E is the kinetic energy of relative motion. It should be noted, that σ_R does not only depend on the quantum states of the reactant molecules but also on that of the product particles. For the number dN of product molecules, say $C(k)$, formed from $A(i)$, $B(j)$ in the volume V and in the velocity intervals $v_{A(i)}\ldots v_{A(i)}+dv_{A(i)}$ respectively $v_{B(j)}\ldots v_{B(j)}+dv_{B(j)}$ scattered in all solid angle in unit time we obtain

$$dN_{A(j),B(j),C(k)}=V\sigma_R\left(\begin{matrix}ij\\kl\end{matrix},\,g_{ij}\right)\boldsymbol{g}_{ij}n_{A(i)}f_{A(i)}(\boldsymbol{v}_{A(i)})n_{B(j)}f_{B(j)}(\boldsymbol{v}_{B(j)})\,d\boldsymbol{v}_{A(i)}d\boldsymbol{v}_{B(j)}\,.$$

From that expression one can immediately calculate the number of particles $C(k)$ formed per unit time and unit volume, *i.e.* the rate of formation of $C(k)$, $dn_{C(k)}/dt$, by dividing through V and integrating over $dv_{A(i)}$ and $dv_{B(j)}$. According to our earlier definition this is

$$\frac{dn_{C(k)}}{dt}=k_{ij}^{lk}n_{A(i)}n_{B(j)}\,,$$

with

$$k_{ij}^{kl}=\int\sigma_R\left(\begin{matrix}ij\\kl\end{matrix},\,g_{ij}\right)\boldsymbol{g}_{ij}n_{A(i)}f_{A(i)}(\boldsymbol{v}_{A(i)})n_{B(j)}f_{B(j)}(\boldsymbol{v}_{B(j)})\,d\boldsymbol{v}_{A(i)}d\boldsymbol{v}_{B(j)}\,.$$

The macroscopic rate, the formation of particles C, independent of the internal quantum states is obtained by summation over all internal quantum states

$$\frac{dn_c}{dt} = \sum_{ij}\sum_{kl} n_{A(i)} n_{B(j)} k_{ij}^{lk} = kn_A n_B$$

and therefore

$$k = (1/n_A n_B)\cdot \sum_{ij}\sum_{kl} n_{A(i)} n_{B(j)} k_{ij}^{lk} .$$

This expression relates the macroscopic rate of a chemical reaction and the cross-sections for reaction. Apparently, even for rather simple reactants this relation is a complex one and one can realize that it is practically impossible to draw conclusions on the microscopic « mechanics » of a reaction from measurements of macroscopic rate constants.

It is useful however to treat a few very simple models. Let us assume

1) the molecules act like hard spheres and the internal degrees of freedom need not be considered;

2) the translational distribution functions are thermal. Then the velocity distributions are given by

$$f(v_\alpha, T) = (m_\alpha/2\pi kT)^{\frac{3}{2}} \exp[-m_\alpha v_\alpha^2/2kT] , \qquad \alpha = A \text{ or } B,$$

and one can transform the integral which defines k_{ij}^{lk} to center of mass and relative velocities v and g by

$$\boldsymbol{g} = \boldsymbol{v}_A - \boldsymbol{v}_B ,$$

$$m_A \boldsymbol{v}_A + m_B \boldsymbol{v}_B = (m_A + m_B)\boldsymbol{v} ,$$

$$d\boldsymbol{v}\, d\boldsymbol{g} = d\boldsymbol{v}_A d\boldsymbol{v}_B .$$

The kinetic energy is

$$\tfrac{1}{2}\mu g^2 + \tfrac{1}{2}(m_A + m_B)v^2 = \tfrac{1}{2}m_A v_A^2 + \tfrac{1}{2}m_B v_B^2$$

with $\mu = m_A m_B/(m_A + m_B)$.

After integration over the center-of-mass co-ordinates one obtains

$$k_\sigma(T) = \left(\frac{\mu}{2\pi kT}\right)^{\frac{3}{2}} \int \sigma g \exp\left[-\frac{\mu g^2}{2kT}\right] d\boldsymbol{g} .$$

In addition integration over the angular part of $\boldsymbol{g}$ can be performed because the integrand depends only on the relative speed. Therefore $d\boldsymbol{g} = 4\pi g^2 dg$.

With $E=\frac{1}{2}\mu g^2$ we obtain

$$k_\sigma(T)=\frac{1}{kT}\left(\frac{8}{\pi\mu kT}\right)^{\frac{1}{2}}\int_0^\infty \sigma_R(E)E\exp\left[-\frac{E}{kT}\right]\mathrm{d}E\,.$$

A very simple model for a reaction is the one where $\sigma(E)$ is given by a step function.

$$\sigma=0 \qquad \text{for } E<E_0,$$

$$\sigma=\sigma_{AB} \qquad \text{for } E\geqslant E_0$$

The expression for the rate constant becomes

$$k(T)=\sigma_{AB}\left(\frac{8kT}{\pi\mu}\right)^{\frac{1}{2}}\left(\frac{E_0}{kT}+1\right)\exp\left[-\frac{E_0}{kT}\right].$$

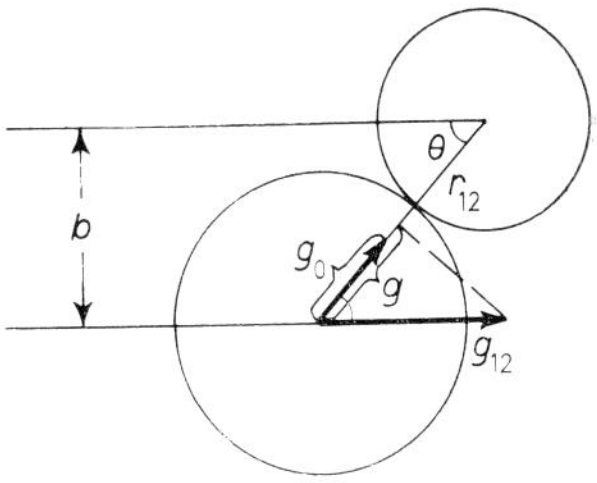

Fig. 1. – Collisional geometry, see text.

A somewhat more plausible hard-sphere model is the one for which reaction is assumed to occur if the kinetic energy along the line of centers exceeds a certain value E_0, or the relative velocity a value g_0 (see Fig. 1). Here $g=g_{12}\cos\Theta\geqslant g_0$; $\cos\Theta=g_0/g_{12}$. The impact parameter is given by $b=r_{12}\sin\Theta$, therefore

$$\left(\frac{g_0}{g_{12}}\right)^2+\left(\frac{b}{r_{12}}\right)^2=1;$$

$$b^2=r_{12}^2\left(1-\left(\frac{g_0}{g_{12}}\right)^2\right)$$

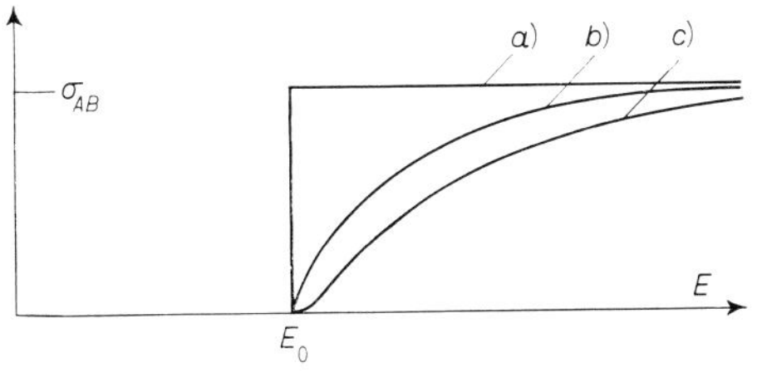

Fig. 2. – Dependence of reaction cross-section on relative energy: *a*) step function, *b*) $\sigma\sim(1-E_0/E)$, *c*) $\sigma\sim(1-E_0/E)^n$ with $n>1$.

and

$$\sigma_{eff}=\pi b^2=\sigma_{AB}\left(1-\frac{E_0}{E}\right) \qquad \text{for } E\geqslant E_0,$$

$$\sigma=0 \qquad \text{for } E<E_0.$$

The energy-dependence of that cross-section is shown in Fig. 2. The resulting rate constant is

$$k=\left(\frac{8}{\pi\mu kT}\right)^{\frac{1}{2}}\frac{1}{kT}\int_0^\infty\sigma_{AB}\left(1-\frac{E_0}{E}\right)E\exp\left(-\frac{E}{kT}\right)\mathrm{d}E=\sigma_{AB}\left(\frac{8kT}{\pi\mu}\right)^{\frac{1}{2}}\exp\left[-\frac{E_0}{kT}\right].$$

It is worth-while looking at the influence of the energy-dependent cross-section and the distribution function on the rate constant. One easily realizes that, as long as $E_0 \gg kT$, the main contribution comes from the domain just above E_0. For the case $\sigma = \sigma_{AB}(1 - E_0/E)$ the integrand has its maximum at $E = E_0 + kT$. If $\sigma \sim (1 - E_0/E)^n$; $n > 1$ the maximum is shifted to higher E values for increasing n. As long as $E \gg nkT$, a good approximation for the position of the maximum is $E = E_0 + nkT$. If the energy-dependence of the cross-section varies, the expression for the rate constant changes too. However, especially if $E_0 \gg kT$ the change is not very large.

A comparison of measured rate constants with those calculated for various σ functions is very difficult. Normally E_0 is not known, in addition the experimental accuracy is not yet high enough to allow the determination of a curvature in the $\ln k$ *vs.* $1/T$ plot and therefore of the temperature-dependence of the pre-exponential factor. It is practically impossible to determine $\sigma(E)$ from measurements of macroscopic rate constants alone, even under the very simplifying assumptions applied in the examples discussed. This becomes also clear, if one realizes, that the rate constant $k(T)$ is the Laplace transform of $\sigma(E)E$ and that the temperature range in which rate constants can be measured is limited.

This shows, that measurements of rate constants do not give information about the detailed microscopic processes which take place during an elementary chemical step. In order to obtain this information one has to perform experiments which measure the σ values under various conditions directly, *e.g.* molecular beam experiments. On the other hand one realizes, that a large amount of data on σ is necessary in order to determine one rate constant, and for most chemical processes not one but many rate constants are important at the same time.

* * *

The author acknowledges the work done by U. BUCK, R. GENGENBACH, and M. v. SEGGERN on a first draft of this lecture.

BIBLIOGRAPHY

For a more detailed treatment see, *e.g.*,

R. FOWLER and E. A. GUGGENHEIM: *Statistical Thermodynamics* (Cambridge, 1949).

M. A. ELIASON and J. O. HIRSCHFELDER: *Journ. Chem. Phys.*, **30**, 1426 (1959).

and especially

J. ROSS, J. C. LIGHT and K. E. SHULER: *Rate coefficients, reaction cross-sections and microscopic reversibility*, in: *Kinetic Processes in Gases and Plasmas* (New York, 1969) p. 281.

Thermal Chemical Reactions and Transition-State Theory.

F. S. Rowland
University of California - Irvine, Cal.

1. – Thermal chemical reactions.

The lectures in this course have so far described in considerable detail the chemical behavior occurring in crossed molecular beams, and the discussion of these reactions in terms of scattering theory seems quite natural to an audience composed largely of physicists. This terminology is, however, quite new in its widespread use in chemical kinetics, and the « alkali age » reactions are a very special, rather limited class in the view of most chemists interested in studying chemical change. To the more traditional chemist, the reaction of $K+Br_2$ would be well described by the term « instantaneous »; chemical systems reacting instantaneously and exothermically are often quite accurately described as « explosions », and as such, are a class of chemical change which experimenters more frequently seek to avoid than to study.

Chemists have lived, and will continue to live for the most part, in a thermal world—a situation in which most chemical reactions under study are made to occur in systems which are essentially in thermal equilibrium. The great bulk of our knowledge of chemical change has been obtained from just such thermal systems, and the concepts and explanations clearly reflect these particular experimental origins. My chief task in the present lecture is to sketch this background of chemical knowledge, attitudes, and explanations—in brief, the normal background of both knowledge and folklore with which persons trained as chemists come equipped—for this audience composed chiefly of atomic physicists [1-4].

The abstraction reaction $H+CH_4 \to H_2+CH_3$ has a cross-section, averaged over the collisions occurring at 25 °C, of $< 10^{-7}$ Å^2; typically, these species are said « not to react at room temperature », and H atoms formed in methane gas at 25 °C almost invariably find a chemical fate different from the abstraction of another H atom from CH_4. Quite commonly, the H atom may undergo numerous ($> 10^5$) nonreactive, scattering collisions with methane, and eventually diffuse long and far enough either to react with a minor impurity in the gas, or else to strike and react with one of the container walls, or something on it.

Molecular collision frequencies are in the range of $(10^9 \div 10^{10})\ s^{-1}$ at one atmosphere pressure, and $\sim 10^{12}\ s^{-1}$ in the liquid phase. In contrast, typical rates of chemical reaction may be expressed in units of ms^{-1}, min^{-1}, h^{-1}, or even d^{-1}. Some extremely slow rates, such as the rate of decarboxylation of amino acids near room temperature (of interest because of the presumed occurrence of the reaction over geologic time in limestone deposits), are estimated to be very much slower, $\sim 10^{-6}\ y^{-1}$ [5]. Despite the enormous disparity between collision frequencies of $10^{12}\ s^{-1}$ and reaction frequencies of $(10^3 \div 10^{-13})\ s^{-1}$, thermal chemical reactions are usually the consequence of bimolecular collisions of A and B, and are described by a rate constant, k, from the formula: reaction rate $= k(A)(B)$, in which (A) and (B) are the concentrations of the colliding species. Obviously, all collisions except one in 10^9 to 10^{25} *fail* to lead to chemical change, and reactive scattering events are numerically negligible compared to scattering events in the usual thermal systems. However, since nonreactive scattering does not create a different chemical species the usual measurement procedure of observing the formation of a new species effectively disregards all collisions *except* those resulting in a permanent chemical change in identity.

The rates of chemical reactions in thermal systems increase very markedly with increasing temperature, much faster than the $T^{\frac{1}{2}}$ dependence of collision frequency. The actual temperature-dependence is found to fit, within experimental uncertainties, the empirical Arrhenius formulation, $k = A \exp[-E_a/RT]$, in which the rate constant, k, is fitted to a two-parameter equation involving E_a, the activation energy, and A, the pre-exponential factor, R being the universal gas constant per mole. Variations in the detailed handling of this equation can include the splitting of A into two terms, pZ, the latter being the collision frequency (*) and the former a « steric factor » reflecting qualitatively that molecular collisions may need to occur in specific orientations for reaction to be a possibility. Unfortunately, p has turned out not to be a quantity capable of calculation, and therefore its quantitative use represents simply a correction factor relating observed and predicted rates of reaction.

The values of activation energies, usually quoted in kcal/mole(**), provide large negative exponents (*e.g.* RT at 298 °K = 0.59 kcal/mole; if $E_a = 11.8$ kcal/mole, the exponential term is e^{-20}), and are critically important in the determination of the rates of reaction. The significance of an activation energy is crudely

(*) Since Z is dependent upon $T^{\frac{1}{2}}$ at constant density, the energy term in the exponential will have a slightly different numerical value than in the first formulation.

(**) 1 eV per molecule = 23.05 kcal/mole.

that chemical change will not take place unless the colliding molecules strike each other with an unusually large amount of energy, an amount sufficient to surmount an energy barrier separating the chemical identity as reaction products from that as reactants. The slow reaction rates *vis-a-vis* collision rates are now readily understood, at least qualitatively, as the consequence of very severe energy requirements on collisions such that only a very rare collision will involve two molecules with sufficient relative energy for chemical reaction to be possibile.

The crucial importance of the activation energy in thermal systems can be illustrated in another way through Fig. 1, in which a hypothetical reactive cross-section is illustrated, rising linearly with velocity above a certain minimum value, or threshold energy. The rate constant for reaction then becomes an integral of $n(v)\sigma(v)v\,dv$, integrated over the $n(v)$ characteristic of thermal systems—the Maxwell-Boltzmann distribution with $n(v) \sim v^2 \exp[-Mv^2/2kT]$, in which $Mv^2/2$ is the kinetic energy of the molecule of molecular weight M. Examination of Fig. 1 indicates that the integral of the Maxwellian distribution over this reactive cross-section receives a negligible contribution from reactions more than $(10 \div 20)\%$ above the threshold velocity. While raising the temperature of the system can greatly increase the magnitude of this integral, the increase does not involve any great change in the range of velocities sampled by the reaction—instead, it merely represents a quite substantial increase in the amount of sampling from the same velocity range as before: the velocities just above threshold. TOLMAN has shown that the activation energy of a reaction is equal to the difference between the average energy of reacting molecules and the average energy of *all* molecules. The activation energy is sensitive to a minor extent to the shape of the reaction cross section above threshold, and hence is not necessarily identical to the threshold energy, although the values will generally be very similar numerically. A change in the form of $\sigma(v)$ above the threshold produces only a small change in the location of the peak in $n(v)\sigma(v)v$; in fact, since the threshold energies are not independently known, no information about the functional form of $\sigma(v)$ for a reaction with an activation energy is available from thermal experiments.

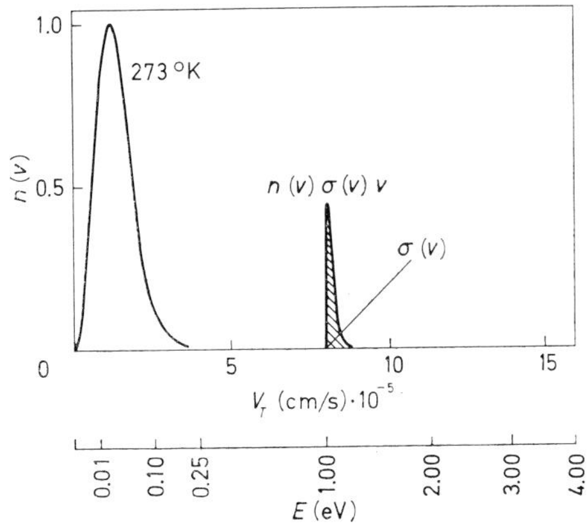

Fig. 1. – Chemical reactions in a Maxwellian system (schematic).

Several points can now be made: the remarkable inefficiency of bimolecular collisions has been rationalized through the imposition of an energy requirement (and, to a lesser extent, an orientation requirement); second, the activation energy is found to be enormously important, virtually determining which reactions are experimentally accessible in the system; third, the disappearance of reagents through reaction is only a minor perturbation on the Maxwell-Boltzmann distribution, and the assumption that thermal equilibrium is maintained is a reasonably good approximation. At very high gas temperatures, for which $E_a \sim RT$, the depletion of high energy molecules is no longer trivial, and appreciable deviations from equilibrium can be expected [5a].

Finally, there is no indication in such a treatment as to what kinds of energy are effective in facilitating reaction, and what kinds do not participate in overcoming the activation barrier. If some kinds of energy do not actually contribute to chemical reaction, this fact is generally concealed in a lower value of the pre-exponential factor, and is thus lumped into the noncalculable p with true orientation factors.

2. – Transition states and reaction mechanisms.

The theoretical treatment of thermal chemical reactions is almost universally carried out through an approach identified as transition state theory. As you may have noticed, this course has proceeded into its second week with only some brief references to this theory, usually to the fact that it *wasn't* being used, and with no details whatsoever of the theoretical formulation. This, of course, isn't really so surprising since the experiments discussed by the lecturers have almost never involved a simple thermal reaction scheme. We shall proceed in our discussion with a particularly simple illustration of a typical thermal chemical reaction.

The two isomers, *cis*- and *trans*-2-butene, differ only in the position of the two methyl groups, and can be converted to one another

$$\underset{\text{cis}}{\begin{array}{c}CH_3 \\ \\ H\end{array}\!\!\!>\!\!C\!=\!C\!\!\!<\!\!\!\begin{array}{c}CH_3 \\ \\ H\end{array}} \underset{k_t}{\overset{k_c}{\rightleftarrows}} \underset{\text{trans}}{\begin{array}{c}CH_3 \\ \\ H\end{array}\!\!\!>\!\!C\!=\!C\!\!\!<\!\!\!\begin{array}{c}H \\ \\ CH_3\end{array}}$$

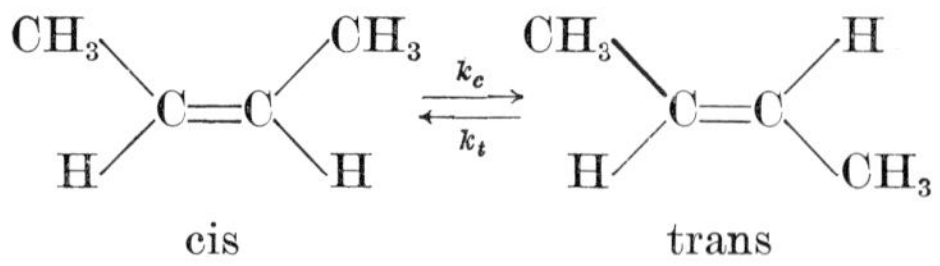

when sufficiently excited. Careful measurement of the heat released when each of these molecules is separately burned (to CO_2 and H_2O) shows that the *trans* form of 2-butene is slightly more stable than the *cis* form ($\Delta H = 1.0$ kcal/mole at 25 °C). Recalling the statistical mechanical formula for the probability of the i-th state as $P(E_i) = g_i \exp[-E_i/kT]$, it is clear that the ratio of the two isomeric forms of 2-butene at chemical equilibrium will be given by the statistical summation of these probabilities. The precise calculation requires knowledge of all of the vibrational frequencies of each molecule, but, because of the close similarity between the two molecules, a crude estimate of the result can be obtained simply by *cis*/*trans* $\sim \exp[-\Delta H/RT]$, which at 25 °C is equivalent to $10^{-\Delta H/1360}$ or $10^{-0.7} \simeq 1/5$. Detailed calculations give an equilibrium *cis*/*trans* ratio at 25 °C of about $\frac{1}{3}$. (A side remark can be interposed here that an energy difference, or uncertainty, of the order of 1 kcal/mole is equivalent to a difference of a factor of 5 at 25 °C. The difficulty in obtaining potential energy surfaces accurate to ± 1 kcal/mole has been well illustrated in the accompanying discussions of the various calculations for the H_3 surface—the surfaces for the more typical reacting systems are far less accurately known than that for H_3.)

If one now considers the actual chemical change of *cis*-2-butene to *trans*-2-butene, or *vice versa*, a reasonable mechanism for this change might be the rotation of one $-\mathrm{C}\langle^{\mathrm{CH_3}}_{\mathrm{H}}$ group through 180° about the C=C axis to $-\mathrm{C}\langle^{\mathrm{H}}_{\mathrm{CH_3}}$, passing through an intermediate stage in which the two $-\mathrm{C}\langle^{\mathrm{CH_3}}_{\mathrm{H}}$ groups are not co-planar, but lie in planes at 90° to one another. A molecule possessing this 90° structure is in a very high energy state relative to either *cis*- or *trans*-2-butene (~ 70 kcal/mole higher), chiefly because the favorable C=C bonding of the stable molecule is partially sacrificed during the rotation. If the potential energy of this system is plotted *versus* rotational angle, one finds that the energy has minima at 0° and 180° (with the *cis* minimum at 0° 1kcal/mole higher than the *trans* minimum at 180°), and a maximum very near 90°. In an equilibrated mixture of *cis*-2-butene and *trans*-2-butene, one would also find the equilibrium amount of the 90° structure (as well as for all other angles, and other states of excitation of both molecules), but the crude estimate of its concentration is, at 25 °C, $\sim 10^{-70000/1360}$, or 10^{-50} relative to the ground state of either stable molecule; the corresponding estimate is still 10^{-25} at 300 °C.

Chemists generally use the terms « transition state » or « activated complex » in a range of meanings from the simple, qualitative identification of the *mechanism* of the chemical reaction, *i.e.* that *cis* isomerizes to *trans* through the 90° intermediate, and not by some other possible route (for example, rupture and then reformation of the $\rangle\mathrm{C}=\mathrm{C}\langle$ bond); to the detailed calculations of the statistical mechanical type, in which the 90° structure is assigned precise bond distances, angles, force constants, vibrational frequences, etc.—essentially the same properties used for the equilibrium calculation for *cis*-2-butene $\leftrightarrows$ *trans*-2-butene. However, while the equilibrium calculations for the two isomeric butenes can be calibrated and based upon extensive physical measurement of their molecular properties, especially spectroscopic frequencies and heat capacities, the corresponding data are totally absent for the transient molecular structures involved in transition states.

The qualitative, mechanistic symbolism of the transition state is not always so straightforward as that indicated for the isomerization of 2-butene. For example, one very thoroughly studied chemical reaction, the isomerization of cyclopropane to propylene, as in (1), has been

$$
(1) \quad \begin{array}{c} CH_2 \\ \diagup \diagdown \\ CH_2\text{---}CH_2 \end{array} \rightarrow \begin{array}{c} CH_3 \\ \diagdown \\ \diagup C=CH_2 \\ H \end{array}
$$

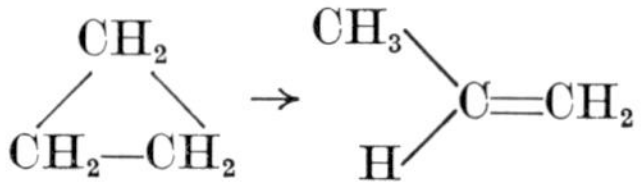

extensively considered in terms of two possible transition states, the C—C elongation of I), and the H atom shift of II) [2].

$$
\begin{array}{cc} \begin{array}{c} CH_2 \\ \diagup \diagdown \\ CH_2\cdots CH_2 \\ I) \end{array} & \begin{array}{c} H\cdots CH_2 \\ \vdots \diagdown \\ CH\text{---}CH_2 \\ II) \end{array} \end{array}.
$$

3. – Calculation of reaction rates.

Quantitative estimation of rates of chemical reaction by the «absolute theory of reaction rates» [1] depends upon the assumption that thermal equilibrium is established and maintained between reactants and the activated complex—for example, between *cis*-2-butene and the 90° intermediate—in the absence of the products of reaction; the methods of statistical mechanics are then applied to this assumed equilibrium. The activated complex is an unusual molecule in some of its spectroscopy because of its basic instability. For example, rotation is perfectly normal near 0° and 180° for the isomeric butene molecules, with a restoring force generated by the resistance of the chemical bonding to distortion. On the other hand, rotation of the $-C{\diagup{CH_3} \atop \diagdown H}$ groups from the 90° position is downhill on the potential energy surface in either direction, and the system is completely unstable against any rotational displacement. For other degrees of freedom, *e.g.* the stretching of a C—H bond in the CH_3 group, the restoring force should be essentially the same at 0°, 90°, and 180°, and the behavior quite normal throughout. The co-ordinates of analysis of the activated complex are therefore divided into two classes—all but one are assumed to be normal, and can be described in the manner applied to all co-ordinates for stable molecules. The remaining co-ordinate contains the energetic instability toward slight displacements, and is designated the reaction co-ordinate. The potential-energy variation in the reaction co-ordinate is usually described by an inverted parabola, with the maximum at the structure of the activated complex, and is then characterized by an imaginary frequency of vibration, $\nu_L^{\ddagger}$. In the specific instance of the isomerization of 2-butene, the reaction co-ordinate can be essentially identified as the rotation of the $-\mathrm{C}\genfrac{}{}{0pt}{}{\diagup \mathrm{CH_3}}{\diagdown \mathrm{H}}$ group. In the decomposition of CH_3—CH_3 into two CH_3 groups, the reaction co-ordinate is crudely the elongation of the C—C bond length. In most cases, however, the reaction co-ordinate is not so readily correlated with a particular, single degree of freedom in the stable molecule.

The further extension of this approach then converts the imaginary frequency—with a potential-energy maximum, no real vibration occurs in the reaction co-ordinate—into a rate of passage of individual complexes across the maximum in the reaction co-ordinate. Taken with the statistically calculated concentration of such complexes, this procedure permits the calculation of the rate at which molecules pass through the potential-energy maximum. The calculation for the reaction co-ordinate becomes a constant factor, kT/h, in which T is the absolute temperature, and k and h are the Boltzmann and Planck constants, respectively. The remaining $3N-7$ vibrational co-ordinates of the activated complex and all of the vibrational co-ordinates of the reactants appear in the rate equation, just as all the co-ordinates appear in the statistical mechanical calculation of a chemical equilibrium. The rate for a bimolecular reaction of A with B is then given by equation (2),

$$v = (A)(B)\frac{kT}{h}\frac{Q^x}{Q_A Q_B}\exp\left[-E_0/kT\right] \tag{2}$$

in which Q_A and Q_B are the partition functions for A and B at temperature T, and Q^x is that for the activation complex with the contribution from the reaction co-ordinate already factored out (*). The Arrhenius rate constant, k, is then equal to $v/(A)(B)$. Two additional factors which are also multiplied onto the right-hand side of (2) are: $\varkappa$, the transmission coefficient, a correction factor to compensate for complexes which do not remain on the product side of the barrier; and the « tunneling » correction, usually neglected except for the isotopes of hydrogen, involving an estimate of the quantum-mechanical penetration of the potential barrier. Many discussions are available about the

(*) Usually chemists will quote activation energies in kilocalories/mole, and use RT in the denominator, with R the universal gas constant per mole. Alternatively, one can use kT involving the gas constant per molecule, and then measure activation energies in energy units per molecule, too.

detailed assumptions underlying this theory, and about the validity of these assumptions [1-4].

The net result of this calculation is a procedure which permits numerical evaluation of the rates of chemical reaction in terms of two definite factors—the activation energy, as obtained from a potential-energy surface; and a set of definite molecular parameters, such as the molecular weights, moments of inertia, and vibrational frequencies. Since the rates of thermal chemical reactions are so sensitive to the value of E_0, and less so to the precise details of the calculation of the Q's, the overall calculation rests its success chiefly upon the knowledge of the difference in potential energy between the reactants and the activated complex along the reaction co-ordinates. Activation energies for chemical reactions are very difficult to estimate with any precision on a theoretical basis—compare, for example, the 8 kcal/mole activation energy for $D+H_2 \rightarrow HD+H$ with the more than 100 kcal/mole required to dissociate either H_2 or HD into an atom pair—and no theory has produced more than semi-quantitative success in such numerical estimates.

The usual practice in rate calculations has therefore been to use the experimentally determined activation energy as a direct parameter in constructing the potential-energy surface. Naturally, the theoretical calculation of a rate from such a surface will give back the correct temperature dependence that was built into it, and any testing of the validity of the theory is confined to the prediction of the Arrhenius A factor. Models have been developed which do throw the empiricism of the calculation onto experimental measurements of a nonkinetic nature, and a comparison of some calculated activation energies with the experimental ones is shown in Table I for the bond energy-bond order method (B.E.B.O.) of JOHNSTON *et al.* [4, 6, 7].

TABLE I. – *Calculated and experimental activation energies for reaction* $A+BC \rightarrow AB+C$.

Reaction	Activation energy (kcal/mole)	
	Observed	Calculated
$H+HI \rightarrow H_2+I$	1.5	1
$H+p\text{-}H_2 \rightarrow o\text{-}H_2+H$	8	10
$Cl+H_2 \rightarrow HCl+H$	5.5	8
$Br+H_2 \rightarrow HBr+H$	18.5	21

4. – Isotope effects on the rates of chemical reactions.

Some of the quantitative calculations of rates of chemical reactions based upon activated complex theory are exceedingly precise, and correspond very closely to the observations made in experiments; probably the best example is the general area of kinetic isotope effects—the calculation of the rates of reaction of isotopic molecules undergoing the same chemical reaction under the same conditions [8–10]. In this case, as in others, the quantitative successes are always obtained in the calculation of *relative* rates of reaction—in a comparison of the rates of reaction of a protonated material and its deuterated counterpart, the calculated value is k_H/k_D and not the value of either k_H or k_D on an absolute basis. Calculations of this kind are, of course, just the ones in which sensitivity to possible errors in the basic assumptions of the theory can be most effectively dulled or cancelled entirely.

The statistical-mechanical calculations of isotope effects on chemical equilibrium can be done very exactly, and have been so done for many systems. The extension of such calculations to kinetic isotope effects is a natural procedure, and follows readily through the statistical calculations of absolute reaction rate theory. Since experimental measurements of relative reaction rates can also be carried out with distinct advantages in precision over absolute measurement, both the theoretical and experimental measurements of kinetic isotope effects are often carried to several significant figures, especially with isotopes of great interest, *e.g.* ^{14}C *vs.* ^{13}C *vs.* ^{12}C; or H *vs.* D *vs.* T.

The comparison of isotopic molecules in the same chemical reaction insures that both molecules are following the same potential energy surface to a very high degree of accuracy; vibrational force constants and interatomic distances are likewise almost identical in value, and will be so in the vicinity of the potential energy maximum, even if the structure near the maximum in not known; vibrational frequencies and moments of inertia are directly related to the isotopic masses. Therefore, when one writes the Bigeleisen-Wolfsberg expression for the rates of reaction of isotopic species 1 and 2, one has an expression, equation (3),

$$\frac{k_1}{k_2} = \frac{\nu^\ddagger_{1L}}{\nu^\ddagger_{2L}} \times \frac{\prod_i^{3N-6} u_{2i}/u_{1i}}{\prod_i^{3N^\ddagger-7} u^\ddagger_{2i}/u^\ddagger_{1i}} \times \frac{\prod_i^{3N-6} [1-\exp[-u_{1i}]]/[1-\exp[-u_{2i}]]}{\prod_i^{3N^\ddagger-7} [1-\exp[-u^\ddagger_{1i}]]/[1-\exp[-u^\ddagger_{2i}]]} \times \frac{\exp\left[\sum^{3N-6} [(u_{1i}-u_{2i})/2]\right]}{\exp\left[\sum^{3N^\ddagger-7} [(u^\ddagger_{1i}-u^\ddagger_{2i})/2]\right]}, \tag{3}$$

which can be calculated very precisely, but which is surely quite insensitive to any specific assumption of the activated complex theory. The detailed calculations are made for the characteristics of a particular activated complex, but comparisons of different possible choices of activated complex indicate that the numerical values are not particularly sensitive to rather substantial changes in the structure of the complex [10]. The isotope effects are a tool for determining force constants at the position of isotopic substitution, and will play the same role with similar magnitudes if assumptions other than those of the activated complex were substituted in the formulation. In the formula, the subscripts 1 and 2 refer to the lighter and heavier isotopic species, respectively; the numerators refer to the isotopically substituted reactant, $\ddagger$ refers to the transition state, $u = h\nu/kT$ with ν a normal vibrational frequency and h and k the usual constants. Statistical symmetry factors, transmission coefficients, and tunneling corrections are omitted.

The calculations based upon this theory are compared to very high precision experimental measurements, of which the decarboxylation of oxalates is one of the best studied. This reaction, written for the manganous compound in (4),

$$MnC_2O_4\,(\text{solid}) \rightarrow MnO\,(\text{solid}) + CO_2\,(\text{gas}) + CO\,(\text{gas}) \tag{4}$$

does not proceed at equal rates for isotopic molecules, but shows slight differences that can be detected by precision mass spectrometry [11]. Some typical data for this reaction, expressed in the standard form illustrated below,

$$\begin{array}{c} {}^{12}COO^- \\ | \\ {}^{12}COO^- \end{array} \xrightarrow{k_1} {}^{12}CO_2 + {}^{12}CO$$

$$\begin{array}{ccc} {}^{13}COO^- & \xrightarrow{k_2} & {}^{13}CO_2 + {}^{12}CO \\ | & & \\ {}^{12}COO^- & \xrightarrow{k_3} & {}^{12}CO_2 + {}^{13}CO \end{array}$$

show that at 270 °C, the ratios of rate constants have the values $(k_2/k_3) = 1.0102 \pm 0.0022$; $(k_1/2k_3) = 1.0080 \pm 0.0023$; and therefore that $(k_1/2k_2) = 0.9977 \pm 0.0018$. In an example taken from organic chemistry, the rate of reaction (5)

$$\begin{array}{l} C_6H_5\diagdown \\ \qquad C{=}O \\ CH_3\diagup \end{array} + H_2N{-}NRH \rightarrow \begin{array}{l} C_6H_5\diagdown \\ \qquad C{=}N{-}NRH \\ CH_3\diagup \end{array} + H_2O\,, \tag{5}$$

is (at 0 °C) 1.119 ± 0.005 faster for $C_6H_5COCD_3$ and 0.9489 ± 0.0020 for $C_6H_5{}^{14}COCH_3$ than for the molecule shown (R is a large organic group = 2,4-dinitrophenyl) [12]. Measurement of such effects and comparison with theoretical expectation is a very useful tool in the study of chemical kinetics; they are not critically dependent, however, upon the absolute validity of the activated complex theory through whose format they have been calculated.

5. – Quantitative kinematic calculations.

How successful has the activated complex theory of chemical reaction rates actually been, and how accurate are its predictions? In one sense, the theory has been very successful and very useful—it has been almost universally adopted in the chemical world as the proper method of approach to all kinetic problems until rather recently. This success is most often somewhat qualitative in nature, and frequently lies more in the realm of the mechanistic understanding of the chemical reactions rather than in the actual calculation of quantitative rates of reaction. Experimental kineticists working with reacting systems are often simultaneously occupied with the problems of *which* products are formed, and *how* did the chemical transformation occur, as well as how fast did it happen. The 2-butene molecule with one 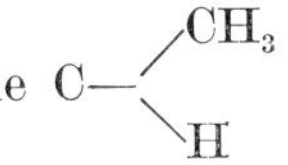 group rotated through 90° aids in the understanding of the qualitative processes occurring in the isomerization of 2-butene; even crude calculations of the effects of this rotation on bond lengths, angles and energies will give further insight into the details of the complicated atomic shifting required in such drastic changes for molecules containing even 12 atoms. This qualitative pattern of organized thinking about chemical change has been very effective, and is responsible for much of the popularity of this approach—perhaps especially when coupled with the feeling that the necessary improvements in quantitative calculations were just around the corner.

In the quantitative sense, no general theory of chemical kinetics has been highly successful in the prediction of rates of chemical reaction without using as an input the knowledge of some other rate of reaction for a similar system. In thermal systems, success depends almost entirely on the accuracy of the potential energy surface, and particularly upon the height of the activation barrier; as we have seen, the usual procedure involves empirical adjustment or testing of a potential-energy surface against a measured activation energy rather than *vice versa*. LAIDLER and POLANYI have recently summarized the situation as follows: « About all that can be concluded at the present time is that comparisons of experimental rates with those calculated on the basis of absolute rate theory do not reveal any discrepancies. None of the tests that have been made are, however, particularly stringent ones; the main reasons for this are that the data are not sufficiently accurate and that the difficulties in calculating activation energies render it necessary to use a considerable amount of empiricism. Accurate data are badly needed for reactions of the type $H + H_2 \rightarrow H_2 + H$, and further effort should be devoted to the *a priori* calculation of potential-energy surfaces » [4].

As the accuracy and reliability of theoretical potential-energy surfaces improves, the empirical content of rate calculations can certainly be reduced. However, at the same time, questions are being increasingly raised about the validity of the underlying assumptions of the theory; comparisons of trajectory calculations with absolute rate theory on the same potential-energy surface are showing discrepancies; the difficulty in calculation of transmission coefficients and the wide spread in tunneling corrections—with the choice usually being settled by comparison with experiment—is still present. And probably most important, more and more of the pertinent kinetic data are being generated in experiments with at least partially state-selected molecular groups, and not from thermally distributed reactants. The study of kinetics is in transition from the integral measurement of rates averaged over the Maxwellian distribution to the differential measurement of reaction cross-sections. In this changing situation, theories designed to treat reactions occurring in thermal equilibrium may be simply inappropriate for a larger and larger body of the experimental data.

REFERENCES

[1] S. Glasstone, K. Laidler and H. Eyring: *The Theory of Rate Processes* (New York, 1941).
[2] S. W. Benson: *Chemical Kinetics* (New York, 1960).
[3] D. Bunker: *Theory of Elementary Gas Reactions* (London, 1966).
[4] K. J. Laidler and J. C. Polanyi: *Progress in Reaction Kinetics*, **3**, 1 (1965).
[5] D. W. Conway and W. F. Libby: *Journ. Amer. Chem. Soc.*, **80**, 1077 (1958).
[5*a*] See this volume p. 249.
[6] H. S. Johnston: *Advan. Chem. Phys.*, **3**, 131 (1960).
[7] H. S. Johnston and C. Parr: *Journ. Amer. Chem. Soc.*, **85**, 2544 (1963).
[8] J. Bigeleisen and M. Wolfsberg: *Advan. Chem. Phys.*, **1**, 15 (1958).
[9] L. Melander: *Isotope Effects on Reaction Rates* (New York, 1960).
[10] M. Wolfsberg and M. J. Stern: *Isotope Mass Effects in Chemistry and Biology* (London, 1964), p. 325.
[11] P. E. Yankwich and P. D. Zavitsanos: *Isotope Mass Effects in Chemistry and Biology* (London, 1964), p. 287.
[12] V. F. Raaen and C. J. Collins: *Isotope Mass Effects in Chemistry and Biology* (London, 1964), p. 347.

The Measurement of Rate Constants.

H. Gg. Wagner
University of Bochum - Bochum

Gas phase reactions of neutral species, *e.g.* combustion reactions, normally involve a large number of parallel and consecutive elementary reaction steps. For the combustion of methane about 100 elementary steps can be written leading to final products. In order to describe the general properties of that process under specific conditions one needs about 8 to 10 steps.

In order to determine the rate of a chemical elementary reaction one would like to find situations over a large range of temperature where only the reaction of interest takes place. This requires among other things that the reaction products are rather inert which is often not the case. For many reactions of interest at least one of the reactants is an active particle, a radical or an atom, which has to be generated in some way. Often this means that another active particle is generated in the same process so that the system cannot be kept clean. Still another problem is the influence of the wall on the reaction. Finally, one has to identify the reaction products and make accurate concentration measurements.

In order to be able to measure a rate constant it is unavoidable in most cases to investigate a mechanism, to postulate intermediates and reaction steps, and then to find out the conditions best suited for the determination of the rate constant [1, 2].

There are some criteria to sort out reaction steps:

1) Thermodynamic properties (often used are bond energies).

2) Chemical analogy.

3) Spin conservation rules.

4) Check, whether closed sequences (chain and chain branching reactions) play a role.

Point 4) needs some explanation:

$$H + Br_2 \rightarrow HBr + Br ,$$

$$Br + H_2 \rightarrow HBr + H .$$

These two reactions form a closed sequence, a « reaction chain », because the active particles consumed are regenerated. There are other reactions, the chain branching steps, where two radicals are formed out of one, *e.g.* $O + H_2 \rightarrow$ $\rightarrow OH + H$, and there are energy chains: $H + F_2 \rightarrow HF^* + F$, $HF^* + F_2 \rightarrow HF + 2F$. Such closed sequences can complicate the investigation of a mechanism. They can be found out by determining explosion limits, the pressure-dependence of the overall reaction, or by adding certain substances which act as radical traps. In the past the determination of explosion limits was a very useful tool for the evaluation of reaction mechanisms and the determination of reaction rates.

The simplest method for measuring reaction rates, which is still in use, is the isothermal static reactor. The mixture is filled into a thermostated vessel and the concentration of one or more species is measured as a function of time together with pressure. The dimensions of the vessel and reaction time have to be chosen such that the temperature can be kept constant, otherwise thermal acceleration of the reaction takes place.

Let us consider one example. It was found [1] that NO_2 and O_3 react and form N_2O_5 and O_2 according to $2NO_2 + O_3 \rightarrow N_2O_5 + O_2$. In the way described, the rate of consumption of O_3 was found to be

$$\frac{d[O_3]}{dt} = -k_3[NO_2][O_3] .$$

Apparently the equation $2NO_2 + O_3 \rightarrow N_2O_5 + O_2$ does not describe the mechanism. From these and other measurements it was found, that the mechanism can be represented by

$$\begin{array}{l} NO_2 + O_3 \xrightarrow{k_3} NO_3 + O_2 \\ NO_2 + NO_3 \rightarrow N_2O_5 \\ \hline 2NO_2 + O_3 \rightarrow O_2 + N_2O_5 \end{array}$$

and at 300 °K: $k_3 = 4.7 \cdot 10^7$ (cm³/mole·s) with $E_a = (7 \pm 0.6)$ (kcal/mole).

If other results would not have been available, one could have postulated another intermediate and another mechanism giving the same rate expression. In general one can say that it is practically not possible to prove a mechanism, it can only be made very probable using all kinds of information related to the chemical system.

A good deal of material available on reactions and reaction rates was obtained in the way just described. The development of new experimental techniques allows us today to apply more favourable conditions for the evaluation of rates of elementary reactions, and to look at many of these rate constants with more confidence. In order to obtain reliable data for rate constants of elementary reactions one should, as already mentioned, try to investigate isolated reaction steps under heat bath conditions. This means: the relaxation times τ (vibration, rotation, electronic) for all species appearing in the system should be very much shorter than the characteristic time for the chemical reaction

$$\tau_{\text{trans}}, \quad \tau_{\text{rot}}, \quad \tau_{\text{el}}, \quad \tau_{\text{vib}} \ll \tau_{\text{chem}} .$$

(One should realize, that these relaxation times are statistical quantities!) In addition the characteristic times for consecutive chemical reactions τ_{CR} should be longer than τ_{chem}.

The various types of elementary chemical reactions are (neglecting the reverse reactions):

$$A \to P \qquad \frac{\mathrm{d}[A]}{\mathrm{d}t} = -k_1[A], \qquad k_1 = \frac{\mathrm{d}\ln[A]}{\mathrm{d}t},$$

$$A + B \to P \qquad \frac{\mathrm{d}[A]}{\mathrm{d}t} = -k_2[A][B], \qquad k_2 = \frac{1}{[B]}\frac{\mathrm{d}\ln[A]}{\mathrm{d}t},$$

$$A + B + C \to P \qquad \frac{\mathrm{d}[A]}{\mathrm{d}t} = -k_3[A][B][C], \qquad k_3 = \frac{1}{[B][C]}\frac{\mathrm{d}\ln[A]}{\mathrm{d}t},$$

$$[B] = f(t), \ [C] = f(t) .$$

If in the two last mentioned cases $[B]$ and $[C]$ can be made so large compared to $[A]$, that their change during reaction remains negligible, one can write

$$k = -K\frac{\mathrm{d}\ln[A]}{\mathrm{d}t} \quad \text{with } K = \text{constant}$$

as long as one is sufficiently far away from equilibrium. One now has to measure $[A(t)]$ at constant pressure and temperature and either evaluate $\mathrm{d}(\ln[A])/\mathrm{d}t = \bar{f}(t)$ directly, or by plotting $\ln[A]$ as a function of time t. From the various $k(T)$ values the activation energy E_a is obtained from $E_a = RT^2 \cdot \mathrm{d}\ln k_a/\mathrm{d}T$, and if k is approximated by $k = A \cdot T^n \cdot \exp[-E_a/RT]$, one finds $E_a = E_0 + nRT$. It is easy to see, that very high accuracy of the measurements is needed in order to determine n, especially because E_0 is unknown.

The range over which rate constants can vary is shown by the following examples:

A typical rate for a chain reaction step is

$k = 10^{13} \exp\left[-\frac{10\,000}{RT}\right]$ (cm^3/mole·s):

T	0 °C	1000 °K
k	10^5	10^{12}
τ_{chem} ($P = 1$ atm)	10^{-1} s	10^{-7} s

There are unimolecular reactions with

$k = 10^{13} \exp\left[-\frac{60\,000}{RT}\right]$ s^{-1}:

T	700 °K	1900 °K	2600 °K
τ_{chem}	days	μs	10 ns

Rate constants can easily vary over many orders of magnitude in a rather limited temperature range. Therefore it becomes often necessary to apply different methods for the investigation of one reaction. In addition, the region of low temperature and long reaction time is very often not accessible due to wall reactions and impurities. An example for the possible influence of impurities is the dissociation of CO_2. (It is well known, that the CO oxidation is strongly influenced by H_2, H_2O and other admixtures.) If the dissociation proceeds via

$$CO_2 + M \xrightarrow{k_1} CO + O + M\,,$$

and impurities act via

$$CO_2 + H \xrightarrow{k_2} CO + OH \ldots \rightarrow H$$

then

$$\frac{d[CO_2]}{dt} = -k_1[CO_2][M] - k_2[CO_2][H] = -k_1[CO_2][M]\left(1 + \frac{k_2}{k_1}\frac{[H]}{[M]}\right).$$

Around 3 000 °K $k_1 \sim (10^7 \div 10^8)$ cm^3/mole·s and $k_2 \sim (10^{11} \div 10^{12})$ cm^3/mole·s therefore $k_2/k_1 \approx 10^3 \div 10^5$. If we now assume $[H]/[M] \approx 10^{-6}$, which means an impurity of 1 ppm, then

$$1 + \frac{k_2}{k_1}\frac{[H]}{[M]} = (1 + 10^{-3}) \text{ to } (1 + 10^{-1})\,.$$

The error in the rate constant under these conditions can be as high as 10%. It increases very rapidly towards lower temperatures, where it can hardly be eliminated because it is extremely difficult to establish a purity of better than a few ppm.

1. – Methods for measuring rate constants.

The methods to be applied to the measurement of the rates of neutral particle reactions in the gas phase depend strongly on the chemical components involved, on pressure and temperature to be applied, and on the order of magnitude of the rate constant.

Depending on how the concentration of a reaction partner $[A]$ is measured at various stages of the reaction, three groups of methods have been applied for the measurement of rate constants.

1) $[A]$ is measured as a function of time:
   a) static system: see the reaction $2NO_2+O_3$ mentioned above, this method is applied for slow reactions;
   b) method of adiabatic compression: reaction conditions are achieved by compression of the gas. This method is useful for medium temperatures up to 1500 °K and pressures up to 50 atm and reaction times above 1 ms;
   c) flash photolysis: reactants generated by light flash. Useful at low temperatures (below a few 100 °C) and low to medium pressures (up to some atm). Reaction times short, depending on system;
   d) relaxation techniques.

2) $[A]$ measured at various places in a stationary flow system:
   a) normal flow system: concentration measurements can be made accurately. Can be used up to more than 1000 °C. Reynolds number of the flow <2300. Lower pressure limit determined by diffusion processes. Reaction time $>10^{-3}$ s;
   b) shock waves: reaction conditions generated by shock wave, reaction time between $10^{-6}<\tau<10^{-3}$ s, reaction temperature up to several 1000 °K, pressure about 10^{-3} to 10^{+4} atm.

3) $[A]$ is measured after chosen time intervals:
   a) stirred reactor;
   b) chemical shock tube and adiabatic compression.

2. – Methods for measurements of concentrations.

For the measurement of concentration as a function of time many methods have been applied. Some of them shall be mentioned.

Direct optical methods: emission $(I \sim [A])$, fluorescence, absorption $(\ln I/I_0 \sim [A])$.

Indirect optical methods: added substances undergo reaction with one of the reaction partners. The reaction products emit light, *e.g.* $O+NO \rightarrow NO_2+h\nu$.

Catalytic probe and metal mirror for the measurement of atomc and radicals. (Requires stationary system.)

Mass spectrometer: requires sampling, preferably beam sampling. The only method which can be used for all species.

Electron spin resonance: requires stationary system, sampling.

Chemical methods: require sampling. If to be used for the measurement of radicals: scavenger technique.

3. – Reacting species.

The species, the reaction of which is to be investigated, can be grouped in two classes: 1) stable species; 2) radicals or atoms. (The excited ones are not considered here.) Stable species are in many cases available commercially. The purity of these substances is always a problem. It should be checked in every case.

For the formation of radicals there are many ways. Very common is the use of electric discharges or of photolytic methods. In both cases, one forms more than one kind of active species (except for homonuclear diatomics), and excited species. Sometimes titration reactions can be used for the generation of active species, *e.g.* $N+NO \rightarrow N_2+O$, or $N+NF_2 \rightarrow N_2+2F$. Sometimes it is possible to apply radical regenerating reactions running parallel to the process of interest. For atom generation the surface dissociation method is a relatively safe one.

Finally a few facts should be mentioned which can drastically obscure the results:

1) Parallel or consecutive reactions can hardly be avoided in most cases. Therefore the mechanism has to be evaluated; sometimes a procedure, rather lengthy but of special interest as far as chemistry is concerned.

2) The influence of wall effects has to be investigated very carefully. The usual method of packing the reaction vessel does not necessarily give the correct answer.

3) The calibration of the concentration measurements. For radicals it is difficult to obtain an absolute value of the concentration. Sometimes titration reactions or chemical equilibria are useful. It should be realized, that in some of the methods for concentration measurement the signal does not only depend on the particles of interest but also on other molecules which may be generated in the course of reaction (*e.g.* NO in ESR).

4. – Some examples.

Let us now consider a few examples and possible errors.

4·1. *Reaction of* N_2O. – At elevated temperatures this molecule dissociates in the gas phase. Final reaction products are N_2, O_2 and NO. It was of interest to know the rate of the reaction [3]

$$N_2O \rightarrow N_2 + O\,(^3P)\,.$$

In order to determine the lifetime of molecules ready for dissociation it was also necessary to reach the high pressure limit of unimolecular reaction. From earlier experiments the step $2N_2O \rightarrow 2N_2 + O_2$ could be excluded. The reaction $N_2O \rightarrow N + NO$ can also be ruled out on the basis of the known reaction enthalpy. One therefore has to worry about two things:

a) How large is the influence of the reaction $N_2O \rightarrow N_2 + O\,(^1D)$, which is not spin-forbidden, but would require more energy?

b) What are possible consecutive reactions and how do they influence the results? Some examples are

$$\begin{aligned} N_2O + O &\rightarrow N_2 + O_2\,, \\ NO + O + M &\rightarrow NO_2 + M\,, \\ NO_2 + N_2O &\rightarrow NO + N_2 + O_2\,, \\ O + O + M &\rightarrow O_2 + M\,. \end{aligned}$$

If the reaction takes place in the way mentioned it should be a unimolecular reaction following a second-order rate law at low pressures

$$\frac{d[N_2O]}{dt} = k[N_2O][M]\,,$$

where conditions are to be applied in which the concentration of an inert collision partner, [M], is large compared to $[N_2O]$. The rate law is checked by wide variations of the concentrations of N_2O and M, by varying P and T and by looking at the rate of disappearance of N_2O in each experiment. If the values $k = -(1/[M])\,d(\ln[N_2O])/dt$ plotted as $\ln k$ *vs.* $1/T$ fall reasonably well together it is very probable that the reaction follows the rate law of a unimolecular reaction in the low pressure range, as long as the Arrhenius parameters are reasonable compared to dissociation energy and collision number.

The search for other species like NO_2 was unsuccessful. It turned out, however, that apparently at very low N_2O concentrations ($< 10^{-3}\%$) the rate started to approach half its initial value. This can be explained by $O + N_2O \rightarrow 2NO$ or $\rightarrow N_2 + O_2$, which will double the rate if it is fast compared to the initial reaction. Measurements of that reaction step confirmed this assumption, so that all facts are in agreement with $N_2O + M \rightarrow N_2 + O(^3P)$.

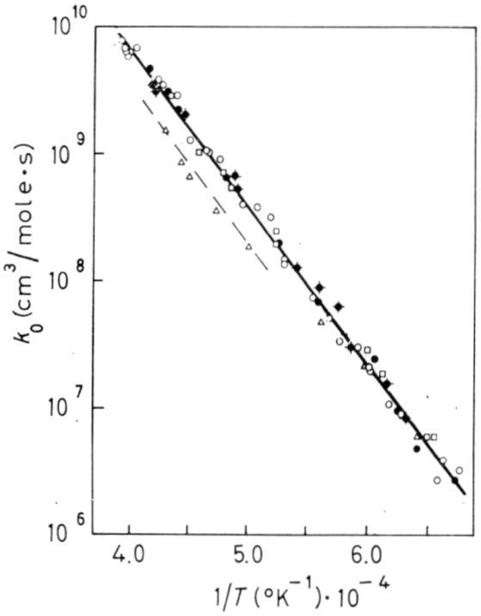

Fig. 1. – Rate constant k for the dissociation of N_2O in argon as a function of temperature at low densities (around 10^{-5} mole/cm^3).

At increased pressures the measured rate constants started to fall below the $\ln k$ *vs.* $-1/T$ curve (Figure 1), and in a plot of $\ln k$ *vs.* $\ln[M]$ the results shown in Fig. 2 were obtained. The dependence of the rate constant on M decreased as to be expected for a unimolecular reaction.

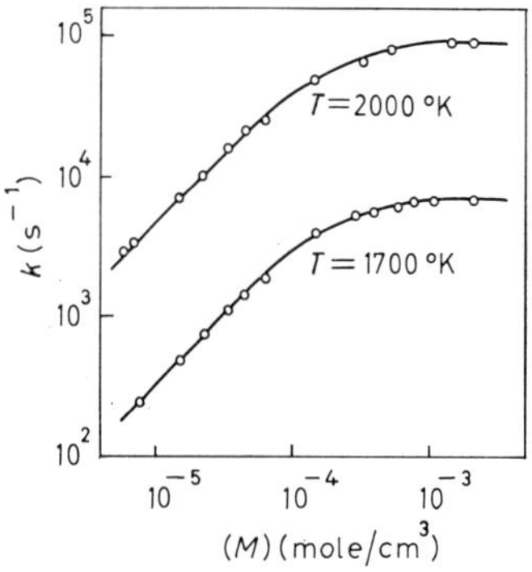

Fig. 2. – Pressure-dependence of the rate constant for the dissociation of N_2O.

Without going into the interpretation of these experiments the experimental procedure and sources of error shall be discussed. The experiments were performed in shock waves at temperatures between 1500 °K and 2800 °K at pressures from 1 to about 400 atm. The essential points in the experiments were:

a) Preparation of mixtures, check of purity.

b) Filling of the shock tube, check of leak rate, temperature, wall adsorption, cleanness of the shock tube walls.

c) Reaction conditions: check of initial pressure, shock velocity and attenuation, check of steady state, especially in reflected shocks.

d) Registration of signals: check of time constant, signal to noise ratio, linearity of system, time scale, quantitative calibration.

e) Data evaluation.

The measured quantities which finally enter into the result, the rate constant, are pressure, concentration, time, and temperature. The rate data obtained are approximated by $k = k_0 \exp[-E_a/RT]$. Pressure and temperature behind the shock front are calculated from its thermodynamic properties and the measured shock velocity and attenuation. Direct measurements of the temperature can hardly be made better than $\pm 50°$ at about 2000 °K. Therefore it might be, that the temperature values show a systematic deviation from the correct temperature which could cause a systematic deviation of the pre-exponential factor of nearly a factor of two in the case of N_2O, but without appreciably influencing the apparent energy of activation. In addition, the normal statistical error has to be taken into account. From the reproducibility of the experiments one estimates it to be about $(30 \div 50)\,\%$. From theory, and from the known characteristic temperatures one would expect a certain curvature of the $\ln k$ *vs.* $1/T$ curve, which one can extend over about 1000 °K if one includes measurements at low temperature and by adiabatic compression. The uncertainty in the measurements, however, allows now firm conclusion on the existence of that curvature.

In other systems, which are not as easy to handle as N_2O the situation is still less satisfying, especially if the reaction products are not known.

4·2. *Reaction of* O *atoms with* COS. – From thermodynamic considerations two reaction paths have to be considered for that system at low temperatures, namely

$$COS + O \xrightarrow{k_1} CO + SO \tag{1}$$

and

$$COS + O \xrightarrow{k_2} CO_2 + S\,. \tag{2}$$

Various methods were applied to investigate the reaction: flash photolysis (the O atoms were produced photolytically from NO_2), and flow methods. In the flow experiments the O atoms must be produced by a titration reaction $N + NO \rightarrow N_2 + O$ because O_2 could react with S or SO.

For the evaluation of the mechanism, the reaction was studied in a plug flow reactor coupled with a time-of-flight mass spectrometer through a molecular beam sampling system. It was important to use this type of sampling, because it allows the application of fairly high pressures in the reactor and avoids the sampling of particles from the boundary layer.

These experiments showed that the essential reaction path is $COS + O \rightarrow CO + SO$, and that $COS + O \rightarrow CO_2 + S$ plays a role only at high temperatures (around 1000 °C). In addition it could be shown that the reaction $SO + O + M \rightarrow SO_2 + M$ is very fast, and that also other reactions, *e.g.* $2SO \rightarrow S_2O_2$,

S_2O, SO_2, S, take place. The determination of these reactions will not be discussed here, it is, however, necessary to know these reactions in order to be able to find out the conditions to be applied for the measurement of k_1.

These measurements were performed in a plug flow reactor coupled to an electron spin resonance spectrometer which allows a very accurate determination of the O-atom concentration. Sampling was done continuously through the ESR cavity into a mass spectrometer for the analysis of stable products. Results are shown in Fig. 3 where $\ln k$ *vs.* $1/T$ is plotted.

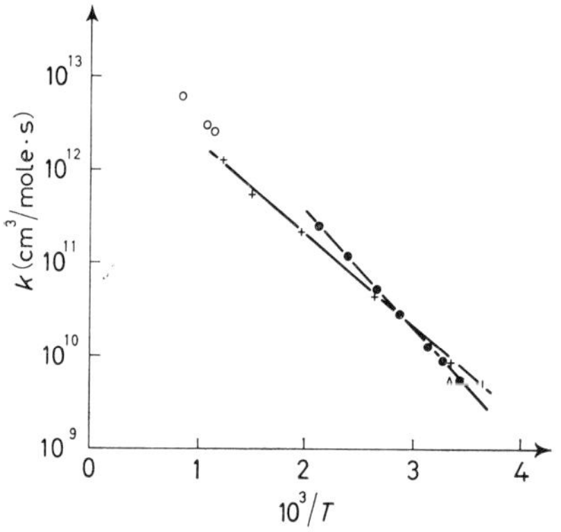

Fig. 3. – Rate constants for the reaction $O+COS \rightarrow CO+SO$ from various sources plotted against T^{-1}. △ [4]; ○ [5]; ● [6]; + [7].

What are the possible errors involved in these measurements?:

a) The purity of the substances: it is very laborious to produce extremely pure COS. Impurities can be CS_2, CO_2, and (if the substance is transported in steel containers) $Fe(CO)_5$.

b) Flow velocity in the reactor: the reaction has to be investigated at low reactant concentrations. The flow consists of carrier gas, COS, O and N_2. The flows of all these components have to be calibrated and the pressure has to be kept extremely constant in order to obtain the real reaction time.

c) Concentration measurements: the concentrations have to be calibrated empirically, and one has to check how the calibrations are influenced by the presence of other components, and by variation of temperature and pressure.

d) The gas temperature.

e) Influence of the wall: during the experiments the wall became coated with some substance which apparently changed its properties with respect to wall reactions.

The errors due to the points 1 to 5 can be kept reasonably small and one can find conditions so that the reaction follows a first-order rate law in the O-atom concentration ($[COS] \ll [O]$). The two limiting rate laws obtained by two groups [5-7] are

$$k_{1,1} = 1.2 \cdot 10^{14} \exp[-(5800/RT)] \text{ cm/mole}\cdot\text{s}$$

and

$$k_{1,2} = 1.9 \cdot 10^{13} \exp[-(4530/RT)] \text{ cm}^3/\text{mole}\cdot\text{s} .$$

Around room temperature the agreement lies within the limits of errors quoted by the two groups. At high temperature, however, the discrepancy becomes larger. The reason for the discrepancy is not yet clear. It could be, that at high temperatures in Westenberg's experiments the influence of the wall is stronger than what he estimates. Our own results in the early state of the experiments, obtained at low pressures and high temperatures, were also lower than the final ones, where the wall influence could be reduced very much.

Nevertheless, compared to rate constants obtained by various authors in the past, the agreement between the results of the two groups is good. It is, however, not yet good enough to draw any conclusion on a temperature-dependent pre-exponential factor.

4·3. *The reaction* $CH_4+OH \rightarrow CH_3+H_2O$. – This reaction may be considered as an example for a reaction which can be favourably investigated by flash photolysis combined with kinetic spectroscopy [8].

Hydroxyl radicals were generated by flash photolysis of H_2O in presence of CH_4: $H_2O+h\nu \rightarrow OH(^2\Pi)+H(^2S)$. The concentration of OH is measured at various reaction times Δt, using the absorption spectrum of OH. As the CH_4 concentration was sufficiently high, k could be obtained from

$$k_1 = -\frac{\Delta \ln[OH]}{[CH_4]\Delta t} = 5.3(\pm 0.2)\cdot 10^9\ \text{cm}^3/\text{mole s}\ .$$

The corresponding value reported by another group is $1.7\cdot 10^{10}$ cm³/mole s. What are the possible errors?

a) One has to be sure that the only photochemical reaction is the photolysis of H_2O. This can be arranged with a spectrosil reaction cell, since CH_4 absorbs only below the transmission limit of the cell walls.

b) Possible secondary reactions have to be considered

$$OH+OH \rightarrow H_2O+O\ , \tag{2}$$

$$OH+OH+M \rightarrow H_2O_2+M\ , \tag{3}$$

$$OH+H+M \rightarrow H_2O+M\ , \tag{4}$$

$$H+CH_4 \rightarrow CH_3+H_2\ , \tag{5}$$

$$OH+CH_3 \rightarrow CH_3OH\ , \tag{6}$$

and less important

$$CH_3+CH_3+M \rightarrow C_2H_6+M\ , \tag{7}$$

$$OH+C_2H_6 \rightarrow C_2H_5+H_2O\ . \tag{8}$$

The decrease of OH concentration caused by the reactions (2), (3), (4) can be determined by runs without CH_4: Using partial pressures of OH between $(10^{-2} \div 10^{-3})$ Torr which are detectable with sufficient accuracy by optical absorption and CH_4 partial pressures between $(10 \div 100)$ Torr, reaction (1) is 10 to 100 times faster than reaction (2). Reaction (5) should not produce appreciable amounts of CH_3 radicals since $k_5 = 10^5$ cm³/mole·s at room temperature. However, H_2O photolysis will produce H atoms with a high amount of kinetic energy, which may react with CH_4 even at 300 °K.

Nothing is known about the reaction (6). Taking a fairly high rate constant for (6) $k_6 = 10^{12}$ cm³/mole·s, a ratio $[CH_4]/[OH]_0 > 10^4$ (which can be chosen as mentioned before) is sufficient to ensure that only k_1 will be observed.

c) Because of the large excess of the reactant CH_4, this has to be very pure since reactions between OH and hydrocarbons, such as C_2H_2, C_2H_4, C_2H_6, are several orders faster than (1).

d) Concentration measurements by absorption involve some problems. During flash duration the concentration of the interesting species is not only determined by reactions but also by photolysis. The concentration of OH is related to its absorption by the well known expression (I = intensity of light, αx absorption coefficient for a cell of length x) $I = I \exp[-\alpha x [OH]]$, and the rate constant is obtained from $\Delta \ln [OH]$. One can easily calculate the error $\Delta \ln [OH]$ for a certain error in I/I_0. Especially if the spectra are recorded on film one has to work very carefully in order to avoid drastic errors.

4·4. *Reaction of* O *with* C_2H_2. – Let us finally consider the reaction of O atoms with C_2H_2 which is of great practical importance in combustion processes [9, 10]. This reaction could proceed in the following ways

$$O + C_2H_2 \rightarrow OH + C_2H\,, \quad HC_2O + H\,, \quad C_2 + H_2O\,, \quad C_2O + 2H\,,$$

respectively $CH_2 + CO$, $HCO + CH$.

Some of these reactions can be ruled out by thermodynamic considerations. However, in every case at least one of the products is a radical and the thermodynamic properties of these radicals are not yet well established. In addition these radicals probably react very fast with O atoms, thus obscuring the results.

For a long time it was assumed, that the first step is $O + C_2H_2 \rightarrow C_2H + OH$.

Results obtained in a flow system with ESR and mass-spectrometric detection showed: if $[C_2H_2]$ was not much higher than [O] it was found *a*) $d[O]/dt = -d[H]/dt$; *b*) for each C_2H_2 consumed two CO appeared but no CO_2.

On the basis of these two results one can rule out some of the above-mentioned initial steps. Further experiments showed that for each C_2H_2 two O atoms were consumed. This leaves

1) $$O + C_2H_2 \rightarrow CO + CH_2\,,$$
$$CH_2 + O \rightarrow CO + 2H\,,$$

2) $$O + C_2H_2 \rightarrow HC_2O + H\,,$$
$$O + HC_2O \rightarrow 2CO + H\,,$$

3) $$O + C_2H_2 \rightarrow C_2O + 2H\,,$$
$$C_2O + O \rightarrow 2CO\,,$$

and

4) $$O + C_2H_2 \rightarrow CO + CH + H\,,$$
$$CH + O \rightarrow CO + H\,.$$

All of these four mechanisms contain steps where three particles are generated in one step, which, due to chemical experience, is not very probable. For the first mechanism a strong rearrangement in the « activated complex » would be necessary.

If the concentration of C_2H_2 becomes very high compared to the O concentration, C_3H_4 can be found as reaction product and the number of H atoms produced per O atom consumed decreases. These observations can hardly be explained by the mechanisms two and three. Unfortunately, CH_2 or CH cannot be measured directly. Therefore the mechanisms one and four remain as main reaction patterns.

The measurement of the O:H ratio at high C_2H_2 concentrations is consistent with mechanism one. This is a weak argument because H recombines rapidly in the presence of C_2H_2. However the first step in mechanism four is endothermic with about 54 kcal/mole, and from experiments a value for the rate constant between 250 and 680 °K of

$$k = 1.2 \cdot 10^{13} \exp[-3\,000/RT]\ \text{cm}^3/\text{mole}\cdot\text{s}$$

was obtained. This confirms, that mechanism one is the most probable one. It is, however, not the only one. Investigations of chemiluminescence of CH and CO in that system show that there are also other reaction channels, through which a small part of the chemical change takes place.

This last example characterizes the situation which normally arises when rate constants and mechanisms are evaluated in systems of chemical interest.

It is a rather simple one, but it clearly shows that it is necessary to apply various experimental methods and to seek for the most favourable conditions for the reaction in order to make a mechanism probable and to measure a rate constant which can be related to a certain elementary reaction step.

* * *

The author acknowledges the collaboration of U. KOLLER and B. SCHIMPKE in preparing a first draft of this lecture.

REFERENCES

[1] H. S. JOHNSTON: *Gas Phase Reaction Rate Theory* (New York, 1966).
[2] S. W. BENSON: *The Foundations of Chemical Kinetics* (New York, 1960).
[3] J. TROE and H. GG. WAGNER: *Ber. Bunsenges. Phys. Chemie*, **71**, 937 (1967).
[4] J. O. SULLIVAN and P. WARNECK: *Ber. Bunsenges. Phys. Chemie*, **69**, 7 (1965).
[5] K. HOYERMANN, H. GG. WAGNER and J. WOLFRUM: *Ber. Bunsenges. Phys. Chemie*, **71**, 603 (1967).
[6] K. H. HOMANN, G. KROME and H. GG. WAGNER: *Ber. Bunsenges. Phys. Chemie*, **72**, 998 (1968).
[7] A. A. WESTENBERG and N. DE HAAS: private communication.
[8] N. R. GREINER: *Journ. Chem. Phys.*, **46**, 2795 (1967).
[9] J. M. BROWN and B. A. THRUSH: *Trans. Faraday Soc.*, **63**, 630 (1967).
[10] K. HOYERMANN, H. GG. WAGNER and J. WOLFRUM: *Zeits. Phys. Chem. N. F.*, **63**, 193 (1969).

Experiments on the H_3 System.

F. S. ROWLAND

University of California - Irvine, Cal.

1. – Introduction.

The atom-molecule chemical reaction most amenable to theoretical calculation is the atomic exchange of H with H_2 through an intermediate involving all three H atoms simultaneously:

$$H + H_2 \rightarrow H_2 + H . \tag{1}$$

The detailed calculations of potential energy surfaces for H_3 and trajectory calculations on such surfaces have been described by other lecturers in this series. Therefore, our attention is given here to laboratory experiments which furnish information about the chemical reactions of hydrogen atoms with molecules. The basic reaction (1), of course, does not give any detectable change in products, and has not been directly investigated. However, variants of reaction (1) have been performed which do lead to measurable chemical change through the choice of suitable labeling techniques: *a*) molecular spin labeling, and *b*) isotopic labeling.

The homonuclear molecular hydrogens, H_2 and D_2, exist in two molecular spin varieties, *ortho* and *para*, in an equilibrium ratio varying with temperature—for H_2, from *ortho*/*para* = 3 at high temperature down to = 0 at 0 °K. Since these spin varieties are found *not* to exchange identity rapidly in the absence of catalytic agents, and do have some measurably different physical properties, nonequilibrium distributions can be produced, maintained, and measured, and reactions such as

$$H + p\text{-}H_2 \rightarrow o\text{-}H_2 + H \tag{2}$$

can be quantitatively studied. In the earliest H_3 experiments, FARKAS followed this reaction through measurement of the heat capacity of the product mixture, the heat capacities of o-H_2 and p-H_2 being different [1].

The other basic labeling approach utilizes the existence of many possible combinations of the three isotopes of hydrogen, H, D, and T. The radioactive isotope, T, cannot be conveniently used as the bulk molecular hydrogen species because of the highly radioactive nature (and accompanying rapid radiation-induced isotopic exchange reactions) of pure HT, DT, or T_2. Thus, the isotopic labeling experiments so far have been largely limited to H_2, HD, and D_2 as molecular species, with H, D, or T as the atomic species. The contributions of triatomic hydrogen species to the exchange reactions of molecular hydrogen can be evaluated through such use of isotopic labeling, and the initial observations of such exchange also relied upon the differences in thermal conductivity between the product HD and the reactants $H_2 + D_2$ [2]. BOATO *et al.* introduced mass-spectrometric analysis of the isotopic composition into their study of the H_2-D_2 exchange reaction near 1000 °K, thereby greatly improving the precision of analysis [3]. Additional new detection techniques have been applied in recent years for the stable isotopes, while radio gas chromatography has been regularly used for the determination of HT and DT.

The total number of H_3 experiments in the three decades after the first experiments of FARKAS was not very large [1-5], and the Farkas and Boato experiments have provided the experimental basis against which various theoretical approaches were usually compared until the early 1960's. In the last few years, however, many additional experiments have been performed, and both the quality and quantity of experimental knowledge has vastly improved. These advances have been accomplished by two basic changes in the experiments: first, the analytical techniques now available permit much greater precision in measurement over much wider temperature ranges by a variety of techniques; and, second, experiments have been extended to nonthermal (*i.e.* hot atom) systems, providing kinetic information of a kind not available from thermal experiments. The advances are so great, in fact, that we shall limit our discussion of quantitative results to the experiments of the last few years. I shall also not discuss molecular beam experiments, since the beam techniques have been covered thoroughly in other lectures.

2. – Experiments in thermal systems.

Thermal experiments of high precision have now been performed for reactions (2) to (5), with extensive sets of data for (4) and (5) by different groups [6-10]:

$$D + o\text{-}D_2 \to p\text{-}D_2 + D\,, \tag{3}$$

$$D + H_2 \to HD + H\,, \tag{4}$$

$$H + D_2 \to HD + D\,. \tag{5}$$

The agreement between these two sets of data is reasonably good, as shown in Fig. 1, although discrepancies do appear at lower temperatures. While the absolute agreement on rate constants at a given temperature is never in gross error, the deviations are systematic and lead to substantial variations when compared in the form of the Arrhenius parameters, *i.e.* A and E in the equation $k = A \exp[-E/RT]$. The values quoted for reaction (5)—the data shown in Fig. 1—are: $k_5 = 4.3_7 \cdot 10^{12} \cdot \exp[-7300/RT]$, ref. [6]; and $k_5 = 4.9 \cdot 10^{13} \cdot \exp[-9390/RT]$, ref. [10]. Clearly, the activation energy for the reaction is in the (7÷9) kcal/mole range, but more accurate estimates will require elimination of systematic errors in one or both of the experiments. The higher rate for reaction (4) than (5) has been rationalized in terms of an expected more rapid decomposition of linear DHH complexes *vs.* that of HDD complexes.

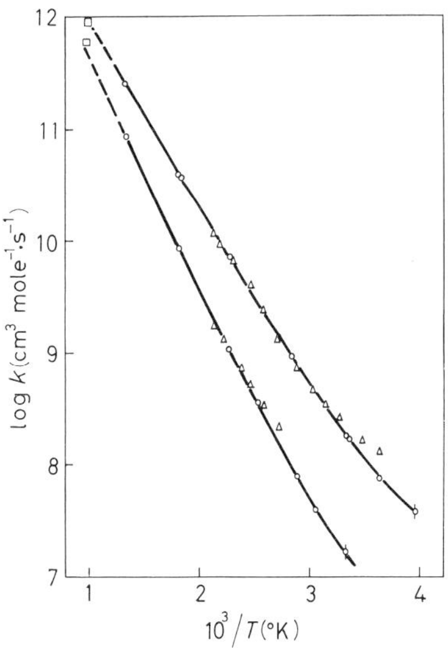

Fig. 1. – From ref. [10]. Arrhenius plots of experimental data for $D+H_2 \rightarrow HD+H$ (upper) and $H+D_2 \rightarrow HD+D$ (lower). o, ref. [10]; △, ref. [6] for $H+D_2$ and [7] for $D+H_2$; □, ref. [3].

The possible presence of systematic upward deviations at low temperature has great theoretical interest, since such a deviation from the straight line can be interpreted as a sign of quantum-mechanical tunneling through the H_3 potential barrier. The EPR measurements of atomic concentration seem less susceptible to complications in the region of discrepancy, and do not indicate any evidence for such tunneling. Additional work in the low-temperature region is obviously called for because of this focus of interest.

3. – Photochemical hot experiments.

Experimental data of a very different kind can be obtained through the study of the reactions of « hot » hydrogen atoms with thermal hydrogen molecules. When light is absorbed by either DBr or DI (or in the isotopic analogs: HBr, HI, TBr, TI), the excess photochemical energy beyond that required to rupture the D-X bond appears as kinetic energy of the two atoms (*);

(*) About half of the absorption in hydrogen iodide molecules leads to electronically excited iodine atoms ($^2P_{\frac{1}{2}}$), with a corresponding reduction in the available kinetic energy. The yield of excited bromine atoms is very small ($< 2\%$) following absorption in hydrogen bromide. The yields of electronically excited halogen atoms are somewhat wavelength-dependent.

conservation of momentum requires that almost all of this kinetic energy be carried by the low-mass hydrogen atoms. Such photolytically formed D atoms are not in thermal equilibrium with the surroundings, and may be capable of reacting with H_2 which is at temperatures far below those shown in Fig. 1, if the « hot » atoms themselves possess sufficient excess kinetic energy [11, 14]. The principle of such experiments is simple, even if the actual experiments are not—if the kinetic energy of the « hot » D atom is insufficient to cause reaction with H_2, no HD will be observed. However, if the kinetic energy *is* sufficient, then reaction (4) will be observed.

The chemical reactions expected for this system are these given in (4*, 6-8), with the I atoms removed by combination to form I_2. (An entirely parallel set of reactions occurs for DBr with H_2, or HBr/HI with D_2).

(6) $$DI \xrightarrow{h\nu} D^* + I\,,$$

(4*) $$D^* + H_2 \rightarrow HD + H\,,$$

(7) $$D^* + H_2 \rightarrow D + H_2\,,$$

(8) $$D + DI \rightarrow D_2 + I\,.$$

Since the H atom from (4*) will react by (9), two molecules of HD are expected for each reaction via (4*), while one molecule of D_2 will be formed by reaction (8) if the hot atom is scattered non-reactively, as in (7). Furthermore since the experiment cannot be carried out without DI, the reactions of D* with DI must also be considered, with the conclusion that some D_2 will be formed by (8*), as well as by (8):

(9) $$H + DI \rightarrow HD + I,$$

(8*) $$D^* + DI \rightarrow D_2 + I\,,$$

(10) $$D^* + DI \rightarrow D + DI\,.$$

The clean distinction between « hot » reactions, forming HD, and thermal reactions forming D_2, is blurred if some of the hot reactions *also* form D_2. This difficulty is neatly circumvented by measurement of the (HD)/(D_2) ratio at many concentrations of (DI)/(H_2), and then extrapolation of the (HD)/(D_2) ratio to the intercept for zero concentration of DI. Such a plot is shown in Fig. 2 for the reactions of 2.9 eV D* atoms with H_2, and readily yields data of the kind summarized in Table I. It is quite clear that an important fraction of these H and D atoms react without thermalization, and that the fraction of hot reaction decreases as the initial energy of the hot atom decreases.

While these experiments involve D* atoms with substantial excess kinetic energy, the choice of still different photolytic wavelengths offers the possibility that excited D atoms might be produced which are below the energy threshold for reaction. Such a threshold experiment has many inherent difficulties—the problems that crop up as one tries to measure the yield of a reaction as the yield approaches closer and closer to zero. Nevertheless, KUPPERMANN and WHITE have made a preliminary report on just such an experiment, and have estimated the threshold energy to be (0.33 ± 0.02) eV [15].

In principle, careful measurement of the fraction of hot reaction *vs.* initial kinetic energy of the reacting D* atom can provide as well information about the reaction cross-section as a function of D* atom energy. The photolytic process itself provides a highly monoenergetic group of D* atoms, and one can contemplate converting a series of measurements of reaction yield *vs.* energy directly into reaction cross-sections. However, the apparent simplicity of this determination of reaction cross-sections from such reactions vanishes upon detailed examination, and a procedure of considerable complexity remains.

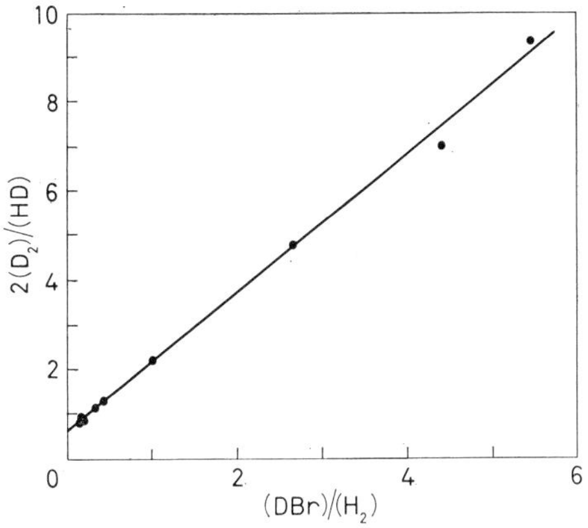

Fig. 2. – Data from ref. [14]. Ratio of $(D_2)/(HD)$ *vs.* $(DBr)/(H_2)$ for reactions of 2.9 eV D atoms with H_2.

The first complication is this—the experiments do not involve a crossed-beam, single-collision arrangement. Consequently, a very energetic atom may

TABLE I.

Atom	Source	Photolytic wavelength (Å)	Initial energy (eV)	Reactant molecule	Fractional yields		Ref.
					Hot	Thermal	
D	DI	2537	1.8, 0.9	H_2	0.42	0.58	[13]
D	DBr	1849	2.9	H_2	0.62	0.38	[14]
H	HBr	1849	2.9	D_2	0.61	0.39	[14]

be scattered in the first, nonreactive interaction to a new energy at which it is still above the threshold energy, and then react in the second interaction. Or in the third; or the fourth. The experimental observations are thus transformed from *a*) an in-principle differential measurement at an energy specified by the photochemistry, into *b*) an integral measurement, with the integration being performed from the initial energy down to the threshold energy. This is, in itself, not critical for the determination of the energy threshold, involving only the parametric form of the approach to zero yield. It is, however, exceedingly important when the interest is extended to the shape and magnitude of cross-section curves above the threshold. The unfolding of a curve of percentage « hot » yield *vs.* initial D atom energy obviously requires accurate knowledge of the non-reactive scattering processes for the D atom [16].

A second major complication in these experiments is the substantial influence of the thermal spread in the energies of the reaction partners. This effect is nontrivial even for energies of the hot atom in the several electron volt range, and for thermal collision partners substantially heavier than the molecular hydrogens. Consider the collisions of 0.36 eV D atoms with H_2 molecules of $\sim$0.04 eV thermal energy. Since the velocities of each are proportional to the square root of the energies (the masses of D and H_2 being equal), the velocities of $0.6k$ for D and $0.2k$ for H_2 will combine to give relative velocities ranging from $0.4k$ to $0.8k$, *i.e.* energies from 0.16 eV to 0.64 eV, *despite* the monoenergetic nature of the D* atom in the laboratory system. The spread in energies is still severe with higher initial D* atom energies (*e.g.* (1.5 ÷ 2.6) eV for 2.0 eV initial laboratory energy), and even the *first* collisions following photolysis have almost completely erased the monoenergetic differential possibilities suggested in the initial description. The presence of *both* the thermal spread in first-collision energies, and the multiple-collision spread for subsequent collisions insures that any appreciable sharpening of the differential nature of these measurements will require simultaneous attention to both.

One important consequence of the thermal spread in first-collision energies is that reactions will be experimentally observed for D* atom laboratory energies below the real threshold. Note for instance, that an appreciable fraction of initial collisions in a 0.36 eV D* atom system would occur at relative energies *above* 0.5 eV—if the threshold energy for stationary H_2 molecules (*i.e.* 0 °K) were 0.5 eV, some reactions would still be experimentally observed in room temperature H_2 for an initial energy of D* considerably below 0.5 eV. The extrapolation to a phenomenological experimental threshold will thus go to a value below the actual threshold in the absence of thermal motion, and corrections for the thermal spread are imperative in this situation. Again, since reaction in collisions after the first is much less important near the threshold, lowering the temperature of the target molecule *can* reduce the discrepancy between the phenomenological threshold and the threshold for hypothetical stationary target molecules.

The actual experiments of KUPPERMANN and WHITE have other complications of a technical nature—HI impurity in DI; HD impurity in H_2; low photon absorption cross-sections for DI at the longer wavelengths; etc. Their preliminary report of the threshold energy for reaction (4*) is illustrated in Fig. 3. A small fraction of thermalized D atoms react with the HI impurity, forming HD, instead of with DI to form D_2. Photolysis of DI-He mixtures, also extrapolated to zero concentration of DI, provides a measure of this thermal DH, and the values given in Fig. 3 represent the difference between experiments in H_2 and in He, expressed as

$$\Delta = \left(\frac{HD}{D_2}\right)^{H_2}_{DI=0} - \left(\frac{HD}{D_2}\right)^{He}_{DI=0}.$$

KUPPERMANN and colleagues have since collected more accurate data over an extended range of energies. The procedures for extracting cross-section information from such data have been discussed, and the work is now in progress [16-17].

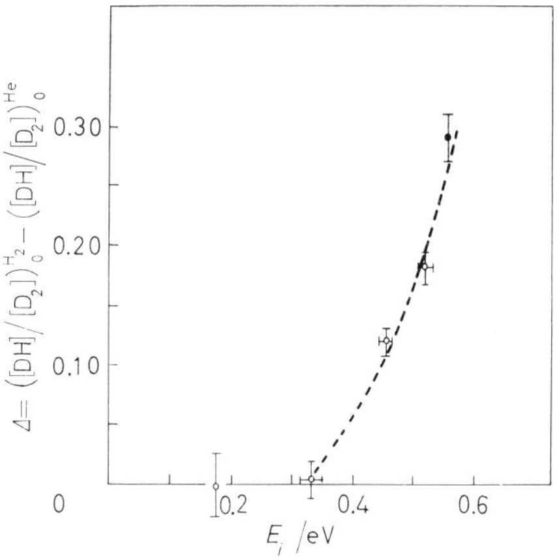

Fig. 3. – From ref. [15]. Effect of average relative initial kinetic energy of reactants of $D + H_2 \rightarrow HD + H$ reaction. Open circles: photolysis of DI mixtures at 3660, 3340, 3130 and 3030 Å, in order of increasing energies. Full circle: photolysis of DBr at 2537 Å, ref. [14].

4. – Experiments with recoil tritium atoms.

Further information about high kinetic-energy reactions in H_3 systems has come from the study of energetic recoil tritium reactions with H_2, D_2, and HD. Such energetic atoms can be formed by the photolysis of TBr, as described above, or by recoil from a nuclear reaction such as $^3He(n, p)^3H$. In the latter case, the recoiling tritium atom is created with 192 000 eV kinetic energy, and must degrade down to the $(10 \div 100)$ eV region before a bond-forming interaction can occur. The experimental and theoretical bases of such experiments have been discussed in detail in the lectures on hot-atom chemistry.

Absolute yield measurements indicate that $> 90\%$ of the tritium atoms from nuclear recoil react by atomic exchange with either H_2 or D_2, forming HT and DT, respectively [18, 19]. Competitive experiments with equimolar H_2-D_2 mixtures show a preference for reaction with the former by a factor of 1.5 ± 0.1 [20, 19].

KARPLUS, PORTER, and SHARMA have calculated reaction cross-sections by trajectory methods for both $T^* + H_2$ and $T^* + D_2$ for energies up to 200 eV, as shown in Fig. 4 [21]. Since the reaction cross-sections with H_2 are larger than for D_2 for all of the higher energies, the trajectory calculations are qualitatively in agreement with the observed ratio of $(HT/H_2)/(DT/D_2)$. Another, perhaps more sensitive, comparison involves the matching of a predicted total hot yield to the experimentally observed values of $(94 \pm 3)\%$. This comparison requires reasonable models for the elastic and inelastic nonreactive scattering cross-sections of T with H_2 and D_2 over the energy range to 200 eV; no results from such a detailed comparison have yet been published.

Close examination of the low-energy segment of Fig. 4 shows that the H_2 and D_2 cross-sections are approximately equal below several eV—and an expanded diagram of this low-energy portion would show a crossover at about $(1 \div 2)$ eV, with the D_2 cross-section *higher* than H_2 at the very lowest energies. This prediction of a low-energy ratio of HT/DT of unity or less has been confirmed by the experimental observation of $(HT/H_2)/(DT/D_2) = 0.99 \pm 0.03$ for photochemically produced 2.8 eV T* atoms [22]. At the same time, an experimental measurement was made of the intramolecular isotope effect in the 2.8 eV T* formation of HT or DT by reaction of 2.8 eV T* with HD, and a preference was found for bonding to the heavier isotope (HT/DT = 0.7) [22]. A similar effect $(HT/DT = 0.62 \pm 0.06)$ has also been found for nuclear recoil tritium [23]. Comparable trajectory calculations agree on the preference for the formation of DT over HT, and are reported to show, in addition, some « structure » in the reaction cross-section curves [24].

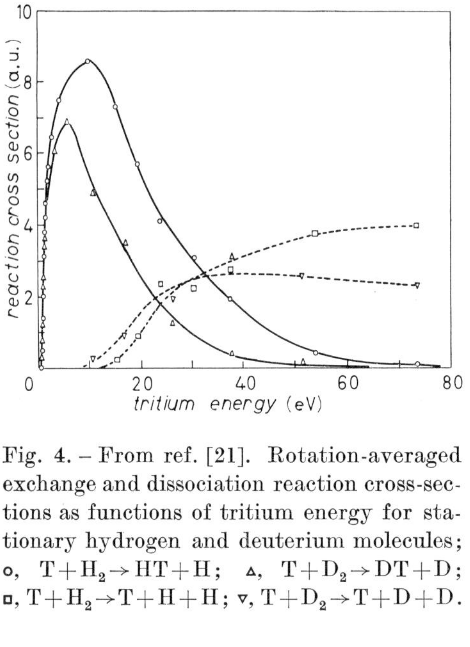

Fig. 4. – From ref. [21]. Rotation-averaged exchange and dissociation reaction cross-sections as functions of tritium energy for stationary hydrogen and deuterium molecules; ○, $T + H_2 \rightarrow HT + H$; △, $T + D_2 \rightarrow DT + D$; □, $T + H_2 \rightarrow T + H + H$; ▽, $T + D_2 \rightarrow T + D + D$.

Finally, we can consider the moderation of the kinetic energy of hot atoms. Progressive dilution of H_2-D_2 mixtures with larger and larger amounts of an inert gas such as argon or neon gradually reduces the hot yield of HT and DT from nuclear recoil experiments. This reduction in hot yield permits a larger percentage of the tritium atoms to survive to lower energies, and thus alters the sampling of different energy ranges and is reflected in the ratio of HT/DT yields. Extrapolation to infinite dilution (and hence zero hot yield) provides a « limiting » estimate of the relative magnitude of the « reactivity integrals » for H_2 and D_2.

A reactivity integral [25], defined as $I = \int_{E_1}^{E_2} p(E)\,d\ln E$ with E_1 and E_2 the ranges of possible chemical reaction, is a useful quantity for the description of probability of reaction curves of the general kind indicated in Fig. 4. The choice of the logarithmic energy scale reflects both the unusually wide range of energies that must be covered for hot atom reactions, and also the fact that fractional energy losses in nonreactive collisions are more nearly constant than absolute energy losses over this energy range. The reactivity integral, so defined, thus represents an attempt to weight the importance of various energy ranges in proportion to the number of collisions likely to occur in that energy range. The experimental observation is that the ratio of HT/DT from H_2-D_2 mixtures approaches the value 1.15 ± 0.04 with progressive dilution in argon in the presence of I_2 scavenger [19]. Direct measurement under the curves of Fig. 4, when plotted on a logarithmic basis and corrected for thermal motion of the target molecules, gave an estimate of 1.37 for the same ratio [21].

This disagreement seems quite substantial and worth further consideration, since the measured isotope effect is only 40% as large as the calculated one. The calculated value would not represent the ratio of reactivity integrals if the cross-section curves are in serious error, and especially if they were subject to a systematic mass-dependent error. The experimentally measured value might not be a valid measurement of the ratio of reactivity integrals *a*) if the experiments are in error, or *b*) if the nonreactive scattering of T atoms in argon were to deviate substantially from the hypothesis of constant logarithmic energy loss. No separate measurements of T-Ar nonreactive scattering from $(0 \div 200)$ eV are available.

Qualitatively, the direction toward agreement is obviously that the real argon-moderated system appears to sample much more heavily from regions of equal reaction probability than does the logarithmic weighting technique applied to the curves of Fig. 4. Both could be in agreement if the fractional energy loss of T in argon changed with energy, being substantially smaller in the low-energy region. Actually, of course, if the logarithmic-energy-loss hypothesis is in serious error in most systems, no useful significance would then be attached to the concept of the reactivity integral as presently defined. Nonetheless, the experimental measurements of the relative magnitudes of these reactivities in excess argon, or in other materials, would still provide a useful basis for intermolecular comparisons.

In summary of all of the experiments described in this lecture, far more data are now available on H_3 systems than were available five years ago, and direct continuation of currently active experiments promises even more information of several varieties in the next several years. The prospects for more and more detailed comparison of theory and experiment in the H_3 system are excellent.

REFERENCES

[1] A. Farkas: *Zeits. Elektrochem.*, **36**, 782 (1930); *Zeits. Phys. Chem. (Leipzig)*, B **10**, 419 (1930); B **14**, 371 (1931).
[2] A. Farkas and L. Farkas: *Proc. Roy. Soc.*, A **152**, 124 (1935).
[3] G. Boato, G. Careri, A. Cimino, E. Molinari and G. G. Volpi: *Journ. Chem. Phys.*, **24**, 783 (1956).
[4] K. Geib and P. Harteck: *Zeits. Phys. Chim.*, Bodenstein Festband, 849 (1931).
[5] M. van Meersche: *Bull. Soc. Chem. Belge*, **60**, 99 (1951).
[6] W. R. Schulz and D. J. LeRoy: *Canad. Journ. Chem.*, **42**, 2480 (1964).
[7] W. R. Schulz and D. J. LeRoy: *Journ. Chem. Phys.*, **42**, 3869 (1965).
[8] B. A. Ridley, W. R. Schulz and D. J. LeRoy: *Journ. Chem. Phys.*, **44**, 3344 (1966).
[9] D. J. LeRoy, B. A. Ridley and K. A. Quickert: *Disc. Far. Soc.*, **44**, 92 (1967).
[10] A. A. Westenberg and N. de Haas: *Journ. Chem. Phys.*, **47**, 1393 (1967).
[11] R. A. Ogg jr. and R. R. Williams jr.: *Journ. Chem. Phys.*, **13**, 586 (1945).
[12] H. H. Schwarz, W. H. Hamill and R. R. Williams jr.: *Journ. Amer. Chem. Soc.*, **74**, 6007 (1952).
[13] R. J. Carter, W. H. Hamill and R. R. Williams jr.: *Journ. Amer. Chem. Soc.*, **77**, 6457 (1955).
[14] R. M. Martin and J. E. Willard: *Journ. Chem. Phys.*, **40**, 2999, 3007 (1964).
[15] A. Kuppermann and J. M. White: *Journ. Chem. Phys.*, **44**, 4352 (1966).
[16] A. Kuppermann, J. Stevenson and P. O'Keefe: *Disc. Far. Soc.*, **44**, 46 (1967).
[17] A. Kuppermann: private communication.
[18] J. W. Root and F. S. Rowland: *Journ. Chem. Phys.*, **38**, 2030 (1963); **46**, 4299 (1967).
[19] D. Seewald, M. Gersh and R. Wolfgang: *Journ. Chem. Phys.*, **45**, 3870 (1966).
[20] J. K. Lee, B. Musgrave and F. S. Rowland: *Journ. Chem. Phys.*, **32**, 1266 (1960).
[21] M. Karplus, R. Porter and R. Sharma: *Journ. Chem. Phys.*, **45**, 3871 (1966).
[22] C. C. Chou and F. S. Rowland: *Journ. Chem. Phys.*, **46**, 812 (1967).
[23] D. Seewald and R. Wolfgang: *Journ. Chem. Phys.*, **46**, 1207 (1967).
[24] R. Porter: private communication.
[25] R. Wolfgang: *Progress in Reaction Kinetics*, **3**, 97 (1965).

Experiments on Unimolecular Reactions.

F. S. ROWLAND

University of California - Irvine, Cal.

1. – Introduction.

The standard experimental kinetic approach to thermal chemical systems involves as one primary goal the determination of the molecularity of the reaction. A finding that the rate of reaction of A with B is proportional to the concentration of A, and to the concentration of B, is a satisfying confirmation of a bimolecular mechanism, first order in each component, and is readily conceived as related to collisions of A with B. The discovery of apparently unimolecular reactions, first order in one species and otherwise independent, has been an important facet of the development of chemical kinetics over the past 50 years, and continues to play a very important role.

The fundamental obstacle to the riddle of a first-order reaction from the collision of *two* molecules was solved by LINDEMANN [1], who formulated the concept of excited molecules, A*, capable either of reaction or of collision-induced stabilization, as in reactions (1) to (3):

$$A + A \xrightarrow{k_1} A + A^* , \tag{1}$$

$$A^* + A \xrightarrow{k_2} A + A , \tag{2}$$

$$A^* \xrightarrow{k_3} P \text{ (Product)} . \tag{3}$$

Assuming that A* is short-lived, and that (dA^*/dt) is therefore $\simeq 0$ at all times, an expression for A* can be obtained, as in (4),

$$0 = \frac{dA^*}{dt} = k_1 A^2 - k_2 AA^* - k_3 A^* \tag{4}$$

and the rate of formation of product will be given by equation (5):

$$\frac{dP}{dt} = k_3 A^* = \frac{k_1 k_3 A^2}{k_2 A + k_3}. \tag{5}$$

At high pressures of A, $k_2 A \gg k_3$ in the denominator and the equation simplifies to (6)

$$\frac{dP}{dt} = \frac{k_1 k_3}{k_2} A = kA, \tag{6}$$

a rate equation that is first order in the macroscopic concentration of A, despite the fact that the initial energizing collision is bimolecular, second order in A.

The proof of the essential correctness of this explanation for the high-pressure unimolecular behavior lies in the observation, at much lower pressures, of deviations from (6). As $k_2 A \simeq k_3$, the denominator of (5) is no longer approximately first order in A, and the observed rate of formation falls off from that expected by (6). Much of the experimentation in unimolecular systems has been concerned with the pressure of onset of this « fall-off » region, and with the detailed shape and significance of such a « fall-off curve » [2].

As with the bimolecular thermal reactions, the experimental kineticist will routinely measure the temperature dependence of the unimolecular reaction, and obtain the activation energy from the Arrhenius plot of log k *vs.* $1/T$. The high-pressure rate constant can then be expressed in terms of the two parameters of equation (7)—the activation energy, E_a, and a temperature-independent factor, k_0:

$$\frac{dP}{dt} = kA = k_0 A \exp[-E_a/RT]. \tag{7}$$

An alternate formulation removes the temperature-dependence of the collision frequency at constant density ($\sim T^{\frac{1}{2}}$), with small numerical changes in the activation energy parameter. The exact details of the manipulation of the parameters are not important for our purposes, but conceptually it is clear that these experiments provide additional information about the sequence (1) to (3), and particularly about (3).

2. - RRKM theory.

While the simple assumptions of reactions (1) to (3) satisfactorily accounted for the qualitative behavior of unimolecular systems, the quantitative details—in particular, the shape of the « fall-off » curves—could not be reproduced in detail. Consequently, the crude classification that molecules were either excited or not excited was then replaced by the concept of a continuum of excited molecules, with the further hypothesis that k_3 steadily increases with increasing energy of excitation. Such a model can fit the experimental data quite well, and leads to the question of the precise relationship between k_3 and energy of excitation. One now finds almost universal usage of the RRKM (RICE, RAMSPERGER, KASSEL, MARCUS) model for this relationship, the quantum-mechanical variant devised by RICE and MARCUS [3, 4] of the original RRK model established more than 35 years ago by KASSEL and by RICE and RAMSPERGER [5-7].

In the original RRK model, the various internal degrees of freedom were considered as classical oscillators, and the procedure essentially involves the hypothesis of distribution and redistribution of excitation energy among these degrees of freedom, until, by random chance, sufficient energy collects in a critical mode, and reaction occurs. Comparison of the RRK theory with actual experiment almost invariably indicated substantial discrepancies with the theoretical lifetimes usually too long. This problem was first solved empirically by treating the number of « active » degrees of freedom more or less as a free parameter, adjusted to fit the experimental data. The connotation of an « active » degree of freedom is that it is a degree of freedom which *is* involved in the energy interchange process; the « inactive » degrees, on the other hand, do not participate in this process. In most cases, the description « active » or « inactive » was not individually specified for actual degrees of freedom, nor were the reasons for « inactivity » often discussed. Most experiments can be correlated rather well with about $\frac{1}{2}$ to $\frac{2}{3}$ of the degrees of freedom counted as « active ».

The arbitrariness of active and inactive counting was subsequently removed in a fundamental manner by RICE and MARCUS through the use of a quantum-mechanical model in which degrees of freedom with high frequencies of vibration were naturally unable to absorb much energy—much as some degrees of freedom are not active in heat capacities at intermediate temperatures. The enumeration of « active » degrees is now in principle simply a question of determining the vibrational frequencies of the excited molecule involved. This is not always without a certain arbitrariness of its own, but the opportunity for large scale variation of parameters has been removed. The details of these models are discussed in Bunker's lecture on the *Theories of Unimolecular Reactions*, cf. also ref. [8].

3. – Molecular lifetimes.

One should pause here and note that the focus of attention in unimolecular reactions has now been shifted from a consideration of the overall macroscopic kinetics and reactions for which it is possible to write $(\mathrm{dP}/\mathrm{d}t) = k\mathrm{A}$, to the reactions of the excited species involved in that kinetics for which unimolecular decay is postulated: $(dP/dt) = k_{3i}A_i^*$. Further, the implication of reaction (3) as written is that these excited molecules have *random* lifetimes, and exhibit the same characteristic exponential decay as that found for other random, unimolecular processes, of which the decay of radioisotopes is both a familiar and excellent example.

Another important corollary concept involved here is that polyatomic molecules are able to exist for appreciable lifetimes despite the possession of excitation energies much higher than that required for bond rupture, or other chemical reaction. Diatomic species have only one vibrational degree of freedom, and, to a good approximation, will either dissociate within one vibrational period if the vibrational energy exceeds the bond dissociation energy, or else will not dissociate at all. Molecules with more than two atoms have, however, many vibrational degrees of freedom—for example, cyclopropane, discussed later, has 21—and it is quite possible for large amounts of excitation energy to be contained in such molecules for substantial lengths of time without the occurrence of a chemical reaction for which only a fraction of this energy would be required.

The experimental observations generally confirm this trend of argument—the more degrees of freedom available, the longer the average lifetime for given excitation and decomposition energies. The decreasing rate constant for the ring-opening reaction (see below) *vs.* increasing molecular complexity is illustrated in Table I for several molecules of the cyclopropane class. All of the

TABLE I (*).

Activated molecule	Methylene source	Olefin	Experimental rate constant s^{-1}
C_3H_6, cyclopropane	CH_2N_2, 4358 Å	Ethylene	$450 \cdot 10^8$
	CH_2CO, 3200 Å	Ethylene	$180 \cdot 10^8$
C_4H_8, methylcyclopropane	CH_2CO, 3100 Å	Propylene	$5.9 \cdot 10^8$
C_5H_{10}, *trans*-1, 2-dimethyl cyclopropane	CH_2N_2, 4358 Å	*trans*-2-butene	$1.4 \cdot 10^8$

(*) Ref. [9].

molecules were excited to approximately the same energy (as indicated, the CH_2 source does have a slight effect on the rate constant) by methylene addition to an olefin [9]—as the molecule becomes larger, the likelihood of concentrating sufficient energy in one critical mode is lessened, and the rate constant decreases.

The general implication might have been drawn from this lecture so far that excited molecular species always deexcite in collisions with other molecules—this is the *strong collision* assumption for reaction (2). A more refined treatment must allow for the possibility of energy transfer of *any* magnitude in collisions, such that excited molecules may retain sufficient energy after collision that further chemical reaction is yet possible. Practically, however, the current precision of experiments is limited to some more general statements about the *average step-size*, or average energy loss, in such de-exciting collisions.

4. - Chemical activation experiments.

Experiments carried out at pressures approaching the high-pressure limit are very near an equilibrium for reactions (1) and (2), and the unimolecular process of reaction (3) removes only a very minor fraction of the molecules excited in the system—most are removed by the de-excitation reaction (2) into an unexcited, nonreactive state. Consequently, one really obtains, as indicated in equation (6), information about the equilibrium—about (k_1k_3/k_2), but not about k_1, k_2, or k_3 individually. Two general kinds of approach have been adopted to avoid these difficulties of the high-pressure region: *a*) experiment in the low-pressure region, measuring the « fall-off » characteristics very precisely; and *b*) obtain molecules sufficiently excited to react through some process other than (1), and then study the decomposition *vs.* stabilization competition of reactions (2) *vs.* (3). Experiments in the low-pressure region can be quite informative, but they are difficult to do for many molecules because the pressures required are often so low as to make homogeneous, gas-phase experiments very hard to attain—at low gas pressures, collisions with, and reactions on, the container walls are often the limiting experimental consideration. When successful, however, the low-pressure experiments are carried out in a region in which the equilibrium is no longer maintained, and at low enough pressures, most of the molecules which are excited in the system actually undergo further chemical reaction, and therefore give direct information about the magnitude of k_1. In this lecture, however, we shall be more concerned with the second procedure for escaping the high-pressure limit—through activation processes, usually called « chemical activation », other than the thermal activation process of reaction (1).

Several general types of chemical reaction have been used in « chemical activation » experiments, all having the characteristic that the energies of the molecules formed in a particular chemical reaction are not thermal but because of the nature of the particular reaction are grouped in a narrow band of energies sufficiently high upon the energy scale that subsequent chemical reactions quite frequently occur unless a de-exciting collision occurs within $(10^{-7} \div 10^{-11})$ s Among the scientists whose names are prominent in the literature of such studies are KISTIAKOWSKY, RABINOVITCH, FREY, TROTMAN-DICKENSON, SETSER, and others.

One important type of chemical activation experiment is the exothermic addition of H atoms to olefins, as illustrated in (8) with *cis*-2-butene [10]:

$$(8) \quad H + \underset{H}{\overset{CH_3}{\diagdown}} C = C \underset{H}{\overset{CH_3}{\diagup}} \rightarrow \underset{CH_2-\dot{C}H}{\overset{CH_3 \qquad CH_3}{\diagdown \qquad \diagup}} .$$

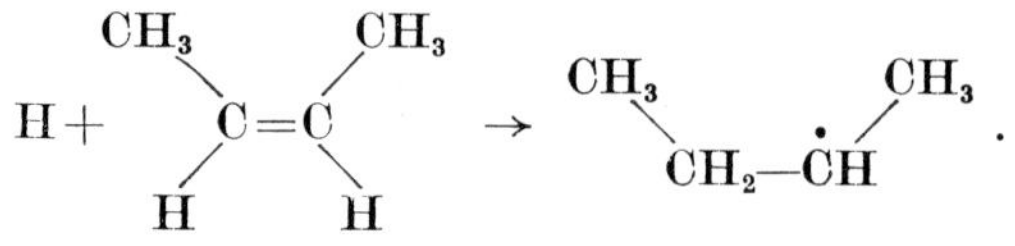

This reaction is exothermic by about 40 kcal/mole, and the excited *sec*-butyl radicals formed all have at least that much energy. The reactants themselves have a thermal spread of energies prior to reaction, and this thermal spread is superimposed upon the exothermicity of the reaction. Nevertheless, the resulting *sec*-butyl radicals have an energy spread of only a few kcal/mole in the region of $(40 \div 45)$ kcal/mole, and are thus crudely « monoenergetic ». The spread of energies from reaction (8) is illustrated in Fig. 1 for two different temperatures, and two isotopic compositions.

Since the subsequent decomposition reaction (9) of excited *sec*-butyl radicals requires only about 33 kcal/mole, *every* one of the radicals has enough excitation energy to decompose if not first stabilized by collisions, as in (10):

$$(9) \quad CH_3CH_2\dot{C}HCH_3^* \rightarrow CH_3 + CH_2=CHCH_3 ,$$

$$(10) \quad CH_3CH_2\dot{C}HCH_3^* + M \rightarrow CH_3CH_2\dot{C}HCH_3 + M .$$

This important difference between thermal and chemical activation should be emphasized, for it is the crux of these experiments: in the chemical activation experiments, the particular molecules under study are formed in the system by exothermic chemical reactions, and *all* are excited enough to react further unless a stabilizing collision intervenes.

The experimental procedure in these H atom plus olefin systems involves measurement of the yields of the decomposition product (D), and the stabilization product (S), and the result of this competition is often expressed in terms of the ratio of these two processes, (D/S). For the particular case of *cis*-2-butene, the decomposition product, propylene, can be directly measured; the *sec*-butyl radical is itself chemically reactive, and is detected after further reactions leading to stable products (*e.g.* $CH_3CH_2CHCH_3 + RH \rightarrow C_4H_{10} + R$).

Small variations in the average energy from (8) can be obtained by changes in several parameters, while maintaining the same basic « hydrogen atom + olefin » reaction. With these small changes, the comparative rates of reaction of the unimolecularly excited species can be evaluated for slightly different average energies of excitation. Among the parameters which are often varied are:

i) *Temperature.* The excitation energy is the sum of the exothermicity and the thermal energy content, and the latter varies by about 2 kcal/mole for 100 °C change in the temperature of the experiment. Such a variation is illustrated in Fig. 1 for 195 °K and 300 °K. (The hydrogen atoms must be produced by a suitable nonthermal source, often a discharge tube, and are then added to the olefin in a separate temperature-controlled gas reservoir).

ii) *Olefinic isomer.* *Trans*-2-butene $\left(\begin{array}{cc} CH_3 & H \\ \diagdown & \diagup \\ C{=}C \\ \diagup & \diagdown \\ H & CH_3 \end{array}\right)$ and *cis*-2-butene differ in energy content by about 1 kcal/mole, and this energy difference is preserved in the *sec*-butyl radical after addition of the hydrogen atom. Still another olefinic alternative is 1-butene ($CH_3CH_2CH{=}CH_2$), from which excited *sec*-butyl radicals are formed by addition to the terminal $=CH_2$ group. When the more drastic step is taken of changing the size of the olefin, the rate constant for decomposition of the radicals again decreases with increasing complexity of the molecule, as shown in Table II [11].

iii) *Isotopic isomer.* Either D or T can be substituted for the H atom in reaction (8), and C_4D_8 or any of the partially deuterated molecules can be substituted for C_4H_8 in each of the olefinic systems listed. Small changes occur in the average energies of the excited molecules when isotopic variants are used— see Fig. 1. However, the rates of reaction are also subject to easily measurable isotopic effects, and the interest in these isotopic systems includes consideration of both factors [10].

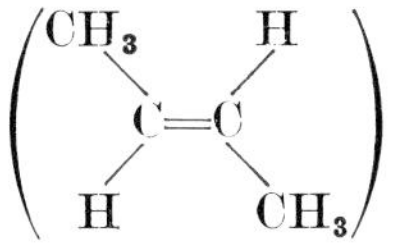

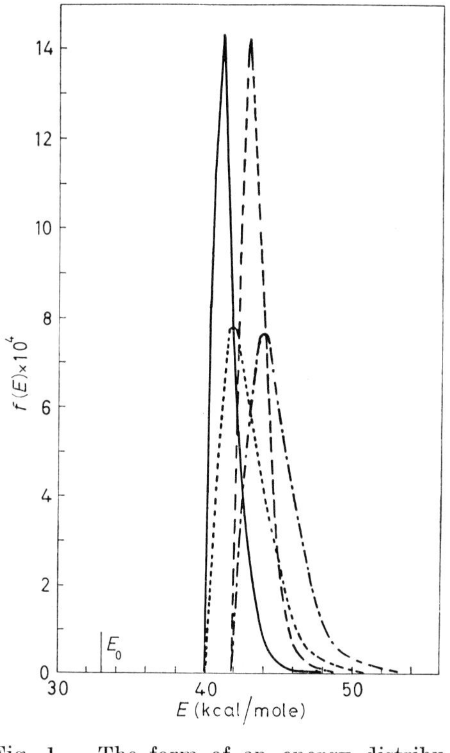

Fig. 1. – The form of an energy distribution for 2-butyl radicals formed by chemical activation: ——— 195 °K, − − − 300 °K; and for 2-butyl-d_1 radicals (D+C_4H_8): − − − 195 °K, —·—·— 300 °K. From ref. [11].

A second general method for chemical activation of molecules is the combination of two free radicals, with the release of the energy generated by the formation of a new chemical bond. A typical example of this method is the

TABLE II. – *Experimental rate constants for alkyl radical decomposition* [11].

Olefin	Radical	k_{exp} (s^{-1})
2-C_4H_8	2-C_4H_9	$2.0 \cdot 10^7$
2-C_5H_{10}	2-C_5H_{11}	$1.2 \cdot 10^7$
2-C_6H_{12}	2-C_6H_{13}	$1.9 \cdot 10^6$
2-C_7H_{14}	2-C_7H_{15}	$3.9 \cdot 10^5$
2-C_8H_{16}	2-C_8H_{17}	$4.8 \cdot 10^4$

radical combination to form ethyl chloride shown in reaction (11), the radicals again being formed by nonthermal processes [12]:

$$(11) \qquad CH_3 + CH_2Cl \rightarrow CH_3CH_2Cl^* .$$

The competitive reactions for $CH_3CH_2Cl^*$ are stabilization, and decomposition by loss of HCl to form ethylene. This elimination process has an activation energy of 56 kcal/mole while the radical combination provides a minimum energy of 88 kcal/mole, and an average energy of 91 kcal/mole, with an energy spread of about ± 3 kcal/mole.

5. – Chemical activation with methylene.

A third major kind of chemical activation reaction involves the chemical reactions of diradicals, of which methylene, CH_2, is the most studied [9]. These species are formed by the photolysis or pyrolysis of molecules such as ketene ($CH_2=C=O$) or diazomethane (CH_2N_2) and are very reactive since they possess only two bonds, instead of the customary four, for carbon atoms. Upon chemical reaction with the formation of new bonds, substantial quantities of energy are again released to the products.

Methylene exists in two electronic states, the singlet with all electrons paired, and the triplet which has two unpaired electrons. While both electronic states are useful in chemical activation studies, the singlet is the more interesting because of its capability for reacting by *insertion* into existing chemical bonds. The singlet reaction with the double bond involves the simultaneous formation of two new single bonds, creation of a three-membered carbon ring, as shown in (12). (The singlet electronic state of methylene is symbolically shown with paired electrons).

$$(12) \qquad :CH_2 + CH_2=CH_2 \rightarrow \underset{CH_2-CH_2^*}{\overset{CH_2}{\diagup\;\diagdown}} \rightarrow CH_3CH=CH_2 .$$

The excited molecule of cyclopropane will isomerize to propylene, as shown, unless first stabilized by collisional deexcitation. The reactions listed earlier in Table I were all initiated by methylene reaction with the olefin, and the listed rate constants in that Table are for the ring-opening reaction shown in (12).

Another kind of insertion reaction introduces the singlet CH_2 group into a C—H bond, as illustrated in (13) for reaction with cyclobutane to form methylcyclobutane:

(13) $\quad :CH_2 + \begin{array}{c} CH_2-CH_2 \\ | \quad\quad | \\ CH_2-CH_2 \end{array} \rightarrow \begin{array}{c} CH_3-CH-CH_2^* \\ | \quad\quad | \\ CH_2-CH_2 \end{array} \rightarrow CH_3-CH=CH_2 + CH_2=CH_2.$

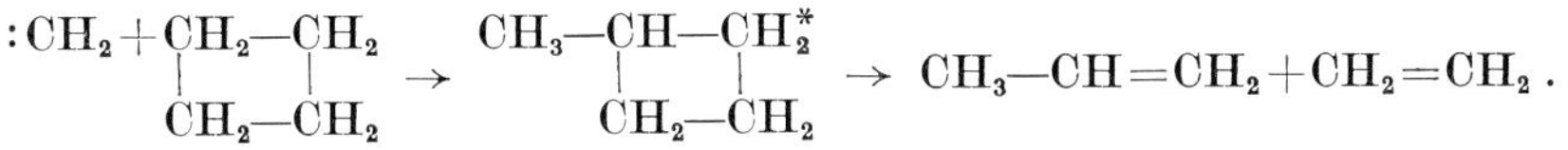

Excited methylcyclobutane is also capable of further reaction—however, molecules containing a four-membered carbon ring such as this do not isomerize, but instead decompose by splitting into two small olefins.

A third reaction of methylene is that observed for the triplet methylene, which reacts with olefins by addition to form an intermediate diradical, as shown in (14):

(14) $\quad \cdot CH_2 \cdot + CH_2=CH_2 \rightarrow \cdot CH_2CH_2CH_2^* \cdot \rightarrow \begin{array}{c} CH_2 \\ / \quad \backslash \\ CH_2-CH_2 \end{array}.$

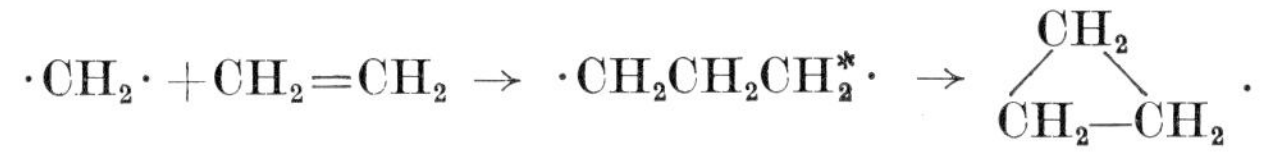

This diradical can then subsequently undergo ring-closure to form the same molecular species indicated in (12). However, since the singlet and triplet methylenes do not have identical energies, and since the intermediate diradical is capable of losing some excess energy by collision *prior* to ring-closure, the cyclopropane formed in (14) is less excited than that formed in (12), and will therefore have different rates for the secondary isomerization to propylene [13].

The introduction of a particular group into a specific position within a molecule provides an especially favorable opportunity for a test of the concept of randomization of energy in unimolecular reactions. For example, if the excess energy of (12) is truly randomized in the molecule, then all three CH_2 groups are equally likely to be involved in each aspect of further reaction, and both C and H atoms should be randomly distributed throughout the product propylene. Just such a randomization test was first carried out by BUTLER and KISTIAKOWSKY, who compared the decomposition reactions of methylcyclopropane formed by two different reaction modes [14]. Reactions (15) and (16) are directly analogous to (12) and (13), forming methylcyclopropane by reaction with propylene and cyclopropane, respectively:

(15) $\quad :CH_2 + \begin{array}{c} CH_3 \\ \backslash \\ C=CH_2 \\ / \\ H \end{array} \rightarrow \begin{array}{c} CH_3 \quad \overset{*}{C}H_2 \\ \backslash \;/\; \backslash \\ C-CH_2 \\ / \\ H \end{array},$

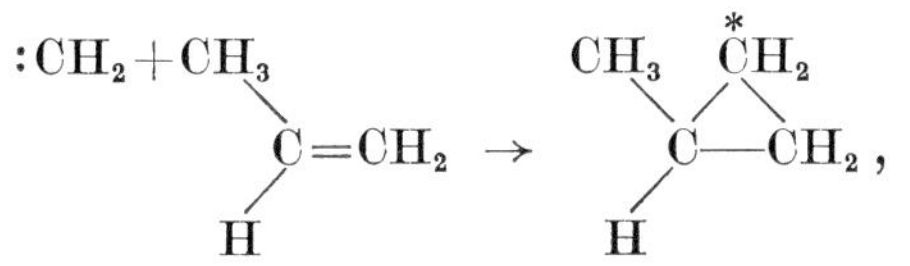

(16) 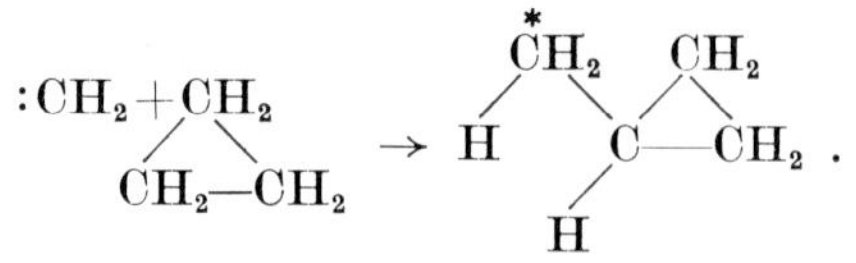

The asterisk designating excitation is placed near the molecular site of the methylene group in the newly-formed molecule in each case. Excited molecules of methylcyclopropane decompose by one of four competing reaction paths to the four butene isomers: *cis*-2-butene, *trans*-2-butene, 1-butene, and isobutene 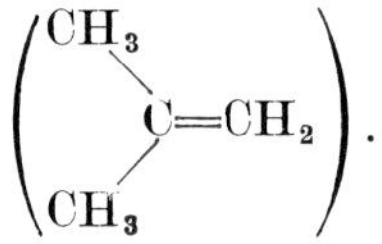. If one postulates that the excess energy accompanying bond formation is initially located in the vicinity of the « new » bonds, differences in the relative probabilities of these four reaction paths might be observed. On the other hand, randomization of energy would effectively remove all trace of the original mode of formation, and give the same distribution of products from either experiment. Neither the Butler-Kistiakowsky experiments, nor any of the subsequent variants of this general type, have indicated any « memory » for the original reaction, and all are consistent with complete randomization of energy.

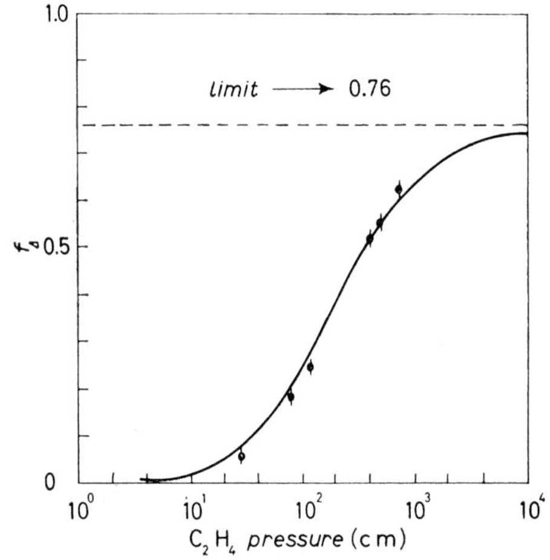

Fig. 2. – Pressure-dependence of cyclopropane-*t* yield from the reaction of tritiated singlet methylene with ethylene. • Experimental points; ——— calculated curve for decrease of a factor of 10 in k_{isomer} in every collision. From ref. [13]. $P_{\frac{1}{2}} = 170$ cm.

The competition between stabilization and further reaction (the isomerization of cyclopropane to propylene in this case) is illustrated in Fig. 2, which shows the results from a typical pressure-dependence experiment [13]. Singlet methylene labeled with tritium was reacted with ethylene over a wide range of pressures, and the two products, cyclopropane and propylene, were measured. The O_2 scavenger in the system prevented the formation of these C_3 hydrocarbons by triplet methylene reactions, while the indicated fractional limit of 0.76 signifies that the residual fraction of singlet methylene (0.24) reacted by direct insertion into the C—H bond of ethylene, and never formed cyclopropane. The points show the measured fraction (f_Δ) of cyclopropane in the sum of the C_3 products, *i.e.* (cyclopropane)/(cyclopropane+propylene). At the higher pressures, collisions occur more quickly on the average, and more molecules of excited cyclopropane are stabilized before they are able to isomerize. A curve calculated on the strong-collision assumption—that one collision always deexcites the excited molecule of cyclopropane—fails to fit the points at lower pressures. The solid curve shown in the Figure is a better fit, and was calculated on the assumption that each collision removes enough energy (about 15 kcal/mole) to lower the rate constant for isomerization by a factor of 10. The results are best correlated then with the assumption that highly excited cyclopropane molecules, formed as in (12), lose an average of (12 ÷ 15) kcal/mole in each collision with a molecule of ethylene. The stabilization/isomerization competition is such that half of the excited cyclopropane molecules will isomerize prior to making one collision when the reservoir pressure of ethylene is 170 cm Hg, corresponding to a rate constant for isomerization greater than 10^{10} s^{-1}. This rate constant is one of the highest observed in chemical activation experiments, and requires measurements at several atmospheres pressure at least. On the other hand, the slower rate constants of Table II will require experiments at pressures of 10^{-4} atmospheres or lower in order to avoid almost total stabilization of the excited species.

The methylene systems introduce two further parameters that can be varied to create small changes in rate constants for reaction [9]: the source of methylene, and the wavelength of absorbed light in the initial photolysis step. In principle, as the wavelength of the absorbed light becomes shorter, the additional energy beyond that required for disruption of the original molecule can appear as additional excitation of the methylene. The amount of excess energy obviously depends upon the energy required for breakup of the original molecule, and therefore upon the chemical identity of the methylene source. At the present time, however, one does not know the distribution pattern for this excess energy—between CH_2 and CO from $CH_2{=}C{=}O$, for instance—or the amount of the excess which remains with the methylene until its actual chemical reaction, and is confined to making measurements of the rate constants for isomerization or decomposition versus wavelength of light in an effort to determine the amount of excess energy carried by the methylene at reaction [9].

I think that the principles of chemical activation experiments as outlined here are reasonably simple and straightforward. However, the details of the applications to particular systems often turn out to be quite complex and intricate, and are not particularly instructive for the consideration of an audience trying to see the role of unimolecular reaction experiments in present day kinetics. The listener wanting to go deeper into the experimental details would do well to start with the review articles of references [9] and [10].

REFERENCES

[1] F. A. LINDEMANN: *Trans. Far. Soc.*, **17**, 599 (1922).
[2] This introductory material is given in most textbooks on chemical kinetics and in many physical chemistry texts. See, for example, S. W. BENSON: *Chemical Kinetics* (New York, 1960).
[3] O. K. RICE and R. A. MARCUS: *Journ. Phys. Colloid Chem.*, **55**, 894 (1951).
[4] R. A. MARCUS: *Journ. Chem. Phys.*, **20**, 359 (1952).
[5] O. K. RICE and H. C. RAMSPERGER: *Journ. Amer. Chem. Soc.*, **50**,1617 (1927).
[6] L. S. KASSEL: *Journ. Phys. Chem.*, **32**, 225 (1928).
[7] L. S. KASSEL: *The Kinetics of Homogeneous Gas Reactions* (New York, 1932).
[8] D. L. BUNKER: *Theory of Elementary Gas Reactions* (Oxford, 1966).
[9] H. M. FREY: *Progress in Reaction Kinetics*, **2**, 131 (1964).
[10] B. S. RABINOVITCH and D. W. SETSER: *Advances in Photochemistry*, **3**, 53 (1964).
[11] M. J. PEARSON and B. S. RABINOVITCH: *Journ. Chem. Phys.*, **42**, 1624 (1965).
[12] J. C. HASSLER, D. W. SETSER and R. L. JOHNSON: *Journ. Chem. Phys.*, **45**, 3231 (1966).
[13] F. S. ROWLAND, C. MCKNIGHT and E. K. C. LEE: *Ber. Bunsengesellschaft f. Phys. Chem.*, **72**, 236 (1968).
[14] J. N. BUTLER and G. B. KISTIAKOWSKY: *Journ. Amer. Chem. Soc.*, **83**, 1324 (1961).

Unimolecular Reactions: Theory.

D. L. BUNKER

University of California - Irvine, Cal.

The R.R.K.M. (Rice-Ramsperger-Kassel-Marcus) theory deals with a certain class of unimolecular processes, defined by the following assumptions:

1) *Strong energizing collisions.* The states of the decomposing molecule before and after a collision are unrelated. This assumption is not inevitable; we make it here primarily for convenience.

2) *Fast relaxation* of vibrational energy on a chemical time scale. The motion of the molecule is so disorderly as to simulate randomness; all internal states at the same energy have equal probability per unit time. This is also called the *random lifetime* assumption, and it is the central one in all present model-type theories of unimolecular reactions.

3) *No isolated states.* At high energies the normal mode description does not apply; all states at the same energy are accessible.

Let the mean collision frequency be ω and the total energy be E. The minimum energy for reaction is the threshold energy E_0. The natural lifetime of a state with respect to reaction is τ. Classical mechanics is used at first.

$$k_{\text{coll}}(E,\,\tau) = \omega P(E) P(\tau)\,. \tag{1}$$

This is the rate of collisional preparation of states $(E,\,\tau)$ with assumption (1). The P's are therefore *equilibrium* probability functions. The decomposition rate constant is

$$k(E,\,\tau) = \omega P(E) P(\tau) \exp\left[-\,\omega\tau\right]. \tag{2}$$

Assumption (2) is that

$$P(\tau) = k_a(E) \exp\left[-\,k_a(E)\,\tau\right]; \tag{3}$$

then

$$k(E) = \omega P(E) \int_0^\infty P(\tau) \exp[-\omega\tau]\, d\tau = \frac{\omega k_a}{\omega + k_a} P(E) \,, \tag{4}$$

and the overall rate constant is

$$k = \int_{E_0}^\infty k(E)\, dE \,. \tag{5}$$

Assumption (3) is that the integral in eq. (4) includes all states.

Next we evaluate $P(E)$ and $k_a(E)$. Let $G(E)$ be the phase volume or (later) the quantum state density *vs.* E; (*) means energized; ($^+$) means critical configuration, that is, in the act of reaction; $E^+ = E^* - E_0$. Then

$$P(E)\, dE = G(E^*) \exp[-E^*/kT]/Q \,, \tag{6}$$

where Q is the reactant partition function.

$$k_a(E_t^+) = \tfrac{1}{2}\nu_D(E_t^+)\, G(E^+ - E_t^+)\, G(E_t^+)/G(E^*) \,. \tag{7}$$

Here E^+ has been partitioned into product translation E_t^+ and $(E^+ - E_t^+)$. Frequency of *outward* passage through ($^+$) is $\frac{1}{2}\nu_D$, with

$$\nu_D = \frac{1}{\delta}\left(\frac{2E_t^+}{\mu}\right)^{\frac{1}{2}}; \qquad \delta \text{ a small distance.} \tag{8}$$

From the energy level density of a one-dimensional boxed particle,

$$G(E_t^+) = \frac{\delta}{h}\left(\frac{2\mu}{E_t^+}\right)^{\frac{1}{2}} \tag{9}$$

and

$$k_a = \frac{1}{hG(E^*)} \int_0^{E^+} G(E^+ - E_t^+)\, dE_t^+ \,.$$

Internal degrees of freedom must now be discussed. Vibrations and internal rotations are *active*; they contribute energy to the reaction without restriction. External rotations are *adiabatic*; they are associated with conservation relations. If a moment of inertia I is approximately constant between (*) and ($^+$) so that the effect of adiabaticity is to keep the corresponding energy constant, the rotation is said to be *inactive*. We take at first the simplest case, that of a classical molecule with inactive rotations. The G's are then vibrational;

$$G = E^{s-1}/(s-1)!\prod^{s} h\nu_i \tag{10}$$

if s is the number of vibration modes, with frequencies ν_i. (This assumption about vibrational modes is purely statistical and does not contradict the third of the ones initially made.) Evaluation of eq. (9) with G's derived from eq. (10) gives

$$k_a(E) = \left(\frac{E^* - E_0}{E^*}\right)^{s-1} \frac{\prod^{s} \nu_i^*}{\prod^{s-1} \nu_i^+} . \tag{11}$$

The overall rate constant, from eq. (5), is

$$k = \int_{E_0}^{\infty} \frac{\omega k_a}{\omega + k_a} G(E^*) \exp[-E^*/kT] \, dE^*/Q , \tag{12}$$

which at $\omega = \infty$ yields the transition state result

$$k^\infty = \frac{kT}{h} \cdot \frac{\prod Q^+_{\text{vib}}}{\prod Q^*_{\text{vib}}} \exp[-E_0/kT] . \tag{13}$$

We should not however, overestimate the degree of connection between transition state theory and the R.R.K.M. treatment. The ($^+$) entity is in no sense an activated complex. The translational motion associated with E_t^+ is separable only over an infinitesimal distance, and there is not necessarily an equilibrium, real or simulated, between ($^+$) and reactants.

The above simplified version of the theory is useful only in an order of magnitude estimate. To approach predictiveness, we must introduce refinements. These are listed in Table I, and will be discussed sequentially.

TABLE I. – *Refinements to basic R.R.K.M. theory.*

Refinement	Direction k changes	Practical for
(1) Quantum statistics	up	all
(2) Adiabatic rotation	up	{ triatomic-exact { others-rough
(3) Correct specification of ($^+$)	down	triatomic
(4) Anharmonicity	down	small molecules

1) *Quantum state counts for vibrational G*: One may adopt a theoretical [1], numerical or semi-empirical approach. A semi-empirical treatment [2] is illustrated here. The integrated form of (say) $G(E^*)$ is written

(14)
$$\int_0^{E^*} G(E^*) = \frac{(E^* + E^*_{zp} - \varepsilon)^s}{s! \prod h\nu_i^*} .$$

The corrections are the zero-point energy E^*_{zp} and the discreteness factor ε.

(15)
$$\varepsilon = \tfrac{1}{2}\alpha h(s-1) \frac{\sum (\nu_i^*)^2}{\sum \nu_i^*} ,$$

where α turns out to be, as might be expected [1, 2], approximately the same tabular function of E^* for all molecules.

2) *Planar adiabatic rotation.* Multiply the former $G(E^*)$ by $G_{rot}(E^*)$ and the former $G(E^+ - E^+_{tr})$ by $I^* G_{rot}(E^+)/I^+$. The G_{rot} are

(16)
$$(2\pi^2 I/h^2 E_{rot})^{\frac{1}{2}}$$

from the state density of a semiclassical rotor. Now k_a is formulated as

(17)
$$k(E^*_{vib}, E^*_{rot}) = \left(\frac{E^* - E_0 + [1 - I^*/I^+]E^*_{rot}}{E_0}\right)^{s-1} \frac{\prod \nu_i^*}{\prod \nu_i^+} ,$$

which may be integrated over a distribution of E^*_{rot}. The result is

(18)
$$k_a(E^*_{vib}) = \frac{\prod \nu_i^*}{\prod \nu_i^+} \left(\frac{E^* - E_0}{E^*}\right)^{s-1} \cdot \left[1 + \left\{\frac{(1 - I^*/I^+)kT}{E^* - E_0}\right\} + \frac{3}{4}\left\{\frac{(1 - I^*/I^+)kT}{E^* - E_0}\right\}^2\right] .$$

The final square bracket is 1 for inactive rotations (cf. eq. (11)), and is sometimes replaced by $(I^+/I^*)^{\frac{1}{2}}$, a 1934 approximation due to RICE and GERSHINOWITZ. This is poor for triatomic molecules, many of which can be treated exactly by extension of the above method, but the error is likely to be less for large molecules because their rotation is more nearly inactive.

3) We should choose the *critical configuration* by $\partial G(E^+_{vib})/\partial r^+ = 0$ if r^+ is the reaction coordinate. Again with a single adiabatic rotation,

(19)
$$E^+_{vib} = E^*_{vib} - U(r^+) + (1 - I^*/I^+)E^*_{rot}$$

where U is the potential along r^+. Note that I^+ and the ν^+ appearing in G are both functions of r^+. Using a roughly defined « rotation barrier » to define $(^+)$, so that this numerical calculation may be avoided, is often a fairly poor approximation.

4) *Anharmonicity corrections* are more important for $G(E^*)$ than for $G(E^+)$, which is not as highly energized above its vibrational ground state. They can be obtained by numerical calculation, *e.g.* by census of randomly placed points in phase space.

Each of the four refinements in Table I may involve as much as a factor of several, so that if not all of them are used the result may or may not be better than the uncorrected prediction. When all important refinements are employed, and the nature of the bonding in the $(^+)$ state is not too uncertain, it is thought that the theory is predictive within a factor of about 2. At present, all known laboratory unimolecular processes conform to assumption (2) given at the beginning. However, many experimentalists are actively looking for exceptions, so we should be careful not to forget the restrictions on the scope of the present theory.

REFERENCES

[1] E. THIELE: *Journ. Chem. Phys.*, **39**, 3258 (1963).
[2] G. WHITTEN and B. RABINOVITCH: *Journ. Chem. Phys.*, **38**, 2466 (1963).

Potential-Energy Surfaces.

M. KARPLUS

Harvard University - Cambridge, Mass.

1. – Introduction.

The study of the microscopic events involved in the atomic and molecular collisions leading to chemical reactions requires, in general, the solution of the Schrödinger equation with the full Hamiltonian of the system. However for the collision energies of primary chemical interest, the nuclear velocities are sufficiently small relative to those of the electrons that the Born-Oppenheimer separation of nuclear and electronic motion is valid. A further simplification applicable to many, though not all, chemical processes is to treat the motion as adiabatic; that is, a single electronic eigenfunction is used to represent the state of the electrons throughout the reactive encounter. With these assumptions, it is possible to divide the problem into two parts: determination of a potential-energy surface parametrically dependent upon the nuclear positions and the calculation of the classical or quantum scattering of the atoms regarded as point particles (with spin) on the surface. In this lecture, we will be concerned with the potential-energy surface; other lectures will discuss the scattering [1].

1·1. *Statement of the problem.* – The determination of a potential-energy surface (or surfaces) requires solution of the electronic Schrödinger equation $H\Psi = E\Psi$ in the Born-Oppenheimer approximation. The Hamiltonian can be written as (in atomic units):

$$H = \sum_i -\frac{1}{2}\nabla_i^2 - \sum_{i,\alpha}\frac{Z_\alpha}{r_{i\alpha}} + \frac{1}{2}\sum_{i,j}\frac{1}{r_{ij}} + \sum_{\alpha,\beta}\frac{Z_\alpha Z_\beta}{R_{\alpha\beta}}, \tag{1}$$

with the following meaning for the terms on the right-hand side:

i) Total electronic kinetic energy (the subscript i designates the i-th electron).

ii) Nucleus-electron interactions ($r_{i\alpha}$ is the distance of the i-th electron from the α-th nucleus with charge Z_α).

iii) Electron-electron interactions (r_{ij} is the distance between electron i and electron j).

iv) Nucleus-nucleus interaction ($R_{\alpha\beta}$ is the distance between nuclei α and β).

The solution of the above equation gives us the energy as a function of the nuclear position vectors $R_1, R_2, ..., R_N$,

$$E = E(\boldsymbol{R}_1, \boldsymbol{R}_2, ..., \boldsymbol{R}_N) = E(\mathbf{R}) \tag{2}$$

and the total electronic wave function Ψ as

$$\Psi = \Psi(\boldsymbol{r}_1, ..., \boldsymbol{r}_n, \boldsymbol{R}_1, ..., \boldsymbol{R}_N) = \Psi(\mathbf{r}, \mathbf{R})\,, \tag{3}$$

where $\mathbf{r}$ designates the electron coordinates and $\mathbf{R}$ the nuclear co-ordinates.

For a given system one can obtain a series of energy surfaces corresponding to the different electronic energy eigenvalues. However, the interest of the chemist is on the lowest-lying one (*i.e.* the lowest eigenvalue of the Hamiltonian (2)) or at most in a few of the lowest-lying surfaces.

Although in principle the problem of solving the Schrödinger equation is well defined, in practice the present state of development of quantum chemistry is such that not even for the simplest reaction systems is the surface available to such an accuracy that it can be regarded as known. The reasons for this derive from a number of computational difficulties:

a) Multicenter problem. All chemical reactions, except special processes like electron transfer (*e.g.* $H+H^+ \rightarrow H^+ + H$), require solution of at least a three-center problem (*i.e.*, three nuclear centers). This requires evaluation of many very complicated and time-consuming multicenter integrals.

b) $E = E(\mathbf{R})$. In contrast to the study of stable molecules, chemical reactions require knowledge of a complete potential surface. For a three-atom system, about 100 points are needed to determine the surface and even more as the number of atoms increases. This implies that very extensive computational work is required in general, though for certain simple systems, approximate analytic expressions can be useful (*e.g.*, H_3 [2]).

c) Accuracy requirement. For many chemical reaction problems, a very high accuracy is essential; *e.g.* a change in the barrier height by $\sim$1.5 kcal can change the rate constant at room temperature by a factor of $\sim$10. Since the total energies are on the order of (1.5 to 150) a.u. ((10^3 to 10^5) kcal) this implies that 0.1 % or better accuracy is desirable for problems where potential barriers (or wells) are important. Of course, certain reactions (*e.g.* purely exothermic with no barrier) and certain aspects of a reaction (*e.g.* differential cross-section, general behavior of total cross-section) can be evaluated with cruder surfaces.

d) Lack of direct checks. In contrast to calculations for stable molecules, for which available dissociation energy, equilibrium distance, and force constant measurements provide checks for approximate treatments, there is very little direct information available on the potential surfaces required for chemical reactions. Furthermore, the kinetic data from gas phase and even from molecular beam experiments cannot as yet be used to obtain quantitative values for surface parameters [3]. It is just this lack of experimental information that makes the theoretical determination of potential surfaces an important part of chemical kinetics.

2. – Outline of computational methods.

For the sake of convenience, it is possible to classify the main ways of approaching the problem into three categories:

a) purely theoretical,

b) semi-theoretical,

c) purely empirical.

These definitions are self-explanatory. We will deal with only the first two, which have many common features; the third is discussed by BUNKER in relation to the trajectory analysis of experimental data [1, 3].

All calculations that have been made so far (although this type of approach is not essential) have involved an expansion of the wave function; that is, the wave function Ψ is written in the form

$$\Psi = \sum_n c_n \Phi_n \, , \tag{4}$$

where

$$\Phi_n = \Phi_n(\mathbf{r}, \mathbf{R}, \boldsymbol{\lambda}) \tag{5}$$

$\mathbf{r}$ and $\mathbf{R}$ being the previously defined co-ordinates, $\boldsymbol{\lambda}$ a set of parameters determining the function Φ_n, and the c_n linear expansion coefficients. A complete set of trial functions would, of course, lead to exact results. For computational reasons, however, the expansion is usually truncated; that is, only a finite (usually a very small) number of terms can be included so that the proper choice of functions is very important.

The most important method for determining the function Ψ by evaluation of the c_n and the $\boldsymbol{\lambda}$, is based on the variational principle; that is, one searches for a minimum of the functional

$$E = \frac{\langle \Psi | H | \Psi \rangle}{\langle \Psi | \Psi \rangle} .$$

Use of a function Ψ of the form (4) generates the well-known secular equation

$$|H_{nm} - ES_{nm}| = 0 , \tag{6}$$

with

$$H_{nm} = \langle \Phi_n | H | \Phi_m \rangle \tag{7}$$

and

$$S_{nm} = \langle \Phi_n | \Phi_m \rangle ,$$

where in general both H_{nm} and S_{nm} are functions of $\boldsymbol{\lambda}$.

The minimum value for the functional is an upper bound to the exact value of the energy for the ground-state surface. Accurate lower bounds to the energy are much more difficult to achieve and have been used much less in calculations [4].

Another way to approach the problem is to minimize the variance $U^2(E)$ [5], where

$$U^2 = \frac{\int (H\Psi - E\Psi)^2 \, \mathrm{d}\tau}{\int \Psi^2 \, \mathrm{d}\tau} . \tag{8a}$$

Substitution of eq. (4) for Ψ into the expression for U^2 yields the secular equation

$$|(H^2)_{nm} - EH_{nm} + (E^2 - U^2) S_{nm}| = 0 , \tag{8b}$$

where

$$(H^2)_{nm} = \int \Phi_n H^2 \Phi_m \, \mathrm{d}\tau$$

and the energy E appearing in eqs. (8*a*) and (8*b*) is an *a priori* estimate for the true energy. This approach coupled with an extrapolation procedure for E has been used extensively by CONROY [5]. It appears to have the advantage over the variation treatment that cruder values for the matrix elements (*e.g.* H_{nm}) are satisfactory for estimating the variance; it has the disadvantage that the energy obtained is not a bound on the true energy.

Finally, we mention the following technique which makes use of the result that, if Ψ and E are the correct Schrödinger wave function and energy, the equation

$$\int X(H\Psi - E\Psi)\,\mathrm{d}\tau = 0$$

is satisfied; here the function X is essentially arbitrary. Introduction of the expansion for Ψ (eq. (4)) again gives rise to a secular equation. The significance of a method of this type, called the « trans-correlated function » method by Boys [6], is that very complicated and accurate forms for Ψ can be used in the equation as long as a function X can be found to « cancel » the complications; that is the required matrix elements remain relatively simple to evaluate. With the variance method, the trans-correlated function approach suffers from the fact that the resulting energy is not a bound.

It is clear from this brief outline that two points are very important for successful potential surface calculations: *a*) the expansion functions Φ_n should be chosen so that rapid convergence of the series is achieved and *b*) the expansion function should be chosen so that the required matrix elements are easy to evaluate. Since these two desirable attributes are complementary, it is their proper balance that determines the success or failure of a particular treatment. In what follows, we describe briefly some of the types of functions that have been employed and the techniques used for calculating the resulting integrals and matrix elements.

2'1. *Special methods for simple systems.* – For two-electron atoms and diatomic molecules, special methods have been used and extremely high accuracy in the resulting energies have been achieved. We mention the results for He and H_2 here, not because they are likely to be extended to potential surface calculations, but because they give us confidence about the ultimate convergence of the approach.

For the ^1S ground state of the He atom, the following wave function was used by Pekeris [7]:

$$\Psi = \exp[-\tfrac{1}{2}(u+v+w)] \sum_{l,m,n} c(l,m,n)\, L_l(u)\, L_m(v)\, L_n(w)\;,$$

where

$$\begin{cases} u = \varepsilon(r_2 + r_{12} - r_1)\;, \\ v = \varepsilon(r_1 + r_{12} - r_2)\;, \\ w = \varepsilon(r_1 + r_2 - r_{12})\;, \end{cases} \qquad \varepsilon = \sqrt{-E}\;, \tag{9}$$

and the $L_l(u)$, $L_m(v)$, $L_n(w)$ are Laguerre polynomials in the indicated variables. Through recursion relations among the coefficients $c(l, m, n)$, Pekeris was able to include up to 1078 terms in the expansion. The total energy obtained for the nonrelativistic Hamiltonian (eq. (1)) is -2.903724376 a.u., which is thought to be accurate to the decimal places listed. If corrections (mass polarization, Lamb shift, etc.) to the Hamiltonian are introduced by perturbation theory, the first ionization potential is calculated to be 198310.687 cm^{-1}, in comparison with the experimental value ($(198310.8_2 \pm 0.15)$ cm^{-1}).

For the $^1\Sigma_g$ ground state of the H_2 molecule, the following wave function has been employed by KOLOS and WOLNIEWICZ [8]:

$$\Psi = \exp[-\alpha(\xi_1 + \xi_2)] \sum_{r,s,\bar{r},\bar{s},\mu} c(r, s, \bar{r}, \bar{s}, \mu)\, \xi_1^r \xi_2^s \eta_1^{\bar{r}} \eta_2^{\bar{s}} \varrho^\mu\,, \tag{10}$$

with the following definitions for the symbols:

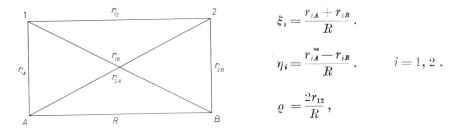

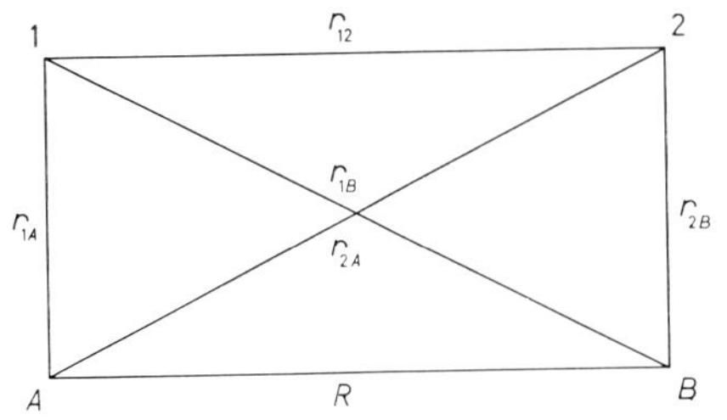

$$\xi_i = \frac{r_{iA} + r_{iB}}{R}\,.$$

$$\eta_i = \frac{r_{iA} - r_{iB}}{R}\,. \qquad i = 1, 2\,.$$

$$\varrho = \frac{2r_{12}}{R}\,,$$

The quantity α is a parameter and r, s, $\bar{r}$, $\bar{s}$, and μ are integers in the power series expansion; the coefficients were determined by the variational method. A calculation with 54 terms in the above expression yielded a dissociation energy $D_e = (38292.9 \pm 0.5)$ cm^{-1} at $R_e = 1.40108$ a.u., which is within the experimental limits of error.

2·2. *Extension to many-electron systems.* – For systems composed of more than two electrons and two nuclei, calculations are much more difficult and the available results are much more limited. However, not quite the same order of accuracy is needed for the potential energy surface of a chemical reaction (~ 1 kcal $\simeq 350$ cm^{-1}) as has been obtained for the He and H_2, so that more approximate methods can be used.

In most studies of systems with more than two electrons [9], the trial functions Φ_n of the expansion (4) are expressed in determinantal form,

$$\Phi_n = \| \varphi_1(1)\varphi_2(2), \ldots, \varphi_n(n) \|\,, \tag{11}$$

where the functions φ_i are *one-electron* spin-orbitals. For the molecular problem, we can distinguish between two limiting choices for the single-particle φ_i's of (11): *a*) atomic functions, where we assume that in each φ_i the individual electrons are in some way centered on one atom of the polynuclear system; and *b*) molecular functions, in which for each φ_i the individual electrons are distributed over all the centers of the polynuclear system.

a) Atomic functions. A variety of different choices have been made for the φ_i of the atomic type. We list some of them below, where we write only the space part (the spin part of the function is suppressed).

1) Exponential-type function on each of the atoms involved (Slater-type orbitals or STO's). Apart from a normalization factor, they can be written for nucleus A as:

$$\varphi_i = \exp[-\alpha r_\mathrm{A}]\,, \qquad \begin{Bmatrix} x_\mathrm{A} \\ y_\mathrm{A} \\ z_\mathrm{A} \end{Bmatrix} \exp[-\beta r_\mathrm{A}]\,, \;\ldots .$$

These functions give a good representation near each of the nuclei, but lead to difficult multicenter integrals (see below).

2) Gaussian-type functions on each center (called GTO's). They have the form for nucleus A

$$\varphi_i = \exp[-\alpha r_\mathrm{A}^2]\,, \qquad \begin{Bmatrix} x_\mathrm{A} \\ y_\mathrm{A} \\ z_\mathrm{A} \end{Bmatrix} \exp[-\beta r_\mathrm{A}^2]\,, \;\ldots .$$

These functions yield rather simple multicenter integrals (see below), but their behavior near the nuclei is poor.

3) Linear combinations of exponential-type, of Gaussian-type, or a mixture of the two. An example is given by the so-called « contracted GTO functions » of CLEMENTI and DAVIS [10], in which a linear combination of Gaussian-type orbitals with coefficients and exponents fixed by atomic calculations are used; this provides one way of overcoming the poor behavior of the GTO's near nuclei.

4) Numerical atomic wave functions (*e.g.* Hartree-Fock atomic functions).

b) Molecular functions. 1)-2) exponential and Gaussian-type functions that are linear combinations of functions on different centers; *e.g.* in a triatomic (ABC) system, a simple example of the exponential type would be

$$\varphi = c_1 \exp[-\alpha r_\mathrm{A}] + c_2 \exp[-\beta r_\mathrm{B}] + c_3 \exp[-\alpha r_\mathrm{C}]\,,$$

where the coefficients c_i are determined by symmetry (symmetry orbitals), by a self-consistent procedure (Hartree-Fock orbitals), or by an alternative method.

3) united atom or one-center expansion where the origins of the functions are fixed on a single center, although they represent the distribution over the entire molecule; this type of function is most useful for cases with a single heavy atom (*e.g.* CH_4).

It is worth mentioning that the above classification is in some way a simplified picture; that is, a variety of intermediate approaches can be found in the current literature. One of these, combining a one-center expansion with a functional form that takes care of the potential singularities at the nuclei is being studied by CONROY [5]. Also, in a « mixed » approach the united-atom method can be used for some of the electrons, whereas, others are described by multicentered wave functions.

The most common method for obtaining well-behaved atomic or molecular functions is the Hartree-Fock self-consistent field (SCF) procedure. For closed shells, we have

$$\Phi = \|\psi_1\alpha(1)\psi_1\beta(2), \ldots, \psi_n\beta(2n)\|, \tag{12}$$

where the ψ's are the spatial orbitals and $\alpha(i)$, $\beta(i)$ the spin functions; in the above restricted HF approximation, $2n$ electrons are accommodated in n orbitals and a single Slater-determinant is used to represent the wave function. Each of the $\psi_j(k)$ is an eigenfunctions of the Fock operator $F(k)$,

$$F(k)\psi_j(k) = \varepsilon_j\psi_j(k), \tag{13}$$

where ε_j is the orbital energy and

$$F(k) = H(k) + \sum_i [2J_i(k) - K_i(k)],$$

with $H(k)$ the kinetic energy and nuclear attraction term in H, $J_i(k)$ the Coulomb operator and $K_i(k)$ the exchange operator,

$$J_i(k) = \int \frac{\psi_i^*(l)\psi_i(l)\,d\tau_l}{r_{kl}}, \qquad K_i(k)\psi_j(k) = \int \frac{\psi_i^*(l)\psi_j(l)}{r_{kl}}\,d\tau_l\,\psi_i(k).$$

In eq. (13), there are a set of occupied ψ_i $(i = 1, 2, \ldots, n)$ plus a set of unoccupied ψ_i $(i > n)$ that form a complete set of eigenfunctions associated with the operator $F(k)$. For most atomic and molecular calculations, the $\psi_i(k)$ are expanded as linear combinations of analytic orbitals

$$\psi_i = \sum_p b_{pi}\chi_p,$$

where the functions χ_p may be all on one atom (atomic Hartree-Fock functions) or on several different atoms (molecular Hartree-Fock functions).

For atoms and diatomic molecules, the primary limitation on accurate calculations is the number of determinants that can be included in the expansion of eq. (4). This limitation results both from the time required for the evaluation of integrals appearing in the individual matrix elements between determinants (H_{nm}, S_{nm}) and from the data handling procedures used in constructing the determinants and solving the secular equation. To obtain an accuracy corresponding to that of the He and H_2 calculations with expansions in terms of one-electron functions is impossible at this time. To demonstrate what can be done, we give results for some simple, four-electron systems (Be, LiH).

First, we present some results for the Be atom for which the experimental value and various computed results for the total energy are listed below.

TABLE I. – Be *calculations*.

	Total energy (e.u.)	Error (a.u.)
Experimental value	— 14.6673	—
Accurate Hartree-Fock ($1s^2 2s^2$)	— 14.57302	0.0943
2 configuration ($1s^2 2s^2$, $1s^2 2p^2$)	— 14.6152	0.0521
37 configurations including up to g-orbitals in the basis set	— 14.6574	0.0099
180 configurations chosen to represent pair excitations and interactions [11]	— 14.6642	0.0031

One can see that the best available result differs from experiment by 0.003 a.u., which is about 2 kcal/mole. This is still a sizable error in terms of accurate energy barrier calculations, in spite of the considerable effort put into the 180 configuration computation.

Second, we consider the LiH molecule for which various computed results at the equilibrium internuclear distance are compared with the experimental energy in Table II. Thus, at best, the error is still about 0.01 a.u. or 6 kcal/mole. As the number of electrons increases, the error in total energy increases; *e.g.*, for F_2, the best value of the total energy is about 0.8 a.u. above experiment [14].

In cases where the molecular calculated energy is higher than the experimental value for the separated atoms, a meaningful purely theoretical dissociation energy D_e cannot be determined. However, if the « same » accuracy wave function is used to compute the energy of both the atomic and molecular systems, D_e can be evaluated with a smaller error than the one involved in each separate computation. This is a simple example of the « semi-theoretical approaches » discussed further below; some results for the dissociation energy of LiH are shown in Table II.

TABLE II. – LiH *calculations.*

	Total energy (a.u.)	Error (a.u.)	D_e (eV)
Experimental value	— 8.0703	—	2.52
Minimal basis set Hartree-Fock	— 7.9667	0.1036	1.39 ([a])
Extended basis set Hartree-Fock	— 7.9860	0.0807	1.44 ([a])
Atomic functions with sizable number of configurations [12]	— 8.0561	0.0142	
Molecular functions with a large number of configurations [13]	— 8.0606	0.0097	

([a]) The value of D_e is obtained by taking the difference between the molecular calculation and a « corresponding » atomic calculation.

2'3. *Integrals in polyatomic calculations.* – For polyatomic systems, the limiting factor for accurate calculations usually is given not by the number of configurations, but by the complexity of the evaluation of the multicenter integrals required for the matrix elements. We therefore consider briefly the kinds of integrals which are involved in molecular computations of the energy by the variational method with determinants of one-electron functions. They are:

Overlap

$$(\mathrm{O}) \quad = \int \chi_p(1)\,\chi_q(1)\,\mathrm{d}\tau\,,$$

Kinetic Energy

$$(\mathrm{KE}) \quad = \int \chi_p(1)(-\tfrac{1}{2}\nabla_1^2)\,\chi_q(1)\,\mathrm{d}\tau_1\,,$$

Nuclear Attraction

$$(\mathrm{NA}) \quad = \int \chi_p(1)\left(-\frac{z_\alpha}{r_{1\alpha}}\right)\chi_q(1)\,\mathrm{d}\tau_1\,,$$

Electron-Electron Repulsion

$$(\mathrm{EER}) = \int \chi_p(1)\,\chi_q(1)\,\frac{1}{r_{12}}\,\chi_r(2)\,\chi_s(2)\,\mathrm{d}\tau_1\,\mathrm{d}\tau_2\,,$$

where p, q, r, s and α can all be different centers. For correlated functions [5], integrals involving more than two electrons can also occur. The overlap and kinetic energy are one-center integrals (p and q on A) or two-center integrals (p on A, q on B); both types can be calculated (for either exponential or Gaussian orbitals) rapidly by straightforward methods [15]. The nuclear attraction integrals can be one-center (p, q, α on A), two center (p, q on A, α on B; p, α on A, q on B) or three-center (p on A, q on B, α on C); of these, the one- and two-center integrals for both exponential and Gaussian orbitals and the three-center integrals for Gaussian orbitals are again straightforward and fast, while the three-center integrals for exponential orbitals are still relatively fast, but require methods corresponding to those used for multicenter two-electron integrals. Among the two-electron integrals, those of the one- and two-center type are rapidly performed for exponential orbitals and all integrals for Gaussian orbitals are straightforward to do. However, the three- and four-center integrals for exponential orbitals are much more time consuming and difficult. In fact, much of the effort spent in recent years in developing techniques for polyatomics has been concerned with this integral evaluation problem.

Although we cannot give a detailed discussion of the methods used for multicenter integrals, we will mention a few important points. From the listing above, it is clear that there are no difficulties for integrals over Gaussian orbitals. The reason for this simplicity is the presence of the r^2 term in the exponent which can easily be transformed from one center to another. Thus, it can be shown directly [15] that

$$\exp[-\alpha_1 r_A^2]\exp[-\alpha_2 r_B^2] = \text{const}\exp[-\beta r_P^2]\,, \qquad \beta = (\alpha_1+\alpha_2)\,, \tag{14}$$

where P is a particular intermediate point between the two centers A and B. Consequently, any three-center one-electron integral or four-center two-electron integral can be reduced to a two-center integral. If we consider the two-electron integral for 1s orbitals on A, B, C, and D, with P and Q the appropriate intermediate points for A, B and C, D respectively, we find [16]

$$\langle 1s_A 1s_B|1s_C 1s_D\rangle_G = \int \exp[-\alpha_1 r_{1A}^2 - \alpha_2 r_{1B}^2 - \alpha_3 r_{2C}^2 - \alpha_4 r_{2D}^2](r_{12})^{-1}\,d\tau_1\,d\tau_2 =$$

$$= 2\pi^{\frac{5}{2}}[(\alpha_1+\alpha_2)(\alpha_3+\alpha_4)]^{-1}\exp\left[-\frac{\alpha_1\alpha_2}{\alpha_1+\alpha_2}\overline{AB}^2 - \frac{\alpha_3\alpha_4}{\alpha_3+\alpha_4}\overline{CD}^2\right]\cdot$$

$$\cdot(\alpha_1+\alpha_2+\alpha_3+\alpha_4)^{-\frac{1}{2}} F_0\,\frac{(\alpha_1+\alpha_2)(\alpha_3+\alpha_4)}{\alpha_1+\alpha_2+\alpha_3+\alpha_4}(\overline{PQ})^2\,,$$

where $\overline{AB}^2$, etc. is the magnitude squared of the line segment connecting A and B, etc. and

$$F_n(u) = \int_0^1 u^n \exp[-ux^2]\,dx = \sqrt{\frac{\pi}{4u}}\,\text{erf}\,(u^{\frac{1}{2}})\,.$$

Integrals involving higher Gaussian orbitals can be expressed correspondingly, so that all of the integrals over Gaussians lead to simple and easily calculable expressions. Present programs for Gaussian two-electron integrals require on the order of ~ 0.025 s/integral for those involving s and p orbitals [10]. As pointed out above, the disadvantage of the method is that a Gaussian function badly describes the electronic orbitals in the neighborhood of the nuclei and for from the nuclei so many more of them may be needed to substitute for the exponential functions.

With exponential orbitals, an approach to the multicenter problem consists of expanding the orbitals from one center about another. In this way, a four-center integral can be reduced to a sum of integrals involving one or two centers. Such a method was pioneered by BARNETT and COULSON [17]. In the original form, the results of the expansion are satisfactory for three-center integrals, but too slow for four-center integral computations in real systems. However, a modified treatment recently suggested by HARRIS and MICHELS [18], in which all orbitals are expanded about a single point, appears to be promising.

Finally, methods based on integral transforms are coming into use. One of these [19] transforms exponential functions to Gaussian functions according to the formula

$$\exp[-\alpha r] = \text{const}\int_0^\infty s^{-\frac{3}{2}} \exp[-\alpha^2/4s] \exp[-sr^2]\, \mathrm{d}s \tag{15}$$

and then makes use of the simplicity of the integration over Gaussians to complete the evaluation of the integral. For a set of exponentials on four centers (A, B, C, D), the electron repulsion integral becomes

$$\langle 1s_A 1s_B | 1s_C 1s_D \rangle_E = \iint \exp[-\alpha_1 r_{1A} - \alpha_2 r_{1B} - \alpha_3 r_{2C} - \alpha_4 r_{2D}] (r_{12})^{-1} \mathrm{d}\tau_1 \mathrm{d}\tau_2 =$$

$$= (16\pi^2)^{-1} \alpha_1 \alpha_2 \alpha_3 \alpha_4 \int_0^\infty \mathrm{d}s_1 \mathrm{d}s_2 \mathrm{d}s_3 \mathrm{d}s_4 (s_1 s_2 s_3 s_4)^{-\frac{3}{2}} \cdot$$

$$\cdot \exp\left[-\frac{\alpha_1^2}{4s_1} - \frac{\alpha_2^2}{4s_2} - \frac{\alpha_3^2}{4s_3} - \frac{\alpha_4^2}{4s_4}\right] \langle 1s_A 1s_B | 1s_C 1s_D \rangle_G .$$

The remaining integrals over the transformation variables s_1, s_2, s_3, s_4 are such that they can be performed relatively easily by suitable numerical or series expansion methods [19]. For s and p orbitals [d orbitals are in progress], the program has been implemented. It requires approximately 1s/integral This is considerably slower (by $\sim 50\times$) than the Gaussian integral computation, but the smaller number of integrals required makes the two approaches competitive. Recent minimum basis set SCF calculations on H_2O required 1 minute and C_2H_6 about 5 minutes on a IBM 7094 [20].

Other transforms (*e.g.* Fourier transforms) have also been suggested [21] and some progress has been made in their utilization for integral evaluation.

2·4. *Semi-theoretical calculations.* – It is useful here to distinguish between the approaches based on the molecular wave function expansion and on the atomic wave function expansion. The main difference between the two types are the same as in the purely theoretical methods (*i.e.* the definition of the φ_i's), but in the present instance some approximations are introduced from experiments to simplify the involved analytic expressions or to improve the rate of convergence of the wave function expansion.

a) Molecular functions. The problem is usually first solved in the Hartree-Fock approximation and the correlation energy (which is defined as the difference between the exact energy for the Hamiltonian of eq. (1) and the Hartree-Fock energy) is introduced after the calculation by some empirical procedure. This has been done successfully for atoms [22], in a preliminary fashion for simple diatomic molecules [23], and so far not at all for polyatomic molecules and potential energy surfaces.

One problem that has to be surmounted in such treatments for potential surfaces is that in the single-determinant (SCF) description, the molecular species usually do not dissociate correctly. For example, in the simple case of the H_2 molecule, the SCF determinant corresponds to the σ_g^2 configuration and dissociates into equal mixture of 2H atoms and of H^+, H^- ions. To obtain the correct dissociation limit into neutral atoms alone, a linear combination of several configurations (C.I.) must be used; in the H_2 case, we have

$$\Phi = c_g \sigma_g^2 + c_u \sigma_u^2 ,$$

where the value of the coefficients c_g and c_u varies as a function of distance [at $R \to \infty$, $c_g = c_u$]. This type of calculation has been done for diatomic molecules (*e.g.* H_2, Li_2, F_2) by WAHL *et al.* [24], who minimized the energy simultaneously with respect to the coefficients of configurational mixing and the form of the orbital expansion. In the simplest treatment, they mixed only the valence configurations required for correct dissociation; in some cases they included additional configurations in their extended Hartree-Fock scheme with promising results. So far, no applications have been made to polyatomic surfaces nor have empirical correlation corrections been introduced.

For certain systems, the relation between reactants and products is such that both limits can be represented by a single-determinant SCF configuration. For instance, RITCHIE and KING [25] studied the proton-transfer reaction

$$H^- + HF \to H_2 + F^-$$

principally in the linear geometry. Here the reactants (H^- and HF) and the products (H_2 and F^-) are closed-shell species and so are properly described by a single determinant. A corresponding calculation has been made by CLEMENTI [26] for the proton-transfer reaction

$$NH_3 + HCl \rightarrow NH_4^+ + Cl^-$$

in a limited set of geometries. In both cases Gaussian orbitals were used to simplify the evaluation of the multicenter integrals. No attempt was made to introduce quantitative correlation corrections in either study, but one can hope that some features of the potential surface are accurate nevertheless.

b) Atomic functions. With this type of wave function expansion, two main semi-theoretical approaches have been introduced: one of these corresponds to the London formula and its extensions, while the other consists of Moffitt's atoms-in-molecules method and its refinements [27]. In the first approach, empirical values are introduced for as many atomic and molecular integrals as possible, whereas in the Moffitt treatment, the choice of VB configurations to be included is guided by chemical intuition and their separated atom energies are determined from experimental data.

As an ultra-simple introduction to the semiempirical atomic expansion, we may mention some calculations for ionic systems. For a molecule such as NaCl, it is known that the ionic model is a good approximation; that is, one considers only the structure Na^+, Cl^- and writes the interaction energy relative to the neutral atoms in the form

$$(15)\qquad \Delta E(R) = \frac{e^2}{R} + I(\text{Na}) + A(\text{Cl}) + \text{Polarization}\left(-\frac{e(\alpha_{\text{Na}^+} + \alpha_{\text{Cl}^-})}{R^4} + \dots\right) + \\ + \text{closed-shell repulsion}\ \left(A\exp[-pR] + \dots\right) + \text{Van der Waals attraction}\ ,$$

as, for example, in the Rittner model [28]. Here I(Na) is the ionization potential of Na, A(Cl) is the electron affinity of Cl, the polarization term corresponds to the charge-induced-dipole interaction (with α the ionic polarizabilities), and the repulsion term (which is usually estimated from experimental data) arises from the Pauli exclusion principle. It is known that the description corresponding to eq. (15) is valid for R in the neighborhood of the equilibrium separation and that additional structures (*i.e.* neutral NaCl) have to be added to obtain the correct ground-state dissociation limit to neutral atoms. If, however, a reaction occurs such as to leave the ionicity of the species unchanged, the simple ionic model may be adequate for the entire potential surface. A case in point is provided by the alkali halide, alkali halide exchange reactions, such as

$$\text{NaCl} + \text{KBr} \rightarrow \text{KCl} + \text{NaBr}\,, \tag{16}$$

which are now being studied experimentally by HERSCHBACH and his students [29] and are being examined theoretically [30]. Here one has the additional advantage that some information is available for intermediate geometries of related species (*e.g.* the $(NaCl)_2$ dimer molecules have been studied [31]).

For the related reactions of which

$$\text{Na} + \text{KCl} \rightarrow \text{NaCl} + \text{K} \tag{17}$$

is a typical example, a somewhat more detailed description is needed since here the neutral Na atom goes to the Na^+ ion, and the K^+ goes to the neutral K atom. Thus, in contrast to the reaction of eq. (16), where no explicit consideration of the electrons is required, the reaction (17) can at most be simplified to that of one electron being transferred from Na to K. This can be done by writing wave functions for the two structures Na, K^+Cl^- and K, Na^+Cl^- and determining their relative contribution as a function of the interparticle distances by the variation principle; *i.e.* the structures Na, K^+Cl^- and K, Na^+Cl^- could be treated simply as $\varphi_{Na}(1)$ and $\varphi_{K}(1)$, respectively, both in the field of Na^+, K^+ and Cl^- cores where $\varphi_{Na}(1)$ and $\varphi_{K}(1)$ should be a flexible atomic function designed to include polarization effects (s, p, ... hybrids). The required matrix elements $H_{Na,Na}$, $H_{K,K}$, and $H_{Na,K}$ can then be determined semi-theoretically by procedures similar to those used for the ionic model combined with a pseudopotential formalism [32]. As has been mentioned by HERSCHBACH in a previous lecture, the most striking attribute of the surfaces corresponding to the reactions (16) and (17) is the presence of a potential well in the transition region.

1) *London formula and its extensions* [2]. The London approach is most simply described by its application to the H_3 problem. We include a $1s$ orbital on each of the three H atoms (a, b, c) and write the two independent, canonical, doublet valence-bond (VB) structures

$${}^2\Phi_{\text{I}} = (ab, c) \qquad \text{and} \qquad {}^2\Phi_{\text{II}} = (a, bc)\,.$$

In ${}^2\Phi_{\text{I}}$, atoms a and b are bonded, whereas in ${}^2\Phi_{\text{II}}$, b and c are bonded; *e.g.* ${}^2\Phi_{\text{I}}$ is

$${}^2\Phi_{\text{I}} = N\{\|1s_a\alpha(1)1s_b\beta(2)1s_c\alpha(3)\| - \|1s_a\beta(1)1s_b\alpha(2)1s_c\alpha(3)\|\}\,,$$

where N is a normalizing factor and we have indicated the main diagonal of the determinants involved (α and β correspond to the spin function associated with the spatial orbitals). If we write the wave function in the form

$$\Psi = c_{\mathrm{I}}\,{}^2\Phi_{\mathrm{I}} + c_{\mathrm{II}}\,{}^2\Phi_{\mathrm{II}}$$

and use the variation principle to determine the coefficients c_{I} and c_{II} and the energy, we find

$$E_{\pm} = \frac{1}{A}[-B \pm (B^2 - AC)^{\frac{1}{2}}] , \tag{18}$$

with

$$\left\{\begin{aligned} A &= (1 - S_1S_2S_3)^2 - \tfrac{1}{2}[(S_1^2 - S_2^2) + (S_1^2 - S_3^2) + (S_2^2 - S_3^2)] ,\\ B &= -(Q - J_{123})(1 - S_1S_2S_3) + \tfrac{1}{2}[(J_1 - J_2)(S_1^2 - S_2^2) + \\ &\qquad + (J_2 - J_3)(S_2^2 - S_3^2) + (J_3 - J_1)(S_3^2 - S_1^2)] ,\\ C &= (Q - J_{123})^2 - \tfrac{1}{2}[(J_1 - J_2)^2 + (J_1 - J_3)^2 + (J_2 - J_3)^2] . \end{aligned}\right. \tag{19}$$

The integrals appearing in eq. (19) are defined as follows:

$$\left\{\begin{aligned} Q &= \text{coulombic integral} = \langle abc|H|abc\rangle ,\\ J_1 &= \text{exchange integral} = \langle abc|H|bac\rangle ,\\ J_2 &= \text{exchange integral} = \langle abc|H|acb\rangle ,\\ J_3 &= \text{exchange integral} = \langle abc|H|cba\rangle ,\\ J_{123} &= \text{exchange integral} = \langle abc|H|cab\rangle \end{aligned}\right. \tag{20}$$

and

$$S_1 = \langle a|b\rangle , \qquad S_2 = \langle b|c\rangle , \qquad S_3 = \langle a|c\rangle .$$

The symbols a, b, and c here refer to the $1s_a$, $1s_b$ and $1s_c$ atomic orbitals, respectively, and

$$\langle abc|H|acb\rangle = \iiint a(1)b(2)c(3)H\,a(1)c(2)b(3)\,\mathrm{d}\tau_1\,\mathrm{d}\tau_2\,\mathrm{d}\tau_3 , \text{ etc..}$$

As indicated in eq. (18), the solution of the secular equation yields two energy values E_+ and E_-. Of these E_- is the surface of lower energy except for the equilateral triangle geometry (D_{3h}) in which the two energies are equal [33].

Up to this point, no approximations have been introduced other than in the ultra-simple choice for the wave function expression. It is primarily to overcome this limitation that empirical information is used. An additional benefit is that the resulting expression for the energy is considerably simplified from the purely theoretical formula. The most elementary approach, which also corresponds to the original London procedure, is to assume that the basis set is composed of orthogonal orbitals (*i.e.* $S_i = 0$ and $J_{123} = 0$). The energy expression (18) then reduces to the London formula [34]:

$$E_{\pm} = Q \pm [\tfrac{1}{2}(J_1 - J_2)^2 + \tfrac{1}{2}(J_2 - J_3)^2 + \tfrac{1}{2}(J_1 - J_3)^2]^{\frac{1}{2}}, \tag{21}$$

where

$$Q = \sum_i Q_i = Q_1 + Q_2 + Q_3, \tag{22}$$

with

$$Q_1 = \int a(1)b(2)\,\mathscr{H}_{ab}(1,2)\,a(1)b(2)\,\mathrm{d}\tau_1\,\mathrm{d}\tau_2, \qquad J_1 = \int a(1)b(2)\,\mathscr{H}_{ab}(1,2)\,b(1)a(2)\,\mathrm{d}\tau_1\,\mathrm{d}\tau_2$$

and

$$\mathscr{H}_{ab}(1,2) = -\frac{1}{r_{1b}} - \frac{1}{r_{2a}} + \frac{1}{r_{12}} + \frac{1}{R_{ab}}.$$

Since the integrals appearing in eq. (21) depend on only two atoms, it is possible to relate them to diatomic energy curves obtained from experimental or theoretical studies. The original suggestion made by EYRING [35] was based on the valence-bond energy formula with zero overlap for the singlet H_2 ground state,

$$^1E_i = Q_i + J_i \qquad (i = 1, 2, 3). \tag{23}$$

He assumed that it was possible to introduce a proportionality constant γ (independent of distance) such that

$$Q_i = \gamma[^1E_i]; \qquad J_i = (1-\gamma)[^1E_i] \qquad (i = 1, 2, 3), \tag{24}$$

at all internuclear distances; if γ is known or determined by some fitting procedure (values in the range $\gamma = 0.1$ to 0.2 commonly have been used), both Q_i and J_i can be evaluated from the experimental energy ^1E_i as a function of R. More recently it has been realized that by use of the lowest triplet energy curve (^3E_i) and its valence-bond zero overlap formula

$$^3E_i = Q_i - J_i \qquad (i = 1, 2, 3) \tag{25}$$

both Q_i and J_i can be determined [36, 37],

$$Q_i = \tfrac{1}{2}[^1E_i + {}^3E_i] \qquad J_i = \tfrac{1}{2}[^1E_i - {}^3E_i] \qquad (i = 1, 2, 3). \tag{26}$$

This treatment is expected to be considerably better than the original Eyring formula (eq. (24)) when accurate singlet and triplet energy curves are available.

If overlap is not neglected, as in the complete formula given in eq. (18), the determination of the various integrals is considerably more complicated. Although the Coulomb integral is still given by eq. (22), the exchange integrals of eq. (20) are the sum of two parts, one being the diatomic molecule exchange integral and the other a residual term composed of triatomic reactions:

$$J_i = J_i^{\text{diat}} + \Delta_i \qquad (i = 1, 2, 3)\,. \tag{27}$$

Inclusion of overlap in the diatomic singlet and triplet formulas gives

$$^1E_i = \frac{Q_i + J_i^{\text{diat}}}{1 + S_i}\,, \qquad ^3E_i = \frac{Q_i - J_i^{\text{diat}}}{1 - S_i} \qquad (i = 1, 2, 3)$$

and

$$Q_i = \tfrac{1}{2}[{}^1E_i + {}^3E_i + S_i^2({}^1E_i - {}^3E_i)]\,, \qquad J_i^{\text{diat}} = \tfrac{1}{2}[{}^1E_i - {}^3E_i + S_i^2({}^1E_i + {}^3E_i)]\,. \tag{28}$$

The residual terms Δ_i can be estimated by comparison with theoretical values. Similarly the double-exchange integral can be approximated by a product of overlap integrals:

$$J_{123} = \varepsilon S_1 S_2 S_3\,, \tag{29}$$

ε being a parameter determined either by comparison with theoretical values of the integral or by fitting some aspect of the surface to a known result.

Such a treatment yields a simple, analytic expression for the potential-energy surface; its application to H_3 is described subsequently. However, it must be remembered that it requires an accurate knowledge of the experimental diatomic curves. Particularly, the triplet curves are difficult to measure and only for H_2 are the theoretical values completely reliable. Moreover, if one wants to introduce higher atomic orbitals (p, d ...) and atomic or ionic excited states, much experimental information, most of which is very poor or non-existent, is required. Some promising attempts at extensions of this type are described by ELLISON and his coworkers [38] and by PEDERSEN and PORTER [39].

2) *Moffit method and its extensions.* An alternative to the London formula (which requires molecular data) is given by the Moffit approach [27], which introduces only atomic information. Moffit's method can be considered to be a means of obtaining dissociation energies by attempting to correct for inaccuracies not only through taking the difference between calculated molecular energies and separated ground-state atom energies, but also by introducing similar differencing corrections for excited states and ionic species.

In the Moffit approach, we write the functions Φ_n of the wave function expansion (4) for the diatomic molecule (AB), for example, in the form

$$\Phi_n = N\mathscr{A}\varphi_n^A(1 \dots n_a)\varphi_n^B(n_a+1 \dots n_b) \dots, \tag{30}$$

where N = normalization factor;

$\mathscr{A}$ = antisymmetrization operator which exchanges each electron between the atoms;

$\Phi_n^{\mathrm{A}}, \Phi_n^{\mathrm{B}}$ = antisymmetrized wave function for atom A and atom B, respectively, in definite states (neutral, ionic etc.) with n_a, n_b electrons; n_a+n_b is the number of electrons in the diatomic molecule.

In principle, an expansion using eq. (30) could include an infinite number of terms, each of which was composed of accurate (and complicated) atomic wave functions. To simplify the calculation, approximations are required. In this method, the number of ionic and excited configurations is restricted to a finite value and the atomic states are evaluated in an approximate way. To counteract the errors introduced by these limitations, the atomic state energies in the Hamiltonian matrix are corrected by the introduction of experimental information. In the original Moffit approach, the Hamiltonian matrix elements H_{nm} of the secular eq. (6) are replaced by

$$H_{nm} \Rightarrow H_{nm} + \tfrac{1}{2}S_{nm}[(E_n - \widetilde{E}_n) + (E_m - \widetilde{E}_m)]\,. \tag{31}$$

The term in brackets constitutes the Moffit correction [27]; the $\widetilde{E}_n$, $\widetilde{E}_m$ terms are calculated for the atom states m and n at infinite distance, while the E_n and E_m values are the corresponding experimental results. By this procedure, there are introduced corrections for each of the atomic energy terms which contribute to the Hamiltonian. The hope is that, since the molecular energy change on bond formation is small (relative to the atomic corrections), the resulting expansion for Ψ should converge more rapidly to the true energy than does the purely theoretical expression.

For the crude atomic wave functions that have been used, the dissociation energies obtained with the Moffit corrections usually are too large. Since it is clear that eq. (31) yields the correct values for all atomic and ionic states at infinite distance ($R \to \infty$), the proper alternative must take account of the fact that for $R < \infty$ the wave functions representing the various states are not completely independent; that is, the set used in the Moffit expansion is overcomplete for finite R because of the nonzero overlap between the different Φ_n. To correct for this, it is appropriate to use instead of these functions a Schmidt orthogonalized set constructed in order of increasing energy [40]. This leads to a

Hamiltonian of the form

$$H_{ij}^{\text{ortho}} \Rightarrow H_{ij}^{\text{ortho}} + (E_i - \widetilde{E}_i)_{ij}\,, \tag{32}$$

where H_{ij}^{ortho} is the Hamiltonian matrix calculated with the orthogonalized functions. Some comparisons of the results obtained with the Moffit method and its orthogonalized alternative are given below.

3. – Illustrative examples.

As an indication of the results that can be obtained with the methods outlined in the previous Sections, we discuss briefly some applications to the H_3 surface, the H_4 surface, and the $Li + F_2 \to LiF + F$ surface, the latter as an example of the series of related surfaces involving different combinations of H, Li and F. The first of these (H_3) is the surface on which most work has been done by different theoretical and semi-theoretical methods and also the one which is most accurately known at present.

3·1. *The* H_3 *potential surface.* – For this system, the general aspects of the potential surface obtained by most methods (excluding the original Eyring formulation) are very similar. Consequently, we use the semi-empirical analytic surface E_- obtained by the generalized London formula with overlap (eq. (18)) to outline the most important features. Figures 1 and 2 show the standard form of potential energy surface for the linear geometry ($\gamma = 0°$)

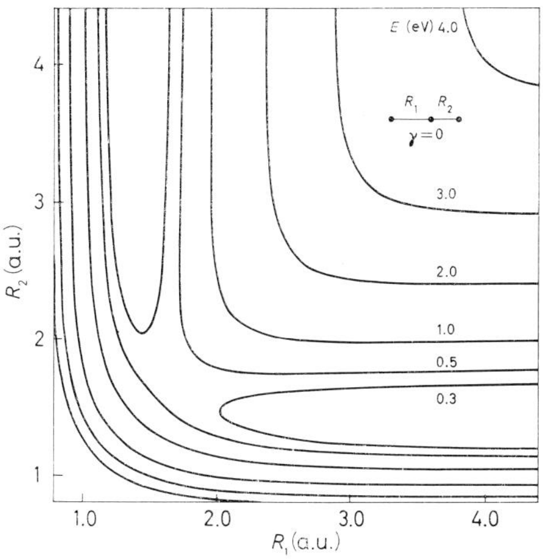

Fig. 1. – Potential-energy contour map for H_3 (relative to H, H_2 as zero) in the R_1, R_2 plane for linear configurations ($\gamma = 0$).

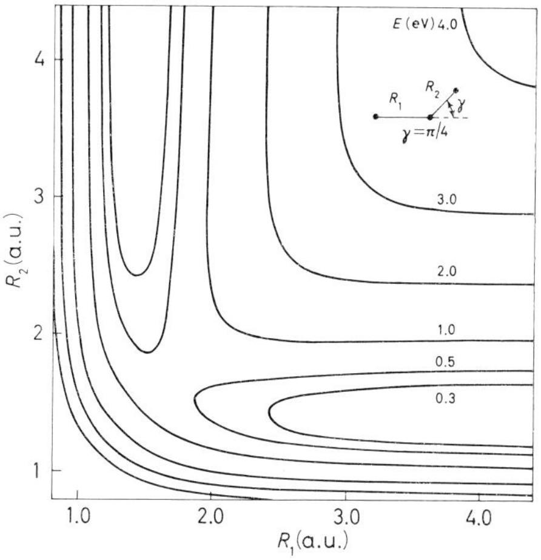

Fig. 2. – Potential energy contour map for H_3 (relative to H, H_2 as zero) in the R_1, R_2 plane with $\gamma = \pi/4$.

of lowest energy and for a slightly bent geometry ($\gamma = \pi/4$). In both drawings the path of minimum energy from reactants to products corresponds to the smooth passage over a saddle point. These minimum energy paths are illustrated for a number of bending angles in Fig. 3. It is to be noted that the

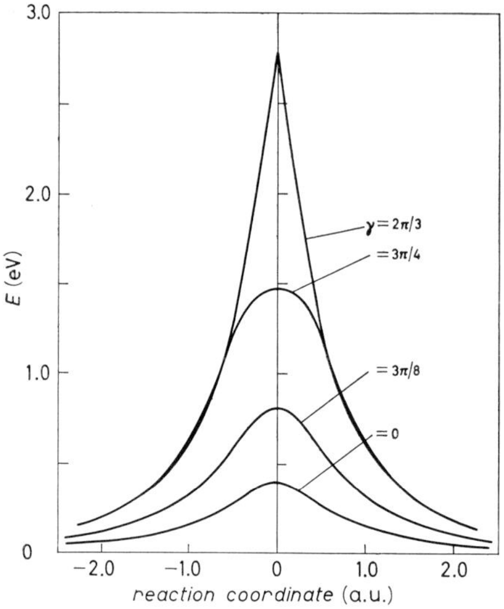

Fig. 3. – Energy of H_3 (relative to H, H_2 as zero) along the reaction path of minimum energy for several values of the bending angle γ.

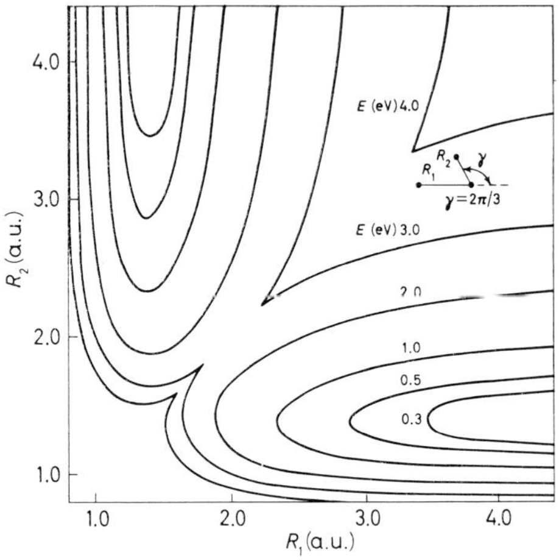

Fig. 4. – Potential energy contour map for H_3 (relative to H, H_2 as zero) in the R_1, R_2 plane with $\gamma = 2\pi/3$.

saddle-point energy increases as γ decreases from 0°. The path corresponding to 60° ($\gamma = 2\pi/3$) is seen to have a cusp for $R_1 = R_2$; this cusp appears along the $R_1 = R_2$ line in the corresponding potential surface shown in Fig. 4. It arises from the degeneracy of the two surfaces E_+ and E_- (eq. (18)) in the equilateral triangle geometry ($\gamma = 60°$, $R_1 = R_2$). A detailed report of this aspect of the surface, which is of considerable importance for studies of the Jahn-Teller effect but of little interest for the H_3 reaction, is given in a separate paper [33].

To indicate the range of surface results obtained by different theoretical and semi-theoretical methods, we list the saddle-point parameters (distance, energy relative to H_2+H, force constants) in Table III. The calculations labelled E, ES, and PK are the semi-theoretical procedures described above and all use the same singlet and triplet H_2 curves (E is the Eyring formulation with zero overlap, ES is the Eyring-Sato formulation with zero overlap, and PK is the Porter-Karplus formulation with overlap). The calculations labelled I and II are theoretical treatments [41] with six basis orbitals ($1s$, $1s'$ on each atom) and fifteen basis orbitals ($1s$, $1s'$, $2p_x$, $2p_y$, $2p_z$ on each atom), respectively; in both cases, optimized orbitals exponents were determined and a

TABLE III. – *Potential-energy surface properties in saddle-point region.*

Surface	R_{sp} (a.u.)	E_{sp} (a.u.) (^{a})	E_0 (kcal) (^{b})	A_{11} (a.u.)	A_{22} (a.u.)	A_{33} (a.u.)
E (^{c})	1.614	−1.6565	11.3	0.323	0.041	(+)0.136
ES	1.781	−1.6497	15.6	0.331	0.028	−0.137
PK	1.701	−1.6600	9.1	0.36	0.024	−0.124
I (1*s*, 1*s*′)	1.788	−1.6305	14.0	0.30	—	−0.054
II (1*s*, 1*s*′, 2*p*)	1.764	−1.6521	11.0	0.31	0.024	−0.061
CB	1.76	−1.6621	7.7	0.32	0.026	~0.00
EK	~1.8	−1.6493	13.5	—	—	—

(^{a}) E_{sp} is the total energy at the saddle point.
(^{b}) E_0 is the barrier height referred to the corresponding H_2, H result.
(^{c}) All results for surface E refer to the bottom of the well ($R_{AB} = R_{BC}$) and not to its rim.

complete set of configurations were included for all geometries (*e.g.*, for the fifteen orbital set, there are 200 configurations for the linear symmetric cases and 402 for the linear asymmetric case). The calculation labelled EK is a theoretical treatment [42] using Gaussian orbitals with seven of *s* type, two of $p\sigma$ type, and 3 of $p\pi$ type on each atom; the calculation labelled CB is the extrapolated theoretical treatment described by CONROY [5, 43]. Except for the Eyring formula, all treatments agree in predicting a symmetric saddle-point geometry although both the internuclear spacing and energy vary by significant amounts. From the theoretical calculations, their range and

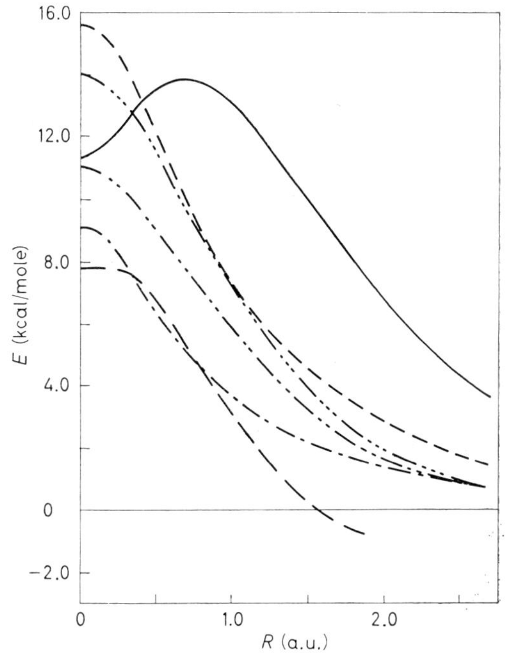

Fig. 5. – Energy profile along the minimum-energy path for linear H_3 from various computations: E (———), EYRING; ES (– – –), EYRING-SATO; PK (—·—·—), PORTER-KARPLUS; I (—· · ·—· · ·—), results using two optimized 1*s* orbitals on each hydrogen; II (—· ·—· ·—), results using two optimized 1*s* and one set of optimized 2*p* orbitals on each hydrogen; CB (— — —), CONROY and BRUNER.

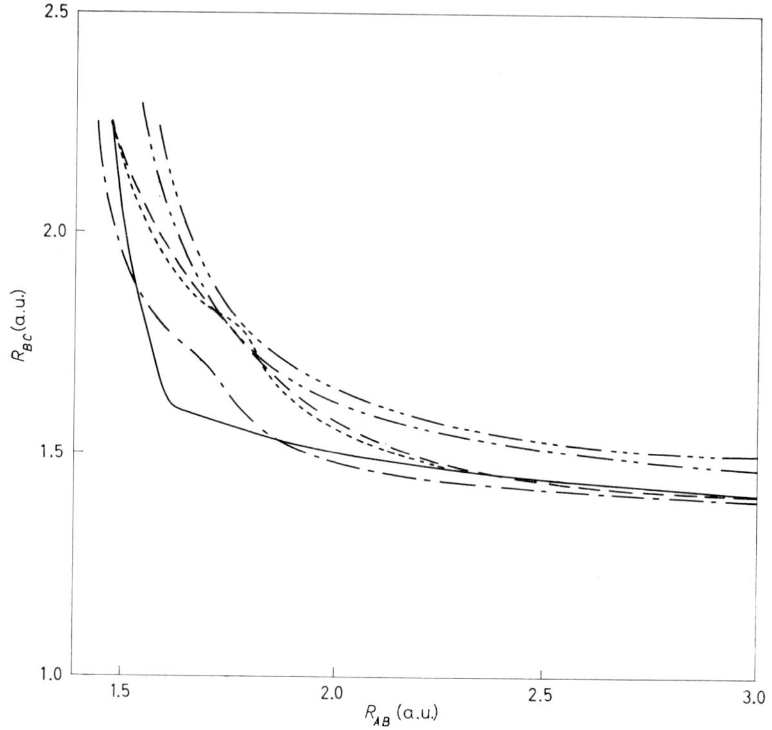

Fig. 6. – Location of the minimum-energy path for various computed linear H_3 surfaces. (See Fig. 5 for explanation of symbols.)

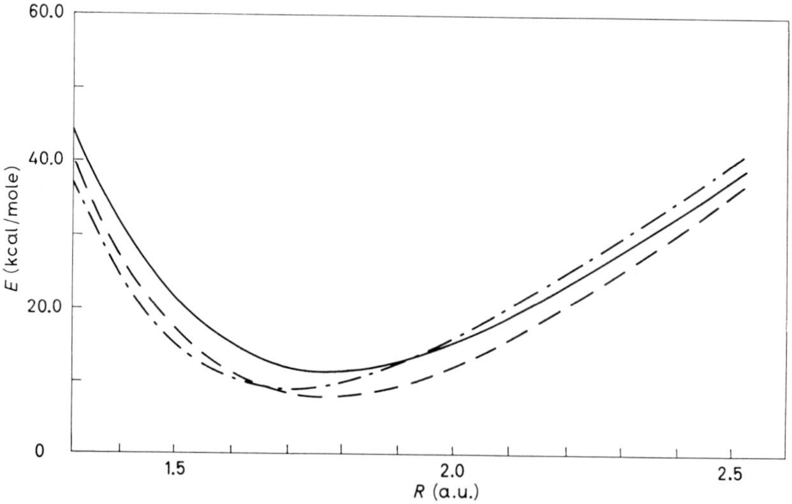

Fig. 7. – Energy as a function of internuclear distance for linear symmetric H_3: PK (—·—·—), results of PORTER and KARPLUS; II (———), results for 15-orbital basis set; CB (— — —), results of CONROY and BRUNER.

estimated errors, it is very likely that the true saddle-point energy is in the range 7.5 to 10.5 kcal above the H_2, H limit; this result ((~9.0±1.5) kcal) still leaves too much uncertainty for unequivocal, absolute, thermal rate-constant calculations [44]. A pictorial comparison of the energy along the path of minimum energy for the various linear surfaces (other than EK) is shown in Fig. 5. The paths themselves are drawn in Fig. 6; they are all similar but there are differences which may be significant for kinetic calculations. Figure 7 shows the energy as a function of inter-nuclear distance for the linear symmetric geometry on surfaces II, CB and PK; the curves represent the potential function for the « symmetric stretching vibration » at the saddle point.

The conclusion to be drawn from the H_3 results is that for this case the best theoretical (II, CB, EK) and semi-theoretical (PK) surfaces are in close correspondence. Moreover, it is likely that in the not too distant future, sufficiently accurate theoretical calculations can be done to reduce the uncertainty in the surface to the more acceptable value of 0.3 kcal/mole.

3·2. *The* H_4 *potential surface.* – For this system, the available results are much more limited than for H_3. However, it is important to mention them because they seem to indicate a serious disagreement between the theoretical and the simplest semi-theoretical formulations. The only semi-theoretical treatment which has been done is of the ES type with zero overlap [45]. It leads to a square-planar saddle-point geometry with side length of 1.97 a.u. and an energy of 61.7 kcal above the H_2, H_2 limit. A preliminary theoretical treatment [45] with eight orbitals ($1s$, $1s'$ on each atom) and twenty orbitals ($1s$, $1s'$, $2p_x$, $2p_y$, $2p_z$ on each atom) and a reasonable, though not complete, configuration list indicates that for the square planar geometry the saddle point is at $R \simeq 2.35$ a.u. with an energy 140 kcal above H_2, H_2, and even 31 kcal above H_2+2H. This value is in significant disagreement with the semi-theoretical result; inclusion of overlap in the latter reduces the discrepancy. Moreover, additional calculations with ($1s$, $1s'$) basis sets for non-square geometries suggest that the minimum energy saddle point is *not a square*; instead distorted configurations of trapezoidal and « kite » form appear to have significantly lower energies [46].

3·3. *The* H, Li, F *potential surfaces.* – The three atoms H, Li, F are sufficiently simple that it is possible to attempt to calculate theoretical or semi theoretical surfaces for the various triatomic reactions that can be formulated by combining them in all different ways, *e.g.* starting with an H atom, we can write

$$\begin{array}{lll} H+H_2 \rightarrow H_2+H\,, & H+LiH \rightarrow H_2+Li\,, & \text{or} \quad HLi+H\,, \\ H+Li_2 \rightarrow LiH+H\,, & H+HF \rightarrow H_2+F\,, & \text{or} \quad HF+H\,, \\ H+F_2 \rightarrow HF+F\,, & H+LiF \rightarrow LiH+F\,, & \text{or} \quad HF+Li\,, \end{array}$$

and similarly for a Li atom and F atom. Various of the reactions are related in that they correspond to different entrance and exit channels on the same potential surface. Many of them, with appropriate isotopic substitution where necessary (*e.g.* H, D or ^{7}Li, ^{6}Li), may be directly amenable to measurement by molecular-beam techniques. Some are very difficult to do because of lack of available isotopes (*i.e.* $F+F_2$ exchange, though here a study of the «non-reactive» scattering would give useful information) or for other reasons (*e.g.* strongly endothermic reactions).

For the simplest cases ($H+H_2$, $H+LiH$, $Li+H_2$, $H+Li_2$, $Li+LiH$, $Li+Li_2$), theoretical treatments by themselves may yield sufficiently accurate results to be of interest. Moreover, these systems are such that *s* orbitals (1*s* in H, 2*s* in Li) are expected to dominate in the binding so that a semi-theoretical approach of the London type may be useful for obtaining additional information. Some studies are already available (*e.g.* LiH_2 and Li_2H [47]). However, as mentioned earlier, the sensitivity of these results to the diatomic triplet state energies which are known only from relatively crude theoretical calculations introduces some doubt as to their reliability.

For the reactions involving fluorine, the presence of additional electrons and the importance of *p* orbitals in the bonding suggest that semi-theoretical approaches of the Moffit type may be most effective. The results are illustrated by applications to the diatomic species LiF, F_2, F_2^- [40] and the potential surface for the reaction $Li+F_2 \rightarrow LiF+F$ [40]. A basis set of 1*s*, 2*s*, 2*p* for both Li and F was used; these basis functions were composed of a linear combination of

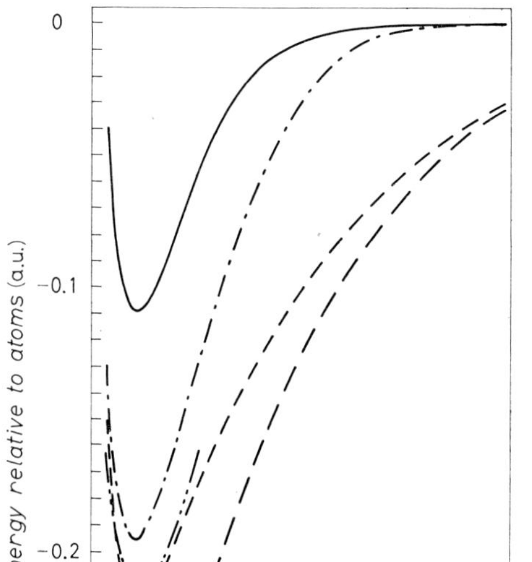

Fig. 8. – Potential-energy curves for LiF: (——) *Ab initio*; (— — —) Moffit's « Atoms in Molecules » method; (—·—·—) Hurley's « Interatomic Correlation Correction » method; (- - -) Orthogonalized Moffit method; (—···—···—) Experiment. The experimental data provide information concerning the shape of the potential-energy curve only in the vicinity of the equilibrium internuclear distance. All the curves tend to zero at large internuclear separations.

Gaussian orbitals obtained by LCAO-SCF calculations for the atoms (*e.g.* the fluorine 1*s* and 2*s* orbitals consisted of five *s*-type Gaussians, while the 2*p* orbitals consisted of three *p*-type Gaussians [48]). From this basis set, atomic functions representing different states of the various species were constructed (*e.g.* for LiF, thirteen different Φ_n corresponding to the possible states of Li, F and Li^+, F^- were included). The results obtained for LiF and F_2 are shown in Fig. 8 and 9. In addition to potential curves from the nonempirical (*ab initio*), Moffit (AIM), and orthogonalized Moffit (OM) calculation, that from Hurley's

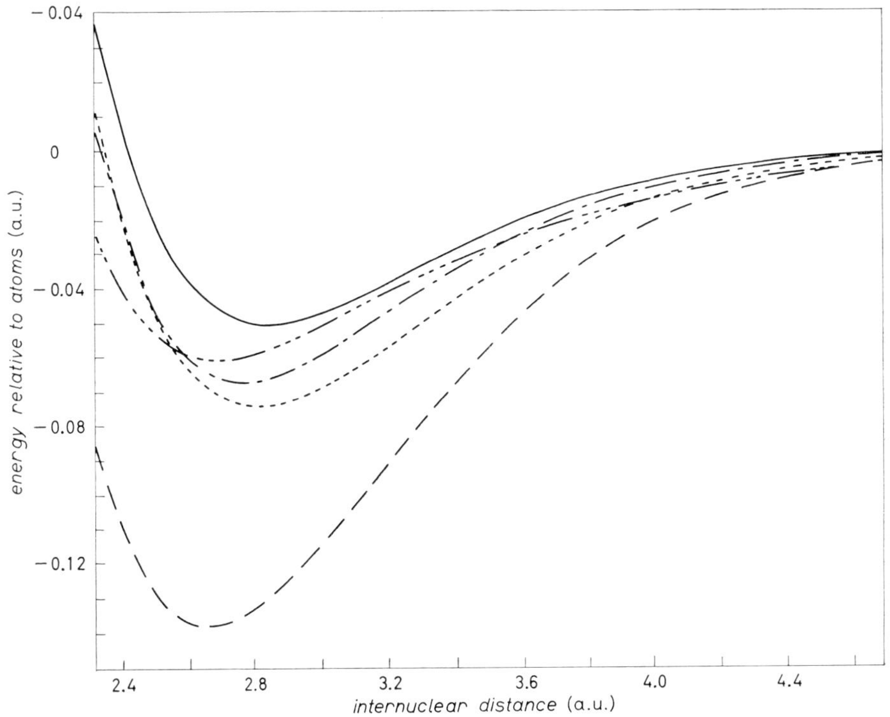

Fig. 9. – Potential energy curves for F_2: All curves tend to zero at large internuclear distances. (See Fig. 8 for explanations of symbols.)

modification (ICC) of the Moffit method is included. It is evident that for both systems, the OM and ICC methods are considerable improvements over the other treatments and give reasonable approximations to the experimental curves, though neither is highly accurate. The choice of the OM method over the ICC correction for potential surfaces was dictated by the fact that the latter does not necessarily go to the correct atomic and ionic limits. For F_2^-, all the calculations agree in indicating that the equilibrium internuclear distance and dissociation energy are larger than those for F_2 itself; *e.g.* the OM values are

	R_e (a.u.)	D_e (a.u.)
F_2	2.8	0.0743
F_2^-	3.3	0.0873

Although the increase of R_e in going from F_2 to F_2^- is expected, that of D_e is rather surprising. Clearly additional calculations with more refined and extended sets are needed to settle the question.

When a corresponding treatment with the OM method and the same basis set is made for Li, F_2, there is obtained a surface of the « attractive » type with the reaction exothermicity being released while the LiF distance is decreasing without a significant increase in the F_2 distance. This behavior is primarily a consequence of the fact that even at rather large Li, F spacings, the ionic states Li^+, F_2^- make the dominant contribution to the ground-state wave function. The other essential feature of the surface is the presence of a « well » of depth $\sim$12.5 kcal relative to LiF (at $R_{F\text{-}F} \simeq 3.3$ a.u., $R_{Li\text{-}F} = 3.0$ a.u.), which arises primarily from the F_2^-, F_2 energy difference. For the perpendicular geometry (Li-F-F $= 90°$), the surface is similar with a well that is somewhat deeper than for the linear case. Comparison of the calculated surface for Li, F_2 with that constructed empirically to obtain the scattering behavior observed for K, Br_2 shows significant differences [49]. The two dominant ones are the presence of the well in Li, F_2 (no such well was introduced for K, Br_2) and the « steeply » rising contours with decreasing F-F distance for LiF near the equilibrium value (the K, Br_2 surface is much softer in the corresponding region). It will be very interesting to have crossed molecular beam measurements for Li, F_2 to see whether the calculated Li, F_2 features are theoretical artifacts or have experimentally verifiable consequences.

* * *

The first draft of this lecture was prepared by P. ECKELT and F. A. GIANTURCO, whose assistance is gratefully acknowledged.

REFERENCES

[1] D. BUNKER: this volume, p. 355; M. KARPLUS: this volume, p. 372 and p. 407.
[2] R. N. PORTER and M. KARPLUS: *Journ. Chem. Phys.*, **40**, 1105 (1964).
[3] For a discussion of this point, see M. KARPLUS: in *Structural Chemistry and Mo-*

lecular Biology, edited by A. RICH and N. DAVIDSON (San Francisco, 1968), p. 837.
[4] For a brief review, see E. B. WILSON jr.: *Journ. Chem. Phys.*, **43**, S 172 (1965).
[5] H. CONROY: *Journ. Chem. Phys.*, **47**, 912, 930, 5307 (1967), and this voulme, p. 349.
[6] S. F. BOYS: *Quantum Chemistry Conference, Montreal, Summer 1967.*
[7] C. L. PEKERIS: *Phys. Rev.*, **115**, 1216 (1959).
[8] W. KOLOS and L. WOLNIEWICZ: *Journ. Chem. Phys.*, **41**, 3663 (1964).
[9] There are some exceptions, such as the Li calculation of E. A. BURKE: *Phys. Rev.*, **130**, 1871 (1963), the work of CONROY [5], and the electron pair treatments (*e.g.*, M. GELLER, H. S. TAYLOR and H. LEVINE: *Journ. Chem. Phys.*, **43**, 1727 (1965)).
[10] E. CLEMENTI and D. R. DAVIS: *Journ. Comp. Phys.*, **1**, 223 (1966).
[11] C. BUNGE: *Phys. Rev.*, **168**, 92 (1968).
[12] J. C. BROWNE and F. A. MATSEN: *Phys. Rev.*, **135**, A 1227 (1964).
[13] C. F. BENDER and E. R. DAVIDSON: *Journ. Phys. Chem.*, **70**, 2675 (1966).
[14] G. DAS and A. C. WAHL: *Journ. Chem. Phys.*, **44**, 87 (1966).
[15] For a recent review of the integral problem, see S. HUZINAGA: *Progr. Theor. Phys. Suppl.*, **40**, 52 (1967).
[16] S. F. BOYS: *Proc. Roy. Soc.*, A **200**, 542 (1950).
[17] M. P. BARNETT and C. A. COULSON: *Phil. Trans. Roy. Soc.*, **243**, 221 (1951).
[18] F. E. HARRIS and H. M. MICHELS: *Journ. Chem. Phys.*, **43**, S 165 (1965).
[19] I. SHAVITT and M. KARPLUS: *Journ. Chem. Phys.*, **43**, 398 (1965).
[20] R. M. STEVENS: unpublished.
[21] M. GELLER: *Journ. Chem. Phys.*, **39**, 84, 853 (1963); M. J. SILVERSTONE: *Journ. Chem. Phys.*, **48**, 4098, 4108 (1968).
[22] V. MCKOY and O. SINANOGLU: in *Modern Quantum Chemistry, Istanbul Lectures*, Part II (New York, 1965).
[23] C. HOLLISTER and O. SINANOGLU: *Journ. Am. Chem. Soc*, **88**, 13 (1966).
[24] For a review see A. C. WAHL, *et al.*: *Int. Journ. of Quant. Chem.*, **15**, 123 (1967).
[25] C. D. RITCHE and H. F. KING: *Journ. Am. Chem. Soc.*, **88**, 1069 (1966).
[26] E. CLEMENTI: *Journ. Chem. Phys.*, **46**, 3851 (1967); **47**, 2323 (1967).
[27] W. MOFFIT: *Proc. Roy. Soc.*, A **210**, 245 (1951).
[28] E. S. RITTNER: *Journ. Chem. Phys.*, **19**, 1030 (1951).
[29] D. R. HERSCHBACH: private communication.
[30] P. BRUMER and M. KARPLUS: work in progress.
[31] J. BERKOWITZ: *Journ. Chem. Phys.*, **29**, 1386 (1958).
[32] A. C. ROACH and M. S. CHILD: *Mol. Phys.*, **14**, 1 (1968).
[33] R. N. PORTER, R. M. STEVENS and M. KARPLUS: *Journ. Chem. Phys.*, **49**, 5163 (1968).
[34] F. LONDON: *Zeits. Electrochem.*, **35**, 552 (1929).
[35] S. GLASSTONE, K. L. LAIDLER and H. EYRING: *The Theory of Rate Processes* (New York, 1941).
[36] S. SATO: *Journ. Chem. Phys.*, **23**, 592, 2465 (1965).
[37] J. K. CASHION and D. R. HERSCHBACH: *Journ. Chem. Phys.*, **40**, 2358 (1964).
[38] F. O. ELLISON: *Journ. Am. Chem. Soc.*, **85**, 3540, 3544 (1963).
[39] L. PEDERSEN and R. N. PORTER: *Journ. Chem. Phys.*, **47**, 4751 (1967).
[40] G. G. BALINT-KURTI and M. KARPLUS: *Journ. Chem. Phys.*, **50**, 478 (1969); and unpublished calculations.
[41] I. SHAVIT, R. M. STEVENS, F. L. MINN and M. KARPLUS: *Journ. Chem. Phys.*, **48**, 2700 (1968).

[42] C. Edminston and M. Krauss: *Journ. Chem. Phys.*, **42**, 1119 (1965); **49**, 192 (1968).

[43] H. Conroy and B. L. Bruner: *Journ. Chem. Phys.*, **42**, 4047 (1965); **47**, 921 (1967).

[44] M. Karplus: this volume, p. 372.

[45] K. Morokuma, L. Pedersen and M. Karplus: *Journ. Am. Chem. Soc.*, **89**, 5064 (1967), and unpublished calculations.

[46] D. Silver, R. M. Stevens and M. Karplus: unpublished calculations; H. Conroy: private communication and ref. [5]; see, also, C. W. Wilson jr. and W. A. Goddard III: *Journ. Chem. Phys.*, **51**, 716 (1969); M. Rubenstein and I. Shavitt: to be published.

[47] A. L. Companion: *Journ. Chem. Phys.*, **48**, 1186 (1968).

[48] The atomic functions used in these calculations were kindly supplied by Dr. D. R. Whitman.

[49] M. Godfrey and M. Karplus: *Journ. Chem. Phys.*, **49**, 3602 (1968); D. Bunker: this volume, p. 355.

Special Results in Potential-Energy Surfaces (*).

H. CONROY

Carnegie-Mellon University - Pittsburgh, Pa.

In this lecture we describe some alternatives to the more common ways of quantum-chemical computation, which have been described in the lecture of KARPLUS [1]. Our treatment has been applied so far to systems up to 4 electrons, especially to H_3 and H_4.

1. – One-electron wave function [2*a*, 4*a*].

The Schrödinger equation for one electron in the field of several nuclei is

$$H\varphi = -\tfrac{1}{2}\nabla^2\varphi + V\varphi = E\varphi\,, \tag{1}$$

where

$$V = -\sum_a Z_a r_a^{-1}\,, \tag{2}$$

the summation being over all nuclei.

The potential V has singularities at each nucleus, which have to be cancelled by the behaviour of φ at those co-ordinates. If we choose

$$\gamma = \sum_a Z_a\,, \qquad \gamma\varrho = \sum_a Z_a r_a\,, \qquad \varphi = \exp[-\gamma\varrho]\,, \tag{3}$$

this cancellation is provided, but the asymptotic behaviour of φ for $r \to \infty$ is wrong. (r, and later θ, ϕ taken from the center of charge.) This can be

(*) Enlarged version of a text prepared after the lecture by P. ECKELT and F. A. GIANTURCO. The responsibility for any errors lies with the editor.

remedied by using

$$(4)\qquad \begin{cases} \sigma^{-1} = (r^2 + s^2)^{\frac{1}{2}}\,, \\ \varepsilon \quad = (-2E)^{\frac{1}{2}}\,, \qquad \alpha = \gamma/\varepsilon - 1\,, \\ \varphi \quad = \sigma^{-\alpha} \exp[-\gamma\varrho + (\gamma - \varepsilon)/\sigma]\,, \end{cases}$$

s is a parameter, which can be fixed in different ways [2*a*, 4*a*]. To add flexibility the final form of φ is a series expansion

$$(5)\qquad \Psi = \sum_n c_n \chi_n = \sum_n c_n A_n(r, \theta, \phi)\varphi(r, \theta, \phi)\,,$$

where the functions A_n are members of a single-center basis set.

Using a 10-term trial function of the type of eq. (5) the results of ref. [3] for H_2^+ have been duplicated.

2. – Polyelectronic wave functions [2*d*, 4*a*].

In principle, it would be possible to use the wave functions obtained in the last Section in an independent-particle model calculation with many configurations. *I.e.*, for two electrons one would use

$$(6)\qquad \Psi_{\text{c.I.}} = \sum_{n,m} c_{nm}[\chi_n(1)\chi_m(2) \pm \chi_m(1)\chi_n(2)]\,.$$

We want to improve convergence by explicitly taking care of the singularity $1/r_{12}$ in the Hamiltonian, and can do this with a multiplicative factor $A(r_{12})$ for which $\nabla_1^2 A = \nabla_2^2 A = A\cdot(1/r_{12})$. In order to provide still more flexibility (mostly for angular correlation) we multiply with an other series B to get

$$(7)\qquad \Psi(1, 2) = A(r_{12})\cdot B(r_{12}, \sigma_1, \sigma_2)\cdot \Psi_{\text{c.I.}}\,.$$

This procedure can be extended to more than two electrons by including all singlet and triplet pair correlations and proper antisymmetrization [4*a*].

3. – Optimization of the wave function [2*c*, 4*c*].

The classical procedure is the minimization of the energy

$$(8)\qquad \varepsilon = \int \Psi H \Psi \,\mathrm{d}\boldsymbol{x} \Big/ \int \Psi^2 \mathrm{d}\boldsymbol{x}$$

(Ritz variational principle). We propose instead to minimize the energy *variance*

$$U^2 = \int (H\Psi - E\Psi)^2 \mathrm{d}\boldsymbol{x} \Big/ \int \Psi^2 \mathrm{d}\boldsymbol{x} \,. \tag{9}$$

With (5) or (7) this leads to a « secular equation » of the type

$$|(H^2)_{nm} - EH_{nm} + (E^2 - U^2)S_{nm}| = 0 \,. \tag{10}$$

The main advantage of our procedure is the reduced sensitivity of the energy obtained with respect to errors in the matrix elements, which in our case come from a multidimensional numerical integration (see below).

Whereas ε is an upper bound to the true energy E_0, neither the minimum of E nor the energy E' which belongs to the minimum of U^2 in eq. (10) are bounds to E_0. It is, however, possible to use some extrapolation procedures in the $[\varepsilon, U^2(\varepsilon)]$-plane in order to approximate well the true energy.

4. – Evaluation of integrals [2*b*, 5].

Since the wave functions described in Sections **1** and **2** are nonseparable, the problem of computing the matrix elements is one of multi-$(3N)$- dimensional numerical integration. This can be solved by Monte Carlo methods [2*b*], but a much better method is « Diophantine » integration [6, 5]. In the latter case systematic (not random) sampling of the co-ordinate space is involved, and one can show that the error decreases as the inverse square of the number of sampling points rather than the inverse square root.

5. – Some results [2*d*, 4*b*].

The methods described in the foregoing have so far been applied to systems of up to four electrons. Most interesting is H_3, because its energy surface is needed in theories for the $H+H_2$ exchange reaction. The energy contour maps for linear and isosceles H_3-arrangements are reproduced in Fig. 1 and 2. In agreement with other treatments the linear surface lies lower than the triangular ones. There is a saddle point at $R_1 = R_2 = 1.76$ a.u. with an activation energy of 7.74 kcal/mole above $H+H_2$ [4*b*].

In contrast, H_3^+ has at least two bound states as an equilateral triangle, whereas H_3^{++} shows no such states. Other 3-electron systems studied include He_3^+ [4*b*] and HeH (unpublished). Among the 4-electron systems He_2 has been computed in essential agreement with PHILLIPSON [7]. Extensive computations on H_4 have been started, considering rectangular, square, trapezoidal and rhombic arrangements. For the square arrangement a saddle point has been found with an activation energy of about 125 kcal/mole, much too high for the observed rate of the exchange reaction $H_2+H'_2 \rightarrow HH'+HH'$, which therefore has to proceed through other conformations [8]. Similar results have

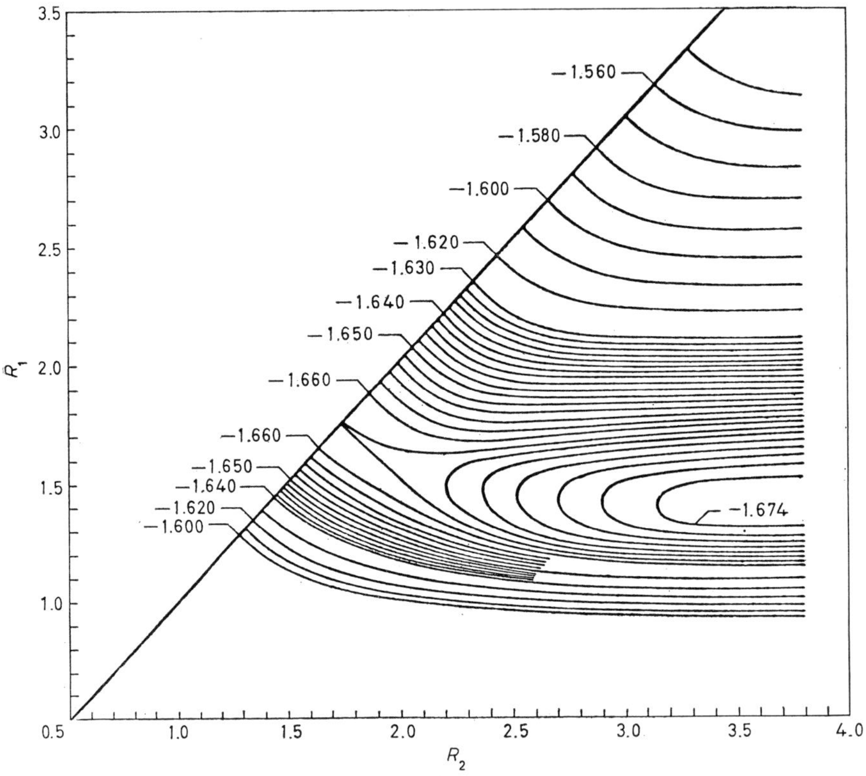

Fig. 1. – Energy surface for linear H_3 in the lowest state ($^2\Sigma_u^+$). From ref. [4*b*].

been obtained for rhombic and trapezoidal arrangements, so that one has to assume that the exchange reaction proceeds via some less symmetric state (cf. also the section on H_4 in ref. [1]).

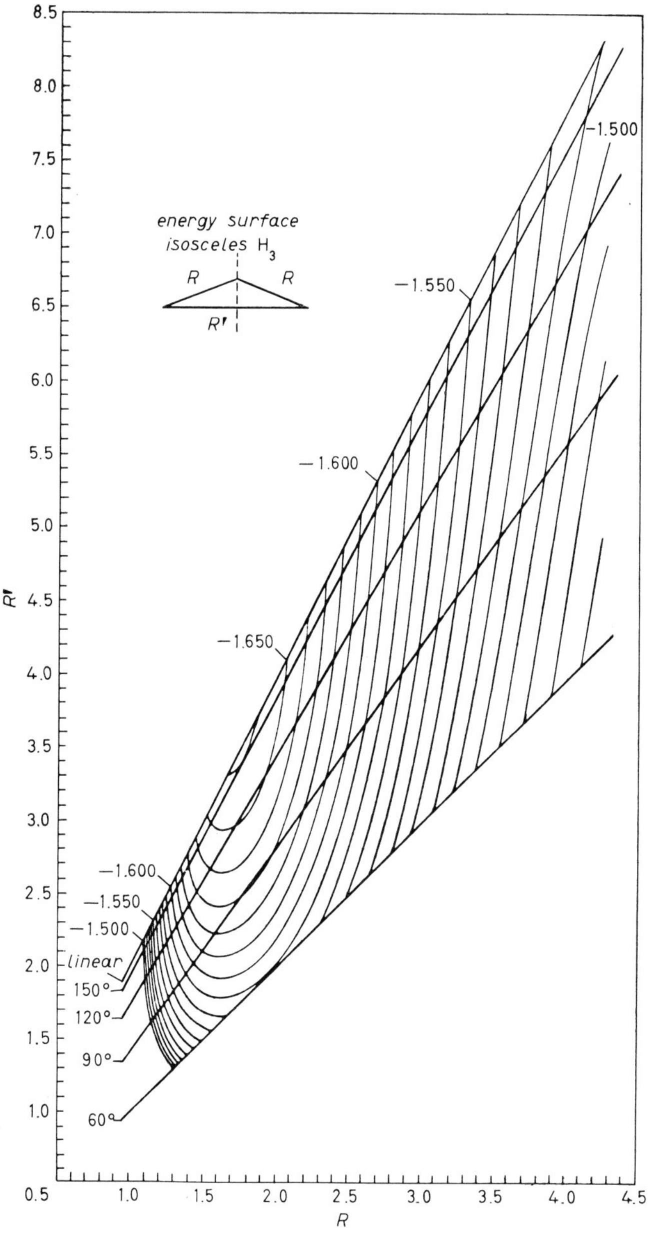

Fig. 2. – Energy surface for isosceles H_3 in the lowest state. From ref, [4*b*].

REFERENCES

[1] M. Karplus: this Volume, p. 320.
[2a] H. Conroy: *Journ. Chem. Phys.*, **41**, 1327 (1964).
[2b] H. Conroy: *Journ. Chem. Phys.*, **41**, 1331 (1964).
[2c] H. Conroy: *Journ. Chem. Phys.*, **41**, 1336 (1964).
[2d] H. Conroy: *Journ. Chem. Phys.*, **41**, 1341 (1964).
[3] D. R. Bates, K. Ledsham and A. L. Stewart: *Phil. Trans. Roy. Soc.*, A **246**, 215 (1953).
[4a] H. Conroy: *Journ. Chem. Phys.*, **47**, 912 (1967).
[4b] H. Conroy and B. L. Bruner: *Journ. Chem. Phys.*, **47**, 921 (1967).
[4c] H. Conroy: *Journ. Chem. Phys.*, **47**, 930 (1967).
[5] H. Conroy: *Journ. Chem. Phys.*, **47**, 5307 (1967).
[6] C. B. Haselgrove: *Math. Computation*, **15**, 323 (1961).
[7] P. E. Phillipson: *Phys. Rev.*, **125**, 1981 (1962).
[8] H. Conroy and G. Malli: *Journ. Chem. Phys.*, **50**, 5049 (1969).

Trajectory Studies.

D. L. Bunker

University of California - Irvine, Cal.

1. – The technology of the method.

As in a more conventional type of experiment, there is an apparatus to be described. This consists of a computer and a specialized kind of program for it. As in the laboratory, there are many interacting parts which have to be fitted together, and once the construction is completed it has to be refined until the results it produces are useful.

The apparatus can be subdivided functionally into three parts—a working definition of the potential energy; a calculation of the classical trajectory for a reacting system; and a method of choosing initial conditions so as to have a representative sample of the events that would occur in the laboratory.

1·1. *The potential energy.* – Even if one knew the potential energy exactly, it would still have to be expressed in a way useful for numerical calculations. It must accurately represent known threshold energies, molecular vibration frequencies, etc., and it should have explicit derivatives for easy computational manipulation. Even in the hypothetical case of an exactly known potential, it is likely that numerical differentiation of a list of numbers will be a less useful procedure than one involving function-fitting. Several approaches to this task are discussed below.

The *modified L.E.P.S.* method is one of the simpler ones. Although it is based on the London equation and began as a semi-theoretical potential, it is here parametrized and treated as if it were wholly empirical. Suppose there are three atoms, *pairs* of which are identified by subscripts 1, 2, and 3. The L.E.P.S. expression is

$$U = \frac{Q_1 + Q_2 + Q_3 - [(J_1 - J_3)^2 + (J_2 - J_3)^2 + (J_1 - J_2)^2]^{\frac{1}{2}} 2^{-\frac{1}{2}}}{1 + S^2}, \tag{1}$$

where Q's are the Coulomb integrals between particles; J's are the exchange integrals; S represents overlap of atomic functions, but it is usually treated as a free parameter.

Empirical information is inserted as follows:

$$Q + J = D(1 - \exp[-\beta\Delta r])^2 , \tag{2}$$

a Morse function representing an attractive potential for a pair of atoms; for the corresponding repulsive state, the « Anti-Morse » function,

$$Q - J = \tfrac{3}{4} D \exp[-2\beta\Delta r] - \tfrac{1}{2} D \exp[-\beta\Delta r] , \tag{3}$$

is used. For fixed S, three each of eqs. (2) and (3) suffice for the evaluation of eq. (1). Selection of S is often accomplished by further appeal to experiment, as by matching a known threshold energy for the reaction.

For general reactions of the A + BC type, eq. (1) does not have enough adjustable parameters. It was therefore modified by J. POLANYI [1] to introduce two more.

$$U = \sum_i \frac{Q_i}{1 + S_i} - \left[\sum_i \frac{J_i^2}{(1 + S_i)^2} - \sum_{i \neq j} \frac{J_i J_j}{(1 + S_i)(1 + S_j)}\right]^{\frac{1}{2}} , \tag{4}$$

where the S_i are to be adjusted. A variety of conceivable potential shapes can be represented in this way.

A competing, entirely empirical procedure is the one employing *Morse functions with switching terms.*

If U = Morse (1) + Morse (2) + Morse (3), then when all the particles are close together there will be a very deep well which is very unlikely to be realistic. Hence the well must be filled in by adding, for example, the following term:

$$D_1[1 - \text{tgh}\,(ar_2 + c)] \exp[-\beta\Delta r_1] , \tag{5}$$

in which the bracket has the shape shown in Fig. 1.

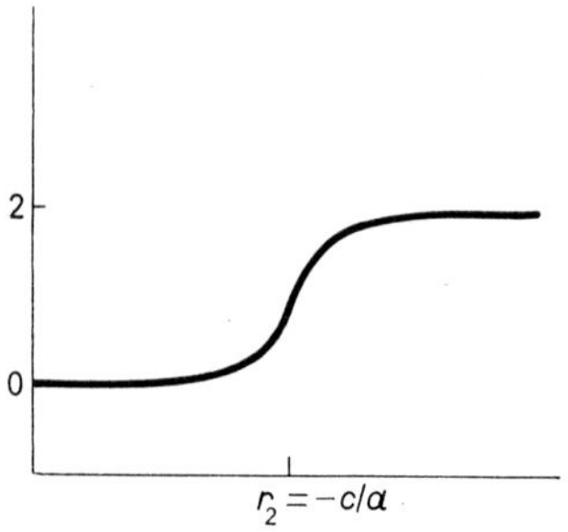

Fig. 1.

The attractive part of Morse(1) is thus switched off when one end of this pair of atoms is approached by the remaining particle. There might be as many as six of these terms, with different parameters a and c, which adjust the position and sharpness of the switching. Repulsive terms, of which a variety are possible, must be included if there are non-bonding atoms.

This is a versatile method. Almost any energy contour along the reaction path can be fitted, although it is more difficult to adjust the shape of the reaction path. Besides the simple abstraction of B by A, two-channel reactions

(6) $$AC + B \leftarrow A + BC \rightarrow AB + C$$

and isomerization reactions

(7) $$ABC \rightarrow ACB$$

have been described in this way.

Although this method is more flexible than the modified L.E.P.S. procedure, the values of parameters required to produce a particular shape are rather unpredictable. A third way of expressing potentials has been developed in order to reduce this difficulty. It may be described as a *map* method, in which the desired potential shape is constructed directly, without recourse to potentials between pairs of atoms.

Two ways that have been used to define a co-ordinate Z along a chosen reaction path are illustrated in Fig. 2 and 3. In Fig. 2 the reaction path is a

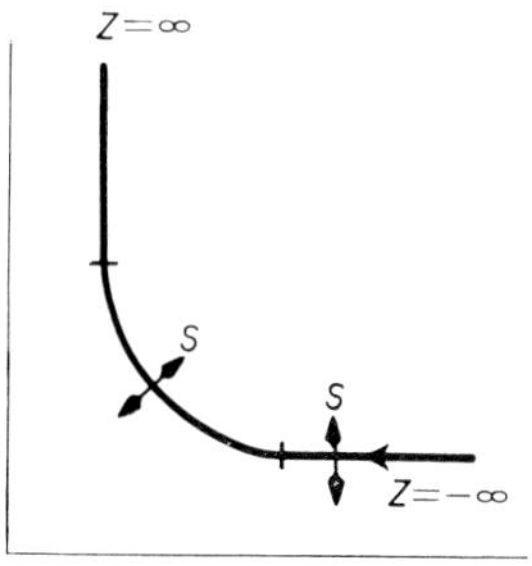

Fig. 2.

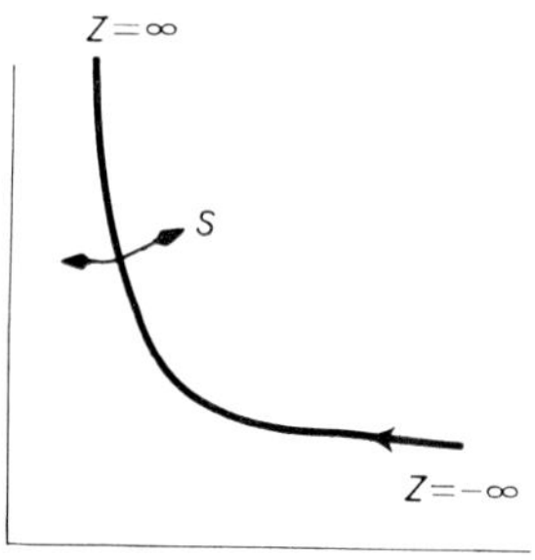

Fig. 3.

quarter circle with a transverse coordinate S defined along its radius [2]; in Fig. 3 both the reaction path and the transverse co-ordinate are (conjugate but not necessarily rectangular) hyperbolas [3]. The shape of the reaction path can be regulated by adjusting the circular radius or the appropriate hyperbolic parameter. The Morse variation is introduced as

(8) $$U = D(1 - \exp[-\beta S])^2, \qquad \beta = \beta(Z), \qquad D = D(Z).$$

A typical Z dependence is

(9) $$\beta = \tfrac{1}{2}(\beta_1 + \beta_2) + \tfrac{1}{2}Y(\beta_1 - \beta_2),$$

which has for $Y=1$, $\beta=\beta_1$; for $Y=-1$, $\beta=\beta_2$. If $Y=Y(Z)$ is now chosen to be some function having the shape of Fig. 1, the potential gradient along the path may be conveniently prescribed.

This method is unusually flexible, but somewhat more expensive than the others. Its use is advisable whenever a systematic variation of potential shape is to be made.

1'2. *The classical trajectories.* – Classical trajectories are calculated because determination of their quantum counterparts is neither practical at present, nor demanded by the nature of the problem. For most purposes a classical treatment is a good approximation for heavy atoms, and even for the interaction of three hydrogen atoms it is far from useless.

The actual equations used are almost invariably those of Hamilton, which give a set of first-order differential equations. This system is briefly illustrated. We define general co-ordinates q_i, with time derivatives $\dot{q}_i$. The kinetic energy will be of the form

$$T=T\,(\text{masses};\ \dot{q}_i^2,\ \dot{q}_i\dot{q}_j,\ q_i)\,. \tag{10}$$

Since the particles are treated as point masses in field-free space, the conjugate momenta are defined by

$$p_i=\frac{\partial T}{\partial \dot{q}_i}=\sum_i \alpha_i\dot{q}_i;\qquad \alpha_i=\alpha_i\ (\text{masses},\ q_i)\,. \tag{11}$$

These linear relationships can be inverted to give

$$\dot{q}_i=\sum_i \beta_i p_i,\qquad \beta=\beta\ (\text{masses},\ q_i)\,,$$

and the kinetic energy may be reconstructed as

$$T=T(\gamma_{ij},\ p_i,\ p_j);\qquad \gamma_{ij}=\gamma_{ij}\,(\text{masses},\ q_i)\,. \tag{12}$$

$$U=U(q_i);\qquad H(p,q)\equiv T+U=E\ (\text{conserved})\,. \tag{13}$$

The Hamilton's equations are

$$\frac{\partial H}{\partial q_i}=-\dot{p}_i\,,\qquad \frac{\partial H}{\partial p_i}=\dot{q}_i\,. \tag{14}$$

It is usual when giving examples to choose one that can be solved exactly. Here the example is instead chosen to require numerical integration, as in a trajectory problem. Consider two atoms in a plane bound by a Morse potential, with polar co-ordinates. Then

$$(15)\quad \begin{cases} T = \frac{1}{2}\mu(\dot{r}^2 + r^2\dot{\theta}^2)\,, \qquad \mu = m_1 m_2/(m_1+m_2)\,,\\ U = D[1-\exp[-\beta(r-r_0)]]^2,\\ p_r = \dfrac{\partial T}{\partial \dot{r}} = \mu\dot{r}\,,\\ p_\theta = \dfrac{\partial T}{\partial \dot{\theta}} = \mu r^2\dot{\theta}\,. \end{cases}$$

The expressions for p_r and p_θ are inverted and the results are inserted in the expressions for T and U.

Then

$$(16)\quad H = \frac{p_r^2}{2\mu} + \frac{p_\theta^2}{2\mu r^2} + D[1-\exp[-\beta(r-r_0)]]^2.$$

$$(17)\quad \begin{cases} -\dot{p}_r = \dfrac{\partial H}{\partial r} = \dfrac{-p_\theta^2}{\mu r^3} + D[1-\exp[-\beta(r-r_0)]]\beta\exp[-\beta(r-r_0)]\,,\\ \dot{r} = \partial H/\partial p_r = p_r/\mu\,, \qquad \dot{\theta} = \partial H/\partial p_\theta = p_\theta/\mu r^2. \end{cases}$$

These three coupled linear first-order differential equations are to be solved numerically.

$$(18)\quad -\dot{p}_\theta = \frac{\partial H}{\partial \theta} = 0\,, \qquad p_\theta = 0\,.$$

Although eq. (18) means that we have an ignorable co-ordinate and hence a constant of the motion, the $\dot{\theta}$ equation must be retained if the θ motion is to be calculated.

The problem described above is too small to be of interest for chemical kinetics. In practice there are between six and thirty coupled equations, and the complexity of each is increased in about the same proportion. But all of the typical features of a real problem occur in this example.

Before the equations of motion may be constructed, it is necessary to select a co-ordinate system. The three kinds most often considered are bond co-ordinates, cartesian, and modified cartesian. Bond co-ordinates are the distances between atoms, the angles between bonds, etc. They are useful only if they greatly simplify U, and have been mostly employed in the study of unimolecular decompositions. In the case of A + BC reactions, bond co-ordinates offer few advantages.

Modified Cartesians are the Cartesian systems with one co-ordinate suppressed in each direction by virtue of the center of mass of the system being fixed. For example, let the positions of A and C be measured from the ABC center of mass; this will specify the position of B implicitly.

Other such arrangements that have been used include A relative to the BC center of mass, B relative to C; A relative to the BCD center, in A + BCD, with C measured from B and D from C; and any number of particles relative to a chosen one, whose location is suppressed. Co-ordinate systems of this type are easy to use because it is always possible to write the kinetic energy in the form given by eq. (12), so that the corresponding part of Hamilton's equations will be simple.

The problem of integrating the equations of motion remains. This is difficult to discuss uniquely because it involves both technique and art. The remarks in Subsection 1·2, up to here, apply to all trajectory studies, but what comes next is merely a description of how things are done in the lecturer's laboratory. Other computers, other types of problem, and other investigators will require variations in what is presented here for illustrative purposes.

There is an analogy with a motor car. We have two gears, a bottom gear for starting and a top gear for speed. The bottom gear is the Runge-Kutta method, illustrated here in its original fourth-order form. In practice various modifications are used. In what follows q will be *either* one of the co-ordinates *or* one of the momenta.

The $(n+1)$-st value of each q is given by

$$(19)\qquad \begin{cases} q_{n+1} = q_n + \dfrac{1}{6}(b_1 + 2b_2 + 2b_3 + b_4)\,, \\ b_1 = \Delta t \left(\dfrac{\mathrm{d}q}{\mathrm{d}t}\right)_{q_n}, \qquad b_2 = \Delta t \left(\dfrac{\mathrm{d}q}{\mathrm{d}t}\right)_{q_n+\frac{1}{2}b_1}, \\ b_3 = \Delta t \left(\dfrac{\mathrm{d}q}{\mathrm{d}t}\right)_{q_n+\frac{1}{2}b_2}, \qquad b_4 = \Delta t \left(\dfrac{\mathrm{d}q}{\mathrm{d}t}\right)_{q_n+b_3}. \end{cases}$$

The second of eqs. (19) is a linear projection of the variation of q at q_n; the others are corrections of higher order. In effect each q is fitted on each interval to a polynomial of fourth degree in t. The time increment Δt must of course be short enough so that q does not have too many extrema to permit this.

The Runge-Kutta is a self-starting method, *i.e.* one only needs to know the q_n before starting each step. No tables of previous values, or their derivatives, are needed. It should be noted that each time t is advanced by Δt the derivative of each q has to be calculated four times.

The top gear is the *Adams-Moulton* procedure, a predictor-corrector method which is also illustrated here in fourth order. We define the predicted value of one of the q's:

$$q'_{n+1} = \frac{\Delta t}{720}\left[1901\left(\frac{\mathrm{d}q}{\mathrm{d}t}\right)_n - 2774\left(\frac{\mathrm{d}q}{\mathrm{d}t}\right)_{n-1} + 2616\left(\frac{\mathrm{d}q}{\mathrm{d}t}\right)_{n-2} - 1274\left(\frac{\mathrm{d}q}{\mathrm{d}t}\right)_{n-3} + 251\left(\frac{\mathrm{d}q}{\mathrm{d}t}\right)_{n-4}\right], \tag{20}$$

and the corrected value

$$q_{n+1} = \frac{\Delta t}{1440}\left[475\left(\frac{\mathrm{d}q}{\mathrm{d}t}\right)_{q=q'_{n+1}} - 1475\left(\frac{\mathrm{d}q'}{\mathrm{d}t}\right)_n - 798\left(\frac{\mathrm{d}q}{\mathrm{d}t}\right)_{n-1} + 482\left(\frac{\mathrm{d}q'}{\mathrm{d}t}\right)_{n-2} - 173\left(\frac{\mathrm{d}q'}{\mathrm{d}t}\right)_{n-3} + 27\left(\frac{\mathrm{d}q'}{\mathrm{d}t}\right)_{n-4}\right]. \tag{21}$$

This method, although it needs a table of previous $\mathrm{d}q/\mathrm{d}t$ in order to start, evaluates derivatives only twice at each step (even in higher order than fourth). It has the same accuracy as the Runge-Kutta method, and is considerably faster.

In numerical integration of trajectories there are four main types of error:

1) *Step size.* This represents the requirement that q between q_n and q_{n+1} must not oscillate so rapidly that it cannot be represented by a fourth degree polynomial. If the step size is too large for this, energy is not conserved.

2) *Intrinsic error.* Even if error 1) is avoided, segments of q are not exactly given by polynomials of any degree. The discrepancy accumulates until finally, if no other error has triumphed, this will.

3) *Truncation error.* This is a result of the fact that a computer can only hold a finite number of digits to represent a number, and hence the number is an approximation. This error builds up in the same way as the intrinsic error.

4) *Functional error.* The manufacturer's routines for evaluation of square root, exponential, etc. are not always as accurate as the computer arithmatic.

If the energy is not conserved then the error is usually of type 1). If the energy is conserved but the completed trajectory cannot be integrated back to its starting conditions, then it is 2), 3), or 4). If increasing the word length improves things, it is 3) or 4), which can be distinguished by further tests.

The example scheme for a complete trajectory is then as follows: Start with the Runge-Kutta method and change gear into the Adams-Moulton for extended periods of operation, shifting down again if (*e.g.*) the step length has to be altered. The calculation is started in a double length mode if this is required for the limiting error to be of type (2). This is continued until the accumulated intrinsic error equals the truncation error for a single word length.

The calculation is completed with single length precision. The best Δt, number of cycles of top and bottom gear, and number of cycles of different precision are all sensitive to the potential energy and to other parameters of the reaction. If optimum speed is to be attained, they must be readjusted for each group of trajectories; in the automobile analogy, the engine is tuned before every race. A factor of as much as ten in efficiency may depend on this.

1·3. *Selecting the initial conditions.* – Since all possible trajectories for a particular reaction cannot be computed, a limited sample must be chosen in some interesting way. Beyond the primitive study, in which only a few trajectories can be calculated and are chosen arbitrarily, there are *scanning* and *statistical* selection methods and mixtures of the two.

In a scan, a systematic sweep through a variable is made in order to show the outcome of the reaction as a function of the variable. For example, in A + BC → AB + C, BC might be fixed in space and the impact parameter of the collision systematically varied. This approach might lead to two kinds of result—smooth, or else noisy, as with a roulette wheel where the result depends in a jerky manner on how the ball was initially placed. A noisy dependence is the usual outcome of any procedure in which more than one or two variables are scanned. Observed smooth variations, in calculations of limited dimensionality or with many parameters fixed, have in the past led to misinterpretation and are to be viewed with suspicion.

In a *statistical* or *Monte Carlo* method the initial co-ordinates and momenta are chosen so that they are a representative sample of the points in the phase space corresponding to a particular type of experiment. Three operationally easy cases can be distinguished:

1) There is a one-dimensional probability distribution for each collision parameter.

2) The dimensionality can be reduced; *e.g.* in A + BCD, the phase of the bending vibration of BCD might be chosen to be zero at the beginning, with or without a compensating variation in the initial position of A.

3) One can neglect some initial parameters; *e.g.* in a molecular beam experiment with H atoms at 10 eV and a cold hydrocarbon, the internal motions of the hydrocarbon would be inconsequential. If none of these conditions is present, a representive sample can still be picked, but the problem becomes much more difficult.

To execute any statistical selection procedure, one needs random numbers in large quantities. One method of obtaining them is as follows: R_0 is a random number $0 < R_0 < 1$ (from tables), odd with N binary digits, and s is the largest such that 5^s has less than N binary digits. Then the remainder of $5^s R_0/2^N$ is taken to be the next pseudo-random number R_1, and so on for R_2, etc. One needs a new R_0 after $2^{s-2} R_i$'s have been generated, so that repeating cycles may be avoided.

Three types of variables occur in A + BC reactions. Some are uniformly random, *e.g.* a planar angle of orientation, which is taken as $\varphi = 2\pi R_i$. An example of an *integrable nonuniform* distribution is that of the impact parameter. The probability of an impact parameter b is

$$P(b) = \frac{2b}{b_{\max}^2}, \qquad 0 < b < b_{\max}, \qquad \text{otherwise } P(b) = 0. \tag{22}$$

The cumulative distribution function

$$C(b) = \int_0^b P(b')\,\mathrm{d}b' = \frac{b^2}{b_{\max}^2}, \tag{23}$$

which is the probability of b smaller than a given value, is equated to a random number.

$$C(b) = R_j; \qquad b = b_{\max} R_j^{\frac{1}{2}}. \tag{24}$$

If the integral in eq. (23) cannot be explicitly done, a rejection method may be used. Suppose we have the probability function.

$$P(v) = v^2 \exp[-\alpha v^2]. \tag{25}$$

Cut off v at some plausibly large value, and take $v = v_{\max} R_m$. This will not generate the distribution we want. We have to throw away some of the v to get something that resembles eq. (25). If

$$\frac{P(v)}{P(\bar{v})} \geqslant R_s, \tag{26}$$

in which $\bar{v}$ is the v for largest $P(v)$, the value of v is accepted; otherwise it is discarded. Clearly v for $P(v) = P(\bar{v})$ are always retained, and those on the wings of the distribution are nearly always rejected. Repetition of this procedure generates the required distribution shape.

In an A + BC reaction, all initial quantities may be chosen by one or another of these methods. They may easily be converted to co-ordinates and momenta in the co-ordinate systems being used, following which the trajectory calculation is ready to proceed.

2. – Results for artificial reactions.

We adopt the reverse of historical order. Originally, trajectory studies had to be closely related to laboratory experiments, so that their significance for the real world could be assessed. General studies not based on a particular laboratory reaction came later. By now, however, the trajectory approach is one of well established validity, and we may use a systematic survey of artificial reactions as a guide to what we may expect to learn in connection with real experiments.

Inquiries into the behavior of reactions in general have fallen into these areas: the character of the motion; the effect of the collision parameters; the influence of the potential surface and of the reactant masses; and the importance of the dimensionality of the calculation.

2·1. *The character of the motion.* – For purely exothermic reactions of the A + BC type, without appreciable basins in the potential surface, there is a standard type of trajectory which dominates the results. It is illustrated in

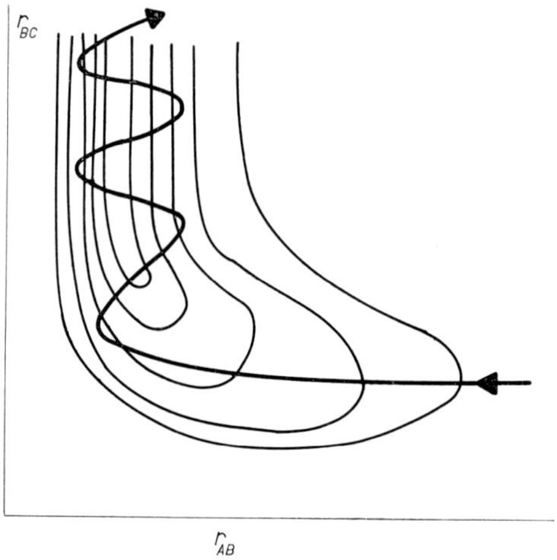

Fig. 4.

Fig. 4, with the angle-dependence of the potential (of which it is more or less independent) suppressed. A characteristic feature is that r_{BC} increases continuously throughout the event—one of the hallmarks of a « direct » reaction.

2·2. *The collision parameters.* – These include impact parameter, orientation, vibration and rotation of BC—all those quantities which are selected by the methods of Subsection **1**·3.

As to which of these produce a recognizable trend in the results of the reaction, the impact parameter b and the initial relative velocity v are the most important, because of their effect on the total angular momentum L. A fairly smooth correlation frequently exists between product scattering angle and L, and sometimes b. This is generated in a straightforward way *via* the rotation period of the intermediate entity ABC, which ordinarily executes only a fraction of a full rotation during the reaction. The velocity v is less well correlated, and the effect of replacing a v distribution with a uniform one is small (in center-of-mass co-ordinates). The same is true of the initial vibrational and rotational variables. With one exception to be noted later, the orientation of BC also has a noisy relationship to the final results.

Another grouping is according to the type of reaction property affected. Including some things yet to be discussed, the overall rough correlation is as follows: b, v, the potential at large r, the direction of favored approach, and the dimensionality of the calculation are associated with the scattering pattern and product rotation; whereas the potential with all distances small largely determines the partitioning of reaction energy between translation and AB internal. This empirical observation is reflected in current methods for reducing molecular beam reactive scattering data.

2·3. *The masses and the potential.* – These two aspects of the problem have not been fully separable from one another. The main question has been whether there exist « pure » mass effects in the sense of their independence of all the other features of the reaction. The well-known special case of the *angular-momentum-limited reaction*, although describable in mass-effect terms, is not a pure one—it is actually a result of conservation laws, as its name suggests.

Ordinarily

$$L \simeq \mu_{A,BC} v b \,, \qquad J_{AB} \hbar = (2 I_{AB} E_{rot})^{\frac{1}{2}} \,, \tag{27}$$

$$(2 I_{AB} \Delta E)^{\frac{1}{2}} \gg L \,, \tag{28}$$

in which μ and I are reduced mass and moment of inertia, J is a rotational quantum number, and E_{rot} and ΔE are the final AB rotational energy and the exothermicty. If eq. (28) holds, there is an upper limit to J_{AB}. Otherwise, we have the angular-momentum-limited reaction, with large J_{AB}, unusually small amounts of AB vibration, and a tendency towards planarity of the trajectories because the orbital angular momentum is forced to be always large. Conditions which favor this are small ΔE, small final r_{BC} relative to b, and small $m_B m_{ABC}/m_{AB} m_{BC}$ from the combined mass dependences of $\mu_{A,BC}$ and I_{AB}. In effect, small m_B is required.

A purer but more elusive mass effect is the light-atom one: Is there anything special about small m_A? Before discussing this, we have to look at the effect of potential energy shape on product internal energy, with which this mass effect may be entangled.

The historically favored potential surface classification has been into *attractive* (Fig. 5) and *repulsive* (Fig. 6) categories, also sometimes called *early-*

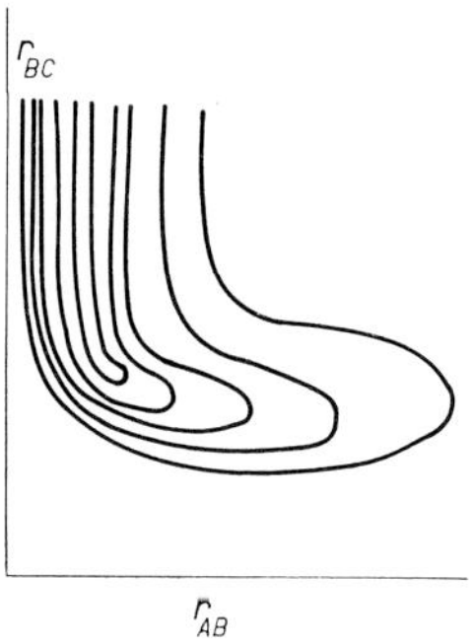

Fig. 5.

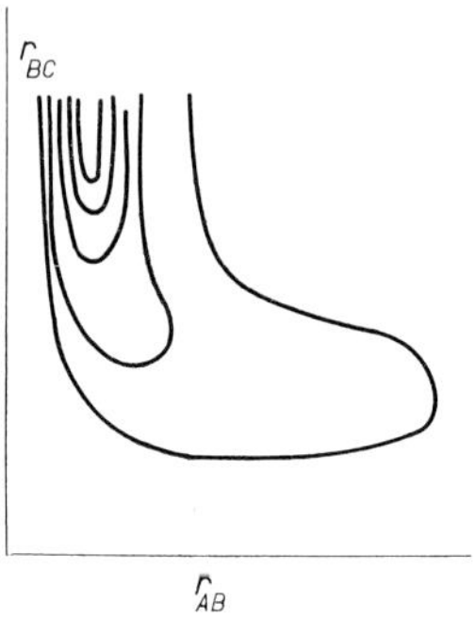

Fig. 6.

downhill and *late-downhill*. These, it is widely agreed, are ordinarily associated with high and low product vibrational energy respectively. Other details of the potential shape, studied [3] with a map-type potential in which there is continuous variation between Fig. 5 and 6, are found to be much less important—the sharpness of the corner, the steepness of the downhill portion, many features of the angle-dependence not shown in the figures, etc.

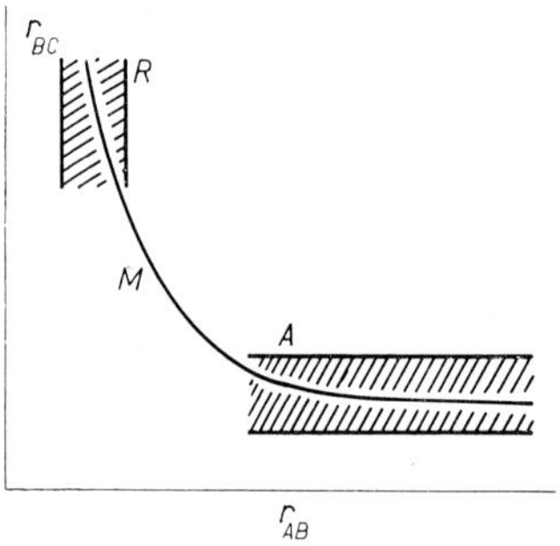

Fig. 7.

An extended classification suggested by J. Polanyi [1] is illustrated in Fig. 7. The shaded regions represent the ranges of diatomic vibration amplitudes at some standard temperature. The surface is described by means of the fraction of exothermicity that is released along the path of minimum energy in regions *A* (attractive), *M* (mixed; possibly absent), and *R* (repulsive).

Returning to the light atom effect: it was first found [4], in a two-dimensional (2D) study on a Morse switching potential, that small m_A produced low AB vibration no matter what the potential. This observation disappeared temporarily with a later 3D study on the map surface, suggesting an interaction with the shape of the path of minimum energy (the map variability could not then reproduce the Morse-switching shape). It then reappeared in the 2D Modified L.E.P.S. study of Polanyi [1], for which the analysis of Fig. 7 was developed. Polanyi finds that although energy release of the *M* type ordinarily behaves as A rather than R, with small m_A there is a much more pronounced variation of the reaction properties with surface detail. In particular, surfaces with much M character are susceptible to a light-atom effect of the sort previously observed.

It seems at the moment reasonable to say that there is a light atom effect (small m_A) leading to lowered product vibrational energy, but that it is not a pure one; and that it is not entirely clear just how it may be best described. If POLANYI replaces the path of minimum energy, in Fig. 7, with a collinear trajectory, he can suppress the effect—in that the energy partitioning now depends only on the amount of A, M and R—but at the cost of making the definition of these mass-dependent.

2'4. *The dimensionality.* – Since this has been discussed in other lectures, a brief summary will suffice here. A reaction simulated by 2D trajectory methods will yield some distribution of scattering angle, $N(\theta)$, where θ is (say) measured between the initial A and final C directions in the center-of-mass system. If the reaction were in fact planar in 3D the corresponding $N(\theta)$ could be constructed on a solid-angle basis by dividing $N(\theta)$ by $\sin\theta$, with some other refinements which need not distract us here. But in fact it is found that although energy partitioning is rather insensitive to dimensionality, the scattering is altered by out-of-plane forces for most potentials; the directions of the orbital-angular-momentum vectors becomes scrambled. As a result an artificial reaction that is isotropic in 2D, $N(\theta) =$ constant, will also be isotropic with $N(\theta)$ proportional to $\sin\theta$ in 3D. A mild exception is the angular-momentum-limited case, which in 3D tends toward planarity for the reasons discussed in Subsect. 2'3.

3. – Results corresponding to real reactions.

All the trajectory features that will arise here have appeared at least once in Sect. 2, but a few laboratory-related examples should be mentioned to illustrate and round out the discussion.

3.1. *Hydrogen and chlorine.* – The reaction

$$H + Cl_2 \rightarrow HCl + Cl \tag{29}$$

has been extensively studied in the laboratory by J. POLANYI [5], who has also investigated it by trajectory methods. For an illustration of the laboratory results, see another lecture in this volume (Fig. 2 of *Special results of theory: Compound-state approaches*). The principal controversy connected with this reaction has concerned whether or not the experimental results demand a repulsive potential surface. This issue is of course occluded to some extent by the problems associated with the light-atom effect (Subsect. 2.3), and also by the fact that the character of the experimental results has changed somewhat in the course of time.

The earliest trajectory work [4] suggested a pure mass effect as the source of the relatively low HCl vibration states observed in these spectroscopic studies. Later studies [1] seemed to support a repulsive surface as the sole explanation, though reinterpretation in view of laboratory refinements [5] might make $50\% A + 50\% R$ equally reasonable. Very recent, preliminary *3D* trajectory studies on the modified L.E.P.S. surface [5] indicate $33\% A$, in Polanyi's classification (Fig. 7), for reaction (29). For the Br_2 analog of eq. (29), 27 to $79\% A$ would fit the product vibrational distribution peak position, with $27\% A$ agreeing best with the *shapes* of the vibrational and rotational distributions. These shapes, however, are relatively minor features of the distributions and may be affected by the minor potential parameters listed in Sect. 2.3, so that the special nature of the L.E.P.S. reaction path may be a factor here.

Since preliminary results have been cited, drawing of firm conclusions would be premature. It seems likely that the final results will turn out to be a not too uneven compromise between the extreme initial positions that were taken.

3.2. *Alkali atoms and alkyl iodide.* – The reaction

$$K + CH_3I \rightarrow KI + CH_3 , \tag{30}$$

well known in the annals of molecular beams [6], has along with its relatives been repeatedly studied by trajectory methods [4, 7, 8]. Here CH_3 is treated as a single particle of atomic mass 15. The main developments, in more or less chronological order, were these:

1) *2D*, attractive Morse potential with one switching term: Too large total cross-section ($\sigma = 200$ Å^2), bad scattering because of the dimensionality. The peak KI internal energy was $E_{int} = 0.7$, expressed as a fraction of the exothermicity. The experimental value, then 0.8, is now 0.6.

2) As 1), but with repulsive potential: The peak E_{int} was zero, which is taken to rule out the repulsive type of surface.

3) As 1), but 3D. The scattering pattern was improved, with essentially no other changes.

4) The 3D map potential: The results were not very different than with 3), except that the peak E_{int} was 0.5 for a fully attractive surface. This would now be considered in good agreement with the experiments.

5) As 3), but with Morse parameters adjusted away from those for stable molecules; 6) as 3), but with two switching terms; 7) as 3), but with an exp-6 K - CH_3I repulsion rather than a simple exponential one between K and CH_3. These were efforts by KARPLUS and RAFF [7] to solve the remaining problems associated with this reaction, especially that of the too large σ. Surface 7) works somewhat less well than the other two. Further preference for surface 5) over 7) is generated by a supplemental study, KARPLUS and GODFREY [9], of the reaction of K with oriented CH_3I.

On the whole, this work provides a reasonably good interpretation of the experimental results. Surface 5) seems to be the most clearly useful of the Morse-switching representations; the other ways of writing potentials have not been adequately explored. Surface 5) has also been used by RAFF [10] in a study of

$$K + C_2H_5I \rightarrow KI + C_2H_5\,, \tag{31}$$

with C_2H_5 treated as two particles of mass 14 and 15. The peak E_{int} is 0.9, of which about 0.14 resides in the C—C bond; $\sigma = 16\ \text{Å}^2$. As far as can be determined at the moment, these are plausible numbers.

There is one loose end. If in reaction (30) we substitute Cs for K, the peak E_{int} drops to near 0.3, according to recent reinterpretations of the experimental results [6]. Only in investigation (1) was this case treated, and the effect did not appear. In view of Subsect. 2˙3, one might question the likelihood of its being accounted for in any of the more advanced studies, *e.g.* (4) or (5).

3˙3. *Potassium and hydrogen bromide.* – This is an angular-momentum-limited reaction, because of its small exothermicity, and behaves as such in both the laboratory [11] and preliminary trajectory calculations. On a map-type potential surface, the smallness of the H atom gives rise to an unusual feature—quasi-diatomic collisions, with protracted orbiting at a critical impact parameter. The relationship of this to experiment is however unclear, since few empirical potentials are very realistic at the distances where the orbiting occurs, and the map surface is not one of the better ones in this respect.

3˙4. *Potassium and bromine.* – Recent unpublished work of GODFREY and KARPLUS, quoted here with permission [12], explains the behavior of the reaction

$$K + Br_2 \rightarrow KBr + Br \tag{32}$$

by means of a novel and interesting switching potential. It is

$$(33)\qquad U = \frac{\exp[\alpha\Delta E]}{1+\exp[\alpha\Delta E]}\,U(\mathrm{K},\mathrm{Br}_2) +$$

$$+\frac{1}{1+\exp[\alpha\Delta E]}\left[\frac{1}{1+\exp[\alpha'\Delta E']}\,U(\mathrm{KBr},\mathrm{Br}') + \frac{\exp[\alpha'\Delta E']}{1+\exp[\alpha'\Delta E']}\,U(\mathrm{KBr}',\mathrm{Br})\right]$$

with

$$(34)\qquad \begin{cases}\Delta E = \min\{U(\mathrm{KBr},\mathrm{Br}'),\,U(\mathrm{KBr}',\mathrm{Br})\} - U(\mathrm{K},\mathrm{Br}_2)\,,\\ \Delta E' = U(\mathrm{KBr},\mathrm{Br}') - U(\mathrm{KBr}',\mathrm{Br})\,.\end{cases}$$

A sufficiently realistic set of pair potentials (*e.g.* Morse for Br_2 and inverse sixth power distance dependence for K-Br_2) makes up $U(\mathrm{K}, \mathrm{Br}_2)$. Terms like $U(\mathrm{KBr}, \mathrm{Br}')$ are similarly constructed but with details appropriate to an alkali halide bond rather than a covalent one. When K is far from Br_2, ΔE is large and positive, and $U(\mathrm{K}, \mathrm{Br}_2)$ dominates; when ΔE becomes negative the balance of eq. (33) takes over. The constant α regulates the sharpness of the switching. For the two (KBr, Br) terms, α' and $\Delta E'$ have corresponding roles.

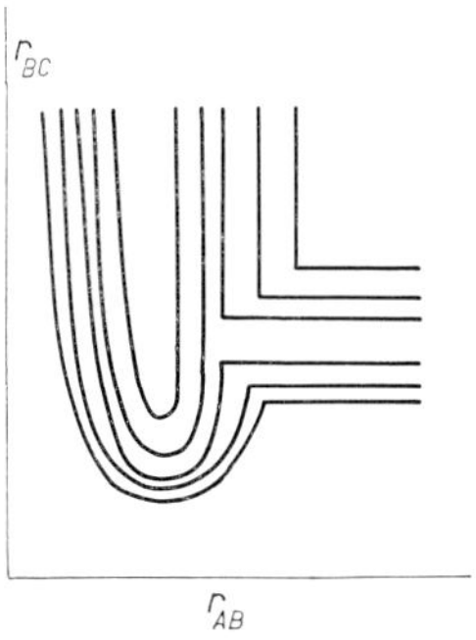

Fig. 8.

It is realistic to make α large, so that the surface may have the appearance qualitatively illustrated in Fig. 8. This produces strongly peaked forward scattering of KBr (*i.e.* same direction as initial K), which has been difficult to attain with other surfaces. Other characteristic features of this class of reaction, such as the very large total cross-sections, are also well reproduced.

3·5. *Tritium and methane.* – This is mentioned merely as an example of a more complicated type of trajectory study now under way. There are six particles, and two reaction paths:

$$(35)\qquad \begin{cases}\mathrm{T}+\mathrm{CH}_4 \to \mathrm{HT}+\mathrm{CH}_3\ \text{(abstraction)}\,,\\ \qquad\qquad \to \mathrm{H}\ +\mathrm{TCH}_3\ \text{(substitution)}\,.\end{cases}$$

Tritium is translationally hot, having in the initial calculations and in some of the key experiments an energy of 2.8 eV. The only potential surface thus far explored is one not having topological features that might be expected to lead to substitution, and only abstraction has been observed. Also, only a

« tight » intermediate state has been considered—one in which the C—H bonds not involved in the reaction are not weakened by the presence of five other atoms near the carbon. On a preliminary basis, it is already possible to conclude that energy deposition in CH_3 will be appreciable, so that the presence of all six particles in the calculation will lead to results appreciably different from those of the more usual three-particle simplification of the reaction.

* * *

The help of C. M. KAPLINSKY and D. DAI-WEI-CHANG in preparing a draft of this lecture is gratefully acknowledged.

REFERENCES

[1] P. KUNTZ, E. NEMETH, J. POLANYI, S. ROSNER and C. YOUNG: *Journ. Chem. Phys.*, **44**, 1168 (1966).
[2] F. WALL and R. PORTER: *Journ. Chem. Phys.*, **36**, 3256 (1962).
[3] D. BUNKER and N. BLAIS: *Journ. Chem. Phys.*, **41**, 2377 (1964).
[4] N. BLAIS and D. BUNKER: *Journ. Chem. Phys.*, **39**, 315 (1963).
[5] K. ANLAUF, P. KUNTZ, D. MAYLOTTE, P. PACEY and J. POLANYI: *Disc. Faraday Society*, **44**, 183 (1967).
[6] E. ENTEMANN and D. HERSCHBACH: *Disc. Faraday Society*, **44**, 289 (1967).
[7] L. RAFF and M. KARPLUS: *Journ. Chem. Phys.*, **44**, 1212 (1966).
[8] N. BLAIS and D. BUNKER: *Journ. Chem. Phys.*, **37**, 2713 (1962).
[9] M. KARPLUS and M. GODFREY: *Journ. Amer. Chem. Soc.*, **88**, 5332 (1966).
[10] L. RAFF: *Journ. Chem. Phys.*, **44**, 1202 (1966).
[11] C. RILEY, K. GILLEN and R. BERNSTEIN: *Journ. Chem. Phys.*, **47**, 3672 (1967).
[12] This has now been published: M. GODFREY and M. KARPLUS: *Journ. Chem. Phys.*, **49**, 3602 (1968).

Special Results of Trajectory Studies.

M. Karplus
Harvard University - Cambridge, Mass.

1. – Introduction.

From a prior lecture [1], we know that the theoretical calculation of reaction attributes can be divided into two parts by means of the Born-Oppenheimer approximation and the adiabatic hypothesis. The first part (*i.e.* the determination of a potential surface parametrically dependent on the nuclear co-ordinates) has been described already [1]; the second part (*i.e.* the determination of the scattering for atoms and molecules treated as point particles moving on such a surface) is the subject of this and a subsequent lecture [2]. Although the calculations of the surface is clearly a quantum-mechanical problem requiring the solution of the electronic Schrödinger equation, the scattering on the surface is a problem at the borderline between classical and quantum mechanics. For most chemical reactions, the deBroglie wavelengths of the particles $[(\hbar/p)\sim 0.1\ \text{Å}]$ relative to the characteristic distances of the potential (0.05 eV/Å) are such that classical mechanics is expected to provide a reasonable approximation to most of the scattering attributes of interest. However, because an unequivocal statement concerning the absence of quantum corrections cannot be made, it is important to have available comparisons of classical and quantum treatments; such a comparison is described in a subsequent lecture [2]. Here we consider the classical or quasi-classical approach which can be carried out to any desired degree of accuracy for potential surfaces of reasonable complexity.

The utility of the classical calculations is twofold. When empirical surfaces are used for the atomic interactions, the classical trajectory analysis serves as the link between the experimental measurement and the surface parameters; this type of approach has been described by Bunker [3]. If instead, reliable semi-theoretical or theoretical surfaces are employed, the trajectory treatment can provide the ultimate details concerning the scattering attributes of the reaction; that is, the consequences of a single collision trajectory with uniquely specified initial conditions can be obtained. For most applications, this is not of interest and averages over suitably chosen sets of initial conditions are introduced. These yield reaction cross-sections as a function of appropriate parameters for comparison with the available measurements. It should be noted that in this type of study, the experimental results serve primarily to check whether or not the theoretical results are likely to be reliable; the significant information concerning the reaction is then obtained with much greater ease and in much greater detail from theory than from experiment.

Since the methods used for calculating trajectories and for performing the desired averages have been described already by BUNKER [3], we limit the present lectures to their application to the $H+H_2$ exchange reaction and its isotopic analogues [4]. For all the calculations presented here, the analytic semi-theoretical surface determined by the generalized London formulation with overlap (PK of ref. [1]) was used to represent the interaction between the three atoms.

2. – Total reaction probability and cross-section.

One property of the collision which can be studied experimentally by an optical model analysis of elastic scattering data [5] is the probability of reaction as a function of impact parameter $[P_r(b)]$. Figure 1 shows $P_r(b)$ calculated for H, H_2 at a fixed initial relative energy ($E_R = 0.48$ eV) and for a fixed

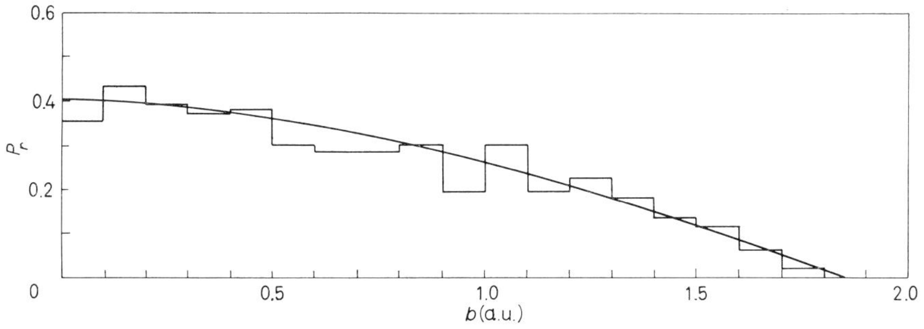

Fig. 1. – Reaction probability P as a function of impact parameter b for $H+H_2$ ($v = 0$, $J = 0$): bar graph from the trajectories; smooth curve is a cosine dependence.

initial quasi-classical molecular rotation-vibration energy ($v = 0$, $J = 0$, $E_{vibrot} = 6.2$ kcal/mole); the other variables defining the trajectories (*e.g.* vibrational phase, molecular orientation) are averaged over by a Monte Carlo sampling procedure [3, 4]. The histogram shown in Fig. 1 represents the actual calculations and the curve, which corresponds to a cosine b dependence, is drawn visually. It appears that $P_r(b)$ is, within statistical error, a smooth, monotonically decreasing function of impact parameter. It is important to note that even for $b = 0$, $P_r(b)$ does not reach unity. This corresponds to the presence of a « steric factor » limiting the reactive collisions to a range of orientations in the transition region. Of course, the « steric factor » is energy-dependent in the sense that $P_r(0)$ is a function of the initial relative energy, increasing from zero for energies below threshold to about 0.6 at energies of 2 eV.

If $P_r(b)$ is integrated over the contributing impact parameters weighted by their probability of occurrence, the total reaction cross-section S_r is obtained; that is

$$S_r(E_R, v, J) = 2\pi \int_0^{b_{max}} P_r(E_R, v, J, b)\, b\, db\,, \tag{1}$$

where b_{max} is the value of b such that $P_r(b)=0$ for $b \geqslant b_{max}$. Figures 2 and 3 show the form of S_r for fixed vibrational energy ($v = 0$), and different rotational

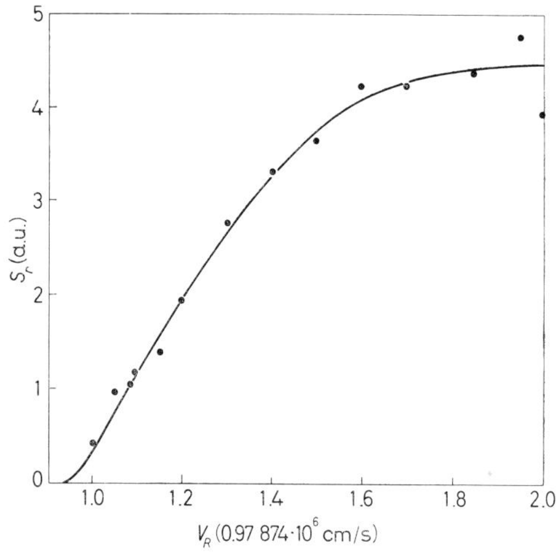

Fig. 2. – Reaction cross-section $S_r(V_R, J, v)$ as a function of V_R for $H+H_2$ ($J=0$, $v=0$).

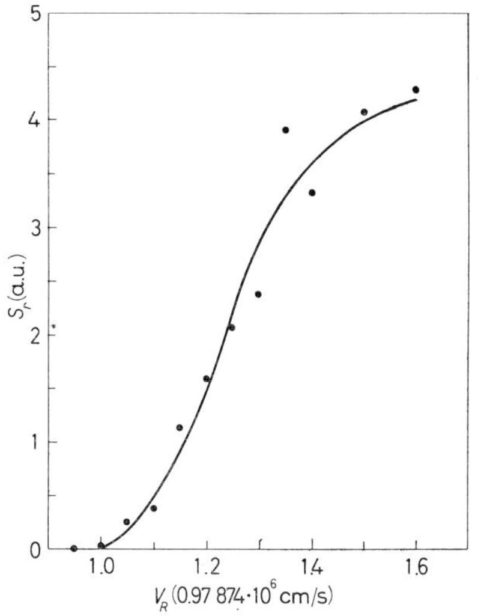

Fig. 3. – Reaction cross-section $S_r(V_R, J, v)$ as a function of V_R for $H+H_2$ ($J=5$, $v=0$).

energies ($J = 0$ and $J = 5$) as a function of the incident relative velocity $V_R(E_R = \frac{1}{2}\mu V_R^2$; with the velocity units used, $0.9787 \cdot 10^6$ cm/s, $E_R(\text{eV}) = \frac{1}{3} V_R^2$ for H, H_2). It is evident that S_r rises monotomically from a threshold value to a plateau for the low-velocity region (but see below). Further the two curves for $J = 0$ and $J = 5$ are very similar, implying that the rotational energy (≈ 5.1 kcal/mole for $J = 5$) is not available for crossing the energy barrier; in fact, a quantitative comparison shows that the $J = 5$ cross-section has a higher threshold value than that for $J = 0$ ($J = 0$, $E_R^0 = 5.7$ kcal; $J = 5$, $E_R^0 = 7.2$ kcal). In contrast to rotational energy, the vibrational energy does contribute significantly to crossing the barrier. Since the barrier height is 9.13 kcal and the $J = 0$ threshold is 5.7 kcal, the zero-point energy of 6.2 kcal is able to provide the additional 3.4 kcal that are required. Thus, about half of the zero-point energy is available in this case, although the exact amount varies with the vibrational state in a way that does not appear to follow any simple model [6]. A first approximation for $v = 0$ is obtained from the assumption that the reaction is vibrationally adiabatic, which corresponds to 3.1 kcal being available.

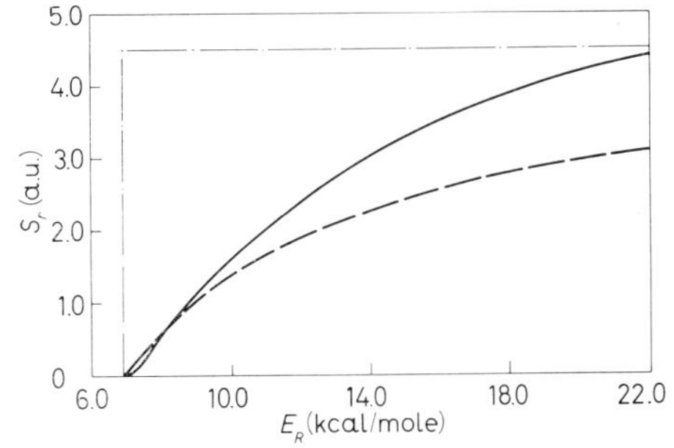

Fig. 4. – Reaction cross-section $S_r(E_R, J, v)$ as a function of E_R: (———) $S_r(E_R, 0, 0)$ from trajectories; (— — —) $S_r(E_R)$ from eq. (2b); (—·—·—) $S_r(E_R)$ from eq. (2a).

The form of the cross-section corresponds to none of the commonly used ultra-simple models (Fig. 4); *e.g.*, it is neither a step function,

$$(2a)\qquad S_r = 0\,, \qquad E_R < E_R^0\,; \qquad S_r = \pi b_0^2\,, \qquad E_R \geqslant E_R^0\,,$$

nor an energy-along-the-line-of-centers curve,

$$(2.b)\qquad \begin{cases} S_r = 0\,, & E_R < E_R^0; \\ S_r = \pi b_0^2\left(1 - \dfrac{E_R^0}{E_R}\right), & E_R \geqslant E_R^0 \end{cases}$$

although the latter gives a reasonable approximate fit near threshold if E_R^0 and b_0 are chosen appropriately.

If the relative velocity is increased significantly beyond the thermal range to that sampled in hot-atom studies, results of the type shown in Fig. 5 are obtained. The specific study [7] was made for $T + H_2$ and $T + D_2$ because these two isotopic species have been investigated experi-

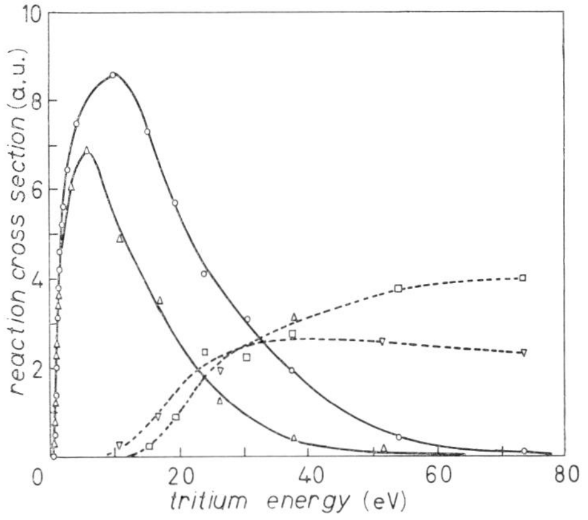

Fig. 5. – Rotation-averaged exchange and dissociation reaction cross-sections as functions of tritium energy for stationary hydrogen and deuterium molecules: ○, $T + H_2$ exchange; △, $T + D_2$ exchange; □, $T + H_2$ dissociation; ▽, $T + D_2$ dissociation.

mentally [8]. We see that as the relative velocity (laboratory T energy is used in the present plots) increases, the cross-section reaches a maximum and then drops asymptotically to zero. It should be noted however that even at very high energies (*e.g.* 50 eV for $T+D_2$, 60 eV for $T+H_2$) nonzero contributions to the cross-section are found. Figure 5 also includes the dissociation cross-sections for these species ($T+H_2 \to T+2H$, $T+D_2 \to T+2D$); dissociation is seen to begin very close to the lower bound set by energy conservation ($T+H_2$, 11.3 eV; $T+D_2$, 7.9 eV), indicating that all three atoms are involved in the dynamics at this point. The general features of both the reaction and dissociation cross-sections can be understood in terms of a simple hard-sphere model with an energy barrier [9]. Although there is no experimental information concerning the shape of the cross-sections, the hot-atom measurements do confirm the yield ratio $([T+H_2]>[T+D_2])$ expected from the relative values given in Fig. 5 [7, 8].

3. – Rate constants and transition-state theory.

From the total reaction cross-sections $S_r(E_R, v, J)$ as a function of E_R, v, and J, the thermal rate constant can be calculated by averaging $S_r(E_R, v, J)$ over Boltzmann distributions for these variables. We have for $K(T)$ in units of $cm^3/(mole\text{-}s)$

$$K(T) = Q_{vJ}^{-1}\left[\sum_{v,J} f_J(2J+1)\exp[-E_{v,J}/kT]\right]\cdot \\ \cdot N_A\left(\frac{2}{kT}\right)^{\frac{3}{2}}(\pi\mu_{A,BC})^{-\frac{1}{2}}\int_0^\infty S_r(E_R, v, J)\exp[-E_R/kT]E_R\,dE_R\,, \tag{3}$$

where Q_{vJ} is the rotation-vibration partition function for H_2, f_J is the statistical weight factor (*e.g.*, in H_2, $f_J=1$ for even J and $f_J=3$ for odd J), $E_{v,J}$ is the vibration-rotation energy, N_A is Avogadro's number, $\mu_{A,BC}$ is the reduced mass for the relative motion ($\frac{2}{3}$ for H, H_2). At temperatures less than or equal to 1000 °K, only $J=0, 1, ..., 5$ and $v=0$ contribute significantly to $K(T)$. Inserting the calculated cross-sections for this set of initial states into eq. (3), we find the results shown as $\log K(T)$ *vs.* $1000/T$ in Fig. 6; the dots are the

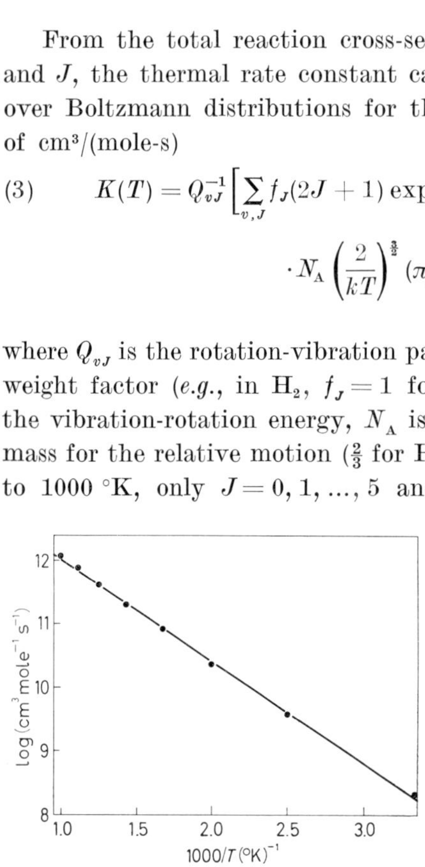

Fig. 6. – Collisional values of $\log K(T)$ *vs.* $1000/T$ for the $H+H_2$ reaction. The dots are the computed results and the straight line corresponds to the equation $K(T)=A\exp[-E_a/RT]$ with $A=4.33\cdot 10^{13}\ cm^3\ mole^{-1}\ s$ and $E_a=7.441$ kcal/mole.

computed values at 300 °K, 400 °K, ..., 1000 °K. Since the straight line drawn through the points is a good approximation to them, we can use it to determine the parameters of an Arrhenius-type rate expression; the result is

$$K(T) = 4.3\cdot 10^{13} \exp[-7.441/RT] \text{ cm}^3/(\text{mole-s}) \tag{4}$$

corresponding to a pre-exponential factor of $4.3\cdot 10^{13}$ cm³/(mole-s) and an activation energy of 7.44 kcal/mole. In considering the very good straight line obtained in Fig. 6, it is important to note that no specific model requiring such an Arrhenius-type temperature-dependence has been assumed for the reaction cross-sections. However, such behavior is not unexpected since for a limited temperature with $E_R \gg KT$, the results depend primarily on the threshold value and are very insensitive to the form of the cross-section [7, 8].

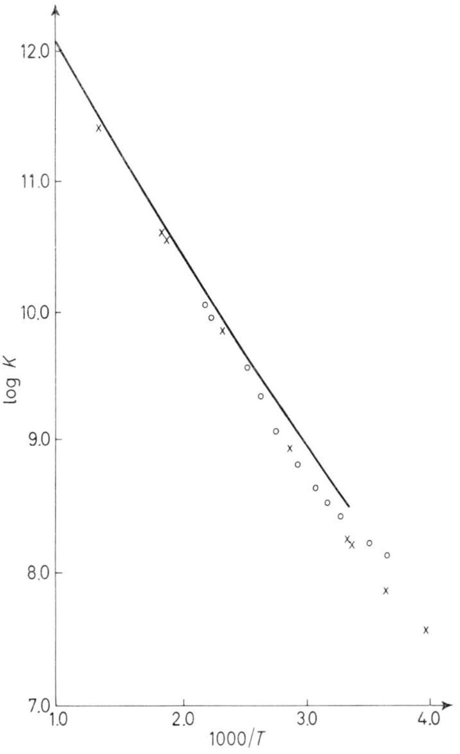

Fig. 7. – Plot of log K *vs.* $1000/T$ for $D + H_2$: (———) calculated rate; ×, experimental values of WESTENBERG and HAAS; o, experimental values of RIDLEY, SCHULZ and LEROY.

To give some indication that the present calculations may be related to the actual behavior of the H, H_2 reaction, we compare in Fig. 7 the calculated temperature-dependence of the $D + H_2 \to DH + H$ rate constant with the recent experimental measurements of SCHULTZ and LEROY and of WESTENBERG and DEHAAS [10]. This reaction is chosen rather than $H + H_2$ itself, because a quantitative interpretation of the ortho → para conversion rate measurements in the latter requires a knowledge of the individual rotational state exchange cross-sections [11]. The reasonable agreement between theory and experiment shown in Fig. 7 is encouraging; the deviations at very low temperatures have been ascribed to quantum effects [10], although the evidence is not unequivocal.

The small deviations from the Arrhenius line appearing in Fig. 6 can be described by a pre-exponential temperature dependence. Writing

$$K(T) = A'\, T^{\alpha} \exp[-E'_A/RT] \tag{5}$$

we find by a least-squares fit

$$A' = 7.867 \cdot 10^9, \qquad \alpha = 1.1792, \qquad E'_A = 6.234 \text{ kcal},$$

for the temperature range under consideration. Examination of the rate constants for the individual rotational states shows that part of the pre-exponential temperature-dependency of $K(T)$ is present in each state and part arises from the temperature variation in the contributions made by the different states.

4. – Significant energies.

It is worth while at this point to emphasize the difference between three different energies that are often referred to in describing a reaction of the $H+H_2$ type. They are *a*) the barrier height E_B, *b*) the threshold energy E_{th}, and *c*) the activation energy E_A. The barrier height E_B (sometimes called the classical activation energy) is a parameter describing the potential energy surface and corresponds to the difference in energy between the separated reactants, each at its equilibrium geometry, and the saddle-point geometry; for H, H_2 on the *PK* surface, the value is 9.13 kcal/mole. The threshold energy E_{th} is a dynamical quantity and describes the relative translational energy at which the reaction cross-section goes to zero. For the quantum-mechanical scattering problem, there is no absolute threshold and it may be necessary to define one in terms of some cut-off; actually a corresponding problem exists in the classical trajectory treatment since its statistical nature makes an exact threshold determination very difficult. Moreover, it must also be remembered that E_{th} in general depends on the internal energy of the reactants (see Fig. 2 and 3), so that the proper quantity to consider is $E_{th}(v, J)$; for H, H_2, $E_{th}(0, 0) = 5.7$ kcal/mole and $E_{th}(0, 5) = 7.2$ kcal/mole. The activation energy E_A is a statistical quantity, which depends on the averages of the cross-section over temperature dependent initial-state distributions. It can be defined only for the systems and temperature ranges for which an Arrhenius equation is a satisfactory fit to the experimental or theoretical results; for H, H_2, $E_A = 7.44$ kcal/mole. Thus, for the H, H_2 reaction, all three energies are similar but clearly far from identical. That $E_A > E_{th}(0, 0)$ is necessary since E_A represents an average over contributing trajectories all of which have relative translational energies greater than or equal to $E_{th}(0, 0)$. This might no longer be exactly correct if higher vibrational states made important contributions since their threshold energies are expected to be significantly lower. The relative value of E_{th} and E_B can vary from reaction to reaction depending in the quasi-classical treatment on the availability of internal energy for crossing the barrier; the result $E_{th} < E_B$ found for H, H_2 is expected to be the most common case. No general statement can be made about the relationship between E_B and E_A; both $E_B > E_A$ and $E_B < E_A$ can be expected to occur.

5. – Transition-state theory.

Because the transition-state or absolute-rate-theory model for chemical reactions provides such a beautifully simple approach for calculating the thermal rate constants (*i.e.* it circumvents the need for a dynamical calculation by a statistical assumption), it is important to investigate its range of validity. In principle, this can be done by comparing calculated rate constants with the experimental measurements. Although useful approximate comparisons can be made in this way [12], the potential parameters for no reaction are known reliably enough to permit a direct experimental test of absolute-rate theory. For the H, H_2 case and its isotopic analogues, such a test may become possible in the near future [1]. However, at present the best procedure is to assume a reasonable potential and compare the rate constants obtained from absolute-rate theory with those from the trajectory analysis; that is, since the same potential surface is used for both calculations, any deviations must represent inadequacies in the theory.

From the absolute rate theory [13], the expression for $K(T)$ has the form

$$K(T) = \varkappa(RT/h)\frac{Q_{\neq}}{Q_A Q_{BC}}\exp\left[-\frac{E_B}{RT}\right]. \tag{6}$$

where $\varkappa$ is the transmission coefficient, $Q_{\neq}$ is the partition function of the activated complex excluding the degree of freedom corresponding to the reaction co-ordinate, Q_A and Q_{BC} are the partition functions for the reactant atom A and molecule BC, and E_B is the previously defined barrier height. Introducing classical partition functions for rotation and quantum-mechanical partition functions for vibration in the standard way, we find for H, H_2

$$K(T) = \varkappa N_A (2\pi)^{\frac{1}{2}}(\hbar^2)(\mu_{A,BC})^{-\frac{3}{2}}\left(\frac{g_{\neq}}{g_A g_{BC}}\frac{I_{\neq}\sigma_{BC}}{I_{BC}\sigma_{\neq}}\right)(kT)^{-\frac{1}{2}}\cdot \frac{\sinh(h\nu_{BC}/2kT)\exp[-E_B/RT]}{4\sinh h\nu_1/2kT)[\sinh(h\nu_2/2kT)]^2}, \tag{7}$$

where g is the electronic degeneracy (2 for A and $\neq$, 1 for BC), I is the moment of inertia, σ is the symmetry number (2 for BC, 1 for $\neq$), ν is the vibrational frequency (ν_1 is the stretching frequency and ν_2 the bending frequency of the complex). All of the quantities required for the numerical evaluation of $K(T)$

from eq. (7) can be determined from the potential surface *PK* [14] except for the transmission coefficient $\varkappa$, which is not given by absolute rate theory except through comparison with experiment. In a certain sense, trajectory calculation can be regarded as a technique for evaluating $\varkappa$. Setting $\varkappa$ equal to unity and fitting the $K(T)$ values from eq. (7) to an Arrhenius equation, we obtain

$$K_{\text{a.r.t.}}(T) = 7.41\cdot 10^{13} \exp[-8.812/RT] \text{ cm}^3/(\text{mole}\cdot\text{s}) . \tag{8}$$

Equation (8) is to be compared with the trajectory result given in eq. (4). We see that the activated complex theory gives an activation energy that is about 1.4 kcal/mole greater than the trajectory treatment. Comparing some of the rate constants, we have the results given in Table I. At the lowest

TABLE I. – *Rate constants from absolute-rate theory and trajectory analysis* ($\text{cm}^3/\text{mole}\cdot\text{s}$).

T (°K)	$K_{\text{a.r.t.}}\cdot 10^{-11}$	$K_{\text{traj}}\cdot 10^{-11}$
300	0.000 3011	0.00845
700	1.260	1.9742
1000	9.640	11.5280

temperature (300 °K), the two values disagree by a factor of six, while at the highest temperature (1000 °K) the disagreement is only a factor of 1.25, the trajectory result being consistently larger. There are two possible sources for the disagreement—first, the divergence between the quasi-classical nature of the trajectory treatment and the quantum-mechanical form used for the vibrational perturbation functions in the absolute rate theory formula [the rotational state distribution is such that use of classical or quantum partition functions leads to very small differences], and second, deviations from absolute-rate theory (corresponding formally to $\varkappa$ not equal to unity). The fact that the agreement between the two treatments is better at high than at low temperatures suggests that the first factor is of some importance. In particular, it appears that although the partly adiabatic character of the vibrational motion in the trajectory calculation leads to behavior corresponding to approximate quantization of the symmetric stretching vibration of the activated complex, the bending vibrations which correspond to rotations in the separated atom, molecule system are not at all « quantized » in the trajectory calculation. Use of classical partition functions for bending in eq. (6) yields values for the rate constat that are in better agreement with $K_{\text{traj}}(T)$ than the $K_{\text{a.r.t.}}(T)$ calculation.

Since an exact quantum-mechanical treatment is not available at the present time [2], the most unequivocal comparison between absolute rate theory and a collision treatment is to be made on a purely classical basis; that is, classical distribution functions are used to determine the initial conditions in the trajectory analysis and the partition functions used in eq. (6) are classical. To simplify the calculation and its analysis, the H, H_2 collision can be restricted to a linear path with two degrees of freedom (*e.g.* in the initial state, they are conveniently chosen as the H_2 vibrational co-ordinate and the distance of H from the center of mass of H_2). For this linear model with the *PK* potential surface, the calculated rate constant between 500 °K and 1300 °K is well represented by the equation [15]

$$K(T) = 10^{5.90} \exp[-9.79/RT] \text{ cm/mole}\cdot\text{s}$$

while that obtained from absolute rate theory is

$$K_{\text{a.r.t.}}(T) = 10^{5.94} \exp[-9.93/RT] \text{ cm/mole}\cdot\text{s}\,.$$

The agreement between these two expressions is excellent, almost within the statistical error of the calculation.

To perform a more refined test of the absolute rate theory, we can analyse the trajectory calculations to determine whether an « equilibrium » energy distribution exists for the degrees of freedom of the complex; that is, we partition the energy at the saddle point for each individual collision between the reaction co-ordinate and the symmetric stretching vibration of the complex and then average over a Boltzmann distribution of initial variables appropriate to a temperature *T*. In this treatment no assumption is made about the existence of a long-lived complex [16] that has time prior to dissociation to achieve an equilibrium distribution for the energy among its internal degrees of freedom. Instead, it is determined whether or not the Boltzmann distribution of initial states is mirrored in the transition state distribution, even though there is no collision complex and the reaction passes smoothly and rapidly over the saddle-point (see below). One way of considering this question is to ask whether Boltzmann distributions describe the energy present in the reaction co-ordinate and the symmetric stretch and, further, if such distributions are appropriate whether they have the same temperatures as that determinging the initial

TABLE II. – *Transition-state temperatures.*

Initial co-ordinates (°K)	Reaction co-ordinate (°K)	Symmetric stretch co-ordinate (°K)
1300	1200	1307
900	842	914
500	490	512

conditions. To a very good approximation, both the reaction co-ordinate and the symmetric stretch correspond to Boltzmann distributions. However, their temperatures are not exactly equal as shown in Table II.

For the linear H, H_2 reaction, the above results show that transition-state theory is a satisfactory approximation. Other systems, with more asymmetric potential surfaces show considerably greater deviations [15].

6. – Details of reaction mechanism.

The most detailed information about the nature of the $H+H_2$ collision is obtained by following the individual trajectories. Some results are shown in Fig. 8*a*)-*c*). Here we plot the three interparticle distances (in a.u.) as a function of time (in units of $0.56\cdot10^{-14}$ s). Figure 8*a*) and *b*) represent the most

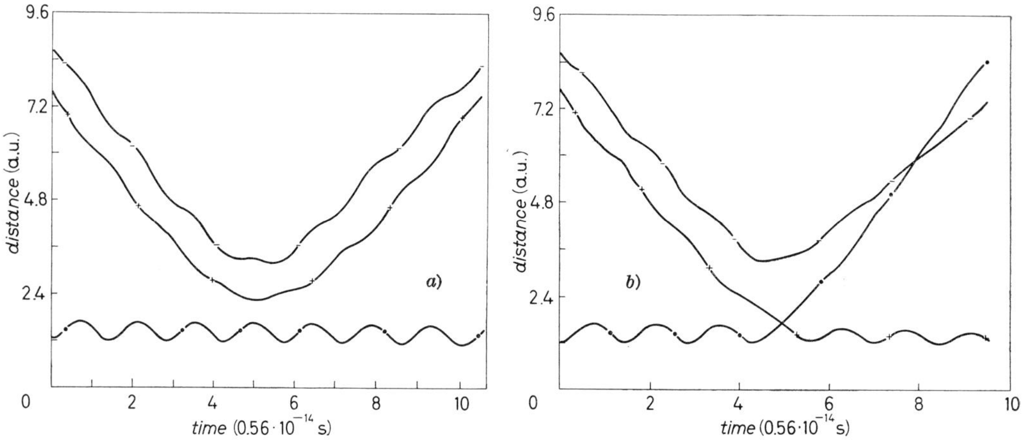

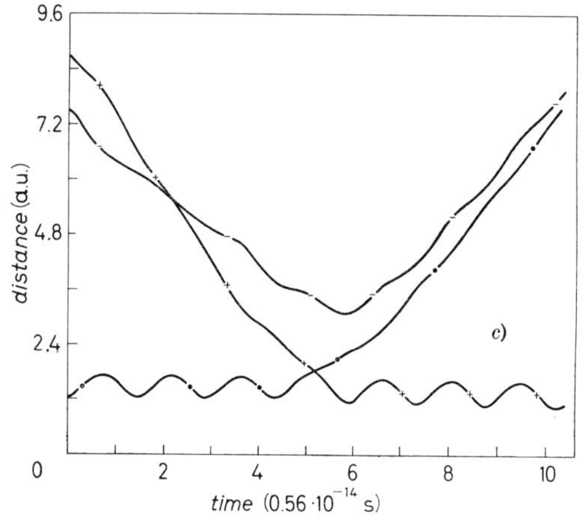

Fig. 8. – Typical $H+H_2$ collision trajectories: (+++) R_{AB}, (———) R_{AC}, (···) R_{BC}. *a*) Nonreactive: $J=0$, $v=0$, $V_R=1.32\cdot10^6$ cm/s; *b*) Reactive: $J=0$, $v=0$, $V_R=1.32\cdot10^6$ cm/s; *c*) Reactive: $J=5$, $v=0$, $V_R=1.18\cdot10^6$ cm/s.

common forms for nonreactive and reactive trajectories, respectively, in the low-energy range (up to 2 or 3 eV). We see that for the two cases, which are very similar in form, there is no indication whatsoever of a collision complex; that is, the collisions are simple and the total interaction time is short, on the order of 10^{-14} s being required for the atom to pass through the distance over which there is a strong interaction with the molecule. Since the vibrational period of H_2 is about $0.4\cdot10^{-14}$ s and the rotational period is generally even longer ($\sim 2\cdot10^{-12}$ s for $J=1$), it is clear that the interaction time is much too short for equilibration of energy among the various degrees of the three-atom system. Moreover, the collisions are not strong in that there appears to be little energy exchange during most trajectories. A few exceptional cases with large energy transfer do occur; a trajectory of this type is shown in Fig. 9*a*). In fact the reactive collisions are slightly more inelastic on the average than are

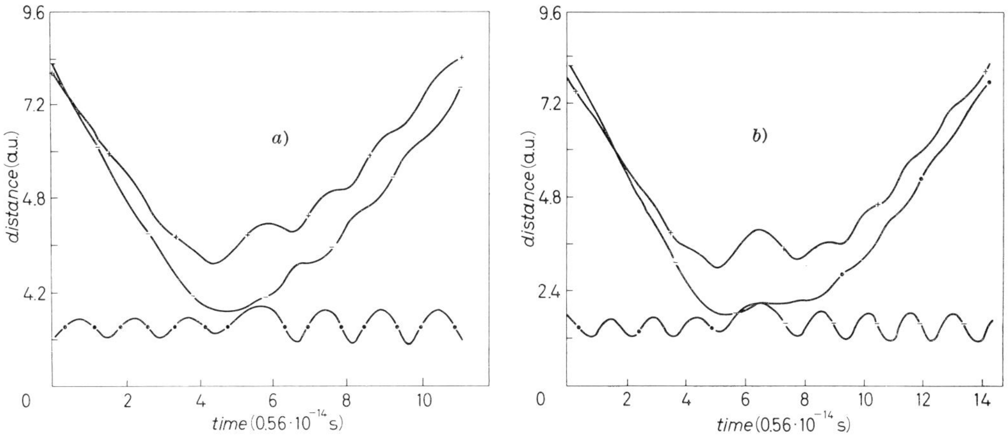

Fig. 9. – Atypical $H+H_2$ collision trajectory: (+++) R_{AB}, (— — —) R_{AC}, (···) R_{BC}. *a*) Nonreactive: $J=5$, $v=0$, $V_R=1.57\cdot10^6$ cm/s; *b*) Reactive: $J=2$, $v=0$, $V_R=$ $=1.32\cdot10^6$ cm/s.

the nonreactive trajectories. In particular, there is significant rotational energy transfer from the orbital angular momentum of the collision to rotational angular momentum of the newly-formed molecule. Figure 10 presents a quasi-classical analysis of this point; the relative probabilities of different final J' are shown as a function of the initial relative velocity [17]. Similar results are found for $J\neq 0$ both as to the form of the trajectories (Fig. 8*c*)) and their rotational inelasticity.

Another type of atypical trajectory is shown in Fig. 9*b*). Here we see that atom and molecule remain together for a time longer than is usually the case; that is, a « complex » is formed for about $3\cdot10^{-14}$ s. However, this period, which was the longest observed in the trajectories studied, is still significantly less than the rotation time. Thus, it is clear that in H, H_2 the mechanism for the reaction is of the simple or « direct » type instead of the « complex » type. Other potentials examined (*e.g.* $H+H_2$ trajectories with an Eyring-type well at the top of the barrier; see the potential marked E in Fig. 5 of [1]) do show contributions from a « complex » mechanism, in that a significant fraction of the trajectories last for many vibrational periods and for a time equal to that required by the entire three-atom system to rotate through 180° or more. If a much deeper well is introduced (*e.g.* in a surface simulating that for H_3^+), very long-lived trajectories ($\approx 10^{-12}$ s) resulting in both reaction and no reaction commonly occur.

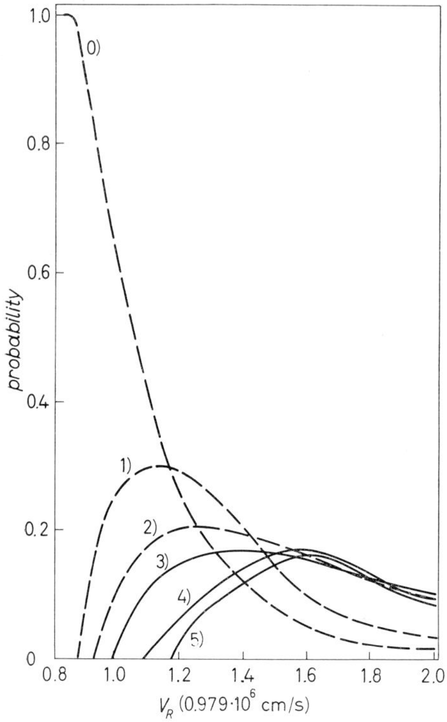

Fig. 10. – Final rotational-state probability J' as function of V_R for $J=0$, $v=0$.

7. – Transition configuration.

In addition to the time-dependence of the interparticle distances in typical trajectories, it is useful to consider the configuration of the three-particle system in the transition region where reaction can be said to take place. Defining the transition point as that at which R_{AB} or R_{AC} becomes shorter than R_{BC} (in most trajectories this occurs only once), we can describe the geometry in terms of a distance ($R_{AB}=R_{BC}$ or $R_{AC}=R_{CB}$) and the corresponding internal angle. Since the internal angle is of primary interest, we limit our discussion to the results obtained for its value in a limited, though representative set of trajectories (Table III). For reactive collisions, the observed angle corresponds to a balance between the volume factor ($\sin\chi$), which makes a 90° orientation most likely, and the form of the potential, which tends to make reaction more likely for near linear geometries particularly at low relative velocites. From Table III, it is clear that a nonlinear geometry is most probable in all cases, the angle χ decreasing (*i.e.* more bent geometries make a greater contribution) as the

TABLE III. – *Internal angle χ for reactive trajectories at transition point.*

V_R (a)	Internal angle χ
~ 1.00	~ 168°
1.10	~ 160°
1.30	~ 148°
1.60	~ 143°

(a) Relative velocity in units of $0.979\cdot10^6$ cm/s.

velocity increases. This result can be described as the increased « excitation » of bending vibrations in a transition-state model of the « *non*complex » type considered above. For a Boltzmann distribution of colliding systems at 1000 °K, the average value of χ for reactive trajectories is about 160°.

8. – Low- and high-energy mechanism.

From Fig. 5 we saw that in $T+H_2$, $T+D_2$ (and similarly in $H+H_2$) the reactive cross-section is nonzero for a large range of initial collision energies. This raises the question whether the mechanism of the reaction remains unchanged over the entire range. For low energies (*i.e.* energies up to the maximum in the S_r curve of Fig. 5), the mechanism is that shown in Fig. 8*b*). It can be characterized as « abstraction », in the sense that the incoming atom « abstracts » from the molecule the atom with which is reacts; that is, the incoming atom combines with the atom of the molecule to which it is closest throughout the trajectory (except for the crossing effect of molecular rotation outside the interaction region, as seen in Fig. 8*c*)). By contrast a typical high-energy trajectory (important for energies greater than that corresponding to the S_r maximum) is shown in Fig. 11. We see here that the incoming atom (A) does not react with the atom (B in this case) to which it is closest during the

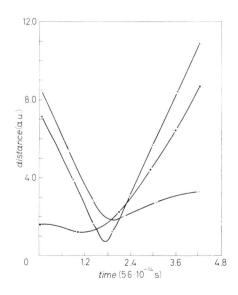

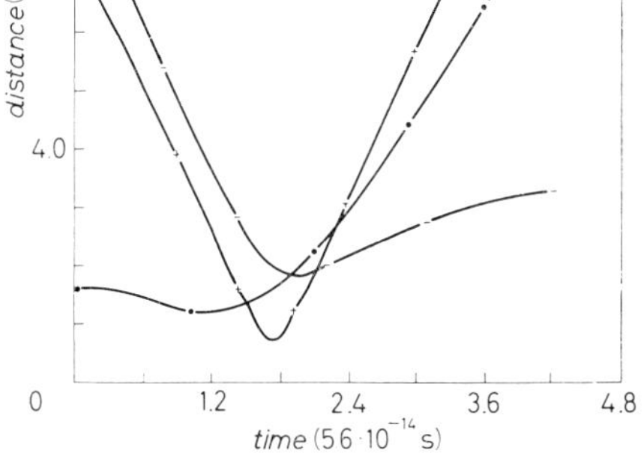

Fig. 11. – Typical reactive high-energy $H+H_2$ collision: $(+++)$ R_{AB}, (———) R_{AC}, $(\cdots)$ R_{BC}. $J=0$, $v=0$, $V_R=4.1\cdot10^6$ cm/s.

initial part of the trajectory. Instead it collides very hard with this atom (the distance of closest approach is very short, corresponding to a strong repulsive interaction), knocks it out of the molecule, and reacts with the other atom (C in this case). This type of reaction mechanism can be termed « displacement », since the incoming atom « displaces » the one with which it collides from the molecule. In the intermediate energy range (the region of the cross-section maximum) both displacement and abstraction contribute significantly, the former increasing in importance relative to the latter as the energy is increased. At very high energies, for which dissociation is dominant and the reaction cross-section is small but still nonzero, a third type of mechanism also contributes. It can be described as a « stripping abstraction »; that is, the incoming atom does react with the atom with which it collides, but this occurs for relatively large impact parameter collisions in which the incoming atom never gets very close to the atom with which it reacts—instead it strips the atom from the molecule. At such high energies, the small impact parameter collisions mainly lead to dissociation.

9. – Differential cross-section.

From the experimental lectures, it is clear that the primary quantity measured in crossed molecular beam studies is the laboratory angular distribution of products or, equivalently, the reactive differential cross-section, ideally as a function of various initial and final state parameters. Considerable analysis is then involved in translating the angular distribution results into quantities related to the dynamics, in contrast to the kinematics, of the reaction. In particular, numerous attempts have been made to extract the center-of-mass angular distribution and the final relative velocity vector which enter importantly into the laboratory measurements. Both of these quantities and the coupling between them (*i.e.* whether the final relative velocity is a function of angle) can be determined directly from the trajectory analysis.

The $H+H_2$ system differs significantly in two ways from most of the systems which have been studied in molecular beams. First, it has a sizeable barrier to reaction and second, it is thermoneutral—there is no heat of reaction or only a very small heat of reaction in some of the isotopic analogues. The latter suggests that rather small changes in relative energy or relative velocity are going to occur in the reaction. Some examination of the trajectories shows that this is true; since the reaction is vibrationally adiabatic to a reasonable approximation and slightly rotationally inelastic ($J' \geqslant J$, see above), the resulting final kinetic energy is somewhat smaller than the initial kinetic energy.

The presence of the energy barrier implies that the form of differential, as well as the magnitude of the total cross-section, are sensitive functions of the initial relative kinetic energy. For $D+H_2$, some preliminary results for the function $q_r(\theta)$, the probability of scattering into an angle θ to $\theta+d\theta$, are shown in Fig. 12. The quantity $q_r(\theta)$ is related to the differential cross-section $\sigma_r(\theta)$ by the equation

$$\sigma_r(\theta) = \frac{q_r(\theta)}{\sin\theta}, \tag{9}$$

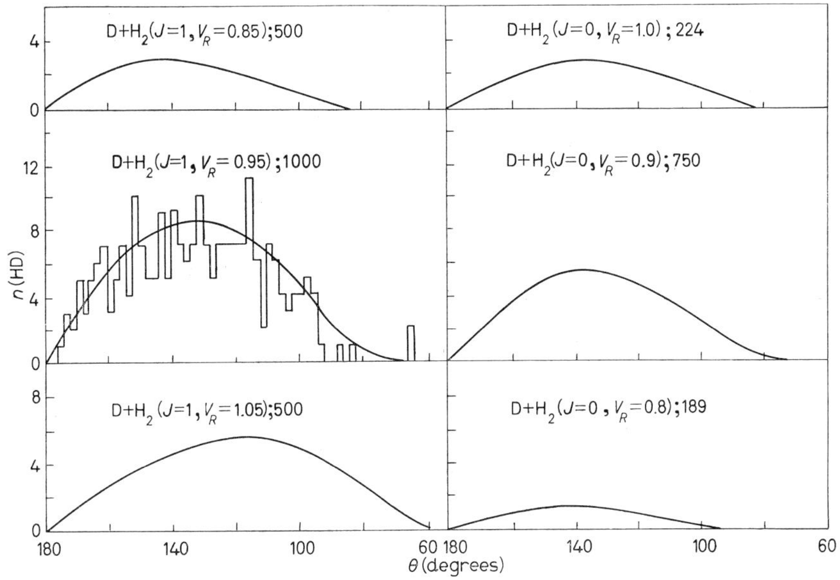

Fig. 12. – Angular distribution function $q_r(\theta)$ for $D+H_2$ reactive collisions with H_2 ($v=0$) and other conditions indicated on the drawings.

where θ is defined as the angle between the final atom-molecule relative velocity vector and the initial atom-molecule relative velocity vector; that is, $\theta = 0$ corresponds to « forward scattering » and $\theta = 180°$ to « backward scattering ». The particular values of relative velocity and rotational states shown in Fig. 12 were chosen to correspond to the most important contributions to the cross-section for a molecular beam experiment with D atoms at 2800 °K and H_2 molecules at 100 °K, the conditions used in the pioneering study of DATZ and TAYLOR [18]. In all cases $q_r(\theta)$ has its maximum in the backward hemisphere ($\theta > 90°$) and $\sigma_r(\theta)$ is peaked at or near 180°. However, even over the small velocity range shown in the Figure it can be seen that the peak shifts forward with increasing velocity. Corresponding results for $T+H_2$ over an energy range between 1 and 8 eV are presented in Fig. 13. Here it is clear that $\sigma_r(\theta)$, which is peaked at $\theta = 180°$ near threshold, is peaked sideways (at $\sim 90°$)

for an energy of 1.7 eV and shifts into the forward hemisphere at even higher energies.

The form of the differential cross-section described above can be understood in terms of a simple model for the reaction. To indicate what is involved, we consider $H+H_2$ at 0.5 eV for which accurate values of $\sigma_r(\theta)$ are shown in Fig. 17. As a preliminary to interpreting the reactive trajectories, we consider the nonreactive scattering of H by H_2. Here we find that the scattering angle θ is approximately a monotonic function of the impact parameter b as would be expected for a purely repulsive spherically symmetric potential. The range of the trajectory results are indicated as the shaded area in Fig. 14. The line, which is close to their mean value, is obtained by the following theoretical argument. We write the three-body PK potential as a function of the H_2 distance r, the distance R from the H atom to the center of mass of the molecule and the external ABC angle γ ($\gamma = \chi - \pi$). Expanding this potential in Legendre functions, we have

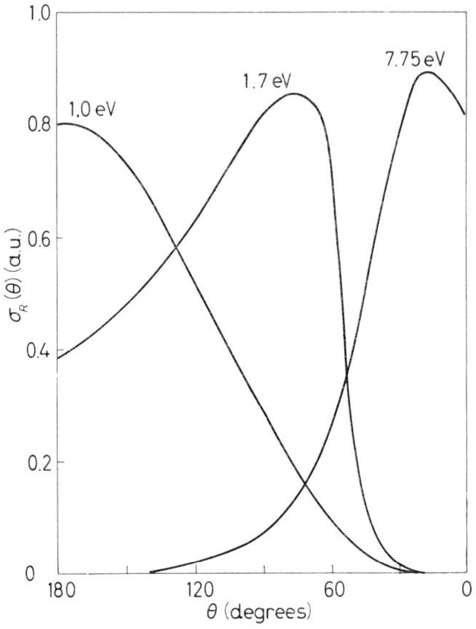

Fig. 13. – Differential cross-section for $T+H_2$ reactive collision ($J=0$, $v=0$) as a function of relative energy.

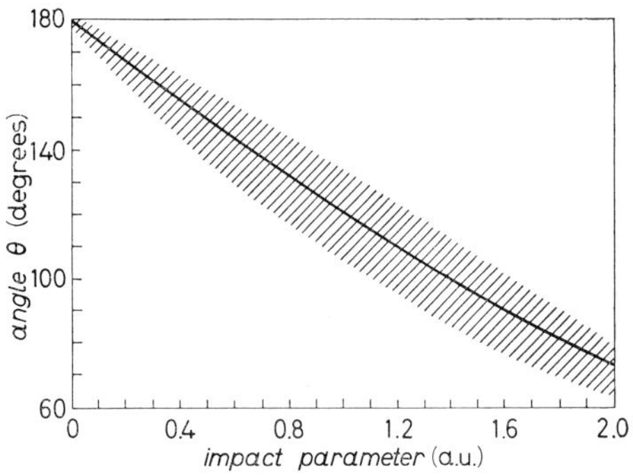

Fig. 14. – Scattering angle θ *vs.* impact parameter b for nonreactive collisions of $H+H_2$ ($J=0$, $v=0$, $E_R=0.48$ eV); the shaded area corresponds to the trajectory results; the line is obtained as indicated in the text.

$$(10) \qquad \mathscr{V}(R, r, \gamma) = \sum_l \mathscr{V}_l(R, r) P_l(\cos\gamma) = \mathscr{V}_0(R, r) + \mathscr{V}_2(R, r) P_2(\cos\gamma) + \dots,$$

where only the even-l terms contribute because H_2 is homonuclear. If we consider eq. (10) with $r = r_e$, the equilibrium internuclear distance, we obtain the curves shown in Fig. 15. It is evident that for the values of R that contribute to scattering for energies less then 1 eV, the spherically symmetric

$\mathscr{V}_0(R, r_e)$ is dominant, $\mathscr{V}_2(R, r_e)$ makes a small contribution, and the higher terms drawn in the Figure are negligible [19]. If the term $\mathscr{V}_0(R, r_e)$ is alone used to calculate the classical elastic scattering, the line drawn in Fig. 14 is obtained. The deviations from this line arise in part from the fact that individual trajectories sample somewhat different potentials deviating from $\mathscr{V}_0$ due to varying atom-molecule orientations, in part from inelastic contributions to the nonreactive scattering, and in part from both. Nevertheless, it is clear that the $\mathscr{V}_0(R, r_e)$ result gives a good approximation to the calculated θ *vs.* b dependence and, therefore, to the elastic differential cross-section

$$\sigma_{el}(\theta) = \frac{b}{\sin\theta}\left|\frac{db}{d\theta}\right| .$$

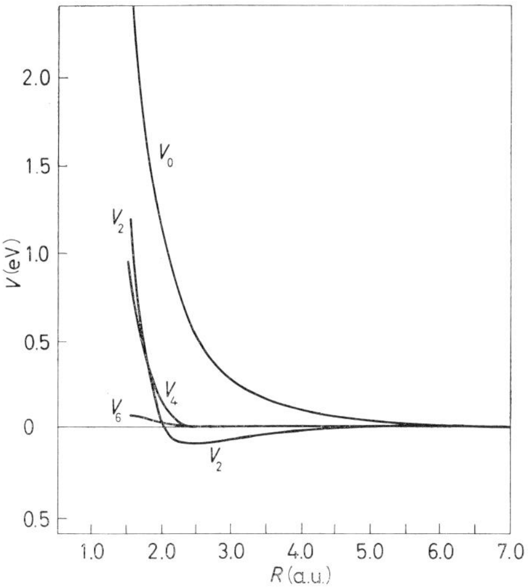

Fig. 15. – Expansion of three-body interaction potential in Legendre polynomials (eq. (10)) with $r = r_e = 1.4$ a.u.

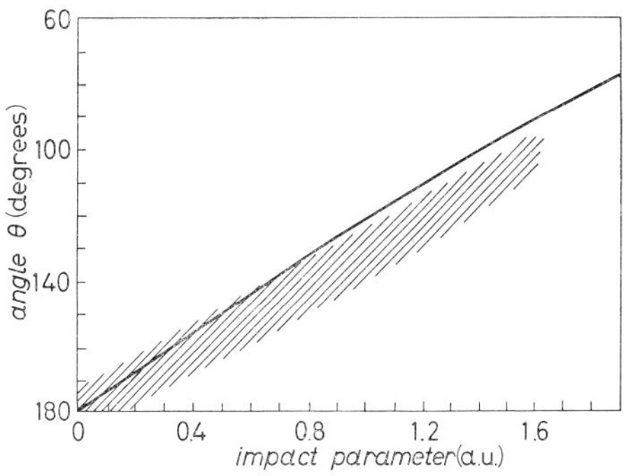

Fig. 16. – Scattering angle θ *vs.* impact parameter b for reactive collisions of $H+H_2$ ($J=0$, $v=0$, $E_R=0.48$ eV); the shaded area corresponds to the trajectory results; the line is obtained as indicated in the text.

For the reactive scattering, we make a corresponding plot of the angle θ *vs.* impact parameter b; the result is shown in Fig. 16, which also includes the θ *vs.* b curve calculated with $\mathscr{V}_0(R, r_e)$. Although the correspondence is not quite as good as for the nonreactive scattering, there is a clear correlation over the range of impact parameters that contribute to the reaction. This result suggests that the angular distribution for the reaction can be obtained by assuming it to occur according to a sudden mechanism, the angle of scattering being determined by a spherical potential (*e.g.* $\mathscr{V}_0$) with a sudden switch from the unreacted A, BC to the reacted AB, C or B, AC. For the symmetric case of $H+H_2 \to H_2+H$, this implies that given a certain impact parameter which corresponds to a scattering angle θ for the reactants, after reaction the same scattering angle and impact parameter are to be found for the products. Thus, from the nonreactive cross-section, we can in this oversimplified model estimate the form of the reactive cross-section by including only that part which arises from impact parameters that lead to reaction; that is

$$\sigma_r(\theta) \simeq P_r[b(\theta)] \frac{b}{\sin\theta} \left| \frac{db}{d\theta} \right| , \tag{11}$$

where $P_r[b(\theta)]$ is the reaction probability expressed as a function of angle by means of the $P_r(b)$ curve from trajectories and the $\theta(b)$ or $b(\theta)$ curve from trajectories or from the spherical-model calculations. The results obtained by this approach are shown in Fig. 17. Although the agreement is only approximate, it is clear that the model gives the dominant feature of the angular distribution. Quantitatively the disagreement is reduced by taking the $\mathscr{V}_0$ potential to be somewhat steeper than that given by eq. (10) [9].

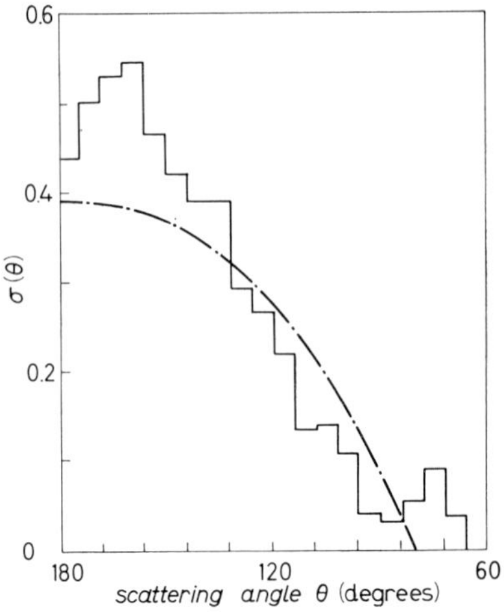

Fig. 17. – Reactive differential cross-section for $H+H_2$ ($J=0$, $v=0$, $E_R=0.48$ eV). (———) from trajectories; (— — —) from eq. (11) with θ *vs.* b from $\mathscr{V}_0$.

* * *

Various aspects of the research reported in these lectures are the result of contributions from a number of different coworkers; they are Drs. K. MOROKUMA, L. PEDERSEN, R. N. PORTER, R. D. SHARMA, K. T. TANG and R. WYATT. It is a great pleasure for me to acknowledge their collaboration. Furthermore, I thank Drs. P. ECKELT and F. A. GIANTURCO for preparing a preliminary draft of this lecture.

REFERENCES

[1] M. KARPLUS: this volume, p. 320.
[2] M. KARPLUS: this volume, p. 407.
[3] D. BUNKER: this volume, p. 355; see also, M. KARPLUS: in *Structural Chemistry and Molecular Biology*, edited by A. RICH and N. DAVIDSON (San Francisco, 1968), p. 837.
[4] M. KARPLUS, R. N. PORTER and R. D. SHARMA: *Journ. Chem. Phys.*, **43**, 3259 (1965).
[5] J. ROSS: this volume, p. 86.

[6] A detailed study of this question has been made in a model calculation for the H_2, H_2 case with a potential that has a square planar saddle-point configuration; see K. MOROKUMA, L. PEDERSEN and M. KARPLUS: *Journ. Am. Chem. Soc.*, **89**, 5064 (1967).
[7] M. KARPLUS, R. N. PORTER and R. D. SHARMA: *Journ. Chem. Phys.*, **45**, 3871 (1966).
[8] F. ROWLAND: this volume. p. 108 and 293.
[9] M. KARPLUS, K. T. TANG and R. WYATT: unpublished calculations. A detailed hard-sphere model study has been made by R. J. SUPLINSKAS: *Journ. Chem. Phys.*, **49**, 5046 (1968).
[10] B. A. RIDLEY, R. W. SCHULTZ and D. J. LEROY: *Journ. Chem. Phys.*, **44**, 3344 (1966); A. A. WESTENBERG and H. DEHAAS: *Journ. Chem. Phys.*, **47**, 1393 (1967); see also D. J. LE ROY, B. A. RIDLEY and K. A. QUICKERT: *Disc. Faraday Soc.*, **44**, 92 (1967).
[11] M. KARPLUS and K. T. TANG: *Disc. Faraday Society*, **44**, 56 (1967).
[12] See, for example, H. S. JOHNSTON: *Gas Phase Reaction Rate Theory* (New York, 1966).
[13] S. GLASSTONE, K. J. LAIDLER and H. EYRING: *The Theory of Absolute Rate Processes* (New York, 1941).
[14] For the numerical values used, see Table X of ref. [4].
[15] K. MOROKUMA and M. KARPLUS: to be published.
[16] D. L. BUNKER: this volume, p. 427.
[17] R. WYATT and M. KARPLUS: unpublished results.
[18] S. DATZ and E. H. TAYLOR: *Journ. Chem. Phys.*, **39**, 1896 (1963).
[19] For a more detailed discussion of the effective potential, see ref. [2].

Quantum Theory of Reactive Scattering.

J. Ross

MIT - Cambridge, Mass.

The purpose of this lecture is to present a short introduction to the formal theory of reactive scattering, to discuss briefly a few approximation methods, and to cite some applications to chemical reactions. There is a wealth of literature on the subject [1]; we shall choose a few topics of interest in molecular scattering and we shall need some of the concepts, equations, and approximations as the course proceeds.

1. – Green's function.

In Beck's lecture [2] he discussed the solution to the time-independent Schrödinger equation in the co-ordinate representation

$$(H(\boldsymbol{r})-E)\,\psi(\boldsymbol{r})=0 \tag{1.1}$$

for the scattering of two particles. This was written

$$\psi(\boldsymbol{r})=\varphi(\boldsymbol{r})+f(\theta)\,\frac{\exp[ikr]}{r} \tag{1.2}$$

for the asymptotic condition of the two particles well separated after the event of the collision. In these equations $\psi(\boldsymbol{r})$ is the wave function, $H(\boldsymbol{r})$ the Hamiltonian, E the energy, $f(\theta)$ the scattering amplitude, and r the distance between the two particles. The Hamiltonian is taken to be a sum of two terms, $H(\boldsymbol{r})=H_0(\boldsymbol{r})+V$, in which $H_0(\boldsymbol{r})$ is the Hamiltonian of the separated particles and V the interaction potential. The function $\varphi(\boldsymbol{r})$ is the wave function describing the motion of the separated particles and hence is a solution of the equation $(H_0(\boldsymbol{r})-E)\varphi(\boldsymbol{r})=0$. Thus the asymptotic form of the wave function

(1.2) is a superposition of the initial incoming wave function of the separated particles and the spherically outgoing scattered wave.

The differential equation (1.1) for $\psi(\boldsymbol{r})$ can be changed into an integral equation which includes the asymptotic boundary conditions expressed by eq. (1.1). For this purpose we introduce Green's function $G(\boldsymbol{r}, \boldsymbol{r}')$ which obeys the equation

$$(1.3) \qquad \big(H_0(\boldsymbol{r}) - E\big)\, G(\boldsymbol{r}, \boldsymbol{r}') = \delta(\boldsymbol{r} - \boldsymbol{r}')$$

and with which $\psi(\boldsymbol{r})$ is

$$(1.4) \qquad \psi(\boldsymbol{r}) = \varphi(\boldsymbol{r}) - \int G(\boldsymbol{r}\boldsymbol{r}')\, V(\boldsymbol{r}')\, \psi(\boldsymbol{r}')\, \mathrm{d}r' \,.$$

The function $G(\boldsymbol{r}\boldsymbol{r}')$ needs now to be evaluated with the correct boundary condition. We indicate the argument [1] briefly for structureless particles. With the completeness relation

$$(1.5) \qquad \delta(\boldsymbol{r} - \boldsymbol{r}') = \left(\frac{1}{2\pi}\right)^3 \int \mathrm{d}\boldsymbol{k}'\, \varphi^*_{\boldsymbol{k}'}(\boldsymbol{r}')\, \varphi_{\boldsymbol{k}'}(\boldsymbol{r}) \,,$$

we have for $G(\boldsymbol{r}, \boldsymbol{r}')$

$$(1.6) \qquad G(\boldsymbol{r}, \boldsymbol{r}') = -\frac{2\mu}{\hbar^2}\left(\frac{1}{2\pi}\right)^3 \int \frac{\exp[i\boldsymbol{k}' \cdot (\boldsymbol{r} - \boldsymbol{r}')]}{k^2 - k'^2}\, \mathrm{d}\boldsymbol{k}' \,,$$

which, on changing to spherical co-ordinates in $\boldsymbol{k}'$ space and introducing the symbol $(\boldsymbol{r} - \boldsymbol{r}') \equiv \boldsymbol{\rho}$, becomes

$$(1.7) \qquad G(\boldsymbol{r}, \boldsymbol{r}') = -\frac{2\mu}{\hbar^2}\left(\frac{1}{2\pi}\right)^2 \frac{1}{2i\varrho} \int_{-\infty}^{+\infty} \frac{\exp[ik'\varrho] - \exp[-ik'\varrho]}{k^2 - k'^2}\, k'\, \mathrm{d}k' \,.$$

If we let k' be a complex quantity, then the integral in eq. (1.7) is along the real k' axis and the integrand has poles at $k' = \pm k$. In order to perform the integration we shift the poles off the axis by changing k^2 to $k^2 \pm i\varepsilon$, where ε is a small positive number. The contour of integration is closed in the upper half of the complex plane for the term containing $\exp[ik'\varrho]$ in eq. (1.7) and in the lower half of the plane for $\exp[-ik'\varrho]$, for then the only contribution to the integral comes from the path along the real axis. In each case the contour encloses one pole so that we have

$$(1.8) \qquad G(\boldsymbol{r}, \boldsymbol{r}') = -\frac{2\mu}{\hbar^2}\, \frac{1}{4\pi\varrho} \exp[\pm ik\varrho] \,.$$

The introduction of $+i\varepsilon$ selects outgoing spherical waves, $-i\varepsilon$ selects incoming waves.

For short-range potentials the condition $r \gg r'$ holds in eq. (1.7) and hence we have approximately

$$\text{(1.9)} \qquad \begin{cases} |\boldsymbol{r}-\boldsymbol{r}'| \approx r - r'\cos\theta\,, \\ k|\boldsymbol{r}-\boldsymbol{r}'| \approx kr - \boldsymbol{k}'\cdot\boldsymbol{r}'\,; \quad \boldsymbol{k}' \equiv k\dfrac{\boldsymbol{r}}{r}\,. \end{cases}$$

For large r, therefore, eq. (1.4) reduces to

$$\text{(1.10)} \qquad \psi^+(\boldsymbol{r}) = \varphi(\boldsymbol{r}) - \frac{\mu}{2\pi\hbar^2}\frac{\exp[ikr]}{r}\int \exp[-i\boldsymbol{k}'\cdot\boldsymbol{r}']\,V(\boldsymbol{r}')\,\psi^+(\boldsymbol{r}')\,\mathrm{d}\boldsymbol{r}'\,,$$

where we have chosen the outgoing boundary condition and have indicated that choice with a plus sign as a superscript on $\psi(\boldsymbol{r})$. If we denote the incident plane wave by $\varphi_i(\boldsymbol{r})$ and the final plane wave $\exp[-i\boldsymbol{k}'\cdot\boldsymbol{r}]$ by $\varphi_f(\boldsymbol{r})$, then we may rewrite eq. (1.10)

$$\text{(1.11)} \qquad \psi_i^+(\boldsymbol{r}) = \varphi_i(\boldsymbol{r}) - \frac{\mu}{2\pi\hbar^2}\frac{\exp[ikr]}{r}\langle\varphi_f(\boldsymbol{r}')|V(r')|\psi_i^+(\boldsymbol{r}')\rangle\,.$$

By comparing eq. (1.11) with eq. (1.2), we see that we have obtained an integral equation for the scattering amplitude $f(\theta)$. This relation can be generalized easily to particles with internal degrees of freedom.

2. – The transition matrix.

It is convenient to have the Schrödinger equation and its solutions in a form independent of the particular representation chosen. We define therefore a state vector $|\psi\rangle$ and a co-ordinate representation vector $\langle\boldsymbol{r}|$ such that the wave function in the co-ordinate representation is $\langle\boldsymbol{r}|\psi\rangle$. Similarly for the Hamiltonian $H(\boldsymbol{r})$ we write $\langle\boldsymbol{r}|H|\boldsymbol{r}'\rangle$, and the Green function $G(\boldsymbol{r}\boldsymbol{r}') = \langle\boldsymbol{r}|G|\boldsymbol{r}'\rangle$. Thus the Schrödinger equation in the state-vector representation (with brackets omitted) is

$$\text{(2.1)} \qquad (H - E)\psi = 0$$

and that equation for the Hamiltonian H_0

$$\text{(2.2)} \qquad (H_0 - E)\varphi = 0\,.$$

Rearrangement of eq. (2.1) yields $(E-H_0)\psi=(E-H_0)\varphi+V\psi$ or

$$\psi=\varphi+\frac{1}{E-H_0}V\psi\,, \tag{2.3}$$

which is the Lippman-Schwinger equation [3]. Since we have $(H_0-E)G=1$ (compare with eq. (1.3)) in the state vector representation, eq. (2.3) is equivalent to eq. (1.4). To see this, we change eq. (2.3) to the co-ordinate representation

$$\langle\boldsymbol{r}|\psi\rangle=\langle\boldsymbol{r}|\varphi\rangle+\langle\boldsymbol{r}|GV\psi\rangle \tag{2.4}$$

and introduce the completeness relation $\int|\boldsymbol{r}'\rangle\,\mathrm{d}\boldsymbol{r}'\langle\boldsymbol{r}'|=1$,

$$\langle\boldsymbol{r}|\psi\rangle=\langle\boldsymbol{r}|\varphi\rangle+\int\langle\boldsymbol{r}|G|\boldsymbol{r}'\rangle\langle\boldsymbol{r}'|V\psi\rangle\,\mathrm{d}\boldsymbol{r}'\,. \tag{2.5}$$

The formal solution of the Lippman-Schwinger equation is obtained from eq. (2.3) by writing

$$(E-H_0)\psi=(E-H_0)\varphi+V\psi+V\varphi-V\varphi\,,$$

or

$$\psi_i^+=\varphi_i+\frac{1}{E-H_0-V+i\varepsilon}V\varphi_i\,, \tag{2.6}$$

with the proper sub- and superscripts.

The transition matrix is defined by the matrix elements

$$T_{fi}\equiv\langle\varphi_f|V|\psi_i^+\rangle \tag{2.7}$$

and a comparison with eq. (1.11) gives the relation of these elements to the scattering amplitude

$$T_{fi}=-\frac{2\pi\hbar^2}{\mu}f_{fi}(\theta)\,. \tag{2.8}$$

An integral equation for the transition matrix T can be derived by introducing the Møller wave propagation operator [4] Ω^+ which transforms the initial state φ_i to ψ_i^+

$$\psi_i^+\equiv\Omega^+\varphi_i\,. \tag{2.9}$$

From eq. (2.6) we have

$$\Omega^+ = 1 + \frac{1}{E - H + i\varepsilon} V \,, \tag{2.10}$$

or

$$\Omega^+ = 1 + \frac{1}{E - H_0 + i\varepsilon} V\Omega^+ \,. \tag{2.11}$$

Again, from eq. (2.7) we may write

$$T_{fi} = \langle \varphi_f | V\Omega^+ | \varphi_i \rangle \tag{2.12}$$

and

$$T = V\Omega^+ \,. \tag{2.13}$$

Therefore the integral equation for T is

$$T = V + V \frac{1}{E - H_0 + i\varepsilon} T \,, \tag{2.14}$$

with the formal solution

$$T = V + V \frac{1}{E - H_0 - V + i\varepsilon} V \,. \tag{2.15}$$

The Born approximation for scattering is obtained by retaining only the first term in eq. (2.15), that is

$$T \simeq V \qquad \text{(Born approximation)} \,. \tag{2.16}$$

The S-matrix, which transforms the initial state φ_i into the final state φ_f, both states being eigenfunctions of H_0, is

$$S = 1 - 2\pi i \delta(E - H_0) T \,. \tag{2.17}$$

3. – Two-potential scattering without rearrangement [5].

For some physical problems it is useful to split up the interaction potential into two or more terms

$$V = V_0 + V_1 \,. \tag{3.1}$$

For instance, in rotational inelastic scattering V_0 might be the spherically symmetric part of the interaction and V_1 the asymmetric part. We consider now the T-matrix for such a case. Let χ be the state vector for the Hamiltonian $(H_0 + V_0)$; the Lippman-Schwinger equations for $\chi^\pm$ are

$$\chi^\pm = \varphi + \frac{1}{E - H_0 \pm i\varepsilon} V_0 \chi^\pm = \varphi + \frac{1}{E - H_0 - V_0 \pm i\varepsilon} V_0 \varphi \tag{3.2}$$

with the subscripts i or f on φ and χ denoting initial or final channels respectively. The T-matrix elements

$$T_{fi} = \langle \varphi_f | V_0 + V_1 | \psi_i^+ \rangle \tag{3.3}$$

are rewritten by substituting for φ_f from eq. (3.2) with the choice χ_f^-

$$T_{fi} = \langle \varphi_f | V_0 | \psi_i^+ \rangle + \langle \chi_f^- | V_1 | \psi_i^+ \rangle - \left\langle \frac{1}{E - H_0 - V_0 - i\varepsilon} V_0 \varphi_f \middle| V_1 \middle| \psi_i^+ \right\rangle , \tag{3.4}$$

or

$$T_{fi} = \langle \varphi_f | V_0 | \psi_i^+ \rangle + \langle \chi_f^- | V_1 | \psi_i^+ \rangle - \langle \varphi_f | V_0 \frac{1}{E - H_0 - V_0 + i\varepsilon} V_1 | \psi_i^+ \rangle . \tag{3.5}$$

In order to combine the first and third term on the right-hand side in this equation, we consider the Lippman-Schwinger equation for ψ_i^+

$$\psi_i^+ = \varphi_i + \frac{1}{E - H_0 + i\varepsilon} (V_0 + V_1) \psi_i^+ . \tag{3.6}$$

We use the identity

$$\frac{1}{E - H_0 + i\varepsilon} = \frac{1}{E - H_0 - V_0 + i\varepsilon} - \frac{1}{E - H_0 - V_0 + i\varepsilon} V_0 \frac{1}{E - H_0 + i\varepsilon} , \tag{3.7}$$

which can be easily verified by multiplying from the left with $(E - H_0 - V_0 + i\varepsilon)$ and from the right with $(E - H_0 + i\varepsilon)$, to give

$$\begin{aligned} \psi_i^+ = \varphi_i &+ \frac{1}{E - H_0 - V_0 + i\varepsilon} (V_0 + V_1) \psi_i^+ - \\ &- \frac{1}{E - H_0 - V_0 + i\varepsilon} V_0 \frac{1}{E - H_0 + i\varepsilon} (V_0 + V_1) \psi_i^+ . \end{aligned} \tag{3.8}$$

Use of eq. (3.6) alters the last term

$$\psi_i^+ = \varphi_i + \frac{1}{E - H_0 - V_0 + i\varepsilon} (V_0 + V_1) \psi_i^+ - \frac{1}{E - H_0 - V_0 + i\varepsilon} V_0 [\psi_i^+ - \varphi_i] \tag{3.9}$$

and substitution of eq. (3.2) yields

$$\psi_i^+ = \chi_i^+ + \frac{1}{E - H_0 - V_0 + i\varepsilon} V_1 \psi_i^+ \,. \tag{3.10}$$

Comparison of this equation with eq. (3.6) is instructive.

We return now to eq. (3.5) to combine there the first and third term with the help of eq. (3.10)

$$T_{fi} = \langle \varphi_f | V_0 | \chi_i^+ \rangle + \langle \chi_f^- | V_1 | \psi_i^+ \rangle \,. \tag{3.11}$$

The first term describes the scattering from V_0 only, and the second term the scattering from V_1, the distortion by V_0 being accounted for. The second term can also be written as

$$\langle \chi_f^- | V_1 + V_1 \frac{1}{E - H_0 - V_0 - V_1 + i\varepsilon} V_1 | \chi_i^+ \rangle \,, \tag{3.12}$$

which shows it to be a matrix element of a T matrix (see eq. (2.15)) in a distorted-wave (χ) representation.

The distorted-wave Born approximation [6] is obtained by approximating ψ_i^+ in eq. (3.11) with χ_i^+, or, equivalently, retaining only the first term, V_1, in the expression (3.12). Thus the distorted-wave method takes full account of the potential V_0 in the absence of V_1, and accounts for V_1 to the first order.

4. – Rearrangement collisions [7].

For rearrangement collisions the Hamiltonian of the separated particles is different in the entrance and exit channel. We write the Hamiltonian as

$$H = H_0 + V \tag{4.1}$$

and

$$H = H_0' + V' \tag{4.2}$$

in the initial (entrance channel) and final (exit channel) state respectively. The wave function ψ_i^+ is, as before, given by

$$\psi_i^+ = \varphi_i + \frac{1}{E - H_0 + i\varepsilon} V \psi_i^+ \,. \tag{4.3}$$

We seek now, however, an asymptotic scattered wave which is an eigenfunction of H_0' and not H_0. Hence we must rearrange the Lippman-Schwinger equation (4.3) to such a form. To do this we use the identity

$$\frac{1}{E - H_0 + i\varepsilon} = \frac{1}{E - H_0' + i\varepsilon} - \frac{1}{E - H_0' + i\varepsilon}(V - V')\frac{1}{E - H_0 + i\varepsilon} \tag{4.4}$$

and substitute it into eq. (4.3)

$$\psi_i^+ = \varphi_i + \frac{1}{E - H_0' + i\varepsilon} V\psi_i^+ - \frac{1}{E - H_0' + i\varepsilon}(V - V')\frac{1}{E - H_0 + i\varepsilon} V\psi_i^+ . \tag{4.5}$$

In the last term we make use of eq. (4.3) again

$$\frac{1}{E - H_0' + i\varepsilon}(V - V')(\psi_i^+ - \varphi_i)$$

to give

$$\psi_i^+ = \varphi_i + \frac{1}{E - H_0' + i\varepsilon}(V - V')\varphi_i + \frac{1}{E - H_0' + i\varepsilon} V'\psi_i^+ , \tag{4.6}$$

or

$$\psi_i^+ = \frac{i\varepsilon}{E - H_0' + i\varepsilon}\varphi_i + \frac{1}{E - H_0' + i\varepsilon} V'\psi_i^+ . \tag{4.7}$$

Only the last term can lead to rearrangement in the limit $\varepsilon \to 0$ and hence, in analogy with the discussion given above, the T-matrix elements for rearrangement are

$$T_{fi} = \langle \varphi_f | V' | \psi_i^+ \rangle , \tag{4.8}$$

or equivalently, by symmetry,

$$T_{fi} = \langle \psi_f^- | V | \varphi_i \rangle . \tag{4.9}$$

If the interaction potential is written as the sum of two terms, $V' = V_0' + V_1'$, and the distortion-wave function χ_f is a solution of the equation $(H_0' + V_0' - E)\chi = 0$, then the transition matrix elements may also be written [6]

$$T_{fi} = \langle \chi_f^- | V_1' | \psi_i^+ \rangle . \tag{4.10}$$

In the distorted-wave Born approximation (DWBA) the wave function ψ_i^+ is again replaced by χ_i^+. A physical reason for the considerable improvement of DWBA over the simpler Born approximation is the exclusion of the incoming wave from much of the interaction region by the distortion potential.

A similar effect can be achieved for direct interaction rearrangement collisions by retaining the simpler substitution of φ_i for ψ_i^+ but limiting the region of integration in the transition matrix element to the exterior of most of the interaction region [8-11].

5. – Faddeev equation for three particles.

The iterative solution of the integral equation for the T-matrix for three-particle collisions may lead to divergences which arise from terms describing the physical situation of two particles colliding while the third proceeds in free motion. FADDEEV has shown how to formulate the three-particle scattering problem without this difficulty, and we outline here briefly a derivation of his equation [12].

Let the Hamiltonian be $H = H_0 + V_{12} + V_{23} + V_{13}$. The Green's function G is now defined

$$(H - E)G = 1 \tag{5.1}$$

and for the kinetic-energy Hamiltonian H_0

$$(H_0 - E)G_0 = 1\,. \tag{5.2}$$

Therefore we have the integral equation for G, $(H_0 - E)G = 1 - VG$, or

$$G = G_0 - G_0 VG\,. \tag{5.3}$$

The integral equation for the transition matrix T is (eq. (2.14))

$$T = V - VG_0 T\,. \tag{5.4}$$

On rearranging this equation to $(1 + VG_0)T = V$ or $(H_0 - E + V)G_0 T = (H - E)G_0 T = V$ we obtain

$$G_0 T = GV\,, \tag{5.5}$$

so that eq. (5.4) becomes

$$T = V - VGV\,, \tag{5.6}$$

which is the same as eq. (2.15). From eq. (5.5) we have

$$VG_0T = VGV; \tag{5.7}$$

therefore the relation

$$VG = TG_0 \tag{5.8}$$

must hold since again on multiplying by V from the right eq. (5.8) yields

$$VGV = TG_0V \tag{5.9}$$

and hence, by comparing eqs. (5.7) and (5.9)

$$TG_0V = VGV = VG_0T\,, \tag{5.10}$$

a result to be needed shortly.

Next we define the transition matrices

$$T_{ij} \equiv V_{ij} - V_{ij}G_0T_{ij} \qquad i \neq j = 1, 2, 3 \tag{5.11}$$

for each pair of particles, but remember that G_0 is the Green's function for three free particles. With the additional definitions

$$T^{(1)} = V_{12} - V_{12}G_0T\,, \qquad T^{(2)} = V_{23} - V_{23}G_0T\,, \qquad T^{(3)} = V_{31} - V_{31}G_0T\,, \tag{5.12}$$

we have for the transition matrix T

$$T = T^{(1)} + T^{(2)} + T^{(3)}\,. \tag{5.13}$$

We proceed next by eliminating the potentials V_{ij} from eqs. (5.11) and (5.12). Multiply eq. (5.12), say the first equation, by $(1 - T_{12}G_0)$:

$$(1 - T_{12}G_0)T^{(1)} = (1 - T_{12}G_0)V_{12} - (1 - T_{12}G_0)V_{12}G_0[T^{(1)} + T^{(2)} + T^{(3)}]\,. \tag{5.14}$$

The first term on the right-hand side may be altered by use of eq. (5.10)

$$V_{12} - T_{12}G_0V_{12} = V_{12} - V_{12}GT_{12}\,, \tag{5.15}$$

which is simply T_{12}, eq. (5.11). Therefore solving eq. (5.14) for $T^{(1)}$ we obtain

$$T^{(1)} = T_{12} + T_{12}G_0T^{(1)} - T_{12}G_0[T^{(1)} + T^{(2)} + T^{(3)}]\,, \tag{5.16}$$

or, on simplifying and adding the other two similar equations,

$$(5.17)\qquad \begin{pmatrix} T^{(1)} \\ T^{(2)} \\ T^{(3)} \end{pmatrix} = \begin{pmatrix} T_{12} \\ T_{23} \\ T_{13} \end{pmatrix} - \begin{pmatrix} 0 & T_{12} & T_{12} \\ T_{23} & 0 & T_{23} \\ T_{13} & T_{13} & 0 \end{pmatrix} G_0 \begin{pmatrix} T^{(1)} \\ T^{(2)} \\ T^{(3)} \end{pmatrix} .$$

These are the Faddeev equations. The iterative solution is free of divergences for well-behaved potentials. The lowest-order solution

$$(5.18)\qquad T \simeq T_{12} + T_{23} + T_{13}$$

is called the impulse approximation [13].

Distortion potentials [6, 14, 15] may be introduced into the Faddeev equations. If we choose as the distortion potential that interaction which reproduces the elastic scattering (optical potential) in the initial channel

$$(5.19)\qquad V_{0i}|\chi_i^+\rangle = V|\psi_i^+\rangle ,$$

where χ_i^+ is the distorted-wave function and V the interaction potential in the initial channel, then the DWBA may be shown to hold

$$(5.20)\qquad \psi_i^+ = \chi_i^+ + 0(T^2)\chi_i^+ \ldots$$

up to second order in the T-matrix elements appearing in Faddeev's equations. For application of the DWBA and similar approaches to chemical reactions see ref. [16-19].

6. – Complex formation.

The formal theory of scattering discussed so far is sufficiently general to include all possible collision mechanisms. Two limiting cases of such mechanisms are direct interaction collisions in which the duration of the interaction is short compared to a characteristic time of motion of the system (rotational period for instance), and complex formation in which the interaction time is long compared to that characteristic time. It is useful to develop a simple approach to complex formation and decay [20, 21]. When generalized, this approach is equivalent to the theories presented in the previous Section.

Consider a molecular aggregate, a complex, with sufficient energy to decompose into two or more fragments [22]. The decay of the complex may be described by assigning a (mathematical) complex eigenvalue to the system,

$E_0 - i\Gamma$, for then the time-dependent wave function is $\exp[-iE_0(t/\hbar)]\exp[-\Gamma t/\hbar]$ and the density changes in time as $\exp[-2\Gamma t/\hbar]$. The asymptotic form of the wave function for the l-th partial wave is written

$$u_l(r) = \frac{C_l}{r}\sin\left(kr - \frac{l\pi}{2} + \eta_l\right), \tag{6.1}$$

or, equivalently,

$$\begin{cases} u_l(r) = \dfrac{iC_l}{2r}\left\{\exp\left[-i\left(kr - \dfrac{l\pi}{2} + \eta_l\right)\right] - \exp\left[i\left(kr - \dfrac{l\pi}{2} + \eta_l\right)\right]\right\}, \\ u_l(r) = \alpha_l(E)\dfrac{\exp[-ikr]}{r} + \alpha_l^*(E)\dfrac{\exp[ikr]}{r}, \end{cases} \tag{6.2}$$

where η_l is the phase shift (see ref. [2], or ref. [1*a*], Sect. **2**). If we have prepared a complex in a single state (a pseudostationary state) with eigenvalue $E_0 - i\Gamma$ then the decay of that complex must be described asymptotically (at large separation of the fragments) by an outgoing wave only. Hence at an energy $E_0 - i\Gamma$ we must have

$$\alpha_l(E_0 - i\Gamma) = 0\,. \tag{6.3}$$

Close to that energy, the asymptotic form of the scattered wave function in the usual scattering experiment in which two molecules collide, interact, and scatter with or without rearrangement, is obtained by a Taylor series expansion of the coefficient α_l

$$\alpha_l(E) = a_l(E - E_0 + i\Gamma) + \ldots, \tag{6.4}$$

where a_l is a constant. The asymptotic wave function eq. (6.2) becomes

$$u_l(r) = a_l(E - E_0 + i\Gamma)\frac{\exp[-ikr]}{r} + a_l^*(E - E_0 - i\Gamma)\frac{\exp[ikr]}{r}\,. \tag{6.5}$$

By comparing eqs. (6.5) and (6.2) we obtain the ratio

$$\exp[2i\eta_l] = \exp[2i\eta_l^0]\left(\frac{E - E_0 - i\Gamma}{E - E_0 + i\Gamma}\right), \tag{6.6}$$

with the definition $\exp[2i\eta_l^0] \equiv -a_l^{-1}a_l^*\exp[il\pi]$ for the phase shift η_l^0. If the quantity Γ is zero, then we have no pseudostationary state or complex formed, and we speak of potential scattering. For $\Gamma \neq 0$ and at E close to $E_0 - i\Gamma$ we say that resonance scattering occurs.

Equation (6.6) may be rearranged to the form

$$\eta_l - \eta_l^0 = \operatorname{tg}^{-1} \frac{\Gamma}{E_0 - E}\,; \tag{6.7}$$

hence for $\Gamma \ll (E, E_0)$ the phase shift η_l increases by π compared to the potential scattering phase shift η_l^0 as the relative kinetic energy varies from E less than E_0 to E greater than E_0. This addition to η_l comes from the addition of a node in the wave function on increasing the relative energy to include the pseudostationary state.

The scattering amplitude $f(\theta)$, eq. (1.2), in the partial-wave expansion is

$$f(\theta) = \frac{1}{2ik} \sum_l (\exp[2i\eta_l] - 1)(2l+1) P_l(\cos\theta)\,. \tag{6.8}$$

Thus, if we write $f^0(\theta)$ for the amplitude evaluated for potential scattering only, we have

$$f(\theta) = f^0(\theta) - \frac{(2l+1)}{k} \exp[2i\eta_l^0] \frac{\Gamma}{E - E_0 + i\Gamma} P_l(\cos\theta)\,. \tag{6.9}$$

If the resonance scattering predominates then the angular dependence of $f(\theta)$ is dominated by $P_l(\cos\theta)$. The total scattering cross-section due only to resonance scattering from a single resonance (pseudostationary state) is

$$\sigma_{\text{resonance}} = \frac{2\pi}{k^2} (2l+1) \frac{\Gamma^2}{(E - E_0)^2 + \Gamma^2}\,. \tag{6.10}$$

The width of the resonance in energy is determined by Γ, called the level width. If Γ is small then the variation of the total cross-section with energy is sharply peaked at E_0, the life-time of the complex $\hbar\Gamma^{-1}$ is long, and the line width associated with the pseudostationary state is small.

The discussion just presented for a single resonance may be generalized to many resonances and to all collisions. This has been developed in a number of different ways [23-28]; we shall describe briefly Wigner's R-matrix method. If two molecules collide during the interaction they are confined to some (arbitrary) region of configuration space. For that region and the Hamiltonian of the interacting molecules, there exists a complete set of eigenfunctions and eigenvalues. The boundary condition on the eigenfunction may be chosen, *e.g.* continuity of the wave function and its derivative on the boundary with the wave functions describing free motion, of either reactants or products, outside the interaction region. The formulation is exact and leads to formal correspondence with the theories discussed above (S-, T-matrix).

Wigner's R-matrix theory has been used [29] to obtain a derivation of the transition-state theory of chemical kinematics and a statistical theory of reactions (see D. BUNKER, lecture on Special Results of Theory: Compound State Approaches). Fano's theory of resonance scattering has been made the basis of a theory of unimolecular decay [30].

* * *

I thank Dr. CHAPMAN for preparing a preliminary manuscript for the lecture.

REFERENCES

[1] A few text books are: *a*) N. F. MOTT and H. S. W. MASSEY: *The Theory of Atomic Collisions*, III ed. (London, 1965); *b*) M. L. GOLDBERGER and K. M. WATSON: *Collision Theory* (New York, 1964); *c*) L. S. RODBERG and R. M. THALER: *Introduction to the Quantum Theory of Scattering* (New York, 1967).
[2] D. BECK: This volume p. 15.
[3] B. LIPPMAN and J. SCHWINGER: *Phys. Rev.*, **79**, 469 (1950); M. GELLMAN and M. L. GOLDBERGER: *Phys. Rev.*, **91**, 398 (1953).
[4] C. MØLLER: *Kgl. Danske Vid. Selsk. Mat.-fys. Medd.*, **23**, 1 (1945).
[5] The presentation follows that of ref. [1 *c*].
[6] K. R. GREIDER: *Adv. Theor. Phys.*, **1**, 245 (1965).
[7] See B. A. LIPPMAN: *Phys. Rev.*, **102**, 264 (1956), and ref. [1].
[8] S. T. BUTLER: *Nuclear Stripping Reactions* (New York, 1957).
[9] W. TOBOCMAN: *Theory of Direct Nuclear Reactions* (London, 1961).
[10] W. TOBOCMAN: *Phys. Rev.*, **115**, 98 (1959).
[11] For an application to some chemical reactions: J. H. HUNTINGTON: Ph. D. Thesis, Brown University (1968); B. C. EU, J. H. HUNTINGTON and J. ROSS: to be published.
[12] L. D. FADDEEV: *Sov. Phys. JETP*, **12**, 1014 (1961); *Mathematical Aspects of the Three-Body Problem in the Quantum Scattering Theory* (New York, 1965); *Sov. Phys. Doklady*, **6**, 384 (1961); **7**, 600 (1963); S. WEINBERG: *Phys. Rev.*, **130**, 776 (1963); **131**, 440 (1963); **133**, B 232 (1964); C. LOVELACE: *Phys. Rev.*, **135**, B 1225 (1964).
[13] G. CHEW: *Phys. Rev.*, **80**, 196 (1950).
[14] K. R. GREIDER and L. R. DODD: *Phys. Rev.*, **146**, 671 (1966).
[15] B. C. EU and J. ROSS: *Journ. Chem. Phys.*, **51**, 159 (1969).
[16] W. THORSON: *Journ. Chem. Phys.*, **37**, 433 (1962).
[17] R. J. SUPLINSKAS and J. ROSS: *Journ. Chem. Phys.*, **47**, 321 (1967).

[18] D. Micha: *Ark. Fys.*, **30**, 425, 437 (1965).
[19] M. Karplus and K. T. Tang: *Disc. Faraday Soc.*, No. **44**, 56 (1967).
[20] G. Breit and E. Wigner: *Phys. Rev.*, **51**, 593 (1937).
[21] L. D. Landau and E. M. Lifshitz: *Quantum Mechanics* (London, 1958).
[22] We follow the discussion in ref. [21].
[23] P. L. Kapur and R. Peierls: *Proc. Roy. Soc.*, A **166**, 277 (1938).
[24] E. P. Wigner: *Phys. Rev.*, **70**, 15, 606 (1946).
[25] E. P. Wigner and L. Eisenbud: *Phys. Rev.*, **72**, 29 (1947).
[26] H. Feshbach: *Ann. of Phys.*, **5**, 357 (1958); **19**, 282 (1962).
[27] J. Humblet and L. Rosenfeld: *Nucl. Phys.*, **26**, 529 (1961).
[28] U. Fano: *Phys. Rev.*, **124**, 1866 (1961).
[29] B. C. Eu and J. Ross: *Journ. Chem. Phys.*, **44**, 2467 (1966).
[30] F. Mies and M. Krauss: *Journ. Chem. Phys.*, **45**, 4455 (1966).

Special Results of Theory: Distorted Waves.

M. KARPLUS

Harvard University - Cambridge, Mass.

1. – Introduction.

The reactive scattering of hydrogen atoms by hydrogen molecules ($H+H_2 \rightarrow H_2+H$) has been studied in detail by quasiclassical trajectory calculations, as described in a previous lecture [1]. Since it has been suggested that quantum effects are important for this reaction [2], it is of interest to have available quantum calculations for comparison with the classical results. For this purpose, we chose to perform a quantum treatment with the realistic, though not exact, semitheoretical potential surface [3] used for the classical calculations. Unfortunately, the complexities inherent in the quantum-mechanical three-body problem are such that for this potential function, as for any not drastically oversimplified function, an exact calculation is very difficult. In this lecture we report the results of some attempts to obtain approximate solutions to the problem [4].

2. – Formulation.

Since Ross [5] has discussed the quantum-mechanical scattering problem, we present only a brief outline to introduce notation and to delineate our method of approach.

We consider the reaction $A+BC \rightarrow AB+C$, where A, B and C are structureless particles (except for nuclear spin) that interact through a known potential surface $\mathscr{V}$. The differential scattering cross-section for rearrangement from the reactant channel i to the product channel f can be written in the centre-of-mass co-ordinate system [6]

$$\sigma_{fi}(\hat{\boldsymbol{k}}_f) = \frac{\mu_i \mu_f}{(2\pi\hbar^2)^2} \frac{k_f}{k_i} |T_{fi}|^2 . \tag{1}$$

Here i corresponds to the quantum numbers for a particular rotation-vibration state of molecule BC and to momentum k_i of A relative to BC, the reduced mass (A, BC) being μ_i; the index f has the same connotation for the final channel. The quantity T_{fi} is the transition or scattering matrix (T matrix) defined by the two equivalent expressions

$$T_{fi} = \langle \Phi_f | \mathscr{V}_f | \Psi_i^{(+)} \rangle = \langle \Psi_f^{(-)} | \mathscr{V}_i | \Phi_i \rangle , \tag{2}$$

where $\mathscr{V}_i$, $\mathscr{V}_f$ are initial- and final-state interaction potentials; *i.e.* $\mathscr{V}_i$ is the part of $\mathscr{V}$ that goes to zero as the initial atom-molecule relative co-ordinate goes to infinity and $\mathscr{V}_f$ is correspondingly defined for the final channel. In terms of these, the total system Hamiltonian $\mathscr{H}$ is written

$$\mathscr{H} = K + \mathscr{V} = \mathscr{H}_i + \mathscr{V}_i = \mathscr{H}_f + \mathscr{V}_f , \tag{3}$$

where K is the total (centre-of-mass system) kinetic-energy operator, and $\mathscr{H}_i$ and $\mathscr{H}_f$ are the noninteracting initial- and final-state Hamiltonians, which have solutions Φ_i and Φ_f (normalized to unit density):

$$\Phi_i = \exp [i\boldsymbol{k}_i \cdot \boldsymbol{R}] \eta_{\text{BC}}(\boldsymbol{r})_i , \qquad \Phi_f = \exp [i\boldsymbol{k}_f \cdot \boldsymbol{S}] \eta_{\text{AB}}(\boldsymbol{s})_f . \tag{4}$$

Here $\boldsymbol{R}$ is the relative co-ordinate, $\boldsymbol{r}$ the molecular co-ordinate, and $\eta_{\text{BC}}(\boldsymbol{r})_i$ the molecular (rotation-vibration) wave function for the initial channel; the quantities $\boldsymbol{S}$, $\boldsymbol{s}$ and $\eta_{\text{AB}}(\boldsymbol{r})_f$ are defined correspondingly for the final channel. The functions $\Psi_i^{(\pm)}$, $\Psi_f^{(\pm)}$ are, respectively, initial- and final-channel eigenfunctions of the entire Hamiltonian with energy E and outgoing (+) or incoming (—) spherical-wave boundary conditions; *e.g.* the $\Psi_i^{(\pm)}$ satisfy the Lippmann-Schwinger equations,

$$\Psi_i^{(\pm)} = \Phi_i + \frac{1}{E - \mathscr{H} \pm i\varepsilon} \mathscr{V}_i \Phi_i , \tag{5}$$

with the positive infinitesimal ε introducing the appropriate asymptotic behaviour. The functions Φ_i, Φ_f, $\mathscr{V}_i$, and $\mathscr{V}_f$ being known, the difficulty in obtaining T_{fi}, and from it the differential cross-section $\sigma_{fi}(\hat{\boldsymbol{k}}_f)$, occurs in the determination of $\Psi_i^{(+)}$ or $\Psi_f^{(-)}$. Since their exact evaluation, which is equivalent to solving the three-body Schrödinger equation with appropriate boundary conditions, is not feasible at present, approximations must be introduced at this point.

a) Born approximation. – The simplest is to replace $\Psi_i^{(+)}$ or $\Psi_f^{(-)}$ in T_{fi} by Φ_i or Φ_f, respectively; *i.e.*

$$T_{fi} = T_{fi}(\text{B}) = \langle \Phi_f | \mathscr{V}_f | \Phi_i \rangle = \langle \Phi_f | \mathscr{V}_i | \Phi_i \rangle , \tag{6}$$

which is the well-known Born approximation. Although it may be useful for some atomic rearrangement problems (*e.g.*, high-energy, low-activation barrier,) $T_{fi}(\mathrm{B})$ would yield incorrect results for the $\mathrm{H}+\mathrm{H}_2$ reaction in the thermal region.

b) Distorted-wave Born approximation. – To account in part for the strong repulsive interaction (high-activation barrier) between the approaching or receding atom and the molecule, it is appropriate to separate $\mathscr{V}_i$ and $\mathscr{V}_f$ into two parts

$$\mathscr{V}_i = \mathscr{V}_i^0 + \mathscr{V}_i', \qquad \mathscr{V}_f = \mathscr{V}_f^0 + \mathscr{V}_f', \tag{7}$$

where $\mathscr{V}_i^0(R)$ and $\mathscr{V}_f^0(S)$ are distortion potentials that cannot produce rearrangement. They are chosen to account for the interaction as completely as possible, subject to the condition that the Hamiltonians $\mathscr{H}_i+\mathscr{V}_i^0$ and $\mathscr{H}_f+\mathscr{V}_f^0$ have solutions $\chi_i^{(\pm)}$ and $\chi_f^{(\pm)}$, respectively, which can be evaluated exactly or, at least, to a high degree of approximation. Here the $\chi_f^{(\pm)}$ satisfy the equation

$$\chi_f^{(\pm)} = \Phi_f + \frac{1}{E-(\mathscr{H}_f+\mathscr{V}_f^0)\pm i\varepsilon}\mathscr{V}_f^0\Phi_f, \tag{8}$$

and a corresponding equation exists for $\chi_i^{(\pm)}$. Introduction of $\chi_f^{(-)}$ and use of eqs. (7) and (8) in the first (post) form of T_{fi} (eq. (2)) yields

$$T_{fi} \simeq \langle \chi_f^{(-)} | \mathscr{V}_f' | \Psi_i^{(+)} \rangle . \tag{9}$$

If now $\Psi_i^{(+)}$ is approximated by $\chi_i^{(+)}$, the so-called distorted-wave Born approximation (DWB) is obtained; *i.e.*

$$T_{fi} \simeq T_{fi}(\mathrm{DWB}) = \langle \chi_f^{(-)} | \mathscr{V}_f' | \chi_i^{(+)} \rangle . \tag{10}$$

As compared with $T_{fi}(\mathrm{B})$ (eq. (6)), eq. (10) should be considerably better for $\mathrm{H}+\mathrm{H}_2$ because it includes distortion of the relative motion wave functions in both the initial and final channels. However, the replacement of $\Psi_i^{(+)}$ by $\chi_i^{(+)}$ in eq. (9) to obtain (10) is generally a serious approximation, the accuracy of the resulting scattering matrix $T_{fi}(\mathrm{DWB})$ depending both on the nature of the problem and on the judicious choice of $\mathscr{V}_i^0$ and $\mathscr{V}_f^0$:

c) Coupled-equation approach. The next step beyond the distorted-wave Born approximation is to introduce some type of expansion to provide an approximation to the functions $\Psi_i^{(\pm)}$ and $\Psi_f^{(\pm)}$. This type of approach [6*b*] has been applied to a variety of scattering problems (*e.g.*, rotational excitation in atom-molecule collisions [7], electron scattering including rearrangement [8]) and clearly is applicable to chemical reactions as well. We write the wave function Ψ (where we drop channel subscripts and boundary-condition superscripts for simplicity) as

$$\Psi = \sum_n \chi_n(\boldsymbol{r}, \boldsymbol{R}) . \tag{11}$$

The functions χ_n can be determined by a variation principle or related methods corresponding to those used in bound state problems [6], though the fact that the variation principle is frequently not a maximum or minimum principle introduces difficulties (but, see ref. [9]). An expansion of the type given in eq. (11) can be made with a finite number of terms contributing, $\chi_n(\boldsymbol{r}, \boldsymbol{R}) = \eta_{\text{BC}}(\boldsymbol{r})_n f_n(R)$, $\chi_n(\boldsymbol{s}, \boldsymbol{S}) = \eta_{\text{AB}}(\boldsymbol{s})_n g_n(\boldsymbol{S})$ where the $\eta_{\text{BC}}(\boldsymbol{r})_n$, $\eta_{\text{AB}}(\boldsymbol{s})_n$ represent the bound states of the molecules BC and AB. Solution for the relative motion functions f_n, g_n then gives rise to a coupled set of integrodifferential equations [8]. Since no such calculation has as yet been made for chemical reactions, we do not present further details here; attempts to apply the method are in progress [10].

3. – Construction of two-body potential.

The work reported in this lecture is primarily concerned with the distorted-wave Born approximation and employs two different assumptions for determining $\mathscr{V}_i^0$ and $\mathscr{V}_f^0$ from the known three-body potential. They represent limiting cases as far as the response of the molecule to the incoming atom is concerned. For the reactant channel in one approximation, it is assumed that the incoming atom does not affect the molecule BC and that the potential $\mathscr{V}_i^0$ between A and BC is simply the total potential averaged over the internal co-ordinates of BC. In the other approximation, the molecule BC adjusts adiabatically to the incoming atom A so that the interaction potential $\mathscr{V}_i^0$ is equal to the change in the eigenvalue of BC produced by the presence of A. The final-channel distorting potentials are determined by corresponding procedures. The « best » two-body potential is between these two extremes and depends both on the atom-molecule relative velocity and the internal state of the molecules; *e.g.*, at high velocities, the free molecule approximation should be appropriate, while at very low velocities, the adiabatically perturbed molecule approximation is better. It would be possible, in principle, to take account of only partial molecular following. This is found to be unnecessary for the elastic scattering, since the two limiting calculations give similar results. For a reliable approximation to inelastic and reactive scattering, however, such a more extended treatment is likely to be important; it involves introduction of some form of the coupled equation approach outlined briefly above.

a) *Free-molecule approximation*. The three-body interaction potential $\mathscr{V}(\boldsymbol{R}, \boldsymbol{r})$ can be expanded

$$\mathscr{V}(\boldsymbol{R}, \boldsymbol{r}) = \sum_l \mathscr{V}_l(R, r) P_l(\cos\gamma) , \tag{12}$$

where γ is the angle between $\boldsymbol{R}$ and $\boldsymbol{r}$ and P_l is the Legendre polynomial of order l in $\cos\gamma$. Only terms with l even contribute for the H, H_2 case because of symmetry; for other isotopes (*e.g.* H, DH scattering) odd terms have to be included because the molecular center of mass is shifted with respect to the center of charge. The terms $\mathscr{V}_l(R, r)$ are obtained from the complete PK potential $\mathscr{V}_T(\boldsymbol{R}, \boldsymbol{r})$ [3] by integration; that is,

$$\text{(13a)} \qquad \mathscr{V}_0(R, r) = \mathscr{V}_T^0(R, r) - \mathscr{V}_{BC}(r),$$

$$\text{(13b)} \qquad \mathscr{V}_l(R, r) = \mathscr{V}_T^l(R, r) \qquad l \neq 0,$$

where

$$\text{(14)} \qquad \mathscr{V}_T^l(R, r) = \frac{2l+1}{2}\int_0^\pi \mathscr{V}_T(\boldsymbol{R}, \boldsymbol{r}) P_l(\cos\gamma)\sin\gamma\, d\gamma\,.$$

and $\mathscr{V}_{BC}(r)$ is the isolated molecule potential. With the assumption that the molecule is unperturbed by the incoming atom, the interaction potential $\mathscr{V}_u^0(R)$ is obtained from eq. (12) by the appropriate average over the molecular wave function $\eta_{BC}(\boldsymbol{r})_i$. For H_2, in its ground state ($v = 0$, $J = 0$), the average yields a result that is very close to that obtained by simply introducing the molecular equilibrium distance r_e for r; that is,

$$\text{(15)} \qquad \mathscr{V}_u^0(R) \simeq \mathscr{V}_0(R, r_e) = \mathscr{V}_T^0(R, r_e) - \mathscr{V}_{BC}(r_e)\,.$$

With the free-molecule assumption, $\chi_i^{(+)}(\boldsymbol{R}, \boldsymbol{r})$ can be written

$$\text{(16)} \qquad \chi_i^{(+)}(\boldsymbol{R}, \boldsymbol{r}) = F_i(\boldsymbol{R})\eta_{BC}(\boldsymbol{r})_i\,,$$

where $F_i(\boldsymbol{R})$ satisfies the equation

$$\text{(17a)} \qquad \left\{\frac{\hbar^2}{2\mu_{A,CB}}\nabla_R^2 + [\mathscr{E}_n - \mathscr{V}_u^0(R)]\right\} F_n(\boldsymbol{R}) = 0\,,$$

with

$$\text{(17b)} \qquad F_n(\boldsymbol{R}) \underset{R\to\infty}{\simeq} \exp[i\boldsymbol{k}_n\cdot\boldsymbol{R}] + \frac{1}{R}\exp[ik_nR]f_n^0(\hat{\boldsymbol{k}})_n$$

and $\mathscr{E}_n = E - \varepsilon_n$. Equation (17) is the standard equation for scattering in a central field, whose solution is well known [6].

b) Adiabatic-perturbation approximation. Since the molecule is adiabatically perturbed by the incoming atom, the potential $\mathscr{V}^0(R)$, labelled $\mathscr{V}_a^0(R)$, is obtained as the change in the molecular eigenvalue $[\varepsilon_n(R) - \varepsilon_n(\infty)]$ in the presence of the incoming atom at R. Thus the separation of relative ($\boldsymbol{R}$) and internal molecular ($\boldsymbol{r}$) motion in this case is analogous to the Born-Oppenheimer separation of nuclear and electronic motion; that is, the molecular wave function and energy levels are obtained as parametric functions of R, with the energy level serving as the potential for relative motion. To find $\mathscr{V}^0_a(R)$, we must solve the molecular problem in the presence of the atom, in contrast to the free-molecule approximation which is based on the unperturbed eigenfunctions. The basic assumption is that we can write $\chi_i^{(+)}(\boldsymbol{R}, \boldsymbol{r})$ in the form (see eq. (16) for comparison)

$$\chi_i^{(+)}(\boldsymbol{R}, \boldsymbol{r}) = G_i(\boldsymbol{R})\eta_{\mathrm{BC}}(\boldsymbol{R}, \boldsymbol{r})_i \,, \tag{18}$$

where the perturbed molecular eigenfunction $\eta_{\mathrm{BC}}(\boldsymbol{R}, \boldsymbol{r})_i$ satisfies the equation

$$\left[-\frac{h^2}{2\mu_{\mathrm{BC}}}\nabla_r^2 + \mathscr{V}_T(\boldsymbol{R}, \boldsymbol{r})\right]\eta_i(\boldsymbol{R}, \boldsymbol{r}) = \varepsilon_i(R)\eta_{\mathrm{BC}}(\boldsymbol{R}, \boldsymbol{r})_i \,, \tag{19}$$

with R a parametric variable. The eigenfunction $\eta_{\mathrm{BC}}(\boldsymbol{R}, \boldsymbol{r})_i$ and eigenvalue $\varepsilon_i(R)$ are chosen such that as $R \to \infty$

$$\eta_{\mathrm{BC}}(\boldsymbol{R}, \boldsymbol{r})_i \underset{R\to\infty}{\simeq} \eta_{\mathrm{BC}}(\boldsymbol{r})_i \,; \qquad \varepsilon_i(R) \underset{R\to\infty}{\simeq} \varepsilon_i \,. \tag{20}$$

Making use of eqs. (18) and (19), we find that the differential equation for $G_i(\boldsymbol{R})$ corresponds exactly to eq. (17) with $[\mathscr{E}_n - \mathscr{V}^0_u(R)]$ replaced by $[\mathscr{E}_n - (\varepsilon_n(R) - \varepsilon_n)]$; *i.e.*, the effective interaction potential $\mathscr{V}^0_a(R)$ is $[\varepsilon_n(R) - \varepsilon_n]$. Thus, the elastic-scattering problem in the adiabatically perturbed molecule approximation can be solved in a straightforward manner, once the eigenvalue $\varepsilon_n(R)$ has been found.

Since the exact solution of eq. (19) is difficult, we simplify the calculation by introducing certain assumptions. The total three-body potential is expanded in Legendre polynominals and the expansion is truncated after the first two nonzero terms; that is, we write

$$\mathscr{V}_T(\boldsymbol{R}, \boldsymbol{r}) \simeq \mathscr{V}^0_T(R, r) + \mathscr{V}^2_T(R, r)P_2(\cos\gamma) \,. \tag{21}$$

Also, we assume that $\eta_{\mathrm{BC}}(\boldsymbol{R}, \boldsymbol{r})_i$ can be expressed in the separable form

$$\eta_{\mathrm{BC}}(\boldsymbol{R}, \boldsymbol{r})_i \simeq \frac{1}{r}\Omega_v(R, r)\Lambda^m_J(R, \gamma, \varphi) \,, \tag{22}$$

with the function $\Omega_v(R, r)$ a solution of the one-dimensional radial eigenvalue equation

$$\left[-\frac{\hbar^2}{2\mu_{BC}}\frac{d^2}{dr^2}+\mathscr{V}_T^0(R,r)\right]\Omega_v(R,r)=\varepsilon_v'(R)\Omega_v(R,r) \tag{23}$$

and the limit $R\to\infty$,

$$\frac{1}{r}\Omega_v(R,r)\underset{R\to\infty}{\simeq}\eta_v(r);\qquad \Lambda_j^m(R,\gamma,\varphi)\underset{R\to\infty}{\simeq}Y_j^m(\gamma,\varphi)\,. \tag{24}$$

The determination of $\Omega_v(R,r)$ and $\varepsilon_v'(R)$ as a function of R from eq. (23) is a straightforward numerical problem; it can be solved by utilizing the Numerov method.

Substituting from eqs. (21) and (22) into eq. (19) and making use of eq. (23), multiplying by $\Omega_v(R,r)$ and integrating over r, and introducing $\lambda=\cos\gamma$ and $\Lambda_j^m(R,\gamma,\varphi)\to(1/\sqrt{2})\Lambda_j^m(R,\lambda)\exp[im\varphi]$, we obtain [11]

$$\left\{\frac{\partial}{\partial\lambda}(1-\lambda^2)\frac{\partial}{\partial\lambda}+\frac{m^2}{(1-\lambda^2)}+A(R)-B(R)\lambda^2\right\}\Lambda_j^m(R,\lambda)=0\,, \tag{25}$$

where

$$A(R)=\frac{2\mu_{BC}}{\Delta\hbar^2}\left[\varepsilon_n(R)-\varepsilon_v'(R)+\frac{1}{2}\int_0^\infty\Omega_v^2(\mathbf{R},\mathbf{r})\mathscr{V}_T^2(R,r)\,dr\right] \tag{26}$$

and

$$B(R)=\frac{2\mu_{BC}}{\Delta\hbar^2}\cdot\frac{3}{2}\int_0^\infty\Omega_v^2(R,r)\mathscr{V}_T^2(R,r)\,dr\,, \tag{27}$$

with

$$\Delta=\int_0^\infty\Omega_v^2(R,r)\cdot\frac{1}{r^2}\,dr\,.$$

Equation (25) is recognized as a spheroidal equation [12] in the variable λ $(-1<\lambda<1)$ with eigenvalue $A(R)$. Since in the present calculation we are studying the ground-state molecule ($j=0$, $m=0$), there is no φ dependence and $\Lambda_0^0(R,\lambda)$ goes to the $j=0$ initial state as $R\to\infty$; for H_2 in excited rotational states, nonzero m values would have to be considered. For the elastic-scattering calculation only $A(R)$ is used; however, for reactive scattering, the function $\Lambda_0^0(R,\lambda)$ is needed as well. Once the eigenvalue $A(R)$ has been determined, the adiabatically perturbed molecule energy is obtained from eq. (26);

i.e., for the ground state

$$(28)\qquad \varepsilon_0(R) = \varepsilon_0'(R) - \frac{1}{2}\int_0^\infty \Omega_0^2(R, r)\,\mathscr{V}_T^2(R, r)\,\mathrm{d}r + \frac{\hbar^2}{2\mu_{BC}}\,A(R)\int_0^\infty \Omega_0^2(R, r)\left(\frac{1}{r^2}\right)\mathrm{d}r\,.$$

c) Results for potentials. The results of the Legendre polynomial expansion for $\mathscr{V}(\boldsymbol{R}, \boldsymbol{r})$ (eq. (12)) with R measured from the center of the H_2 molecule and $R_e = 1.40$ a.u., the equilibrium distances for H_2, are shown in Fig. 1. It is evident that for the most significant energies ($E < 1$ eV) the strongly repulsive potential $\mathscr{V}_0$ [$\mathscr{V}_0(R, r) \simeq \mathscr{V}_u^0(R)$] is dominant with $\mathscr{V}_2$ making a small additional contribution; for the three atoms on a line and $R > 2$ a.u., $\mathscr{V}_2$ decreases the magnitude of $\mathscr{V}$ ($\mathscr{V} \simeq \mathscr{V}_0 + \frac{1}{2}\mathscr{V}_2$, with $\mathscr{V}_2 < 0$), corresponding to the fact that the linear geometry is the one of minimum energy in the present calculation. Thus, for energies of 1 eV or less, the effective potential in the free-molecule approximation is nearly spherically symmetric and the neglect of higher terms should be a very good approximation, particularly for elastic scattering.

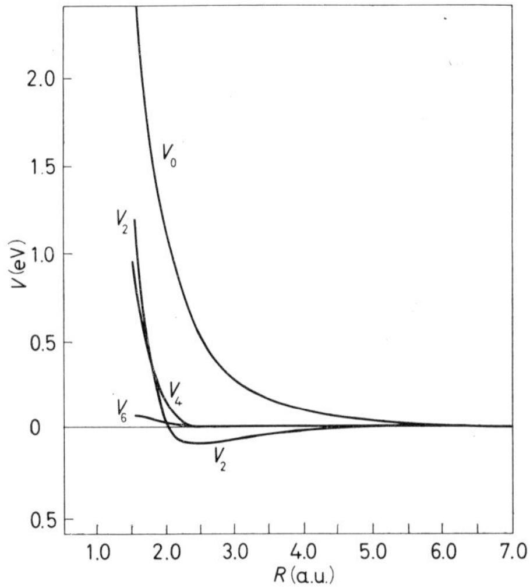

Fig. 1. – Expansion of three-body interaction potential in Legendre polynomials (eq. (12)) with $r = r_e = 1.4$ a.u.

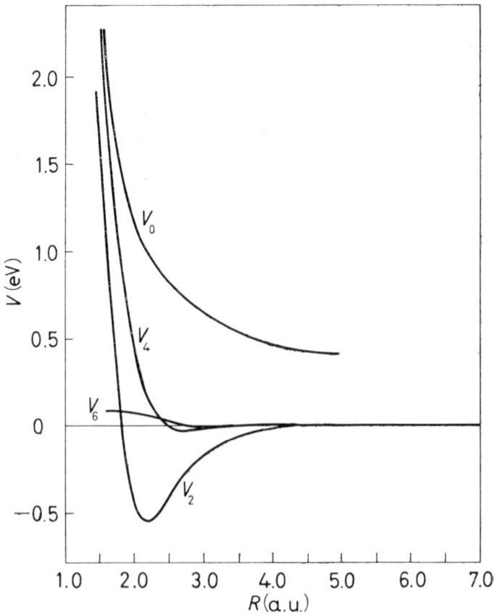

Fig. 2. – Expansion of three-body interaction potential in Legendre polynomials (eq. (12)) with $r = 1.7$ a.u.

To provide an indication of the dependence of $\mathscr{V}(\boldsymbol{R}, \boldsymbol{r})$ on the value of r, we plot in Fig. 2 the corresponding $\mathscr{V}_l(R, r)$ coefficients for $r = 1.7$ a.u., the value of r associated with the minimum energy geometry for the reactive scattering problem (the saddle point of the linear three-atom system with $r = 1.7$ a.u., $R = 2.55$ a.u.). The most significant difference between Fig. 1 and 2 is that for the «stretched» molecule system, the anisotropic term $\mathscr{V}_2$ is much more important than for the molecule at its equilibrium separation.

The effective potential corresponding to the adiabatically perturbed molecule is $\mathscr{V}_a^0(R) = [\varepsilon_0(R) - \varepsilon_0]$, with $\varepsilon_0(R)$ obtained from eq. (28); it is shown in Fig. 3 with $\mathscr{V}_u^0(R)$ for comparison. Although the two potentials are similar in shape, $\mathscr{V}_a^0(R)$ is somewhat «softer» than $\mathscr{V}_u^0(R)$ in the region $1.8 < R < 3.5$ a.u. Two factors are responsible for this difference: the first is the change in the molecular vibrational eigenfunction and eigenvalue (eq. (23)) and the second, which is in part a consequence of the first (see below), is the change in the rotational wave function due to the orienting effect of the anisotropic potential term $\mathscr{V}_T^2(R, r)P_2(\cos\gamma)$. To visualize the vibrational effect, we plot in Fig. 4 the potential $\mathscr{V}_T^2(R, r)$ as a function of r for several different fixed R. At each given R, this is the effective vibrational potential for the adiabatic molecule wave function $\eta_{BC}(\boldsymbol{R}, \boldsymbol{r})_i = \eta_0(R, r)$ and eigenvalue $\varepsilon_0'(R)$ (eqs. (22) and (23)). The most important changes that occur in $\mathscr{V}_T^0(R, r)$ as the atom approaches are that the value of the potential at $r_e(R)$ is raised—*i.e.* there is a repulsive interaction corresponding to $\mathscr{V}_T^0[R, r(R)]$, and that the force constant is lowered—*i.e.* the potential broadens and the zero-point energy

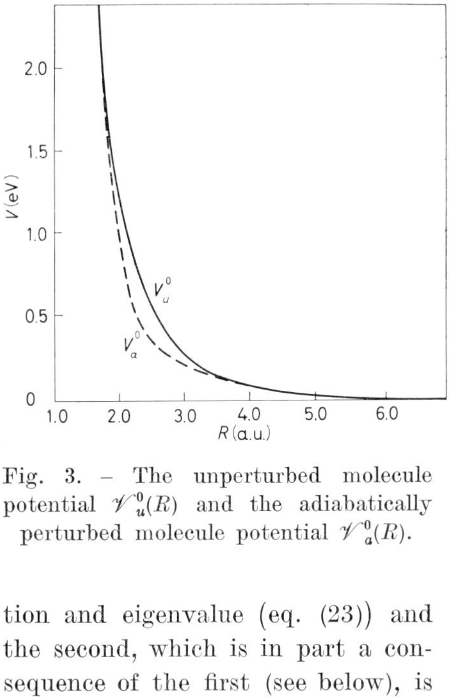

Fig. 3. – The unperturbed molecule potential $\mathscr{V}_u^0(R)$ and the adiabatically perturbed molecule potential $\mathscr{V}_a^0(R)$.

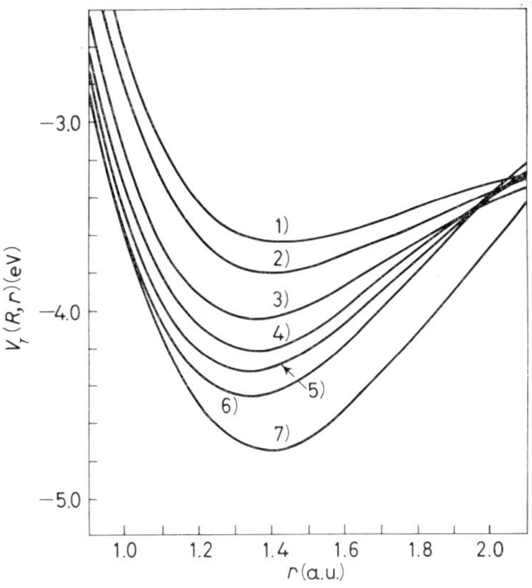

Fig. 4. – Effective vibrational potential $\mathscr{V}_T^0(R, r)$ for the molecule in the adiabatic approximation; each curve gives $\mathscr{V}_T^0(R, r)$ as a function of r for fixed R. 1) $R = 2.0$; 2) $R = 2.1$; 3) $R = 2.2$; 4) $R = 2.5$; 5) $R = 2.7$; 6) $R = 3.0$; 7) $R = \infty$.

$\varepsilon_0'(R)$ decreases. Moreover, the broadening of the molecular potential alters the vibrational wave function so that the average of $\mathscr{V}_T^2(R, r)$ over $\Omega_0(R, r)$ has significant contributions from values of r larger than r_e. From a comparison of Fig. 1 and 2, it is clear that, as the molecule stretches, the anisotropic term $\mathscr{V}_T^2(R, r) = \mathscr{V}_2(R, r)$ increases in magnitude. Thus, for the adiabatic case the vibrational average used for the orienting potential (eq. (28)), leads to a much larger effect than that produced by $\mathscr{V}_T^2(R, r_e)$. A significant lowering of the interaction energy results from the fact that the adiabatically perturbed molecule turns toward the atom (*i.e.* toward the linear geometry of minimum energy). The actual molecular alignment corresponding to the orienting potential is shown in Fig. 5, where we plot the spheroidal wave function Λ_0 $(R, \cos\gamma)$ as a function of γ for several values of R. At first as R decreases, the wave function becomes concentrated in the small-angle region, in contrast to the uniform angular distribution for the undistorted molecule ($R \to \infty$). For small distances ($2.0 < R < 2.4$ a.u.; see Fig. 2), the orienting $\mathscr{V}_2$ term decreases in magnitude (it finally becomes positive) and the distortion in the rotational wave function is less; this is seen explicitly in Fig. 5 by comparing the $R = 2.2$ a.u. function (5) with that at $R = 2.4$ a.u. (4).

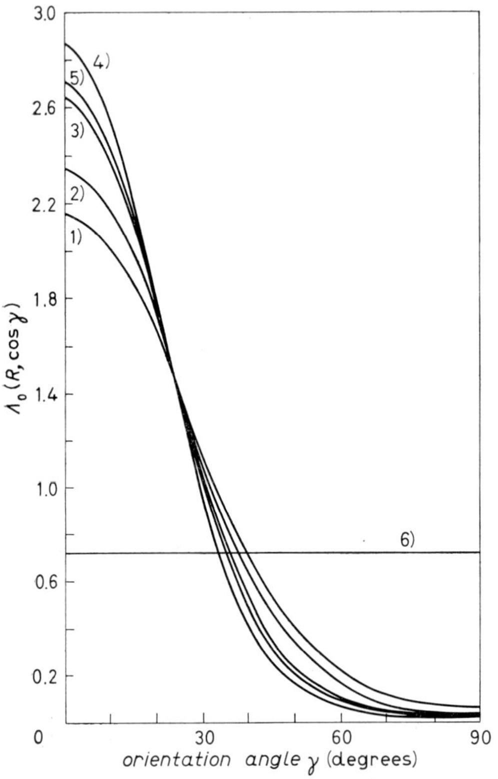

Fig. 5. – Spheroidal function $\Lambda_0(R, \cos\gamma)$ as a function of γ for fixed R; the spheroidal function is normalized. 1) $R = 3.0$; 2) $R = 2.8$; 3) $R = 2.6$; 4) $R = 2.4$; 5) $R = 2.2$; 6) $R = \infty$.

An important point about the effective two-body potentials $\mathscr{V}_u^0(R)$ and $V_a^0(R)$ is that they are both purely repulsive. This is as expected, since the semi-empirical potential $\mathscr{V}_T(R, r)$ from which they were obtained does not include any attractive Van der Waals (dispersion) term. The neglect of the very small Van der Waals energy is unimportant for the reactive (H, H_2) scattering, which occurs at large collision energies ($\gtrsim 0.3$ eV) and small impact parameters; however, for the elastic scattering, particularly in the important small-angle, large impact-parameter region, the Van der Waals term is significant. Consequently, modified forms of $\mathscr{V}_u^0(R)$ and $\mathscr{V}_a^0(R)$ with a Van der Waals « tail » had to be constructed. Although there exist good theoretical values for the Van der Waals energy for H, H_2 [13], there is no unique procedure for joining the result valid only for large R, to the calculations for the inner region, where semi-empirical techniques or independent theoretical methods have been used. One approximate formula given by DALGARNO, HENRY, and ROBERTS [14] is

$$\mathscr{V}_p^0(R) = 511.088 \exp[-1.9R] - \frac{251.546}{R^6}$$

for $\mathscr{V}_p^0(R)$ in eV and R in a.u. In Fig. 6, we plot $\mathscr{V}_p^0(R)$, $\mathscr{V}_u^0(R)$, and $\mathscr{V}_a^0(R)$.

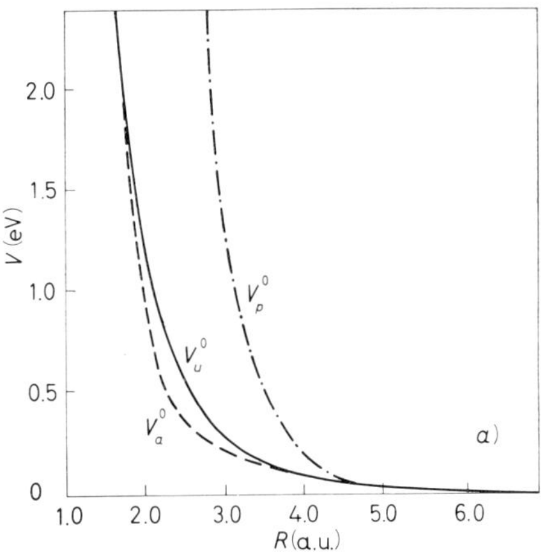

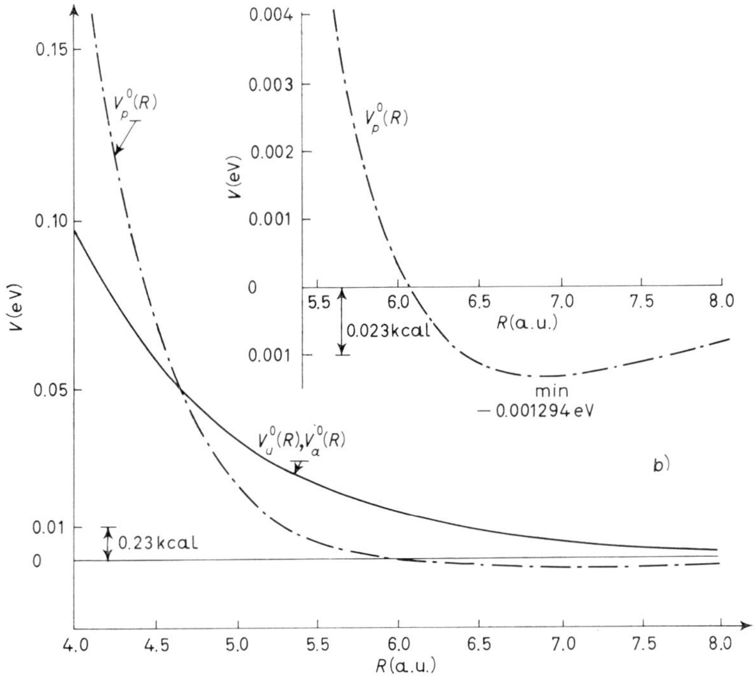

Fig. 6. – Comparison of effective spherical potentials. *a*) Inner region: (——) $\mathscr{V}_u^0(R)$, (— — —) $\mathscr{V}_a^0(R)$, (—·—·—) $\mathscr{V}_p^0(R)$; *b*) outer region: (——) $\mathscr{V}_u^0(R)$ and $\mathscr{V}_a^0(R)$, (—·—·—) $\mathscr{V}_p^0(R)$. For definitions, see text.

Figure 6*a*) shows that for the repulsive part of the potential, $\mathscr{V}^0_p(R)$ begins to rise steeply at a significantly larger R value than do either $\mathscr{V}^0_u(R)$ or $\mathscr{V}^0_a(R)$, which are very similar in behavior. Since a potential based on a wave function corresponding to that used in obtaining $\mathscr{V}^0_p(R)$ leads to the activation barrier for the exchange reaction that is much too high, while $\mathscr{V}_T(\boldsymbol{R}, \boldsymbol{r})$ yields a more reasonable value, it is likely that the $\mathscr{V}^0_u(R)$ or $\mathscr{V}^0_a(R)$ repulsion is considerably nearer the correct behavior. The outer region of the potentials is illustrated in Fig. 6*b*), which shows that the purely repulsive forms $\mathscr{V}^0_u(R)$ and $\mathscr{V}^0_a(R)$ have a long positive exponential tail, while $\mathscr{V}^0_p(R)$ drops rapidly to zero and becomes slightly negative due to the presence of the attractive Van der Waals term. Since in this region $\mathscr{V}^0_p(R)$ is almost certainly more accurate, hybrid potentials were constructed by joining $\mathscr{V}^0_u(R)$ and $\mathscr{V}^0_a(R)$ at or near the crossing point; the resulting potential has a Van der Waals minimum equal to $-0.001\,294$ eV at 6.9 a.u. (see insert of Fig. 6*b*)).

4. – Elastic scattering.

The differential and total scattering cross-sections for the H, H_2 system were calculated for the various interaction potentials as a function of the collision energy by standard methods [6]. In Fig. 7 we present differential cross-

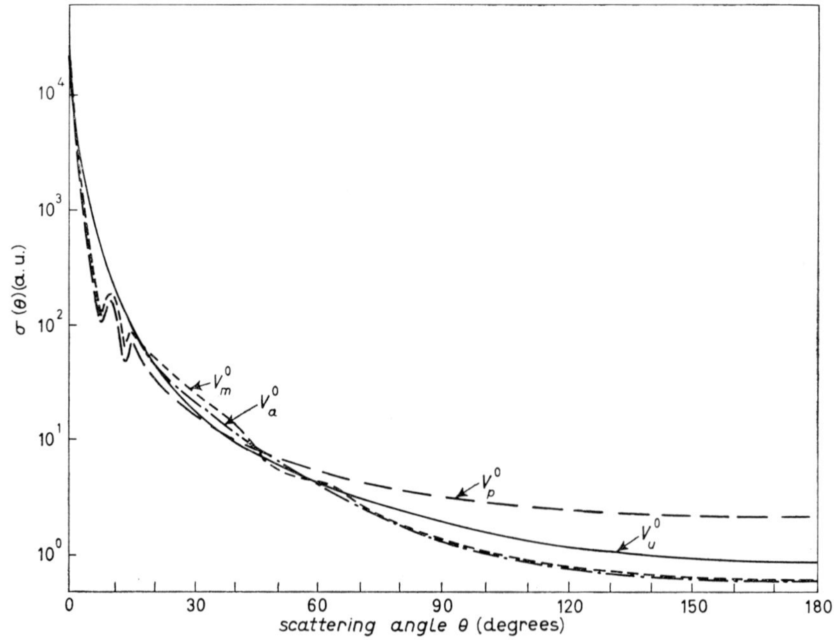

Fig. 7. – Elastic differential cross-section at energy $E = 0.5$ eV; (———) from potential $\mathscr{V}^0_u(R)$, (—·—·—) from potential $\mathscr{V}^0_a(R)$, (– – –) from potential $\mathscr{V}^0_m(R)$, (———) from potential $\mathscr{V}^0_p(R)$.

sections at a relative energy of 0.5 eV. The four potentials all yield different results. Comparing $\mathscr{V}_u^0(R)$ and $\mathscr{V}_a^0(R)$, we see from Fig. 7 that for scattering angles less than 60°, the two differential cross-sections are very similar, while for large scattering angles (60° to 180°) there is a greater difference. The large-angle collisions, corresponding to small impact parameters (small l values), sample the potential at the distance ($R \simeq 2.5$ a.u.) where $\mathscr{V}_a^0(R)$ is most « softened » relative to $\mathscr{V}_u^0(R)$ (see Fig. 3). By contrast, the smaller scattering angles associated with larger l values are most affected by the identical outer portions of the potentials $\mathscr{V}_u^0(R)$ and $\mathscr{V}_a^0(R)$. Although the small l collisions make only a minor contribution to the total elastic cross-section, the reactive scattering is dominated by them. Consequently, we expect the reactive cross-sections to be significantly different for a distorted-wave calculation based on $\mathscr{V}_u^0(R)$ *vs.* one employing $\mathscr{V}_a^0(R)$ (see below). Comparing the shape of the 0.5 eV differential cross-section from $\mathscr{V}_p^0(R)$ with those from $\mathscr{V}_a^0(R)$ and $\mathscr{V}_u^0(R)$, we see that for large scattering angles ($\theta > 60°$), the former has a considerably higher value, in correspondence with the fact that the strongly repulsive (hard core) portion of $\mathscr{V}_p^0(R)$ is larger than that of $\mathscr{V}_u^0(R)$ and $\mathscr{V}_a^0(R)$ (Fig. 6*a*)). For smaller scattering angles, the differential cross-section from $\mathscr{V}_p^0(R)$ is below that from $\mathscr{V}_u^0(R)$ and $\mathscr{V}_a^0(R)$ because the primary effect of the Van der Waals term in the former is to decrease the overall range of the potential as compared with the long exponential tail of the latter. This is also evident from the smaller total cross-section resulting from $\mathscr{V}_p^0(R)$ as compared with $\mathscr{V}_u^0(R)$ and $\mathscr{V}_a^0(R)$ (Fig. 8).

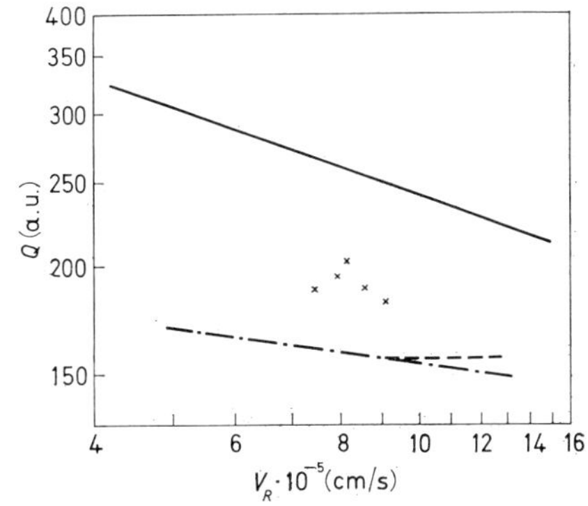

Fig. 8. – Total elastic cross-section as a function of relative velocity; (———) from potential $\mathscr{V}_u^0(R)$ and $\mathscr{V}_a^0(R)$, (—·—·—) from potential $\mathscr{V}_p^0(R)$, (– – –) from potential $\mathscr{V}_m^0(R)$. The crosses correspond to the measurements of HARRISON (ref. [17]).

For the potential $\mathscr{V}_m^0(R)$, constructed by joining the long range Van der Waals interaction to $\mathscr{V}_a^0(R)$ at the crossing point (Fig. 6*b*), the 0.5 eV elastic cross-section approximates that from $\mathscr{V}_a^0(R)$ for large scattering angles and that from $\mathscr{V}_p^0(R)$ for small angles. This is expected from the form of the potential, which is like $\mathscr{V}_a^0(R)$ for small distances ($R \leqslant 4.5$ a.u.) and like $\mathscr{V}_p^0(R)$ for $R > 4.5$ a.u. Correspondingly, because the large scattering angles dominate the total cross-section, $\mathscr{V}_m^0(R)$ yields a value near that of $\mathscr{V}_p^0(R)$. Since the construction of $\mathscr{V}_m^0(R)$ is somewhat arbitrary (as it is for $\mathscr{V}_p^0(R)$), checks were made on the cross-sections for potentials which differed slightly (by 0.15 a.u.) in the particular point at which $\mathscr{V}_p^0(R)$ was joined to $\mathscr{V}_a^0(R)$. For energies of 0.1 eV and 0.2 eV, the total cross-sections were identical to 1 part in 10^4 and the differential cross-sections were very similar, the maximum differences (~ 1 part in 10^2) occuring for large scattering angles.

It is of interest to note that the differential cross-sections from $\mathscr{V}_u^0(R)$ and $\mathscr{V}_a^0(R)$ are monotonically decreasing functions with increasing scattering angle; that is, there are no oscillations. This simple behavior is a consequence of the form of the interaction potential, which decreases smoothly and monotonically with an approximately exponential dependence on distance. Corresponding behavior of the differential cross-section is obtained, for example, from a Coulomb potential. By contrast, the results from $\mathscr{V}_m^0(R)$ and $\mathscr{V}_p^0(R)$ show oscillations in the angular dependence. This is the expected result for a potential which is not a monotonic function of distance [15].

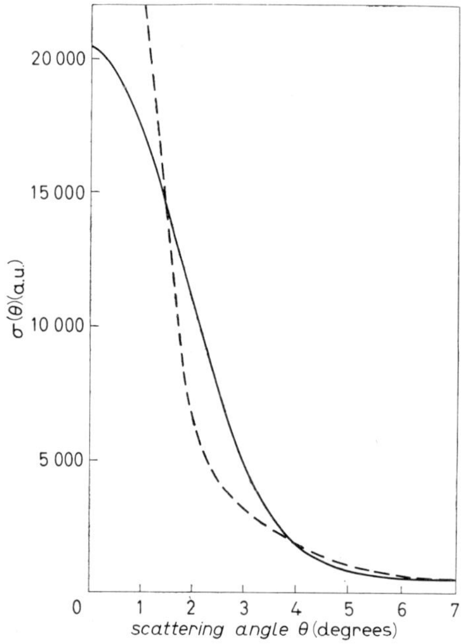

Fig. 9. – Comparison of classical and quantum differential cross-sections for potential $\mathscr{V}_u^0(R)$ at $E = 0.5$ eV; (———) quantum calculation, (— — —) classical calculation.

The classical differential cross-section for potential $\mathscr{V}_u^0(R)$ at an energy of 0.5 eV is shown in Fig. 9 together with the corresponding quantum cross-section. For large angles ($\theta \gtrsim 7°$), the classical and quantum results are in agreement, but for smaller angles deviations appear. For $\theta \lesssim 2°$ the classical values rise sharply, in correspondence with the fact that the total classical cross-section is infinite. Quantum corrections to the classical results in the small-angle region could be obtained by the use of the appropriate limiting formulae [15]. For the potentials $\mathscr{V}_p^0(R)$ and $\mathscr{V}_m^0(R)$ the overall comparison with classical calculations is expected to be similar to what was found above. However, quantitative deviations are expected since the classical formulation cannot reproduce the oscillations in the quantum differential cross-section for these nonmonotonic potentials. The appropriate semi-classical techniques would be adequate to introduce the necessary corrections [15], which become less important as the scattering energy increases.

Although there are no differential cross-section measurements for H, H_2 elastic scattering, Fite and Brackmann [16] have examined the nonreactive scattering of a thermal H beam at 3000 °K crossed with a D_2 molecule beam at 77 °K. Their results are shown in Fig. 10. The open circles represent an earlier curve which is normalized to yield a total cross-section equal to the Harrison measurement (see below, [17]); the solid circles (with error bars) represent a revised interpretation of the same experimental data, apparently without adjustment of the total cross-section to agree with Harrison's value. Also plotted in Fig. 10 are calculated laboratory differential cross-sections obtained from potential $\mathscr{V}^0_u(R)$ and $\mathscr{V}^0_m(R)$. They were determined from the center-of-mass system results (normalized to Harrison's value) by assuming that D_2 is stationary and that the H atom has a laboratory energy of 0.26 eV, corresponding to the most probable value for a thermal beam at 3000 °K. The general form of all of the differential cross-sections is similar, but there appears to be better agreement between theory and experiment for one set of data than for the other. More measurements as a function of velocity would be helpful here. Also, if velocity selection were possible, quantitative data concerning the oscillations in the differential cross-section would be of interest for comparison with the calculated results. In particular, an investigation of the changes in the oscillations due to anisotropy of the potential could be very informative.

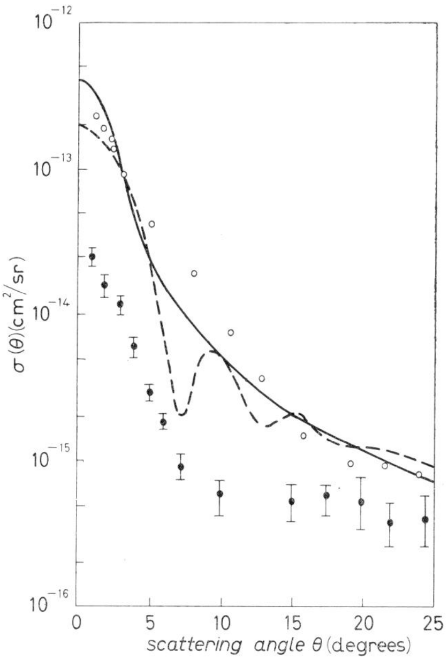

Fig. 10. – Laboratory elastic differential cross-section for $H+D_2$: (———) calculated from $\mathscr{V}^0_u(R)$, (— — —) calculated from $\mathscr{V}^0_m(R)$; open circles and solid circles (with error bars) from experiments of FITE and BRACKMANN (ref. [16]).

The total cross-sections from the different potentials are shown as a function of velocity in Fig. 8. The results from potential $\mathscr{V}^0_u(R)$ and $\mathscr{V}^0_a(R)$ are the same over the velocity range considered. The $\mathscr{V}^0_p(R)$ and $\mathscr{V}^0_m(R)$ cross-section values are very similar, though there is some difference at the higher velocities, which correspond to energies where the two potentials differ most. The magnitudes of the $\mathscr{V}^0_u(R)$, $\mathscr{V}^0_a(R)$ cross-sections are 1.5 to 1.75 times as large as the $\mathscr{V}^0_p(R)$, $\mathscr{V}^0_m(R)$ cross-sections. This is not unexpected (see above) since for the energy range studied the main effect of the Van der Waals term appearing in the latter potentials is to provide a « cut-off » of the long exponential tail present in the former. On the log-log scale of Fig. 8 all of the cross-sections are approximately linear throughout the velocity range considered. This suggests that the Landau-Lifshitz or the Massey-Mohr relationships [15] can be used to relate the logarithm of the total cross-section Q to the logarithm of the relative velocity V_R

$$\log Q = \text{const} - \left(\frac{2}{n-1}\right) \log V_R ,$$

where n is the exponent of the inverse power or r of an effective potential of the form (A/r^n). For $\mathscr{V}_u^0(R)$ and $\mathscr{V}_a^0(R)$, the slope is 0.35, which yields $n \simeq 6.7$, while for $\mathscr{V}_p^0(R)$ and $\mathscr{V}_m^0(R)$ the slope is 0.14 with $n \simeq 15$. For all of the potentials, the velocities considered in Fig. 8 are such that the repulsive part of the potential dominates the scattering; thus, the n values obtained from the slopes correspond to an inverse power potential that approximates the repulsive interactions for that velocity range. The velocity dependence of the total cross-section has been studied recently by FLUENDY, MARTIN, MUSCHLITZ, and HERSCHBACH [18] for the velocity range 3 to $11\cdot10^5$ cm/s ($\sim$0.03 to 0.4 eV). On a $\log Q$ *vs.* $\log V_R$ plot, the results fall on a straight line with slope 0.18 ± 0.03. This is in good agreement with the $\mathscr{V}_m^0(R)$, $\mathscr{V}_p^0(R)$ value (0.14) and in disagreement with the $\mathscr{V}_u^0(R)$, $\mathscr{V}_a^0(R)$ value (0.35).

HARRISON [17] has made absolute total cross-section measurements by determining the attenuation of an H atom beam that traversed a scattering chamber filled with H_2 gas. The experimental values range from 175 to 210 a.u. (Fig. 8) without any discernable velocity dependence for V_R between 7.5 and $9\cdot10^5$ cm/s. Since there are some uncertainties in the experiment (*e.g.* pressure measurement, lack of velocity selection in beam), it would be very useful to have a redetermination of the total cross-section.

5. – Reactive scattering.

With the distorted wave functions from the effective two-body potentials, the DWB scattering matrix and cross-sections were evaluated. Only the reaction with both the initial and final molecule in the lowest rotational state ($J=0$, $J'=0$) was considered. The total cross-section S_{fi} obtained by integrating $\sigma_{fi}(\hat{\boldsymbol{k}}_i)$ over all angles and multiplying by two are listed in Table I

TABLE I. – *Total cross-section* ($J=0$, $J'=0$) *by* DWB *approximation.*

Relative energy ([a]) (eV)	Free molecule approximation (I) (a.u.)	Perturbation molecule approximation (II) (a.u.)
0.5	0.009	0.20
0.33	—	0.027
0.21	—	0.0001

(*a*) The barrier height is 0.396 eV.

for a few incident energies. At 0.5 eV there is a profound difference between the two approximate models, the adiabatic perturbation of the molecule by the incoming atom yielding a 20 fold increase in S_{fi} over that corresponding to an unperturbed molecule. From the collision time for this relatively low energy, it appears that the adiabatically perturbed model should be the better approximation. Certainly, this is true for the vibrational distortion, although it is more questionable for the rotational reorientation.

The differential cross-section (in arbitrary units) obtained from the adiabatic model at an energy of 0.5 eV is shown by the solid line in Fig. 11. It corresponds to « backward scattering » in the centre-of-mass system; *i.e.* the incoming atom strikes the molecule, picks up an atom, and the newly formed molecule goes back dominantly in the direction from which the atom came. The differential cross-section for the free-molecule model is similar in shape to the adiabatic result although the magnitude is much smaller. Also shown in Fig. 11 by a dashed line is the differential cross-section (in arbitrary units) determined from the quasi-classical trajectory treatment at 0.5 eV($J = 0 \to$ all J'). The form is almost identical to the adiabatic model result. Such strongly backward peaked cross-sections are expected when the quantum-mechanical wave function or classical path of the incoming atom is strongly distorted by a repulsive barrier. They contrast sharply with the Born approximation, which yields an oscillating cross-section with its maximum in the forward direction.

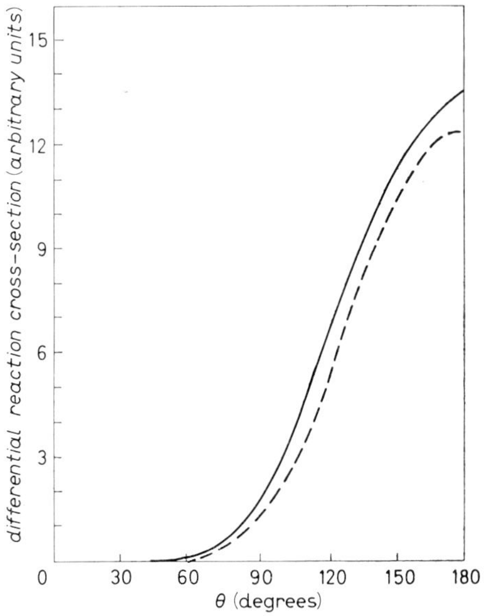

Fig. 11. – Differential reaction cross-section as a function of the scattering angle θ for an incident energy of 0.5 eV: (———) from adiabatic model (II) for ($J = 0$, $J' = 0$); (— — —) from exact quasi-classical calculation for ($J = 0$, all J').

Figure 12 presents a plot of the variation with incident energy of the form of the ($J = 0$, $J' = 0$) differential cross-section (in arbitrary units) [19]. As the energy increases from 0.4 to 1.5 eV, the primary peak in the cross-section gradually shifts in the forward direction and a secondary peak appears. Thus, the incoming atom is strongly repelled by the barrier for low incident energies, but tends to « remember » its initial direction of motion as the incident energy becomes greater than the barrier height (0.39 eV). This trend can be obtained

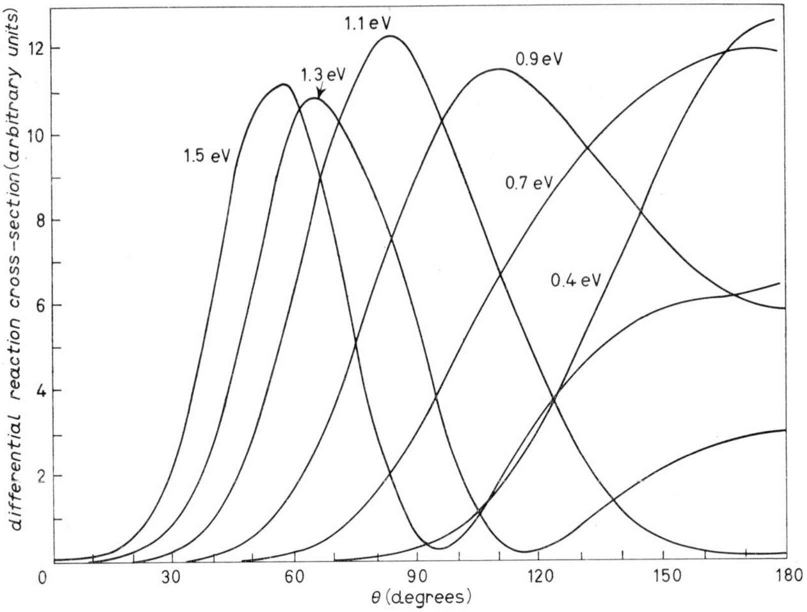

Fig. 12. – Differential reaction cross-section as a function of the scattering angle θ for $J=0$, $J'=0$ at a series of incident energies from the linear model.

from a simple semi-classical model which relates the elastic and reactive scattering. Although it has not yet been observed in the molecular case, exactly corresponding results are found for the (d, p) stripping reaction in a Coulomb field as the energy increases from a few MeV to a few hundred MeV [20].

To obtain an idea of the configurations of the three nuclei which make the dominant contributions to reaction, we used the free-molecule approximation and considered a quantity $T_{fi}(\mathrm{DWB}, \tau)$ defined by

$$T_{fi}(\mathrm{DWB}, \tau) = \langle \chi_f^{(-)} | \mathscr{V}'_f H(\tau - \gamma) | \chi_i^{(+)} \rangle \,, \tag{29}$$

with $H(x)$ the Heaviside function $[H(x)=0,\ x<0;\ H(x)=1,\ x>0]$ and $0 \leqslant \tau \leqslant \pi$; thus, $T_{fi}(\mathrm{DWB}, \pi) = T_{fi}(\mathrm{DWB})$. The corresponding total cross-section $S_{fi}(\tau)$ provides a semi-classical measure of the contribution to reaction for atom, molecule orientations with γ in the range between 0 and τ. The quantity $S_{fi}(\tau)/S_{fi}$ for the $J=0$ to $J'=0$ reaction at an energy of 0.5 eV is plotted as a function of τ in Fig. 13. Only small angles contribute; *i.e.* 80 % of the cross-section is obtained with $\gamma \leqslant 40°$. In the quasi-classical calculation [1], the average value of τ is 24° at the same relative energy.

By an expansion of the total reaction cross-section in terms of contributions from individual partial waves l, a comparison with the classical impact parameter (b) dependence can be made. For $J = 0$ to $J' = 0$ and $E = 0.5$ eV the final state (exit) l values, which are essentially the same as those for the initial state, were found to make contributions that decrease smoothly with increasing l and approach zero for $l \simeq 10$ ($b \simeq 2$). This behaviour is very similar to that of the classical reaction probability [1], which goes to zero at $b = 1.85$ a.u. In Fig. 14 are given the sum of the linear model cross-section ($J = 0 \to \to J' = 0, 1, 2$) [19] and for comparison the quasi-classical result ($J = 0 \to$ all J'). The most important point is that the quantum calculations yield a significantly higher effective threshold than does the quasi-classical treatment. This is not unreasonable when one considers that the same initial molecular zero-point energy is present in both approaches, but that the quantum constraints in

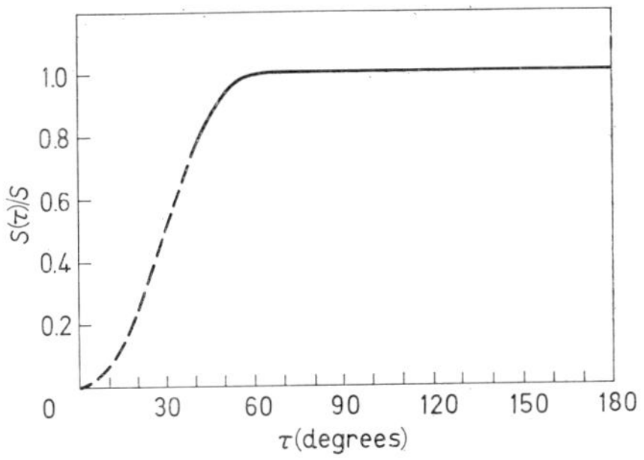

Fig. 13. – Fractional contribution of H_3 configurations to the total reactive cross-section ($J = 0$, $J' = 0$) for model (I) at 0.5 eV. For each value of τ, all configufation angles γ less than τ are included (see text).

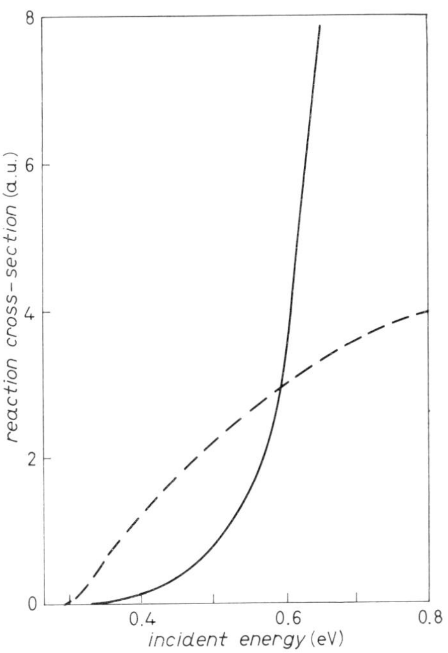

Fig. 14. – Total cross-sections as a function of incident energy: (———) from linear model [19] for $J = 0 \to 0$, 1 and 2, (— — —) from quasi-classical calculation for $J = 0 \to$ all J.

the saddle-point region may provide a limit on the vibrational energy available for crossing the barrier which does not exist for the classical trajectories. The much higher values reached by the quantum cross-sections at large energies may result from the breakdown of the adiabatic approximation, which invalidates an energy-independent choice of the strength parameter. Also, since the DWB method is a perturbation procedure, it becomes less valid as the magnitude of the total cross-section increases. To determine whether the differences are real or the consequence of the approximation in the quantum formulation will require further study.

REFERENCES

[1] M. KARPLUS: this volume, p. 372.
[2] See, for example, H. S. JOHNSTON: *Gas Phase Reaction Rate Theory* (New York, 1966).
[3] M. KARPLUS: this volume, p. 320.
[4] M. KARPLUS and K. T. TANG: *Disc. Faraday Society*, **44**, 56 (1965).
[5] J. ROSS: this volume, p. 392.
[6] See, for example, *a*) A. MESSIAH: *Quantum Mechanics* (New York, 1962); *b*) R. G. NEWTON: *Scattering Theory of Waves and Particles* (New York, 1966).
[7] A. M. ARTHURS and A. DALGARNO: *Proc. Roy. Soc.*, A **256**, 540 (1960); W. A. LESTER jr. and R. B. BERNSTEIN: *Journ. Chem. Phys.*, **48**, 4896 (1968).
[8] B. H. BRANSDEN: *Adv. Atom. and Mol. Phys.*, **1**, 85 (1965); R. PETERKOP and V. VELDRE: *Adv. Atom. and Mol. Phys.*, **2**, 263 (1966).
[9] W. H. MILLER: *Journ. Chem. Phys.*, **50**, 407 (1969).
[10] G. WOLKEN, W. H. MILLER and M. KARPLUS: work in progress.
[11] For details, see K. T. TANG and M. KARPLUS: *Journ. Chem. Phys.*, **49**, 1676 (1968).
[12] P. M. MORSE and H. FESHBACH: *Methods of Theoretical Physics*, Chapter 11 (New York, 1953).
[13] A. DALGARNO: *Rev. Mod. Phys.*, **35**, 522 (1963); M. KARPLUS and H. J. KOLKER: *Journ. Chem. Phys.*, **41**, 3955 (1964).
[14] A. DALGARNO, R. J. W. HENRY and C. S. ROBERTS: *Proc. Phys. Soc.*, **88**, 611 (1966).
[15] See, for example, H. PAULY and J. P. TOENNIES: *Adv. in Atomic and Molecular Physics*, **1**, 195 (1965).
[16] W. L. FITE and R. J. BRACKMANN: in *Atomic Collision Processes*, edited by M. R. C. McDOWELL (Amsterdam, 1964), p. 955; *Journ. Chem. Phys.*, **42**, 4057 (1962).
[17] H. HARRISON: *Journ. Chem. Phys.*, **37**, 1164 (1962).
[18] M. A. D. FLUENDY, R. M. MARTIN, E. E. MUSCHLITZ jr. and D. R. HERSCHBACH: *Journ. Chem, Phys.*, **46**, 2172 (1967).
[19] This calculation was made with a simplified «linear» model that approximates the distorted wave Born approximation; see ref. [4] for a description.
[20] J. R. ERSKINE, W. W. BEUCHNER and H. A. ENGE: *Phys. Rev.*, **128**, 720 (1962); L. C. BIEDENHARN, K. BOYER and M. GOLDSTAIN: *Phys. Rev.*, **104**, 383 (1956).

Special Results of Theory: Compound-State Approaches.

D. L. BUNKER
University of California - Irvine, Cal.

In this lecture we consider processes of the type

(1) $$A + BC \rightarrow (?) \rightarrow AB + C\,,$$

in which A, B and C are monatomic.

One way to classify simplified models for (?) is according to the amount of logical connection between the reactants and the products of the reaction. One of the extremes is loosely described as the *adiabatic* model. In this the vibration quantum number of BC is the same as that for AB, and a similar but somewhat more detailed restriction relates the rotational state of BC to the angular momenta of the products. The opposite approach (which is the subject of this lecture) is the *statistical* or *compound-state* approach of LIGHT [1-3]; for alternate theoretical routes see also EU and ROSS [4]. In this, except for requirements imposed by conservation laws, there is as little connection as possible between the states of AB and those of BC.

Define first the *statistical complex*. Classically two particles colliding subject to a potential $U = -(C/r)^n$ would have a critical impact parameter

(2) $$b^* = \left[\frac{n}{2(n-2)}\right]^{\frac{1}{2}} \left(\frac{n-2}{2}\right)^{1/n} \left(\frac{C}{E_R}\right)^{1/n},$$

where E_R is the initial relative energy. This corresponds to a critical angular momentum $L^*(E_R)$ for which orbiting occurs. Extending this idea in an approximate way to 3 particles, and assuming that only $L_R \leqslant L^*$ results in a complex, we find the cross-section for its formation to be

(3) $$\sigma_F(E_R) = \pi b^{*2}\,, \qquad \text{classically}\,,$$

(4) $$\sigma_F = \frac{\pi \hbar^2}{2\mu_R E_R}(L^* + 1)^2\,, \qquad \text{quantum-mechanically;}$$

also

$$\sigma_F(L_R, E_R) = \pi\hbar^2(2L_R + 1)/2\mu_R E_R , \tag{5}$$

whose sum over L_R yields eq. (4).

Let the angular momentum of the complex be K, and that of the reactant BC be J_R. The probability of forming a complex with K, given L_R and J_R, is

$$P_K = \frac{2K+1}{(2J_R+1)(2L_R+1)} ; \quad |L_R - J_R| \leqslant K \leqslant |L_R + J_R| . \tag{6}$$

Next construct $(2J_R + 1)/\mu_R E_R \sigma_{RP}$, in which σ_{RP} is the total cross-section for formation of products from reactants.

Substitute

$$\sigma_{RP} = \sum_K \sigma_{RK} P_{KP} , \tag{7}$$

in which σ_{RK} is the cross-section for formation of a complex in state K, and P_{KP} is the probability that this will decompose into products. Next substitute

$$\sigma_{RK} = \sum_{L_R, v_R, J_R} \sigma_F(L_R, E_R, v_R) P_K , \tag{8}$$

in which dependence on the vibrational state of BC has now been explicitly included. The result is

$$(2J_R + 1)\mu_R E_R \sigma_{RP} = \tfrac{1}{2}\pi\hbar^2 \sum_{K, v_R L_R, J_R} (2K+1) P_{KP} = \tfrac{1}{2}\pi\hbar^2 \sum_K (2K+1) P_{KP} n_{KR} , \tag{9}$$

in which n_{KR}, the number of reactant states accessible from K, arises because P_{KP} is independent of v_R, J_R and L_R; the summation over these is equal to the number of its terms.

The expression that we have constructed in eq. (9) is the one which must be the same for both directions of the reaction if detailed balance is to be satisfied:

$$\sum_K (2K+1) P_{KP} n_{KR} = \sum_K (2K+1) P_{KR} n_{KP} , \tag{10}$$

a condition fulfilled if $P_{KP}/P_{KR} = n_{KP}/n_{KR}$. This is the way the essential idea of the model is usually verbalized: Any outcome of the reaction is probable in simple proportion to the number of states that correspond to it.

This allows the reaction cross-section to be written as

$$\sigma_{RP}(E_R, J_R, v_R) = \sum_K \sigma_{RK} \frac{n_{KP}}{n_{KP} + n_{KR}} . \tag{11}$$

If we successively substitute eqs. (8), (5) and (6), the final result is

$$\sigma_{RP}(E_R, J_R, v_R) = \frac{\pi\hbar^2}{2\mu_R E_R(2J_R+1)} \sum_K (2K+1) \frac{n_{KP} n_{KR}}{n_{KP}+n_{KR}} . \tag{12}$$

With somewhat more work, we could alternatively have derived σ_{RP} with a detailed dependence on the states of the reactants and products, and summed to obtain eq. (12).

The n's must still be evaluated, usually numerically. The following is one possible procedure, not necessarily the one used by LIGHT. Let E be the total energy (E_R + exothermicity + BC internal). To evaluate n_{KR}:

1) For each (K, E) scan through quantum number v_P and J_P, counting 1 for each way to partition K into J_P and L_P consistent with:

a) Energy conservation.

b) $|L_P - J_P| \leqslant K \leqslant |L_P + J_P|$.

c) $L_P \leqslant L_P^*$ corresponding to final b_P^*.

2) Sum over K to get $\sigma_{RP}(E_R, v_R, J_R)$ and the corresponding product energy distributions.

3) Sum over E_R, v_R and J_R if desired.

The following example applications are taken from the tests performed by LIGHT [1]. A favorable case is K+HBr. This is an angular-momentum limited reaction on account of its small exothermicity. Its possible behavior is strongly restricted by conservation laws. Large J_{AB} and small v_{AB} are to the expected. These features (qualitatively illustrated in Fig. 1) are reproduced in this theory and in most other theoretical treatments, and also in trajectory studies. The cross-sections *vs.* b and E_R are also well predicted by the compound state approach.

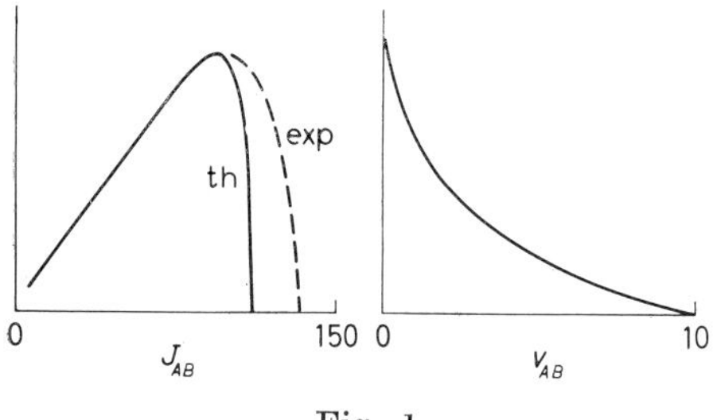

Fig. 1.

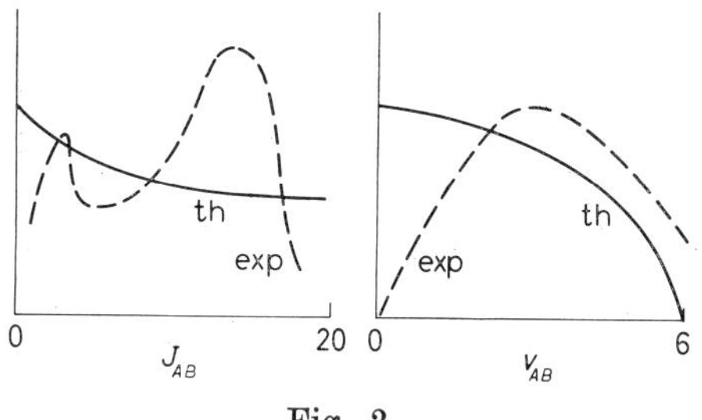

Fig. 2,

A less favorable case is $H+Cl_2$ (Fig. 2). Others that have been evaluated are: Neutral reactions with results similar to the above, *viz.*, Cl+KH, $Cl+Na_2$, and Na+NaCl; other kinds of reaction such as Hg*+CO (quenching), $H_2O^* \rightarrow \rightarrow H + OH$, $He + H_2^+$.

LIGHT's summary of the overall success of this model is:

σ_{RP}: good.

J_{AB} distribution: good.

v_{AB} distribution: poor (it always descends).

The following simplified picture may help to show why the theory always predicts that O is the most probable v_{AB}. It is of the semi-classical (v, J^2) -plane of the products (Fig. 3); we ignore the vibrational zero-point energy and the difference between J^2 and $J(J+1)$. The upper diagonal line in the Figure, which is for fixed (K, E), arises from E conservation;

$$E > v_{AB} h\nu + J^2_{AB} \hbar^2 / 2I \,. \tag{13}$$

If we exclude angular-momentum-limited reactions, the line $J_{AB} = K$ will cross the triangle. Since angular momentum must be conserved with a reasonable value of the final impact parameter b^*_P, we have for $J_{AB} \leqslant K$:

$$L_P \hbar \leqslant \mu_P u b^*_P ; \qquad \tfrac{1}{2} \mu_P u^2 = E - v_{AB} h\nu - J^2_{AB} \hbar^2 / 2I \,, \tag{14}$$

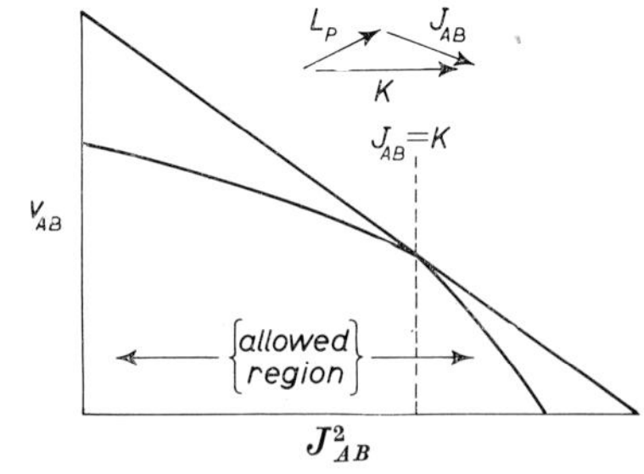

Fig. 3.

if the vector relationship illustrated at the top of Fig. 3 is to be feasible. (u is outgoing relative velocity.) The equation of the line bounding the allowed region is then

$$2\mu_P b^{*2}_P (E - v_{AB} h\nu - J^2_{AB} \hbar^2 / 2I) \geqslant (K - J_{AB})^2 \hbar^2 \,. \tag{15}$$

A similar argument holds for $J_{AB} \geqslant K$.

The weight associated with an area element in the allowed region is not uniform, but it depends only upon the number of ways a collection of angular momenta may be composed. It is independent of v. Thus for any K, or for any combination of them, the v distribution always descends. Therefore no auxiliary statement solely about angular momenta can remedy the difficulty. Since the theory includes no information about the shape of the potential energy surface, one possible direction of modification of the model might be to insert at least one bit of such information, in such a way as to weigh against the states near the origin of the diagram in Fig. 3.

REFERENCES

[1] J. Light: *Disc. Faraday Soc.*, **44**, 14 (1967).
[2] P. Pechukas and J. Light: *Journ. Chem. Phys.*, **42**, 3281 (1965).
[3] P. Pechukas, J. Light and C. Rankin: *Journ. Chem. Phys.*, **44**, 794 (1966).
[4] B. C. Eu and J. Ross: *Journ. Chem. Phys.*, **44**, 2467 (1966).

SUBJECT INDEX

PROCEEDINGS OF THE INTERNATIONAL SCHOOL OF PHYSICS
« ENRICO FERMI »

Course XIV
Ergodic Theories
edited by P. CALDIROLA

Course XV
Nuclear Spectroscopy
edited by G. RACAH

Course XVI
Physicomathematical Aspects of Biology
edited by N. RASHEVSKY

Course XVII
Topics of Radiofrequency Spectroscopy
edited by A. GOZZINI

Course XVIII
Physics of Solids (Radiation Damage in Solids)
edited by D. S. BILLINGTON

Course XIX
Cosmic Rays, Solar Particles and Space Research
edited by B. PETERS

Course XX
Evidence for Gravitational Theories
edited by C. MØLLER

Course XXI
Liquid Helium
edited by G. CARERI

Course XXII
Semiconductors
edited by R. A. SMITH

Course XXIII
Nuclear Physics
edited by V. F. WEISSKOPF

Course XXIV
Space Exploration and the Solar System
edited by B. ROSSI

Course XXV
Advanced Plasma Theory
edited by M. N. ROSENBLUTH

Course XXVI
Selected Topics on Elementary Particle Physics
edited by M. CONVERSI

Course XXVII
Dispersion and Absorption of Sound by Molecular Processes
edited by D. SETTE

Course XXVIII
Star Evolution
edited by L. GRATTON

Course XXIX
Dispersion Relations and Their Connection with Causality
edited by E. P. WIGNER

Course XXX
Radiation Dosimetry
edited by F. W. SPIERS and G. W. REED

Course XXXI
Quantum Electronics and Coherent Light
edited by C. H. TOWNES and P. A. MILES

Course XXXII
Weak Interactions and High-Energy Neutrino Physics
edited by T. D. LEE

Course XXXIII
Strong Interactions
edited by L. W. ALVAREZ

Course XXXIV
The Optical Properties of Solids
edited by J. TAUC

Information about Courses I-XIII may be obtained from the Italian Physical Society.

Course XXXV
High-Energy Astrophysics
edited by L. Gratton

Course XXXVI
Many-Body Description of Nuclear Structure and Reactions
edited by C. Bloch

Course XXXVII
Theory of Magnetism in Transition Metals
edited by W. Marshall

Course XXXVIII
Interaction of High-Energy Particles with Nuclei
edited by T. E. O. Ericson

Course XXXIX
Plasma Astrophysics
edited by P. A. Sturrock

Course XL
Nuclear Structure and Nuclear Reactions
edited by M. Jean

Course XLI
Selected Topics in Particle Physics
edited by J. Steinberger

Course XLII
Quantum Optics
edited by R. J. Glauber

Course XLIII
Processing of Optical Data by Organisms and by Machines
edited by W. Reichardt